34th Center for Chemical Process Safety International Conference 2019 (CCPS)

Topical Conference at the 2019 AIChE Spring Meeting and 15th Global Congress on Process Safety

New Orleans, Louisiana, USA
31 March - 4 April 2019

ISBN: 978-1-5108-8687-2

Printed from e-media with permission by:

Curran Associates, Inc.
57 Morehouse Lane
Red Hook, NY 12571

Some format issues inherent in the e-media version may also appear in this print version.

Printed by Curran Associates, Inc. (2019)

For permission requests, please contact AIChE
at the address below.

AIChE
120 Wall Street, FL 23
New York, NY 10005-4020

Phone: (800) 242-4363
Fax: (203) 775-5177

www.aiche.org

Additional copies of this publication are available from:

Curran Associates, Inc.
57 Morehouse Lane
Red Hook, NY 12571 USA
Phone: 845-758-0400
Fax: 845-758-2633
Email: curran@proceedings.com
Web: www.proceedings.com

TABLE OF CONTENTS

547777 Using Process Safety Tools to Find and Sell an Inherently Safer Design: A Case Study

Using Process Safety Tools to find and sell an Inherently Safer Design: A Case Study

By

Claire Fluegeman and Margery Kosch

3M Company

This paper will focus on the use of PSM tools to communicate risk and influence decision making during the design process.

As our facility began the design and installation of a new large-scale process, we identified numerous challenges. A significant processing culture shift came when the new continuous operation was introduced into a production structure that is largely used to working with batch process operations. Included in this was a shift from using drumstock Acrylic Acid to using bulk tank storage and delivery to the process equipment. Receiving bulk Acrylic Acid in tank trucks and railcars presented new challenges, and new opportunities for error, that had not yet been considered.

During the detailed design process hazard analysis that was conducted for the project, a potentially catastrophic scenario was identified involving the unloading of Acrylic Acid. Because of the potential impact of the event, a Layers of Protection Analysis (LOPA) was conducted. Ultimately, the only true way to reduce the risk of the event to an acceptable level was to evaluate numerous modifications to the unloading process. However, this was met with a fair amount of resistance from the design team as it could significantly impact the project cost and implementation timetable.

The site PSM Coordinator, who facilitated both the PHA and the LOPAs for the project, had the responsibility to persist with those that make the decisions and ensure that the true risk of the scenario was properly communicated. Using the qualitative and semi-quantitative results of the PHA and LOPA, site safety personnel were able to intervene and convince the decision makers that the right thing to do was develop an inherently safer design for the unloading of this highly hazardous material. Even though there were additional costs and some redesign was involved, the final product resulted in an inherently safer design for unloading and delivering bulk acrylic acid, and the repurposing of a facility that was currently not in service.

548217 Should Chemical Engineers Have More Influence on Safety Than They Do Now?

This presentation will focus on the question of whether chemical engineers and other health, safety, and environmental professionals within the corporation possess enough influence and resources to effectively prevent major process safety hazards and will discuss what increases in authority may be appropriate.

548395 Courageous Process Safety Leader – a Bad Guy

An honest competent process safety engineer is considered to be the trouble maker and a job stopper. This is one of the most thankless role in the high hazard industries. If this role is performed honestly without taking into consideration the impact on the relationship with the management and employees, then the results are certainly positive and encouraging for a long run.

The author will share three real life examples with challenges and success stories. The first example is from a corporate world, he influenced senior management about the benefits of process safety, where none of their operations were covered by OSHA-PSM. The second example is from the major projects, an engineering and design world, where he influenced project directors, projects managers and design engineers to implement inherent safety in design in projects. The last example is from an operating refinery, where he changed the process safety culture at the management as well as shop floor level by implementing effectively a work permit system which was completely broken. The author faced several challenges and threats which impacted his personal life in addition to the professional relationships.

This paper will be concluded by sharing the Do's and Don'ts for transforming the process safety culture from corporate world to engineering and design and down to the shop floor in complex high hazard industries.

A Leadership Vision for Accelerating Learning from Past Lessons to Create a Future Free of Incidents

David A. Jones
JCL Risk Services LLC
16225 Park Ten Place, Suite 500
Houston, Texas 77084
djones@jclsc.com

James C. Lefler, CSP
JCL Service Company LLC
12626 E. 60th Street, Suite 101
Tulsa, OK 74146
jlefler@jclsc.com

Eric Biggs, CSP
JCL Safety Services Company LLC
12626 E. 60th Street, Suite 101
Tulsa, OK 74146
ebiggs@jclsc.com

Prepared for Presentation at
American Institute of Chemical Engineers
2019 Spring Meeting and 15th Global Congress on Process Safety
New Orleans, LA
March 31 – April 3, 2019

A Leadership Vision for Accelerating Learning from Past Lessons to Create a Future Free of Incidents

David A. Jones
JCL Risk Services LLC
16225 Park Ten Place, Suite 500
Houston, Texas 77084
djones@jclsc.com

James C. Lefler, CSP
JCL Service Company LLC
12626 E. 60th Street, Suite 101
Tulsa, OK 74146
jlefler@jclsc.com

Eric Biggs, CSP
JCL Safety Services Company LLC
12626 E. 60th Street, Suite 101
Tulsa, OK 74146
ebiggs@jclsc.com

Keywords: leadership, risk, lessons learned, incidents

Abstract

The practice of process safety management (PSM) involves the analysis and understanding of what can go wrong, what the effects could be, and what should be done to prevent or mitigate an incident. Design, development and effective implementation of a robust risk management system, driven by a proactive culture of leadership to reduce operational risks and improve performance, enables organizations to avoid future incidents. Lessons of incident investigations are full of learning opportunities, yet 15-20 years is an unacceptable period for gradual learning. A conventional approach of longer-term continual improvement of operational risk performance could easily be outpaced by rapid changes in technology, organizational and human factors, and dynamics of the market. This paper poses a challenging question: *"Knowing what we already know, why would we want to take years to gradually reduce our risks, when we could accelerate the learning from past lessons and prevent incidents using available best industry practices?"*.

Given that the global industry has made remarkable progress toward developing proven solutions aimed at prevention, organizations must now commit to accelerating the adoption of these solutions before at-risk operations experience near future incidents. The focus of the paper is the challenge to accept a leadership position to: (1) resolve existing but perhaps unidentified risks, (2) to commit action to proactively identify hazards and their associated risks, and (3) urgently implement risk reduction measures before further incidents occur with losses. Practical takeaways include how individual management system elements can be improved with "leadership designed in" to control the process safety risks at three opportune times: before an incident (prevention), during an incident (emergency management) and after an incident (incident investigation, root cause analysis and lessons learned).

1. Introduction

The practice of process safety management (PSM) involves the analysis and understanding of what can go wrong, what the effects could be, and what should be done to prevent or mitigate an incident. Design, development and effective implementation of a robust risk management system, driven by a proactive culture of leadership to reduce operational risks and improve performance, enables organizations to avoid future incidents. Lessons learned from past disasters, as well as successes by industry leaders in preventing such incidents, form the basis for a practical framework for a PSM system as a holistic management tool for managing hazards and their associated risks.

However, a conventional approach of longer-term continual improvement of operational risk performance could easily be outpaced by the rapid changes in technology, organizational and human factors, and the dynamics of the market. Operating companies and regulators alike are challenged by such changes and their potential impacts. Lessons of disaster investigations from the process industry have been full of learning opportunities. In fact, many of the lessons and their root causes are common across most process safety incidents ... yet, 15-20 years has proven to be an unacceptable period for gradual learning.

This paper poses a challenging question: *"Knowing what we already know, why would we want to take years to gradually reduce our risks, when we could accelerate the learning from past lessons and prevent incidents using available best industry practices?"*.

This paper will present some common lessons as a priority for establishing improvement strategies as organizations strive for operational excellence. Available proven risk management methodologies and best practices for operational risk reduction will be summarized. Given that the global industry has made remarkable progress toward developing proven solutions aimed at prevention, organizations must now commit to accelerating the adoption of these solutions before at-risk operations experience near future incidents from human error and equipment malfunctions.

The focus is the challenge to accept a leadership position to: (1) resolve existing but perhaps unidentified risks, (2) commit action to proactively identify hazards and their associated risks, and (3) urgently implement risk reduction measures before further incidents occur with losses. This paper addresses using a framework for a robust risk-based management system, incorporating all of the necessary elements based on lessons learned and including supporting leadership and management processes to rapidly achieve risk reduction performance objectives.

Examples demonstrate how management system elements can be improved with "leadership designed in" to control the process safety risks at three opportune times: before an incident (prevention), during an incident (emergency management) and after an incident (incident investigation with root cause analysis). This paper draws upon the authors' extensive experiences in applying the principles and practices of effective leadership, organizational and human factors, as well as designing and developing comprehensive PSM systems, conducting PSM system audits, facilitating process hazard analyses (PHAs), and leading incident investigations.

To help us on the journey of accelerating the learning process from past lessons, the format for this paper follows a standard strategic planning process: (1) Where are we today … and why? (2) Where do we want (need) to be … and when? (3) How do we get there … and faster?

2. Where Are We Today … and Why?

> *"**Knowing what we already know**, why would we want to take years to gradually reduce our risks, when we could accelerate the learning from past lessons and prevent incidents using available best industry practices?"*

Over the past 30+ years since the founding of the Center for Chemical Process Safety (CCPS) in 1985, the U.S. chemical process industry has been remarkable progress in improving the culture and leadership, the methods and tools, and the overall performance for process safety. However, experience has shown that we still have a way to go to create an incident-free work environment.

2.1 What We Already Know

Where a particular organization is today, as a starting point for their continuing (or new) journey in process safety, depends on their knowledge of the current: (1) status of maturity for their process safety culture and PSM systems as leading indicators; and (2) level of process safety risk and incident performance with an evaluation of their tolerability as lagging indicators. Several methods and tools are available to assist organizations assess where they are today in terms of safety culture and various leading and lagging performance indicators [16,18,32]. Figure 1 illustrates where organizations may be today (i.e. at time zero). Simply, an organization should consider accelerating their learning based on a determination of where the levels of their risks and incidents, and the status of their current culture and systems, are:

- Acceptable and mature – sound culture and vibrant systems delivering positive results; but could benefit from accelerated learning to get even better, faster.
- Intolerable and inadequate – maturing culture and evolving systems delivering mixed results; and should benefit from accelerated learning to get in better shape, faster.
- Unacceptable and deficient – delivering poor results; and would benefit from accelerated learning to get in good shape, as fast as possible.

Besides a knowledge of the starting point for their journey, organizations can evaluate their rate of progress based on past experience and projected progress (see Figure 1). Simply, an organization should consider accelerating their learning based on a determination of whether they are:

- Making steady progress – you're getting where you're headed, but have a potential for further improvement; you can accelerate learning to get even better, faster.
- Making no net progress – you're not getting anywhere, but holding firm against challenges; you should accelerate learning to get on a more positive track, especially given new risks.
- Losing ground – you're going backwards, and increasing the risk of incidents; you need to accelerate learning to gain control, get back on track, and turn performance around.

Finally, we know that many leading organizations have already been on a successful journey of improvement, having learned the hard way; and we also know that the methods and tools for improvement are now more readily available. Therefore, we should be accelerating the learning from lessons; firstly because we can, and secondly because it's needed and it's responsible. Society (the shareholders, the workforce, the public, and the regulators) will expect more in the future.

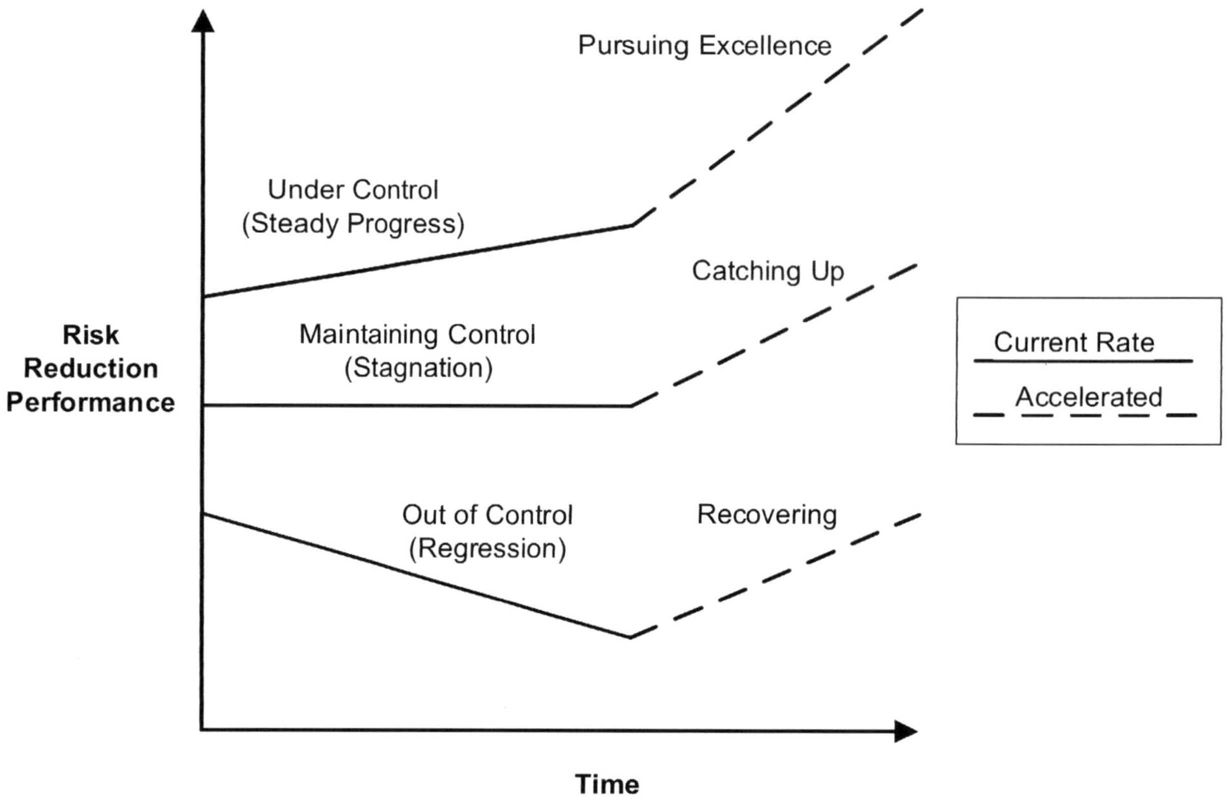

Figure 1: Where Organizations May Be Today and Rates of Progress

2.2 What More We Should Know

Using what we already know is essential to determining where we are today. However, if we don't know, then we are severely handicapped in determining whether and/or how to improve our management systems and their performance. Further, if we don't know that we don't know, then it makes the task even more difficult, leaving gaps in our knowledge base that could increase the risk of incidents.

If we don't know, then we need to go find out. Of course, this assumes that someone in the organization recognizes gaps and how to close them. For example, a formal assessment based on requirements or expectations can be used to fill a knowledge deficiency (e.g. training assessment). Also, audit checklists serve to document assessments against requirements which serve as a basis for implementing actions to improve the management systems [9]. In a broader context, CCPS *Vision 20/20* contains an Assessment Tool to be used by organizations as a starting point for planning process safety improvements towards achieving "great process safety performance" [18].

Organizations often commit to managing compliance with legal or regulatory requirements for process safety as a minimum expectation. Unfortunately, a performance-based regulation such as OSHA's Process Safety Management Standard does not spell out all of the details for achieving compliance, let alone performance [27]. While some requirements are stated, others are only implied; an example being that a training process is needed to assure the training of operators, but there are no requirements for such a process stated in the OSHA PSM Standard. In fact, there are literally hundreds of informational gaps that require the organization to know (or define) and communicate internally what is needed. Beyond compliance, organizations need to know how they plan to manage process safety commensurate with the complexity and risks of their operations and other organizational factors regarding how resources and activities are used to deliver their products. In other words, how is ongoing process safety performance to be achieved?

Finally, to increase the chances of success in managing process safety, organizations should: (1) know (and continually monitor) and analyze how external factors, as well as internally-driven "changes", can affect the risk picture (i.e. where there are possible bumps in the road); and (2) be prepared to manage such changes which could adversely affect process safety performance.

2.3 Why We Are Where We Are

Understanding why we are where we are is crucial in establishing a strategy for accelerating learning of lessons for improvement, whether it be sustaining a proven course or plotting a new course. Simply, even when we've made improvements, we may not be where we need to be. Based on what we've seen and understand that works well, we need to leverage that success and then continue to work on other areas for improvement. The more we learn, the more we can apply, and the more we can accelerate the rate of risk reduction to decrease the number of potential incidents.

Commitment of leadership, allocation of sufficient resources, and application of best practices to a large extent answers the question about why we are succeeding. However, let's recognize that much of the success is the result of correcting what was not in place, was not adequately designed, or was not effectively implemented.

To answer why PSM systems and performance aren't where they need to be, there are several possible reasons based on knowledge, attitudes, behaviors and other factors of the organization and individuals, depending on the maturity of the process safety culture:

- Don't care or share values (i.e. resistance).
- Don't have the resources (i.e. lack of commitment).
- Don't know, and may be uniformed (i.e. ignorance, lack of awareness).
- Don't comprehend what they do know, nor know why (e.g. lack of understanding).
- Don't know how to apply what they learned (e.g. lack of skills).
- Don't think they need to know, so don't ask questions (e.g. lack of perceived benefits).
- Don't see the forest for the trees (e.g. organizational myopia, competing priorities).
- Don't recognize the need or urgency (i.e. lack of perceived risk or underestimated risk).
- Don't think it'll happen to me/us (i.e. complacency or "not invented here" syndrome).
- Don't think it's a system or process problem (i.e. someone else is to blame)

Regardless of the root causes for the above road blocks, what's important going forward is that the organization needs to know why it is where it is at before it can expect to plan a meaningful strategy for improvement. This places emphasis on the ability to analyze their own situational internal strengths and weaknesses, as well as external threats and opportunities. Later we'll focus on the application of best practices for compliance and performance, and for identifying and addressing system deficiencies related to risk controls, including identifying what's worked and what hasn't in the control processes, and what went right and what went wrong as observed from the results.

3. Where Do We Want (Need) to Be ... and When?

> *"Knowing what we already know,* ***why would we want to take years to gradually reduce our risks****, when we could accelerate the learning from past lessons and prevent incidents using available best industry practices?"*

The previous question of *Where Are We Today ... and Why?* involved not only an assessment of performance and capability (i.e. what's working and what's not?) but also an analysis of why we are where we are (i.e. what are the causal factors?). As such, it represents the starting point for the journey, with the ultimate destination being a future free of incidents.

3.1 A Future Free of Incidents

That is, the overall aim of *Where We Want (Need) to Be* is the prevention of process safety incidents. An incident can be defined as "an unplanned sequence of events with the potential for undesirable consequences" [17]. Process safety incidents also include what are commonly referred to as *near misses*, defined as "an unplanned sequence of events that could have caused harm or loss if conditions were different or if the events were allowed to progress, but actually did not" [17]. The commitment to achieve a future free of incidents (with no harm to people, property or the environment) as a goal for process safety management is a moral and noble cause.

However, reporting and tracking incidents is only an indirect or lagging performance indicator that a process hazard has been fulfilled as a result of a cause (e.g. equipment malfunction, human error, or external event) and one or more failures of safeguards (e.g. administrative or engineering risk controls designed to prevent, detect, control, mitigate and/or respond to the incident) [28].

3.2 In Better Control of Risks

Better control of process safety risks is the means to the end of effectively preventing incidents. That is, we can and should more directly manage risk reduction in order to reduce the likelihood and severity of incidents. Leading indicators for process safety performance involve various risk reduction measures associated with the health of the risk controls or safeguards [17,34].

With a goal of continually reducing risks, we can set specific objectives for managing the quality of the resources and activities associated with the management processes for assuring that the appropriate risk controls are in place, are adequately designed, and are effectively implemented [9,34]. Based on established performance standards, each of these risk controls can be managed and continually improved through a robust management process (e.g. PDCA life cycle) [29].

3.3 As Soon As Reasonably Practicable

The simple answer to the question of *when?* is ... as soon as reasonably practicable; that is, with all due speed within the constraints of the organization. The longer the time delay in reducing risks before we achieve performance targets, the sooner it will be for the next incident. It's not enough to simply finish the race; we must win the race against time, or else risk undue harm. Fortunately, accelerating the learning of past lessons and incidents can help to close the time gap.

4. How Do We Get There … and Faster?

> *"Knowing what we already know, why would we want to take years to gradually reduce our risks,* ***when we could accelerate the learning from past lessons and prevent incidents using available best industry practices?****"*

To quote from *Alice in Wonderland* [5],

"'Would you tell me, please, which way I ought to go from here?' said Alice.
'That depends a good deal on where you want to get to,' said the Cat.
'I don't much care where -' said Alice.
'Then it doesn't matter which way you go,' said the Cat.
'- so long as I get SOMEWHERE,' Alice added as an explanation.
'Oh, you're sure to do that,' said the Cat, 'if you only walk long enough.'"

Depending on the path taken for process safety management, an organization can experience higher risks and repeated incidents, contrary to the goal of prevention. But using best roads with the best vehicles and the best drivers is likely to get you where you need to be… faster and safer!

4.1 Borrowing from Lessons

What Went Wrong? – Within Your Organization

A valuable source of lessons is from those incident investigation findings and recommendations within your own organization. Typically treated as company-confidential, this information can range from lessons at a particular facility to shared common lessons across the entire organization. Besides the obvious lessons learned derived from conducting a basic incident investigation to determine what went wrong and why and how to correct identified system deficiencies, there are even more benefits for organizations who adopt rigorous and proven methodologies that address: ***near misses, corrective actions and preventive actions***.

Near Misses – Those organizations who conduct investigations of near miss incidents can further broaden and accelerate their learning from two aspects: (1) increasing the number of incident investigations, thus expanding the number of lessons as opportunities for improvement; and (2) decreasing risks sooner, thus preventing the recurrence of incidents. Although the OSHA PSM Standard does not explicitly use the words "near miss" in the Incident Investigation element, it does require investigation of all incidents, including those "… which could reasonably have resulted in a catastrophic release of highly hazardous chemical in the workplace" [36].

Corrective Actions – Employing an investigation methodology (e.g. with cause and effect trees) for identifying ALL causal factors and analyzing corresponding root causes related to identified system deficiencies, enables an organization to deepen and accelerate their learning from two aspects: (1) maximizing the number of causal factors, thus expanding the number of lessons as opportunities for improvement; and (2) decreasing risks sooner, thus preventing recurrence of incidents [12]. Although the OSHA PSM Standard does not explicitly use the words "root" or "cause" in the requirements for the Incident Investigation element, it does require documentation of "corrective actions". This implies that a root cause analysis (RCA) is required based on quality standards where a corrective action is defined as the *action to eliminate the cause of an identified deficiency to prevent recurrence* [23]. Interestingly, the Mechanical Integrity element of the OSHA PSM Standard requires correction of deficiencies in equipment that are outside acceptable limits, without requiring that corrective actions be taken to eliminate the causes of these deficiencies [36]. Yet RCA and resultant corrective actions for mechanical integrity programs is recognized as a best practice in the industry. As such, this practice is another example of how an organization can further accelerate their learning of lessons beyond the minimum of regulatory compliance.

Preventive Actions – Based on industry quality standards, a preventive action is defined as the *action to eliminate the cause of a potential deficiency or other potential undesirable situation to prevent occurrence* [23]. Although not required by the OSHA PSM Standard, "… a preventive action extends the corrective action to other potential deficiencies, perhaps in other management system elements, other equipment or process units, or even other facilities" [29]. Organizations who apply the lessons from corrective actions to develop preventive actions to eliminate similar causes can further accelerate the learning process.

Based on the authors' experience in practicing and auditing the management of quality and process safety, many in the industry confuse and misuse the above terms for actions and their intents. In fact, many so-called corrective action systems do not actually address the elimination of the causes of deficiencies. Others assume (or use) the term preventive action to mean prevent the occurrence or the recurrence of deficiencies, while not even analyzing their causes in many cases.

While such confusion and misapplication leads to lower process safety performance, as measured by both leading and lagging indicators, it has been recognized that leaders who apply available best practices to "challenge the process" can accelerate an organization's learning and offer opportunities for future improvements to break the cycle of recurring incidents [31].

What Went Wrong? – Within Our Industry

The chemical process industry is replete with examples of process safety incident investigation reports and case histories from which we can borrow lessons and learn to apply to the improvement of PSM systems and performance [29,30,32,33,40]. The benefit is that we can accelerate our learning from others without having to actually experience such incidents.

Unfortunately, all too often some organizations are either unaware of the availability of incident investigation information, or perhaps cannot see the direct application of particular lessons to their own situation, and therefore miss out on these free opportunities for improvement.

One of the most valuable public sources of lessons is available from the findings and recommendations published in incident investigation reports by the U.S. Chemical Safety and Hazard Investigation Board (CSB). From their website:

> "The CSB conducts root cause investigations of chemical accidents at fixed industrial facilities. Root causes are usually deficiencies in safety management systems, but can be any factor that would have prevented the accident if that factor had not occurred. Other accident causes often involve equipment failures, human errors, unforeseen chemical reactions or other hazards." [20]

A summary of lessons below from ten publicly available incident investigation reports by the CSB (that were addressed in more detail during the 2016 GCPS Conference) shows example opportunities for improvement affecting the entire PSM system, as follows [29]:

1. The Employee Participation management process was in place and periodic safety culture assessments conducted, but actions were not effectively taken to improve identified weaknesses in the Stop Work Authority (SWA) program and therefore contributed to the incident.
2. The Process Hazard Analysis (PHA) management process was in place, but a recommended action for installing remotely-operated shut-off valves (ROSOVs) made during the initial PHA eleven years prior was never implemented, was subsequently inaccurately closed out as "complete", and was not discovered during a 5-year PHA revalidation.
3. An Operating Procedures management process was in place, but the operating procedures did not reflect actual practice, were not corrected over time, were not reviewed for changes as required before multiple start-ups, and were routinely deviated from during operations.
4. An Operator Training management process and related verification processes for managing risks, including PHA and MOC, were not effectively implemented to prevent injuries from an ink dust explosion and flash fire in a new dust collection system.
5. Contractors and Hot Work Permit programs were not in place, but could have provided for safe work practices and emergency procedures, as well as a means for verification of the welding task (e.g. hot work permit) and verification of contractor safety management (e.g. evaluating contractor safety performance when selecting contractors, informing contract employers of the known potential fire, explosion, or toxic release hazards related to the contractor's work, and periodically evaluating on-the-job performance) prior to the incident.
6. A Mechanical Integrity management process with periodic inspections was in place, but was not effectively implemented to determine causes and to correct deficiencies of process equipment outside acceptable limits before a process safety incident occurred.
7. Management of Change (MOC) and Pre-Startup Safety Review (PSSR) management processes were in place, but did not adequately address the deficiencies associated with the safety effects of a procedural change from water flushing to nitrogen purging for removal of hazardous chemicals as a precursor step to steam purging, nor the identification of low-point traps from a previous piping change, by the time a flash fire incident occurred.
8. The Incident Investigation management process was in place, but did not adequately identify causes or related hazards for previous incidents and no effective measures had been developed to prevent recurrence of numerous fires or tank overfills, nor was a review system in place to detect trends by the time a major process safety incident occurred.
9. An Emergency Planning & Response management process was not in place, which would have provided employees a means for planning, training, and verification to prepare for emergencies such as the release of heptane vapor from a heated mixing tank which exploded.
10. A Compliance Audit management process was in place, but the latest audits carried out by the facility missed many deficiencies that became apparent as a result of the incident.

The following section focusses on available sources of lessons (both internally and externally) based on *what went right*, addressing many of the above elements of an integrated PSM system.

What Went Right? – Within Your Organization & Our Industry

Although we tend to focus heavily on improvements based on *what went wrong*, incident investigations, as well as the results of other PSM processes, offer the benefits for accelerating learning from those activities, resources and results that are working well. For example, we often hear that an incident could have been so much worse had it not been for the successful function of certain safeguards or the proper actions of trained personnel; that is, *what went right*? Besides lessons from the internal strengths and successes of an organization, industry as a whole also offers a wealth of positive information regarding lessons to enable the acceleration of learning.

Your Organization – Capitalizing on *what went right* within a successful PSM system is about recognizing and extracting value from PSM system process activities and their deliverables. Tangible benefits can be derived from the activities of individual PSM processes themselves, as well as from their outputs (e.g. results). A detailed treatment of practical applications for the PHA process was presented during the 2014 GCPS Conference, with the following summary (see Table 1) offered as food for thought as individual organizations can undertake similar exercises to recognize and extract value from their PSM system processes [28]:

Table 1: Example Benefits from Both the Process Hazard Analysis Process and Its Results

Benefits from the PHA "Process"	Benefits from the PHA "Results"
1. Increased Workforce Participation	6. Purposeful Risk Reduction
2. Improved Interface Management	7. Supplemental Design Verification & Validation
3. Unit-Specific Awareness and Understanding	8. Assurance for Compliance and Conformance
4. Shared Ownership for Risk Management	9. Enhanced Management Systems
5. Strengthened Culture of Prevention	10. Continual Improvement for Operability

On a broader scale, "for each of the management processes for the PSM system, the aim is to take appropriate actions on a continual basis to improve the quality of the PSM system and its effectiveness in reducing and/or controlling process safety risks"[29]. With an understanding of the design intents for each PSM process as performance improvement objectives (see Table 2), an organization can take inventory of what has worked well in the past and can evaluate such lessons for accelerating their learnings. Additional lessons are offered by recognizing potential value "added" and value "derived" between a single process and other PSM processes (see Table 3).

Table 2: Example Performance Improvement Opportunities [29]

PSM Element	Performance Improvement Objectives
Employee Participation	Improve participation and ownership
Process Safety Information	Improve information quality
Process Hazard Analysis	Improve the process design & safeguard effectiveness
Operating Procedures & Safe Work Practices	Improve operational control of process and job tasks
Operator Training	Improve competence and performance
Contractors	Improve qualifications and performance
Pre-Startup Safety Review	Improve the adequacy of PSM elements before startup
Mechanical Integrity	Improve the asset integrity
Management of Change	Improve the control of changes
Incident Investigation	Improve the system and prevent incident recurrence
Emergency Planning & Response	Improve response capability
Compliance Audits	Improve the effectiveness of the PSM system

Table 3: Examples of Valued Added/Derived for the PHA Process [26]

Element	PHA Inputs – Value "Added"	PHA Outputs – Value "Derived"
Employee Participation	• Leadership to drive/enable risk reduction improvements to design & operation • Management commitment of resources to conduct PHAs • Empowerment and active participation of knowledgeable and experienced team	• Increased awareness of process safety and the process design and its operation • Experience in predicting hazard scenarios and evaluating risks of potential incidents • Cultural change of "believers" (e.g. I learned something new; golden nuggets)
Process Safety Information	• Accurate design and operating data • Verification of availability and effectiveness of safeguards (e.g. administrative and engineering controls)	• Recommendations for improvements to the Process Safety Information process • Supporting documentation for PHA archives (e.g. drawing mark-ups)
Process Hazard Analysis	• Robust PHA process to deliver quality results • Previous Hazard Analysis reports as input for revalidation	• Recommendations for risk reduction • Recommendations for improvements to the PHA process • Use of PHA results for future revalidations
Operating Procedures & Safe Work Practices	• Verification of availability and effectiveness of safeguards (e.g. administrative controls)	• Recommendations for improvements to the Operating Procedures and Safe Work Practices processes and related documentation • Use of hazard scenarios for develop operating procedures and consequences of deviations
Operator Training	• Verification of availability and effectiveness of safeguards (e.g. administrative controls)	• Recommendations for improvements to the Training process and related documentation • Use of hazard scenarios to raise awareness and provide basis for training
Contractors	• Active participation by contractors and equipment suppliers • Verification of vendor design data and operating experience	• Recommendations for improvements to the Contractors process • Use of PHA results for contractor training, pre-startup reviews
Pre-Startup Safety Review	• Verification of safeguards as input to the PHA process	• Recommendations for improvements to the PSSR process and records • Use of PHA results (e.g. hazard scenarios) for planning future PSSRs • Use of PHA to close out PSSR actions
Mechanical Integrity	• Accurate quality assurance information • Verification of availability and effectiveness of safeguards (e.g. administrative controls) • Understanding of equipment history and performance (e.g. equipment failures)	• Recommendations for improvements to QA & Mechanical Integrity process and related data • Use of PHA results for preventive maintenance, reliability centered maintenance, and risk-based inspection programs
Management of Change	• Verification of changes to process safety information as input to the PHA process • MOC records for PHA revalidations	• Recommendations for improvements to the MOC process and records • Use of PHA results for MOC reviews
Incident Investigation	• Lessons learned from previous investigations for hazard scenario development • Incident records for PHA revalidations	• Recommendations for improvements to the Incident Investigation process • Validation of risk controls for prevention • Recommendations for prevention of incidents • Use of PHA results for incident investigations
Emergency Planning & Response	• Verification of availability and effectiveness of safeguards (e.g. administrative and engineering controls)	• Recommendations for improvements to the Emergency Response process and plans • Use of hazard scenarios for developing emergency response plans and planning drills
Compliance Audits	• Corrective actions for improvements to the PHA process from previous audits	• Use of PHA process and reports for future audits (i.e. in place, adequately designed, effectively implemented)

Our Industry – We can borrow lessons from other organizations within our industry (who share their successes), as well as from academia (who research new methods), regulators (who promote good practices), trade organizations (who evolve industry standards), technical societies (who facilitate collaboration), and service and product providers (who offer subject matter expertise and technology). We can also borrow from the successes brought about by the global implementation of international standards for formal management systems (e.g. quality, environmental, occupational health and safety, risk, security), which have been adopted over the past 30+ years and have often been adapted to PSM systems, particularly the well-established continual improvement life cycle involving the steps: Plan-Do-Check-Act (PDCA) [29].

Of particular note is the role that industry organizations such as CCPS play in promoting the sharing of lessons learned, knowledge, and practices by member organizations and others:

> **CCPS Vision:**
>
> "In order to protect people, property and the environment, CCPS is committed to bringing the best process safety knowledge and practices to industry, academia, the government and the public around the world through collective wisdom, tools, training and expertise." [7]

Relevant to our discussions on accelerating learning from past lessons, in 2012 CCPS developed and rolled out the initiative *Vision 20/20* which "looks into the not-too-distant future to describe how great process safety is delivered when it is collectively and fervently supported by industry, regulators, academia, and the community worldwide; driven by the five industry tenets; and enhanced by the four global societal themes" [35]. Core principles within those *Vision 20/20* industry tenets and societal themes **bolded** in Table 4 have been reinforced throughout this paper.

Table 4: CCPS Vision 20/20 [35]

Industry Tenets	Societal Themes
• **Committed Culture** • **Vibrant Management Systems** • **Disciplined Adherence to Standards** • **Intentional Competency Development** • **Enhanced Applications & Sharing of Lessons Learned**	• Enhanced Stakeholder Knowledge • Responsible Collaboration • Harmonization of Standards • **Meticulous Verification**

As part of this initiative, benchmarking results from the *Vision 20/20* Assessment Tool can provide an indication of *what went right* as a basis for accelerating an organization's learning for applying lessons; thus translating a vision with strategy into a game plan for practical action [18].

4.2 Applying Learnings

A well-defined strategy for applying learnings (i.e. from borrowed lessons) leverages an organization's strengths and opportunities against weaknesses and threats. By capitalizing on what works best for us (e.g. proven methods and tools) and by eliminating what's holding us back (e.g. system deficiencies), we can accelerate our journey of risk reduction.

Learning to Drive, Driving to Learn – Acquiring Competence

All employees and contractors within an organization who are responsible for designing, constructing, commissioning, starting up, operating and maintaining process units for a facility, are potential drivers for managing process safety performance. Within a "committed culture", learning to drive faster can result in driving the learnings towards excellence [18]. Learning from process safety incidents and other lessons is all about how to improve; the more we learn, the more we improve, and the more we can accelerate not just the learning itself but also risk reduction.

A dictionary definition of learning is "the acquisition of knowledge or skills through study, experience, or being taught" [37]. Much has been studied and reported on individual learning (e.g. adult learning [4]) and organizational learning [41]. In fact, international standards for formal management systems (e.g. quality, environmental, occupational health and safety) have gone beyond the typical minimum regulatory requirement for training by placing broader emphasis on competence [34]. Competence refers to the understanding and interpretation of needed knowledge and the experience and skills to be able apply such knowledge to perform tasks correctly and safely [10]. Industry standards and guidelines have developed competencies as training criteria [3,21,22]. In fact, competence is considered a "personal" performance influencing factor (PIF) when aiming to reduce human error that could lead to process safety incidents [34].

With the global labor force in transition, previous experience and proven skills is increasingly retiring or leaving the industry, creating an urgency to fill the corporate knowledge gap. A new generation (e.g. managers, engineers, operators, mechanics) is coming along that may work in a plant that is already designed and operational (with the bugs worked out), and therefore doesn't need to re-invent the wheel (unless there are changes or incidents). New recruits may be taught to know the ways things work or the skills for how to perform a task, but they may not necessarily understand why. Inexperienced personnel cannot (and should not have to) wait for a generation to "learn" what they need to know in order to apply lessons for process safety already available.

A well-designed training process is crucial for acquiring competence within the workforce [3]. From an organizational design perspective [1], a competence assurance system addresses recruiting, training, and development of human assets. The *recruiting* process is intended to ensure that potential employees have been attracted, pre-qualified to minimum competency performance standards, and hired. The *training* process is intended to ensure that appropriate knowledge is imparted to, and necessary skills are developed by, all employees in order to qualify them as competent to perform their assigned job tasks proficiently. From implementation of the training process, competent people can become qualified to perform. The *development* process is intended to ensure that all employees continue to be involved in their own personal development (e.g. through job enrichment, etc.).

Further, the benefits of improved training and communications extend throughout the entire PSM system, where employees participate in managing their work and managing risks, either directly or indirectly. This includes several human factors issues: written operating procedures that provide clear instructions for safely conducting activities; receiving and understanding training in these procedures and safe work practices; refresher training to assure understanding of and adherence to operating procedures and safe work practices; being informed of and/or trained in hazards, changes, and actions. The bottom line here is: competent people can drive process safety performance; but highly competent drivers can accelerate that performance.

Navigating the Hazards – Assessing and Controlling Risks

Process hazards are the driving conditions on the journey towards improved risk reduction. A standard with established principles, framework and a generic process for managing risk is well suited for describing the key steps for managing process safety risks (see Figure 2) [24].

Risk Assessment – The risk assessment process can be considered the engine for identifying, analyzing and evaluating process hazards and associated risks [11]. Drivers must be able to "see" the hazards if they are to have any control of the risks potentially being fulfilled as future incidents.

- Firstly, if we cannot or do not identify all process hazards, then we will be unable to adequately analyze and evaluate the process safety risks. A thorough PHA requires a well-defined scope and objectives, as well as complete and correct process safety information.
- Further, if we cannot or do not fully understand: (1) the causes of how these process hazards (based on credible hazard scenarios) can be initiated, nor (2) the physics and chemistry that lead to consequences with harmful impacts, nor (3) the design intent and effectiveness (i.e. availability, sufficiency, reliability) of existing risk controls; then we will underestimate the risks and miss opportunities for accelerating risk reduction. A meaningful PHA requires a suitable methodology (e.g. Hazard and Operability (HAZOP)) and a competent team to analyze the process hazards, as well as their causes, consequences and risk controls [28].
- Finally, if we cannot or do not properly estimate the severities, likelihoods and resultant risks for each hazard scenario against a practical set of criteria for judging risk tolerability, then we will be unable to evaluate the importance of existing risk controls and will miss more opportunities for improving risk controls to achieve process safety risk reduction.

Risk Control – The risk treatment process can be considered the vehicle for controlling such risks. This powerful vehicle includes both the engineered systems (i.e. engineering risk controls) and the whole PSM system (i.e. administrative risk controls). This process applies to designing new vehicles, analyzing existing vehicles, and changing the design features of existing vehicles for improved performance. A PHA is an independent check on the adequacy and effectiveness of the risk controls (i.e. safeguards or layers of protection) of the chemical process, which are designed for prevention, detection, control, mitigation and emergency response, e.g.:

- Engineering Risk Controls – reliable equipment designs, indicators with alarms, equipment safety trips, pressure safety valves, secondary containment, fire suppression systems, etc.
- Administrative Risk Controls – operating procedures, safe work practices, operator surveillance, response to alarms, training/drills for emergency response, etc.

A quality PHA can determine whether the controls are: (1) in place (i.e. available), (2) adequately designed (i.e. reliable), and (3) effectively implemented (i.e. in use, reviewed, tested, maintained). From the PHA, if existing safeguards are not adequate or effective for the level of risk evaluated, then additional and/or improved risk controls may be warranted. Thus, PHA recommendations drive risk reduction performance. However, if needed recommendations are not being generated, or if recommendations are not being properly resolved, or if recommended actions are not being closed out, then risk reduction performance decelerates. The presence of any of these conditions deserves immediate attention to determine why the engine and/or vehicle is running poorly. You cannot depend (or wait) on a compliance audit every few years to safely run your daily operation.

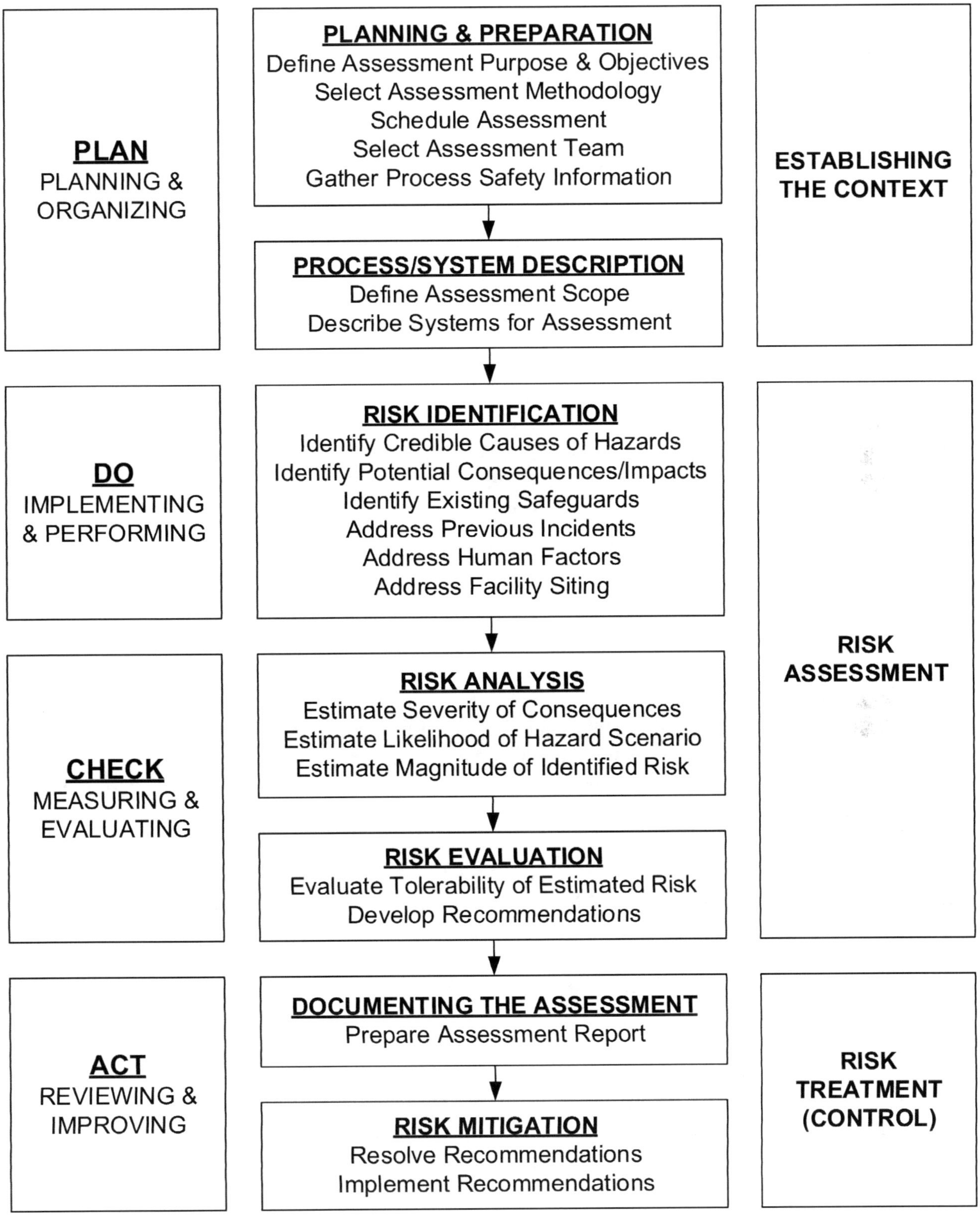

Figure 2: Risk Management Process Adapted to PSM [24]

Maintaining the Vehicle – Auditing PSM System Effectiveness

Even when we can see the hazards and are equipped to handle the road conditions, drivers for the journey also need to perceive the assessed risks and the consequences of their actions (or inactions) in order to make the journey safely. That is, they also need to address the behavioral aspects of managing process safety, beyond the technical execution of the risk management process [25]. One of the most important leadership actions is commitment to a periodic assessment of the PSM system; that is, the vehicle used for maintaining risk control and for driving continual risk reduction performance. Such an assessment is often referred to as a management system audit, or where regulated, a compliance audit. Since the OSHA PSM Standard is a performance-based regulation, the compliance audit is in fact a PSM system assessment, or a check on its health with actions to address system deficiencies to maintain that health and kept it functioning [9].

Much has been written about designing a PSM system [17,27]. In general, we know that robust management systems that integrate all of the resources and activities of a facility can be a powerful tool in reducing incidents by controlling process safety risks. We also know, from these same systems, where we could have weaknesses and how we can improve them to deliver better risk reduction results. Using the basic premise of auditing, we can verify (and validate) the quality of our resources and activities as they relate to each individual management system process for process safety by asking three critical questions:

- Are they in place? Do they exist?
- Are they adequately designed? Can they get the job done?
- Are they effectively implemented? Are they getting the job done?

Auditing PSM system components (e.g. policies, processes, programs, procedures, plans, etc.) for effectiveness follows a similar methodology as the PHA in determining whether the administrative controls are: (1) in place (i.e. are established and documented); (2) adequately designed (i.e. to address who, what, when, where, how and why); and (3) effectively implemented (i.e. have been rolled out and put into practice for managing the continual improvement life cycle, including: Plan (e.g. assignments, training), Do (e.g. use), Check (e.g. reviews), Act (maintenance, improvement)).

Thus, accelerating learning from past lessons applies not only to incident investigations, but to the learnings from results of PSM system audits, including the audit process itself. Increasing the frequency and/or quality of PSM audits can accelerate improvements to the PSM system. This in turn can improve efficiency and effectiveness for other PSM system processes to directly accelerate reducing risks and therefore prevent the occurrence/recurrence of process safety incidents. Maintaining PSM system effectiveness through the auditing process as a management tool enables safe driving at higher speeds. Some examples include: running the PSM system in a sustained race (i.e. as a process) versus a shorter-term sprint (i.e. as a project); driving with a clean windshield to see the benefits of heading towards performance instead of stopping at compliance; looking under the hood to check the vehicle's condition by utilizing proven verification methods for auditing (document/record reviews, inspections, interviews) and also design reviews, pre-startup safety reviews, equipment inspections/testing, evaluation of contractors, etc.

A robust PSM audit process can identify numerous system deficiencies which can then turn the PSM system vehicle into a high-performance race car, where acceleration is possible!

4.3 Creating and Sustaining Momentum

During the journey toward incident-free performance, the risk of incidents remains very real; hence, there is an urgent need for swift action to accelerate the learnings from past lessons to achieve risk reduction in order to prevent future harm. Urgency is synonymous with importance, thus requiring a further (or revitalized) commitment by an organization's management to make process safety a higher priority and to sustain a culture to take process safety performance to a higher level. Anything that slows (or stops) progress (e.g. time delays from gradual learning or re-learning) to the destination is considered a bump in the road (or worse, a roadblock).

Keeping It In High Gear – Efficiency with Timing

By definition, accelerating learning by applying best practices implies shifting gears, picking up speed and gaining momentum. Creating that momentum requires overcoming inertia (unless the organization is already making good progress and their risks are generally under control). Demonstrated leadership and commitment of resources are necessary to drive the acceleration, especially where the culture needs improving. Where an organization starts from in terms of overall risk performance, and how well they are already progressing (if at all) in reducing their risks, should influence the rate of acceleration needed to reach the target for acceptable levels of risk. Further down the road, the challenge of sustaining that momentum also needs to be addressed.

To prevent incidents in a timely manner, we must be efficient if we hope to accelerate the process of reducing risks to acceptable levels. Reducing process cycle times and non-value added tasks helps us to speed up, while eliminating waste and other obstacles helps to clear the path and reduce delays. In many cases, improvements to management processes can achieve both objectives simultaneously. Why waste time and precious resources to re-invent the wheel when you are able to use a best-in-class engine with better training, better information, better communications, better methodologies, and better tools to drive performance improvement faster with higher ***efficiency*** (defined as the "relationship between the results achieved and the resources used") [23]?

For a PSM system that utilizes a formal Plan-Do-Check-Act (PDCA) continual improvement life cycle approach, every management process and sub-process within the PSM system is a candidate for improved efficiencies through the adoption of best management practices. These are aimed at utilizing quality resources (e.g. competent personnel, accurate information, reliable tools) and carrying out quality work activities (e.g. detailed planning & organizing, rigorous implementing & performing, meticulous measuring & evaluating, and continual reviewing & improving) [34]. Generated savings from efficiency gains can then be reinvested for cost-effective acceleration.

Efficiency efforts in process safety can benefit from applying proven methods and tools that are already being employed in Lean and Six Sigma initiatives by organizations pursuing operational excellence. "Six Sigma focuses on reducing process variation and enhancing process control, whereas lean drives out waste (non-value added processes and procedures) and promotes work standardization and flow" [2]. A recommended strategy for the PSM system includes reducing management process review cycle times. That is, shorter-term cycles for reviewing process safety information and operating procedures, delivering refresher training, managing changes, and conducting periodic PHAs and compliance audits, can be beneficial for improving those verification activities necessary for reducing process safety risks. For such cases, there can be greater assurance in the overall robustness of the PSM system as the vehicle for managing risks.

Keeping It Fueled – Effectiveness with Actions

To enable the acceleration of learnings, we must continually get results along the way to get where we are headed. To prevent incidents, we must be effective in eliminating or controlling risks to tolerable levels. Individual management processes and their integration within the PSM system serve to produce results (especially actions) that directly or indirectly achieve risk reduction. Within the context of the PDCA design for each management process, the **actions** produced from the "act" step for these processes are thus the key driver for continual process safety performance improvement [29]. By setting an example and inspiring others to create a "bias for action" [38], leaders at the top and at all levels of a "leaderful" organization [39] (with full employee participation in managing process safety) can engage in active decision making and problem solving. Without such dedication to action, we risk running out of gas and stalling.

A review of ten previous process safety incidents investigated by the CSB (that were addressed in more detail during the 2016 GCPS Conference) shows that improperly managed deficiencies and associated actions in the Plan, Do, Check, and/or Act steps for eleven elements of PSM systems contributed to their occurrence, as follows [29]:

- For instance, some incidents involved missing or ineffectively implemented policies and programs that would have required in the "Plan" step the need for identifying deficiencies, and therefore prevented any progress in the subsequent "Do", "Check", and "Act" steps.
- Next, some incidents which may have had successful planning also inadequately identified deficiencies in the "Do" step, which did not allow for meaningful causes to be determined in the subsequent "Check" step.
- Further, other incidents included no root causes or improperly determined causes of the identified deficiencies in the "Check" step, which did not allow for suitable actions to be proposed for the "Act" step.
- Finally, one incident demonstrated that despite apparent success in the "Plan", "Do", and "Check" steps, ineffectively managed action within the action system prevented improvement and actually contributed to the incident.

Since incidents often occur as a result of a lack of actions, one the highest priorities to maintain the overall effectiveness of the PSM system is to rigorously manage these actions. A best practice in process safety is to "implement a system to manage the life cycle of actions in order to correct identified deficiencies, eliminate the causes of existing deficiencies through corrective actions in order to prevent their recurrence, and eliminate the causes of potential deficiencies through preventive actions in order to prevent their occurrence in the future" [29].

Within such action systems, as well as throughout the PSM system of management processes, proper planning, implementing, and monitoring of progress toward completion of corrections, corrective actions, and preventive actions are required for managing process safety ***effectiveness*** (defined as the "extent to which planned activities are realized and planned results are achieved" [23]). Thus, accelerating the learnings of lessons necessitates that we keep the vehicle for managing risks (i.e. the PSM system) full of fuel during the journey of risk reduction for incident prevention. We can achieve higher performance by ensuring that ALL needed actions are effectively managed and closed out in a timely manner (i.e. with all due speed), thus minimizing their backlog.

Keeping It Tuned Up – Sustainability with Changes

To sustain the momentum during the journey towards the incident-free goal, we need to be in control of changes. All process safety incidents are the result of unplanned or uncontrolled changes. Changes can either decrease risk (if managed) or increase risk (if not controlled). Lessons from many process safety incidents involving the MOC process have shown that changes can affect both the engineered and management processes, as well as the organization itself [32,40]. In response to these incidents, both regulations and industry guidelines have provided sound principles for identifying and controlling changes to process chemicals, technology, equipment, facilities, and procedures [13,36]. Organizational changes can occur internally through the normal course of running a business, or can be externally driven (e.g. the labor pool). Recruitment, training, and continuity planning are necessary, but can also introduce changes that can affect the effectiveness of the work performed and the PSM system processes being operated. From its first publication, CCPS stated in their *Challenge to Commitment* brochure sent to 1,500 CEOs that "particular attention must be paid during organizational transitions, such as down-sizing or mergers, acquisitions or divestitures"[8]. More recently, industry guidelines and some regulations have addressed management of organizational change related to PSM [6,14]. One of the biggest opportunities for improvement is associated with the proper design and implementation of the MOC management process. Based on the authors' experiences, some systems are very effective, while others are often complex and cumbersome to use. Further, unless the MOC process is robust enough to adequately evaluate hazards that may be introduced by changes, the risks of such changes may go undetected until future PHA revalidations, or perhaps never.

Other changes (i.e. those that are unplanned but that may be expected) often occur as the result of degradation in either the engineered systems or the management systems. For example:

- Commitment to Asset Integrity – degradation through normal wear and tear, corrosion, and especially operating outside of the design envelope can affect the integrity of engineered systems. This degradation increases the likelihood of equipment malfunctions as causes for hazard scenarios, as well as failures of engineering controls required in safeguarding an event. A progressive Asset Integrity process with verification steps for inspection, testing and maintenance, including preventive maintenance programs (PM), reliability-based inspection (RBI), and reliability-centered maintenance (RCM), offers improvement opportunities for risk reduction over a poorly executed "run to failure" philosophy [15,17].
- Culture of Operational Discipline – degradation of human performance through skills becoming rusty, knowledge being forgotten, or "normalization of deviance" from operating procedures can affect the operational discipline needed within the PSM system [19,34,42]. This degradation increases the likelihood of human factors as causes for hazard scenarios, as well as inadequate human responses required for managing an event. Refresher training to maintain competence, rigorous review of changes to maintain accurate operating procedures, and reinforcement of strict adherence to procedures, including safe work practices and permit-to-work systems, offer improvement opportunities for risk reduction [18]. This is especially important during a shutdown operation or during a startup following a turnaround, where the highest potential for incidents exists.

Keeping the vehicle tuned up and its drivers operating at their highest level requires a commitment to anticipating and controlling changes affecting system degradation, and by maintaining the ongoing integrity of the physical resources and the operational discipline of the human resources.

5. Conclusions

This paper has posed a challenging question: *"Knowing what we already know, why would we want to take years to gradually reduce our risks, when we could accelerate the learning from past lessons and prevent incidents using available best industry practices?"*.

The simple answer is that we don't, and shouldn't. The industry already has available a wealth of knowledge and experience of what can go (and has gone) wrong, and a powerful toolkit of information, methods and tools to reduce process safety risks and prevent future incidents.

To accelerate process safety performance, leadership can ask (and get answers to) the following:

- If you are having incidents do you know why? – If so, you can prevent future incidents by accelerating improvements before the next incident. If not, you should fix the incident investigation process to know what to improve to avoid unknown risks, since you don't know what you don't know.
- If you are not having incidents, do you know why? – If so, is it that you're doing something right, or that you've been lucky so far? If you do know, then sustain your successes and improve further. If not, then you should find out, since what you don't know can cause harm. Better investigations of near miss incidents can be very helpful as a starting point.

This paper has shown that we have an excellent opportunity to accelerate risk reduction performance by learning from borrowed lessons and applying available best industry practices. The leadership challenge is to aim for an incident-free industry, nurture the culture, and empower all drivers in the race against time. Towards this end and with these means, we can live up to our obligations to better protect people's safety and health, the environment, and our communities. The *roadmap* below summarizes some of the key topics from this paper for your journey:

Sections	Key Topics
1. Introduction	
2. Where Are We Today ... and Why?	
2.1 What We Already Know	Knowing Your Current Rate of Progress
2.2 What More We Should Know	Assessing System Requirements and Gaps
2.3 Why We Are Where We Are	Understanding Strengths & Weaknesses
3. Where Do We Want (Need) to Be ... and When?	
3.1 A Future Free of Incidents	Preventing All Process Safety Incidents
3.2 In Better Control of Risks	Setting Performance Standards for Control
3.3 As Soon As Reasonably Practicable	Racing Against Time
4. How Do We Get There ... and Faster?	
4.1 Borrowing from Lessons	
What Went Wrong? – Within Your Organization	Identifying Your Roots Causes
What Went Wrong? – Within Our Industry	Learning from Others
What Went Right? – Within Your Organization & Our Industry	Leveraging Successes and Best Practices
4.2 Applying Learnings	
Learning to Drive, Driving to Learn – Acquiring Competence	Maintaining Knowledge/Skills/Experience
Navigating the Hazards – Assessing and Controlling Risks	Driving Risk Reduction Performance
Maintaining the Vehicle – Auditing PSM System Effectiveness	Assuring It's In Place, Adequate, Effective
4.3 Creating and Sustaining Momentum	
Keeping It In High Gear – Efficiency with Timing	Reducing Process Cycle Times & Waste
Keeping It Fueled – Effectiveness with Actions	Generating and Closing Out Actions
Keeping It Tuned Up – Sustainability with Changes	Assuring Integrity & Controlling Changes

6. References

[1] Ackoff, Russell L. *Re-Creating the Corporation: A Design of Organizations for the 21st Century.* ISBN-13: 978-0195123876, ISBN-10: 9780195123876, Oxford University Press, Inc., New York, New York, 1999.

[2] ASQ. "What is Lean Six Sigma?", American Society for Quality, Milwaukee, Wisconsin. Retrieved on January 15, 2019 from: https://asq.org/quality-resources/six-sigma

[3] ASSE. ANSI/ASSE Z490.1-2016, *Criteria for Accepted Practices in Safety, Health, and Environmental Training*, American Society of Safety Engineers (ASSE), Park Ridge, Illinois, January 2016.

[4] ATD. Association for Talent Development. Retrieved on January 15, 2019 from: https://www.td.org/about

[5] Carroll, Lewis. Alice in Wonderland, Quotes, Retrieved on January 15, 2019 from: https://www.goodreads.com/quotes/225938-would-you-tell-me-please-which-way-i-ought-to

[6] CCHS. "Industrial Safety Ordinances (ISO) Risk Management ", Contra County Health Services, Retrieved on January 15, 2019 from: https://cchealth.org/hazmat/iso/risk-management.php

[7] CCPS. "CCPS Vision", Center for Chemical Process Safety (CCPS) of the American Institute of Chemical Engineers (AIChE). Retrieved on January 15, 2019 from: https://www.aiche.org/ccps/about/mission-vision

[8] CCPS. "Chemical Process Safety Management: A Challenge to Commitment" (brochure), Center for Chemical Process Safety (CCPS) of the American Institute of Chemical Engineers (AIChE), New York, New York, 1985.

[9] CCPS. *Guidelines for Auditing Process Safety Management Systems*, 2nd Edition, ISBN 13: 978-0-470-28235-9, Center for Chemical Process Safety (CCPS) of the American Institute of Chemical Engineers (AIChE) and John Wiley & Sons, New York, NY, 2011.

[10] CCPS. *Guidelines for Defining Process Safety Competency Requirements*, 1st Edition, ISBN 13: 978-1-118-79522-4, Center for Chemical Process Safety (CCPS) of the American Institute of Chemical Engineers (AIChE) and John Wiley & Sons, New York, New York, 2015.

[11] CCPS. *Guidelines for Hazard Evaluation Procedures*, 3rd Edition, ISBN: 13: 978-0-471-97815-2, Center for Chemical Process Safety (CCPS) of the American Institute of Chemical Engineers (AIChE) and John Wiley & Sons, New York, New York, 2008.

[12] CCPS. *Guidelines for Investigating Chemical Process Incidents*, 2nd Edition, ISBN: 13: 978-0-8169-0897-4, Center for Chemical Process Safety (CCPS) of the American Institute of Chemical Engineers (AIChE) and John Wiley & Sons, New York, New York, 2003.

[13] CCPS. *Guidelines for Management of Change*, 1st Edition, ISBN 13: 978-0-470-04309-7, Center for Chemical Process Safety (CCPS) of the American Institute of Chemical Engineers (AIChE) and John Wiley & Sons, New York, New York, 2008.

[14] CCPS. *Guidelines for Managing Process Safety Risks During Organizational Change*, 1st Edition, ISBN 13: 978-1-1183-7909-7, Center for Chemical Process Safety (CCPS) of the American Institute of Chemical Engineers (AIChE) and John Wiley & Sons, New York, New York, 2013.

[15] CCPS. *Guidelines for Mechanical Integrity Systems*, 1st Edition, ISBN 13: 978-0-8169-0952-0, Center for Chemical Process Safety (CCPS) of the American Institute of Chemical Engineers (AIChE) and John Wiley & Sons, New York, New York, 2006.

[16] CCPS. *Guidelines for Process Safety Metrics*, 1st Edition, ISBN 13: 978-0-470-57212-2, Center for Chemical Process Safety (CCPS) of the American Institute of Chemical Engineers (AIChE) and John Wiley & Sons, New York, New York, 2009.

[17] CCPS. *Guidelines for Risk Based Process Safety*, 1st Edition, ISBN 13: 978-0-470-16569-0, Center for Chemical Process Safety (CCPS) of the American Institute of Chemical Engineers (AIChE) and John Wiley & Sons, New York, New York, 2007.

[18] CCPS. "Vision 20/20", Center for Chemical Process Safety (CCPS) of the American Institute of Chemical Engineers (AIChE). http://www.aiche.org/ccps/resources/vision-2020 (Accessed January 15, 2019).

[19] CSB. "Investigation Report – Refinery Explosion and Fire" (BP, Texas City, Texas, March 23, 2005), No. 2005-04-I-TX, U.S. Chemical Safety and Hazard Identification Board, March 2007.

[20] CSB. U.S. Chemical Safety and Hazard Identification Board. Retrieved on January 15, 2019 from: https://www.csb.gov/about-the-csb/mission/

[21] EPSC. "Process Safety Competence – How to Set Up a Process Safety Competence Management System", Report Number 35, European Process Safety Centre (EPSC), UK, 2013. Retrieved on January 15, 2019 from: https://epsc.be/Products/Reports/EPSC+Reports+Available/_/ReportNo35.pdf

[22] ISC. "ISC Process Safety Competency Guidance", IChemE Safety Centre (ISC), Edition 2, 2018. Retrieved on January 15, 2019 from: https://www.icheme.org/media/1094/0007_18-competency_brochure-final.pdf

[23] ISO. ISO 9000:2015, "Quality management systems – Fundamentals and vocabulary", International Organization for Standardization (ISO), Geneva: ISO, 2015.

[24] ISO. ISO 31000:2018, "Risk management – Guidelines", International Organization for Standardization (ISO), Geneva: ISO, 2018.

[25] Jones, David A. "Addressing the Behavioral Aspects of Human Factors in Process Safety Management", 2007 International Symposium, Mary Kay O'Connor Process Safety Center, College Station, Texas, October 23-24, 2007.

[26] Jones, David A. "Enhancing HAZOP Studies – Adding Value by Integrating the Hazard Analysis Process into the Process Safety & Environmental Management System (PSM/SEMS)", Offshore Process Safety Management Conference, Houston, Texas, September 11-12, 2013.

[27] Jones, David A. "How to Design Effective EH&S Management Systems for Continual Improvement of Risk Reduction Performance", ASSE Safety 2002, American Society of Safety Engineers (ASSE), Nashville, Tennessee, June 2002.

[28] Jones, David A. "Tangible Benefits Derived from the Application of the Hazard and Operability (HAZOP) Methodology for Process Safety Risk Assessments of Mobile Offshore and Land Drilling Units", American Institute of Chemical Engineers (AIChE), 10th Global Congress on Process Safety, New Orleans, Louisiana, March 30-April 2, 2014.

[29] Jones, David A. and Kay Modi. "A Call to Action: Understanding the Key Driver for Continual Process Safety Performance Improvement", American Institute of Chemical Engineers (AIChE), 12th Global Congress on Process Safety, Houston, Texas, April 10-13, 2016.

[30] Kletz, Trevor. *What Went Wrong? Case Histories of Process Plant Disasters and How They Could Have Been Avoided*, 5th Edition, ISBN-13: 978-1856175319, ISBN-10: 1856175316, Butterworth-Heinemann/IChemE, 2009.

[31] Kouzes, James M. and Barry Z. Posner. *The Leadership Challenge*, 6th Edition, ISBN-13: 978-1119278962, ISBN-10: 1119278961, John Wiley & Sons, Inc., Hoboken, New Jersey, 2017.

[32] Mannan, Sam (Editor). *Lees' Loss Prevention in the Process Industries,* Volumes 1-3, Fourth Edition, ISBN 10: 0123971896, ISBN 13: 9780123971890, Butterworth-Heinemann /Elsevier, Oxford, UK, 2012.

[33] Marsh. "The 100 Largest Losses 1978-2017: Large Property Damage Losses in the Hydrocarbon Industry", Marsh Report, 25th Edition, Marsh Ltd., 2018. Retrieved on January 15, 2019 from: https://www.marsh.com/uk/insights/research/100-largest-losses-in-the-hydrocarbon-industry.html

[34] Mathur, Minu Shikha and David A. Jones. "How to Design Human Factors into a PSM System to Ensure Effectiveness and Deliver Risk Reduction Performance", American Institute of Chemical Engineers (AIChE), 12th Global Congress on Process Safety, Houston, Texas, April 10-13, 2016.

[35] McCavit, Jack, Scott Berger, Cheryl Grounds, Louisa Nara. "A Call to Action: Next Steps for Vision 20/20", American Institute of Chemical Engineers (AIChE), 10th Global Congress on Process Safety, New Orleans, Louisiana, March 30 - April 2, 2014.

[36] OSHA. "Process safety management of highly hazardous chemicals", 29 CFR 1910.119, (e), Process Hazard Analysis, Occupational Safety and Health Administration (OSHA), *Federal Register*, February 24, 1992 (57 FR 6356).

[37] Oxford. "Definition of Learning", English Oxford Living Dictionary. Retrieved on January 15, 2019 from: https://en.oxforddictionaries.com/definition/learning

[38] Peters, Thomas J., Robert H. Waterman Jr. *In Search of Excellence: Lessons from America's Best-Run Companies*, First Edition, ISBN-10: 0446378445, ISBN-13: 9780446378444, Grand Central Publishing, New York, New York, 1984.

[39] Raelin, Joseph A. *Creating Leaderful Organizations: How to Bring Out Leadership in Everyone*, First Edition, ISBN-13: 978-1-57675-233-3, Berrett-Koehler Publishers, Inc., San Francisco, CA, 2003.

[40] Sanders, Roy E. *Management of change in chemical plants: learning from case histories,* Fourth Edition, ISBN-13: 978-0-12-397189-0, Butterworth-Heinemann, Oxford, UK, 2015.

[41] Senge, Peter M. *The Fifth Discipline: The Art & Practice of The Learning Organization*, ISBN-10: 0-385-51725-4, Doubleday Publishing, New York, New York, 1990.

[42] Vaughan, Diane. *The Challenger Launch Decision: Risky Technology, Culture, and Deviance at NASA*, ISBN-13: 9780226851754, University of Chicago Press, Chicago, Illinois, 1996.

548252 Dow's Life Critical Standards - Making Them Easier to Follow

Dow's Life Critical Standards: Making Them Easier to Follow

Dow Chemical is making it easier for employees to follow our Life Critical Standards.

Dow has eight Life Critical Standards:

1. Safe Work Permit
2. Isolation of Energy Sources
3. Line & Equipment Opening
4. Elevated Work
5. Hot Work
6. Electrical
7. Confined Space
8. Hydroblasting

We know that when employees follow these Standards—we reduce our process safety incidents.

Punishing employees for violating the Standard is not the best way to get compliance. We have large plants, all over the world, usually with 24-hour shifts. We are not going to "police" our way into compliance. And, we know the research: punishment accounts for less than 10% of the variance in employee rule following.

Instead of punishment, we asked: How could we make these Standards easier to follow?

Our solution was to make a cover page for each Standard. That cover page visually communicates the most critical tasks inside each Standard.

To learn these critical tasks, we asked our in-the-field safety representatives: "Inside this Standard, what critical tasks are your employees most likely to ignore or forget?" Once we knew the critical tasks—usually around six critical tasks for each Standard—we represented each task in an illustration.

We want our cover pages to communicate visually. So, the cover pages are mostly illustrations. There are very few words. Again, we knew the research: illustrations have about seven times higher recall than words.

Finally, we made frontline supervisors our first priority in receiving the new cover pages. We know that close to 70% of frontline employees will follow a Standard if their supervisor specifically asks them to do so.

These new cover pages make it easier for supervisors to ask for this compliance.

We know the importance of our Life Critical Standards. So, our goal is to make them easier to follow.

END

548244 Process Safety Risk Prioritization and Auditing 3rd Party Manufacturers and Suppliers

Many companies utilize 3rd party manufacturers, often called tollers, to allow for flexibility in production for a wide variety of products and package sizes. From a business standpoint this model is very reasonable but it can add safety and environmental challenges when the right oversight and management system is not in place. 3rd party tollers needs to be evaluated based on business and HSE risk and continually assessed to improve HSE performance as part of a commitment for a healthy and safer world along with continued strong business relationships.

Most companies utilize 3rd party raw material (chemical) suppliers for additives for chemical manufacturing processes. Raw material suppliers can present business and supply challenges. Health and Safety of chemical suppliers should be prioritized and assessed to help ensure a continued supply chain but also sustaining our commitment for a healthy and safer world.

- Ways to prioritize toller & supplier risk
- HSE assessment and audit framework

Don't Let Your Safe Operating Limits Leave You S-O-L (Out of Luck)

Todd Stauffer
Exida Consulting, LLC
Sellersville, PA
tstauffer@exida.com

Denise Chastain-Knight
Exida Consulting, LLC
Sellersville, PA
dchastainknight@exida.com

Prepared for Presentation at
American Institute of Chemical Engineers
2019 Spring Meeting and 15th Global Congress on Process Safety
New Orleans, LA
March 31 – April 3, 2019

Don't Let Your Safe Operating Limits Leave You S-O-L (Out of Luck)

Todd Stauffer
Exida Consulting, LLC
Sellersville, PA
tstauffer@exida.com

Denise Chastain-Knight
dchastainknight@exida.com

Keywords: Safe Operating Limits, Design Limits, Never Exceed Limits, SOL Exceedances, Tier 3 Leading Indicators, Process Safety Information,

Abstract

Collection and Utilization of process safety metrics is an important tool for driving improved safety. Tier 3 leading indicators (challenges to safety system) indicate failures of process safety management systems and highlight areas that should be improved to prevent a more serious event. Safe Operating Limit (SOL) exceedances are a commonly used Tier 3 leading indicator. Surprisingly, there are many different approaches used in industry to calculate safe operating limits and to apply them. This inconsistency potentially diminishes the usefulness of SOL exceedances as an effective indicator.

This paper discusses current industry practices around the determination and application of safe operating limits as established by a recent benchmark survey of over 150 safety practitioners from around the world. Areas explored in the survey of SOLs include; methodology for calculating, how / where information is stored, how / when established values are reviewed and audited, usage as a Process Safety Management Leading indicator, integration with operations (training, documentation), identification and tracking of when exceedances have occurred, and actions taken on exceedance. Key results and conclusions will be presented as well as recommendations on where industry should focus on improvement.

1 Introduction

Determination and documentation of Safe Operating Limits (SOLs), the limits beyond which would be considered upset conditions, is required per the OSHA Process Safety Management (PSM) Regulation 1910.119. SOLs and associated consequences of deviation that could occur if these limits were violated are considered part of the Process Safety Information (PSI) [1]. Exceeding Safe Operating Limits are considered a Tier 3 Process Safety Management Leading Indicator per API RP-754 [2]. SOLs are typically documented in operating procedures and should be well known and understood by the operations team, including how to respond to prevent the potential consequence (the consequence of deviation). SOL values should be firmly established to support HAZOP analysis and evaluation of process changes as part of management of change (MOC).

Despite the broad applicability across industry, the determination and application of SOLs often varies from company to company. An industry benchmark study was conducted amongst 152 functional safety practitioners from around the world. The survey respondents were asked both multiple choice and short answer questions.

2 Notation

API	American Petroleum Institute
CCPS	Center for Chemical Process Safety
DCS	Distributed Control System
ISA	International Society of Automation
I/O	Input/Output
IPL	Independent Protection Layers
KPI	Key Performance Indicator
LOPA	Layer of Protection Analysis
LOPC	Loss of Primary Containment
MAWP / MAWT	Maximum Allowable Working Pressure / Temperature
MOC	Management of Change
OSHA	Occupational Safety & Health Administration
PHA	Process Hazard Analysis
PRV	Pressure Relief Valve
PSI	Process Safety Information
PST	Process Safety Time
RT	Response Time
SIS	Safety Instrumented System
SOL	Safe Operating Limit

3 What are Safe Operating Limits and Why Are they Important

Although the requirement to determine and document Safe Operating Limits is defined in OSHA PSM, there is no prescriptive instruction provided. This leaves the definition and methodology for determination open to interpretation. Industry guidelines are often inconsistent and contradictory in methodology and terminology. Following are the definitions for SOL from CCPS, ISA, and API industry guidelines.

Safe Operating Limit (SOL) - Limits established for critical process parameters, such as temperature, pressure, level, flow, or concentration, based on a combination of equipment design limits and the dynamics of the process [3,4].

Safe Operating Limit (SOL) – A value for a critical operating parameter (e.g., low/high pressure, level, temperature, pH, composition, and flow) that defines the equipment or process unit safe operating envelope beyond which a process will not intentionally be operated due to the risk of imminent catastrophic equipment failure or loss of containment. Operational or mechanical corrective action ceases and immediate predetermined actions are taken at these critical operating parameter values in order to bring equipment and process units to a safe state [5].

Safe Operating Limit (SOL) – value for a critical operating parameter outside of the normal operating limits defining a threshold of abnormal process condition beyond which predetermined actions are taken in order to avoid a hazardous event and return the process to a safe state.

NOTE 1: The definition of "safe state" can be different for different protection layers, such as returning the process to the normal operating limits or safely shutting down the equipment.

NOTE 2: Exceeding a safe operating limit is indicative of an abnormal operating condition.

NOTE 3: The IPL setpoint can coincide with a SOL in some cases [6].

Figure 1 below provides a means to visualize Safe Operating Limits as applied to API-RP 754, where SOL exceedances are defined as a Tier 3 Process Safety Performance leading indicator.

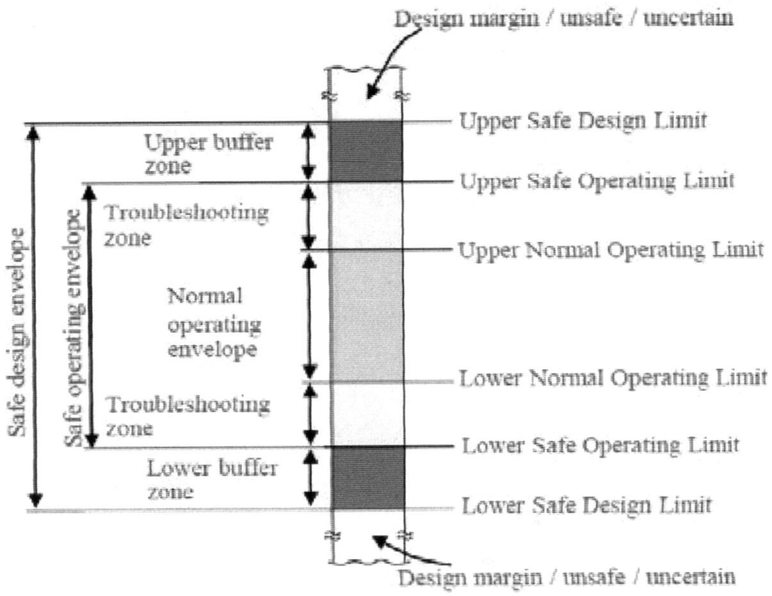

Figure 1. Safe Operating Limits [7]

Safe Operating Limits represent the point beyond which troubleshooting ends and pre-determined action occurs to return the process to a known safe state. Pre-determined actions may range from manually executed operating procedures to a fully automated safety instrumented system [8]. Operating outside the Safe Operating Limits results in a higher probability of Loss of Primary Containment [7].

Consolidating key aspects of the definitions above yields the following elements for discussion of safe operating limits. Safe operating limits:

- Are values of critical operating parameters (e.g., low/high pressure, level, temperature, pH, composition, and flow) outside of the normal operating limits.
- Are established based on a combination of equipment design limits and the dynamics of the process.
- Define the equipment or process unit safe operating envelope beyond which a process will not intentionally be operated due to the risk of imminent catastrophic equipment failure or loss of containment.
- Represent the point at which troubleshooting ceases and immediate predetermined actions are taken in order to bring equipment and process units to a safe state. Pre-determined actions may range from manually executed operating procedures to a fully automated safety instrumented system.

4 Survey Demographics

Over 150 functional safety practitioners from around the world participated in the survey. The table below shows the major countries that were represented.

Table 1. Top Respondents by Country

#	Country	% of Respondents
1	United States	30.9%
2	Canada	6.6%
2	India	6.6%
2	United Kingdom	6.6%
5	Australia	4.6%
6	Netherlands	4.0%
7	Brazil	2.6%
7	Germany	2.6%
	Other	25.3%

Respondents were asked whether their site / company has one or more processes that are covered by the OSHA PSM regulation (CFR 1910.119) or equivalent regulation (e.g., COMAH, Seveso III). As shown in Figure 2 below, 57% answered Yes.

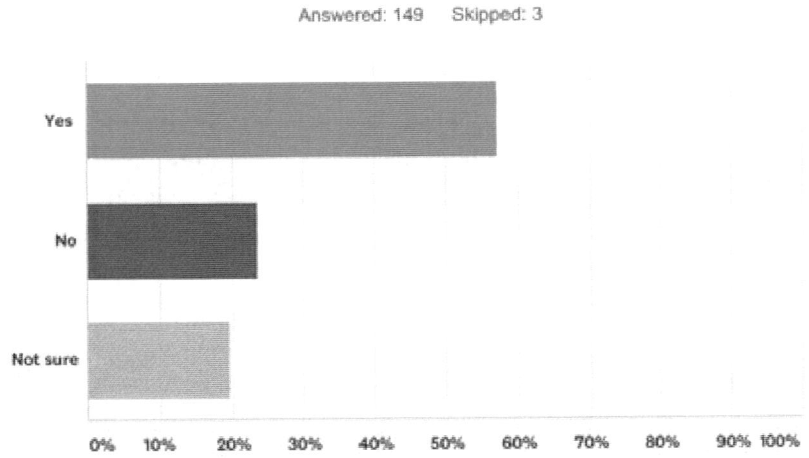

Figure 2. Respondents having processes covered by a Process Safety Regulation

5 Survey Results

Operational Limits

The survey respondents were asked what types of operational limits were used or defined in their companies. Figure 3 shows that 84.3% of the respondents define Safe Operating Limits (or equivalent) at their company. Almost as many (75.6%) define process control operating limits, such as normal operating limits. Some respondents indicated that they define separate and distinct environmental limits for emissions and effluent.

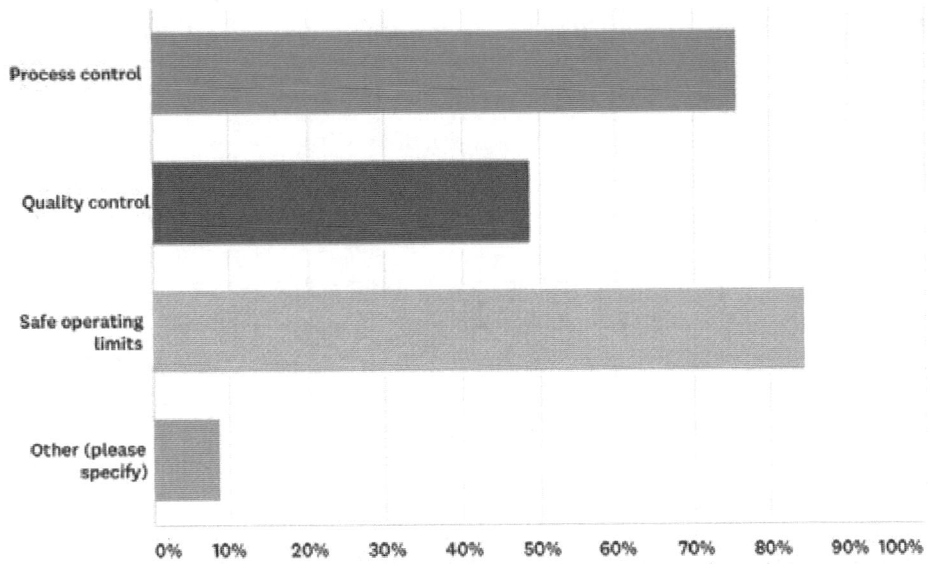

Figure 3. Types of Operational Limits that are Defined

Key Information for Establishing Safe Operating Limits

Respondents were asked what information was most important to establishing Safe Operating Limits. As shown in Figure 4, the most popular answer (but not the majority) was "Design Constraints" at 38%. Almost as many respondents (36.4%) answered "Safety Instrumented System trip points and relief valve settings"; while another 15% answered "Normal operating limit". None of the respondents selected "Mechanical Integrity Limit" as the most important information.

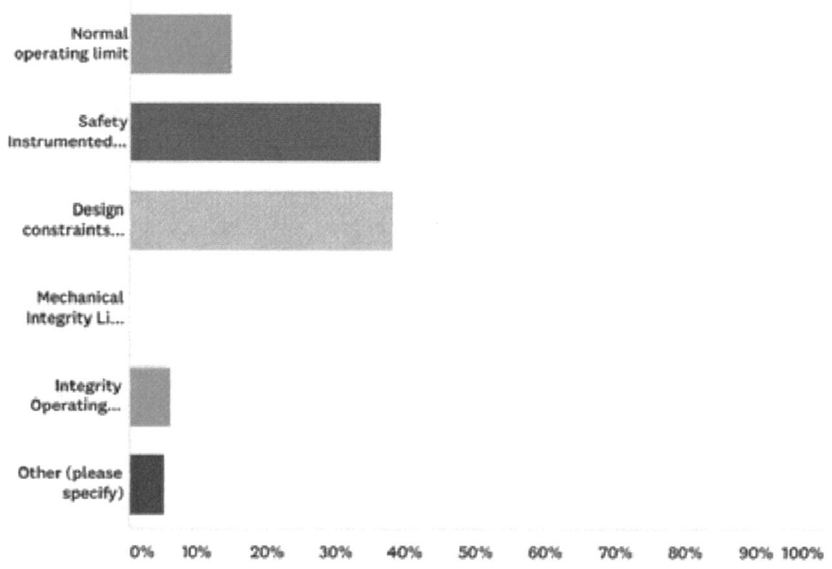

Figure 4. Important Information for Establishing Safe Operating Limits

Based on the definitions provided earlier, the expected answer to this question was "Design Constraints (Maximum allowable working pressure)". Strictly speaking, the Safe Operating Limit is not a function of the SIS setpoint; it is a function of the process hazard and the normal operating limit [6].

Free form input different from the multiple choice answers could also be entered. Responses of note included the following:

- A combination of following factors:
    - Mechanical integrity (maximum and minimum pressures and temperatures, effects of corrosion and erosion, effects of vibration, fatigue and differential thermal expansion, safety factors in design.
    - Process specific safety requirements
    - Response time and integrity of protective systems
    - Response time and preparedness of personnel to abnormal situations and human limitations
    - Systematic risk assessment with respect to safety, environment and asset/production loss.

- Process safety time - which depends on the relevant design constraints (not just pressure) and on the dynamic behavior of the process.
- In our systems, the process media for a specific parameter e.g. pressure, or temp. has different layers as below:
    - Never Exceed (Low/ High)
    - Safe Operating Limit (Low/ High)
    - Interlock Levels (Low/ High)
    - Alarm Levels (Low/ High)
    - Operating Range (Low/ High)

The Process for Establishing Safe Operating Limits

As a follow up to the previous question, respondents were asked to describe, at a high level, the process for defining SOLs. Responses of note included the following:

- Mechanical design
    - Maximum Allowable Over Pressure is the determination for most safe operating limits in the facility
    - Safe Operating Limit are defined by taking into consideration the design constraints (design values) of the operating equipment. Based on the metallurgy and other factors the mechanical integrity values are determined. The safe operating limits are then defined at these mechanical integrity values
    - Based on mechanical design limits, process design limits for safe operation, response time and reliability limitations of protective systems/measures and systematic risk assessment.
    - Limit is set at 90% of equipment's MAWP and/or Temperature.
    - The safe operating limits are usually below the maximum allowable working limits.
- Process consideration
    - Safe Operating Limits are governed by the process design requirements. As per the intended process, Safe operating limits are defined and then mechanical design limits are identified.
    - process flow in & out from the equipment in question and the time required to get to high level (alert level), generally it is 5-10%of the higher limit
    - as per the normal operating values of the process flow diagrams and material and heat balances
    - The process dynamic analysis is used to set limits based on inventories, flowrates, pressures, temperatures, flow or pressure relief capacity, design safety margin, etc.
    - For high level, threshhold value is the level at which entrainement can take place is given by the Process engineer which is configured as SIS set point. Safe operating limit : Depending on the volume and time taken to reach the SIS level from a high level alarm is generated Example : 40-55% safe operating level and 75% is SIS set level. margin of 20% approximately has process safety time of 30 minutes
- Process and mechanical consideration

  - It's a combination of equipment design constraints and process technology information; the more restrictive of the two becomes the SOL
  - Inferred from equipment design limits or process simulation and defined such that safety systems can respond fast enough if the process exceeds the SOL .
  - Safe operating limits are defined based on RBI (Risk based inspection), design and operational constraints.
  - During the plant design (plant subject of a project) Safe Operating Limits are established to avoid: damages to equipment with consequence of injuries, loss of containment and loss of production mainly
  - Safety margin applied to materials and process values
  - We apply a safety factor on Design limits. This factor is based on mechanical integrity and operational constraints
  - Process engineering defines safe operating limits based upon equipment design, system capacity, static/dynamic operating conditions.
  - A range outside of the normal operating limits beyond which LOPC or other serious process safety events can occur (runaway reactions, etc.). SOLs still provide a buffer between the boundary of the SOLs and the equipment limits (MAWP, etc.) to provide response time in the process shutdown before a catastrophic loss of containment should occur.
  - Safe operating limits is defined as limits based on flow, temperature, pressure, and levels in accordance with the mechanical equipment and process conditions.
- Set to match (e.g. interlock trip, relief set point)
  - In many cases, it is not available or assumed to be equal to the interlock / relief valve settings.
  - Currently SOLs are the design constraints or process safety time.

Storing Official Record of Safe Operating Limits

Respondents were asked about the location for the official repository of SOL information. Figure 5 shows that the results ranged from 32% for a "Documentation System", to 30% for the "Unit Operating Procedure" to 18% for a "Spreadsheet". The main documentation systems mentioned included Documentum, SAP, Dymenzions, and MS Sharepoint. Other responses included P&IDs, Equipment & Piping data sheets, SRS, and SIF datasheets.

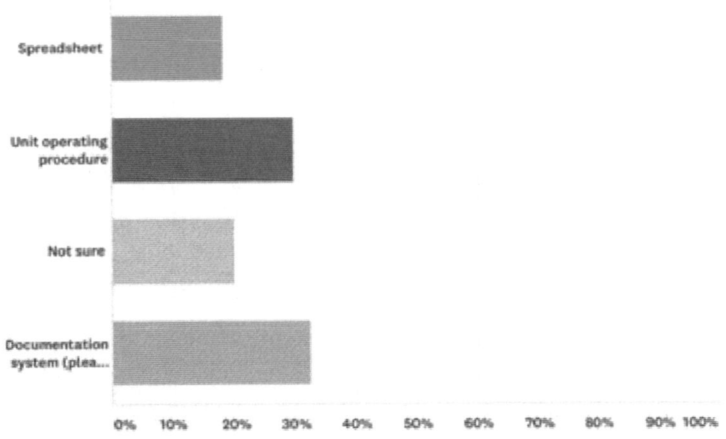

Figure 5 – Storage Location for Safe Operating Limits

Periodic Review of Safe Operating Limits

Figure 6 shows that 62.5% of respondents answered that Safe Operating Limits were reviewed periodically.

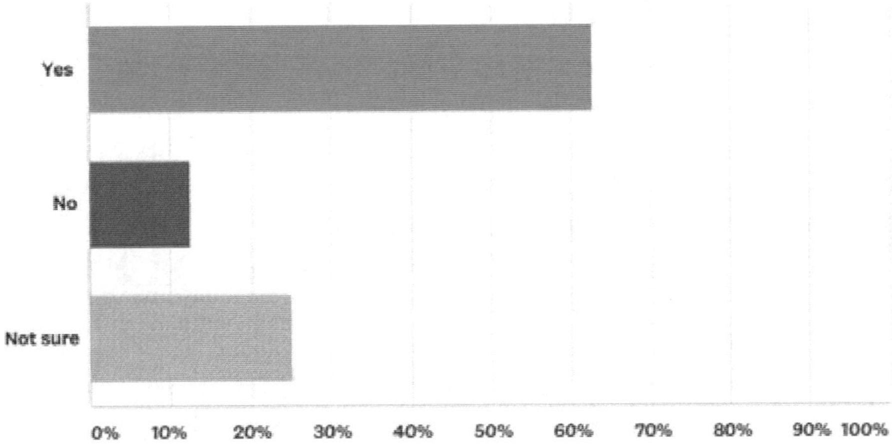

Figure 6 – Periodic Review of Safe Operating Limits

Using Process History to Validate Safe Operating Limits

When respondents were asked whether they review process history from a data historian to validate safe operating limits, 54.8% responded that they do. As shown in Figure 7, 45% said that they did not, with 24% responding that although they do not, they would like to be able to do this. Providing industry practitioners with the ability to review SOLs against actual operating envelopes (and other design limits) appears to be an area that could be improved.

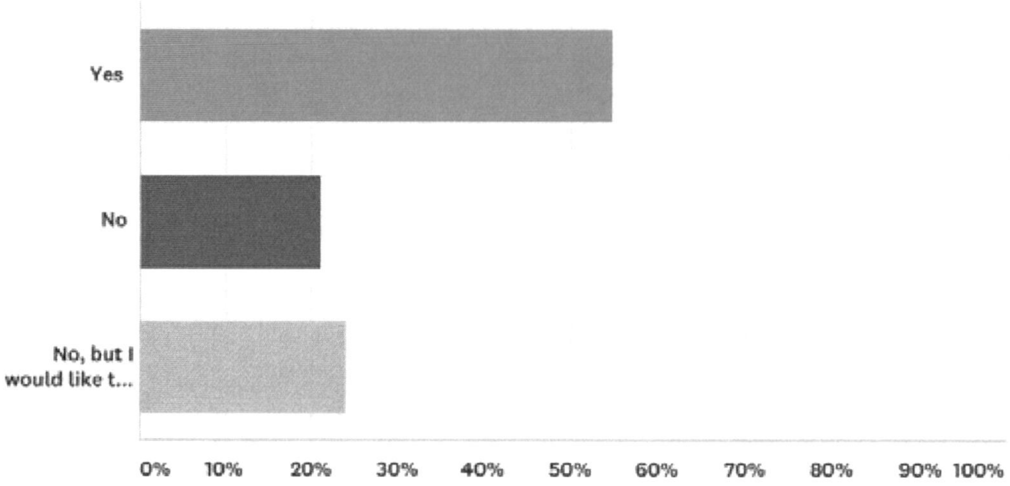

Figure 7 – Use of Process History to Validate Safe Operating Limits

Actions Taken When a Safe Operating Limit is exceeded

Respondents were asked what actions were taken when a Safe Operating Limit was exceeded. Following are some of the notable responses:

- Alarm, event recording, incident investigation, project revision (during project startup)
- root cause failure analysis
- Investigation and if change is required the Management Of Change procedure must be followed.
- An alarm is generated and automatic protective actions, if any, are initiated. Manual response is based on standard operating procedures for the plant. Alarms are generally prioritized based on an alarm rationalisation study to focus attention of operators on the most important issues.
- Our clients have formal procedures for incident investigation. All trips are investigated. Alarm frequency and duration are also monitored and investigated if too high. We have challenged some of our clients who have misunderstood the basis of the limits and raised change orders to increase limits when the first action should have been to review the cause of frequent alarms.
- Alarm then Trip
- depends on the client; however, part of clients unfortunately do not treat activation of SIFs or PRV as a potential catastrophic incident. Reset-carry-on because "nothing happend"
- Alarms are set before SOL. SOL point is usually when a trip occurs. Depending on severity, incident investigation may occur.
- initially alarm generated, if followed by an automated shutdown (SIS) a formal incident investigation is conducted.
- Actions as defined in operating procedures of equipment/plant are normally undertaken

- An investigation is done why it has occurred and depending on the outcome e.g. instrument failure, real process failure, decisions is made how to prevent that in the future.
- Alarm generated when a safe operating limit is exceeded. KPI system to capture the number of critical alarms generated are recorded as Lagging indicators in Process Safety KPI
- Process review will be carried out and alarm setting reviewed and process incident will also be reviewed. Result of the review will be analysed and measures put in place to forestall reoccurrence. This will be documented and cascaded to relevant persons
- "Near miss" event recorded and possibly investigated.

85% of respondents also reported that Safe Operating Limit exceedances needed to be reported to management.

Frequency of Exceeding Safe Operating Limits

Respondents were asked how often they thought their site experienced an exceedance of an SOL. As shown in Figure 8, the majority (60.9%) indicated that exceedances occur once per year or less. 32% of the respondents indicated that exceedances occur once per month or more frequently.

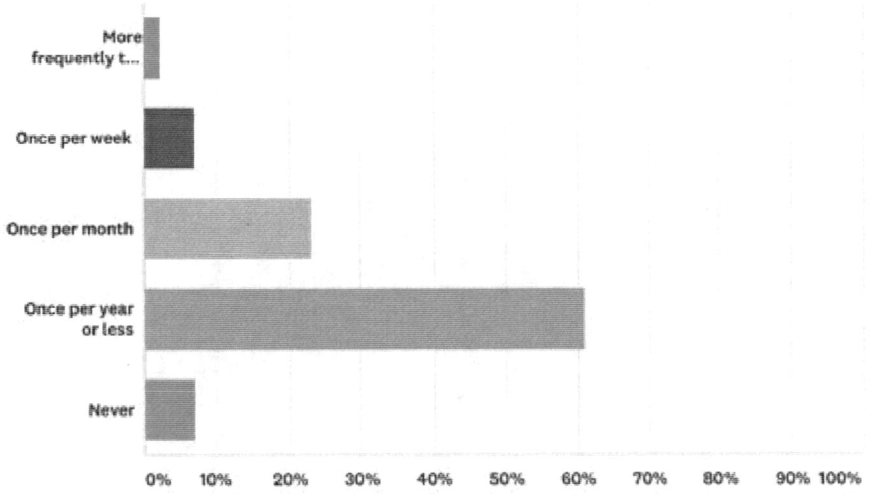

Figure 8. Frequency for Exceeding Safe Operating Limits

Use of SOL Exceedances as Process Safety Leading Indicators

The survey asked whether SOL exceedances were counted as Process Safety Leading Indicators. Figure 9. shows that 62.9% indicated “Yes”, while 21.4% responded that they were “Not sure”. It seems notable that such a large percentage of the respondents (21.4%) were not sure; this indicates a potential area where companies can improve in the communication of their Process Safety Management plans.

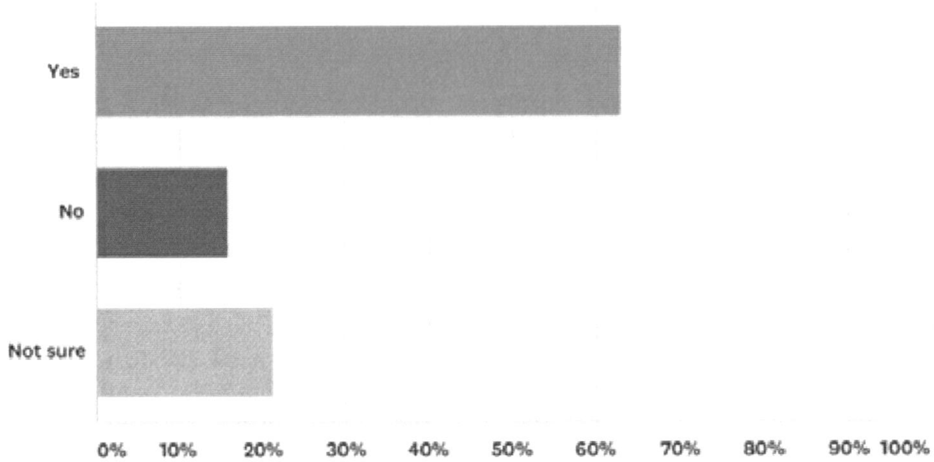

Figure 9 – Counting SOL Exceedances as Leading Indicators

According to ANSI / API-RP 754 "Process Safety Performance Indicators or the Refining and Chemical Industries" and as shown in the figure below, performance indicators are divided into four Tiers.

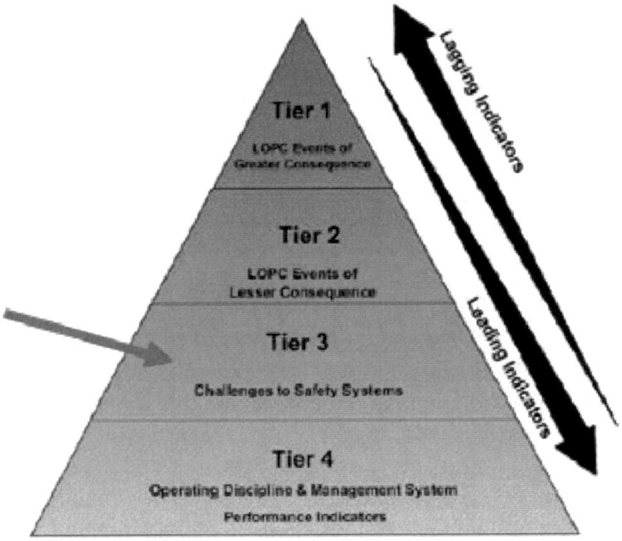

Figure 10 – Process Safety Leading and Lagging Indicators (4 Tier Structure) [7]

SOL Exceedances are one of four recommended Tier 3 Leading Indicators.

- **Safe Operating Limit Exceedances**
- Demands on Safety Systems
- Primary Containment Inspection of Testing Results Outside Acceptable Limits
- De minimis LOPC Events [7,8]

For each valid SOL exceedance, the following information should be measured and reported:

- peak value,
- duration of exceedance, and
- description of the initiating cause.

Frequent and/or prolonged SOL exceedances should be investigated to determine the initiating cause and the barrier failure [7].

Operator Training and Awareness of Safe Operating Limits

Respondents were asked whether their operators are trained and made aware of the Safe Operating Limits for the process units they monitor and control. The majority of respondents (82.6%) answered "Yes", while 3.3% answered "No" and 14.1% were "Not Sure".

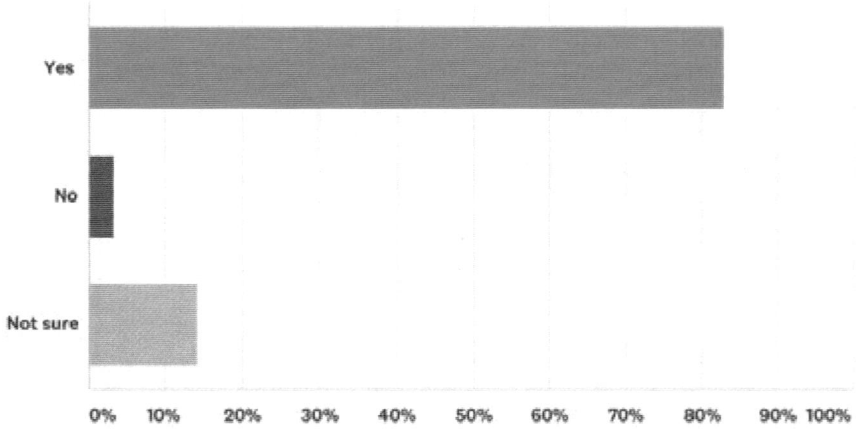

Figure 11 – Operator Training / Awareness of Safe Operating Limits

Per the OSHA PSM regulation, operating procedures are required to document (safe) operating limits, consequences of deviation, and the steps required to correct or avoid deviation; therefore, it should be expected that 100% of operators are trained and aware of relevant safe operating limits. SOL-related alarms will likely be some of the highest priority alarms in the system; operators should be trained to understand their importance [3]. This might be a relevant learning that industry can get better at ensuring operators are trained on the Safe Operating Limits for their areas of responsibility. Additional recommendations on how to optimize operator response to SOL-related alarms can be found in previous works [9,10].

Making Safe Operating Limit Information Available during PHA, LOPA, and Alarm Rationalization

Respondents were asked whether it would be beneficial to have Safe Operating Limit information available during a PHA / LOPA and during alarm rationalization. Approximately 95% of the respondents answered "Yes" to both of these questions.

This result also introduces an area that could be improved. PHA / LOPA and alarm rationalization tools could potentially be improved to allow Safe Operating Limit information to be easily entered and then displayed in context as necessary. During alarm rationalization, evaluation of alarm limits for safety alarms or SOL pre-alarms should reference the Safe Operating Limits.

Defining Alarms to Indicate SOL Exceedances

Respondents were asked whether they define alarms in the control system to indicate Safe Operating Limit exceedances. As shown in Figure 12 below, 69.2% of the respondents answered "Yes" to this question. This question probably requires / required additional information. There could be applications where an alarm and operator response is the required action to prevent the consequence of deviation from occurring. If the response to an SOL exceedance is a SIS trip or PRV event, then there might not be an alarm generated (SIS trip takes the process to a safe state and no operator action is needed); thus the essence of the question is whether an alarm is generated for the purpose of capturing the SOL exceedance. Per the definition of an alarm in ANSI/ISA-182.-2016, if no operator action is needed, then the notification would not be considered an alarm [11].

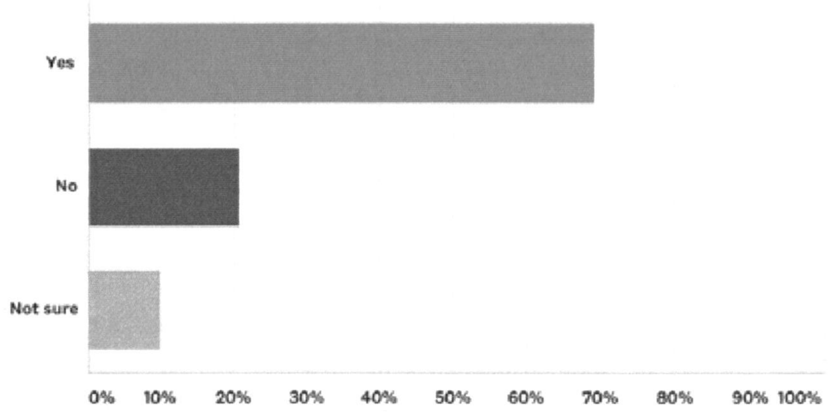

Figure 12 – Defining Alarms to Indicate SOL Exceedances

Recording SOL Exceedances

Respondents were asked whether they keep records of SOL exceedances in a database for the purpose of analyzing performance. As shown in Figure 13, 52.3 % answered "Yes", 31.4% answered "No" and 16.3% answered that they "Plan to in the future". Adding the ability to record and document SOL exceedances for performance analysis is another area where tools could be enhanced to better support PSM practitioners.

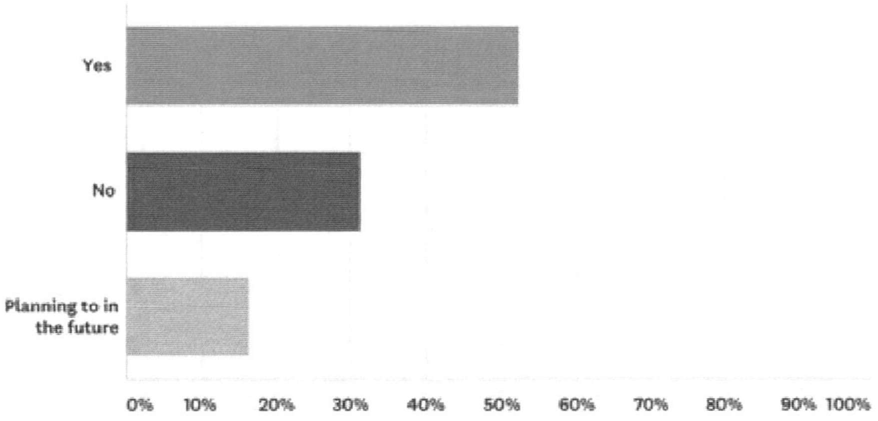

Figure 13 – Recording of SOL Exceedances in Database for Analyzing Performance

6 How are Safe Operating Limits Determined

SOLs, which are typically identified for each hazardous process unit, should meet the following requirements:

- SOLs are established on critical operating parameters only if it is physically possible to exceed the limit and if exceeding the limit could lead to a catastrophic failure of process equipment or loss of containment [5].
- SOLs are set based on a combination of equipment design limits and process dynamics [3] and must consider the process safety time.
- SOLs are determined by identifying design limits of equipment within a system. The most limiting elements will establish the SOLs for the system.
- SOLs are often set by code or by operating experience (particularly when these values were not provided as part of the design documentation).
- The technical basis for the SOL should be documented as part of the PSI package (Process Technology Information).
- Safety and environmental consequences of pressure relief system activation to the atmosphere (relief valve, rupture disc, etc.) should be considered in the determination of SOL settings [5].
- The Safe Operating Limit is not a function of the SIS setpoint; it is a function of the process hazard and the normal operating limit [6].

Safe Operating Limits in relation to Design Limits

As mentioned in the beginning of the text, one of the main issues with definition and application of safe operating limits is terminology. In Figure 14 below taken from the draft ISA-84 TR4, the Safe Operating Limit for an increasing process variable is shown in relation to the Design Limit, Never Exceed Limit, IPL Setpoint, and Normal Operating Limit. Clear and consistent definition of these other terms is important to be able to understand SOLs.

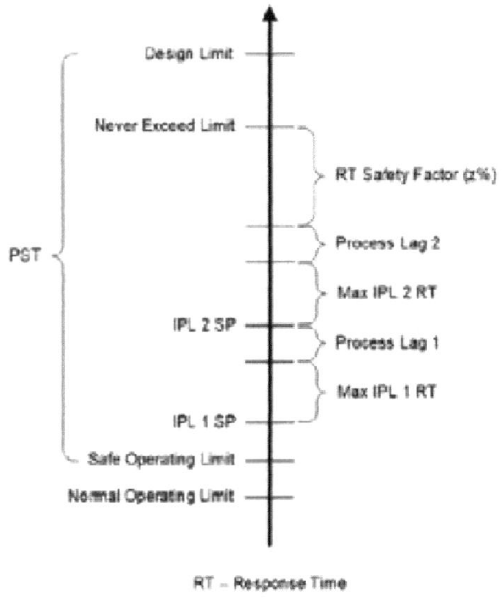

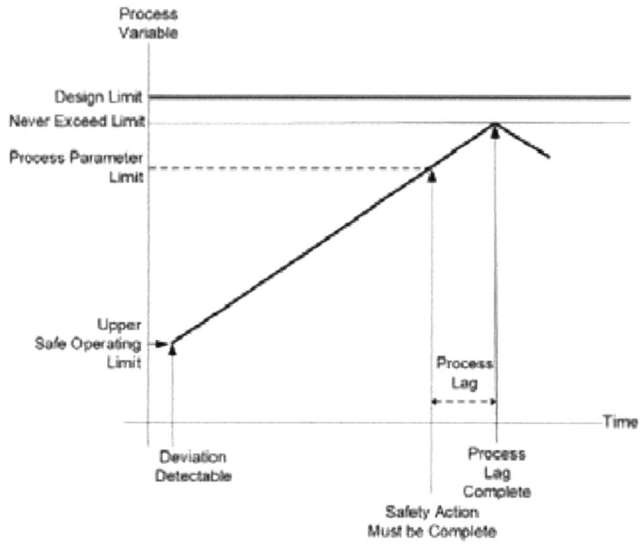

Figure 14 – Safe Operating Limit in relation to Design Limit, Never Exceed Limit, and IPL Setpoints [6]

Design limit (DL): the extreme value of a process variable that protects the asset integrity of the process equipment.

NOTE Factors to be considered in establishing the design limit typically include, but are not limited to: equipment design specification values (e.g., MAWP/MAWT) and standards used for testing the mechanical integrity of the equipment upon manufacture or modification [6].

Never Exceed Limit (NEL): the closest approach value to the design limit, allowing for operational and asset integrity uncertainties.

NOTE 1: These values are sometimes referred to as Safe Upper and Lower Limits.

NOTE 2: Factors to be considered in establishing the never exceed limit typically include, but are not limited to: design limit for the parameter, age of the equipment, severity of the operating environment and internal process, observed trend in condition of the equipment (e.g., results of periodic inspections), and Integrity Operating Windows (IOW) for the process equipment and piping.

NOTE 3: The NEL and the Design Limit can coincide in some cases [6].

The definitions and figures above hopefully provide additional clarity around the definition and application of Safe Operating Limits. For example, it shows that SIS trip points (shown as IPL 1 SP or IPL 2 SP in Figure 14) would typically be set above the Safe Operating Limits, but below the Never Exceed and Design Limits.

One discrepancy that has been discovered in the literature concerns how Safe Operating Limits relate to Never Exceed Limits. In some references, the Never Exceed Limits are shown inside of the Safe Operating Limits [5]. This disagrees with the portrayal of Never Exceed Limits being outside of the SOLs as shown above in Figure 14 above.

7 Conclusion

Consistent definition and application of Safe Operating Limits is important to maintaining the process safety of hazardous plants and is integral to the following:

- Unit Operating Procedures and Emergency Shutdown Procedures
- Calculation of Tier 3 Process Safety Events (Leading Indicators)
- Management of Change - Evaluation of whether a proposed change takes the process conditions outside of its safe operating limits [12]

As shown in this paper, the definition (calculation) of Safe Operating Limits has significant variability. Some of the key takeaways regarding definition of SOLs include:

- When evaluating Design Limits it is important to include the effects of Mechanical Integrity and Integrity Operating Windows in addition to design constraints such as MAWP or MAWT.
- Many practitioners set SOLs based on Design Limits such as MAWP. Some set SOL values equal to MAWP, others include a Safety Factor (e.g., SOL = 0.9 MAWP).
- The rate at which a scenario develops should be considered when setting SOL to assure there is sufficient time to take corrective action and return the process to a safe state.
- Industry would benefit by having more published examples of SOL calculations, such as those shown by Sutton [13].
- Inconsistent terminology makes it difficult to apply SOLs; for example, some references show SOLs falling outside of Never Exceed Limits.

Some of the key takeaways regarding application of SOLs include:

- Industry can improve operator training and visibility to the SOLs in their area of responsibility.
- Significant variability exists in where the official record of SOLs are stored, from unit operating procedure to Excel spreadsheet to Documentation system under access control.
- Industry can improve / provide the ability to compare actual process history from a data historian to safe operating envelopes for validation of the SOL values.
- Use of SOL Exceedances as a Process Safety Management Leading Indicator could / should be increased.
- Use of alarms to help prevent SOL exceedances and to capture SOL exceedances for review could be applied more rigorously.

8 References

[1] OSHA, Process Safety Management of Highly Hazardous Chemicals, CFR 1910.119

[2] "ANSI/API Recommended Practice 754: Process Safety Performance Indicators for the Refining and Petrochemical Industries," 2nd ed., Amer. Petrol. Inst., Washington, D.C. (2016).

[3] CCPS. *Guidelines for Risk Based Process Safety.* Center for Chemical Process Safety, American Institute of Chemical Engineers, Hoboken, NJ, 2007.

[4] AICHE process safety glossary. https://www.aiche.org/ccps/resources/glossary/process-safety-glossary/safe-operating-limits (accessed January 15, 2019)

[5] CCPS. *Guidelines for Engineering Design for Process Safety Systems.* Center for Chemical Process Safety, American Institute of Chemical Engineers, Hoboken, NJ, 2012.

[6] ISA–TR84.00.04-2019, Part 1. Guidelines for the Implementation of ANSI/ISA-61511-1:2018, Edition 4 – Revision D, Draft, 2019.

[7] "ANSI API RP-754: Quarterly Webinar", Amer. Petrol. Inst., Washington, D.C. (September 13, 2016).

[8] "ANSI/API RP-754: Process Safety Performance Indicators for the Refining & Petrochemical Industries, Part 3 – Tier 3 and 4 Process Safety Indicators" Webinar, Amer. Petrol. Inst., Washington, D.C. (September, 2010).

[9] Stauffer, T., Clarke, P., Using Alarms as a Layer of Protection – AICHE 8th Global Congress on Process Safety, April 2012.

[10] Stauffer, T., Sands, N., Strobhar, D., Closing the Holes in the Swiss Cheese Model – Maximizing the Reliability of Operator Response to Alarms, AICHE 13th Global Congress on Process Safety, April 2017

[11] ANSI/ISA–18.2–2016. Management of Alarm Systems for the Process Industries. June 2016.

[12] "Safety Moment #58: From "Change" To "Change"", https://iansutton.com/safety-moments/safety-moment-58-change-change (accessed February 1, 2019).

[13] "Process Safe Limits: Defining safe limits quantitatively" https://suttonbooks.wordpress.com/article/process-safe-limits-2vu500dgllb4m-4/ (accessed February 1, 2019).

Metrics Driven Decision Making Improves Performance at a Complex Process Facility

Elise Diaz
Chevron Australia Pty LTD
250 St Georges Tce
Perth, WA 6000
ediaz@chevron.com

Murray Watts
Chevron Australia Pty LTD
250 St Georges Tce
Perth, WA 6000
Murray.Watts@chevron.com

Prepared for Presentation at
American Institute of Chemical Engineers
2019 Spring Meeting and 15th Global Congress on Process Safety
New Orleans, LA
March 31 – April 3, 2019

Metrics Driven Decision Making Improves Performance at a Complex Process Facility

Elise Diaz
Chevron Australia Pty LTD
250 St Georges Tce
Perth, WA 6000
ediaz@chevron.com

Murray Watts
Chevron Australia Pty LTD

Keywords: process safety, leading and lagging metrics, performance indicators, process safety culture

Abstract

The importance of using leading and lagging metrics to drive improvement in process safety performance at complex process facilities has been well documented. Organizations are challenged to develop an appropriate set of metrics that will provide a holistic view of process safety management system health to allow for corrective action against poor performance. This paper will share a new complex process facility's journey in establishing leading and lagging metrics, the metric review cycle applied to drive improvement, and examples of these improvements.

1 Introduction

The importance of identifying and managing leading and lagging process safety metrics, also referred to as performance indicators, was boosted in the aftermath of the 2005 Texas City Isomerization Unit explosion. This major accident was a catalyst for the focused development of industry guidance on leading and lagging process safety metrics through the following Baker Panel recommendation [1] "…develop a consensus set of leading and lagging indicators for process safety performance for use in the refining and chemical industries." ANSI/API Recommended Practice 754 [2], Process Safety Performance Indicators for the Refining and Petrochemical Industries, was released in 2010 to address this recommendation. CCPS concurrently released

Guidelines for Process Safety Metrics [3] and the International Oil and Gas Producers (IOGP) followed a year later with Report 456, Recommended Practice on Key Performance Indicators [4] aimed to support API RP 754s applicability for upstream activities. In addition to industry guidance already mentioned, the volume of literature covering process safety metrics continues to grow.

This paper is not intended as an academic dive into process safety metrics. The paragraph above is included to frame the challenges faced by new facilities to implement an appropriate set of leading and lagging metrics that will be used to drive improvement. Finding the right balance of leading and lagging metrics is essential for understanding and managing process safety performance at complex process facilities. Too many metrics will create an excessive burden to collect and analyze data detracting attention from more meaningful metrics. With too few metrics a facility will not get a holistic view of process safety management system health. Similarly, choosing the wrong type of metric, such as a low-value metric with no impact on high consequence activities, will also waste time and effort. This paper will discuss how an appropriate set of metrics and their performance expectations were established for a new facility by involving the right people and taking a “less is more” approach.

However, an appropriate set of metrics alone will not drive improvement. This paper continues with detail on the metric review cycle put in place to effectively monitor and improve the facility’s process safety performance. Owned by the plant leadership team and supported by a site process safety lead, the review cycle includes collecting quality data, analyzing performance, generation and follow up of actions, and continuous data monitoring. Managing metrics by exception focused the plant leadership team’s attention to deteriorating performance and allowed for robust conversation and finite resource allocation to drive process safety improvements.

This paper provides examples to demonstrate how using process safety metrics has led to improvements in process safety performance at this facility, as evidenced by improvements in leading and lagging metric trends supported by a strong process safety culture.

2 Facility Process Safety Metrics

2.1 Metric selection

Process Safety metrics measured during the facility's first year of operation were based on industry recommended practices API RP 754 [2] and IOGP 456 [4], as well as industry guidance from CCPS [3], Energy Institute [5], and UK HSE [6]. A number of these metrics and their performance expectations were chosen at a Corporate and local Business Unit level to support Company monitoring, reporting and benchmarking. The remainder of these metrics, called site stewardship metrics, were developed to provide the plant leadership team visibility on the performance of each process safety element during the transition from construction and commissioning to operations, as well as area-based performance within the facility. Site stewardship metrics were selected after consideration of recommendations from plant leadership team members assigned as process safety element champions and metric reporters supporting this activity. Some of these metrics were intended to be leading indicators to the Corporate and local Business Unit metrics. Where there was enough data, performance expectations were also recommended to indicate whether the metric was in or out of tolerance using red-amber-green traffic light indicators. Final endorsement and overall accountability of site stewardship process safety metrics and performance expectations was granted by the plant leadership team.

2.2 Metric review cycle

Process safety metrics are reviewed in a four-step monthly review cycle, beginning with collecting quality data, followed by analyzing, acting and continuously monitoring the data (Figure 1). Using and repeating this process provides visibility on the facility's process safety management systems. Ultimately, it is a tool to help improvement in process safety performance.

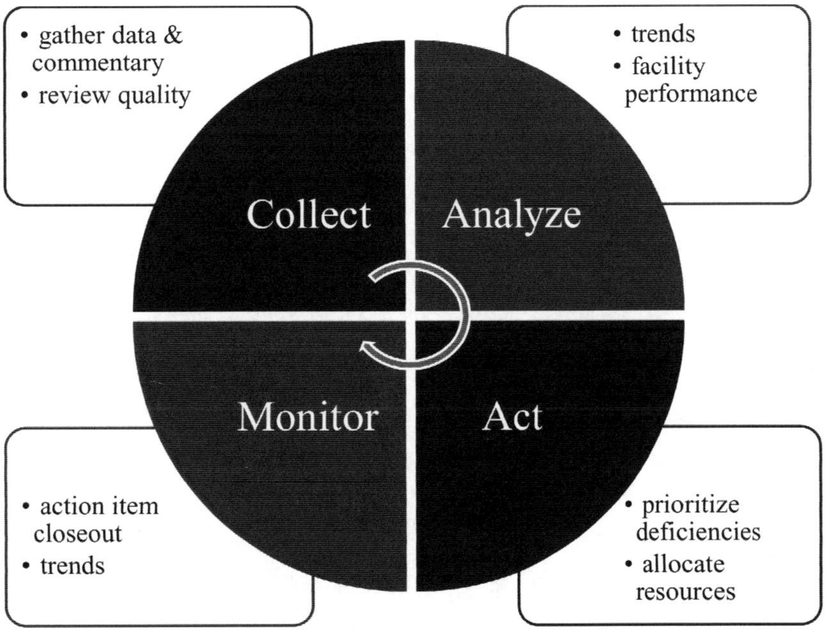

Figure 1. The four-step monthly review cycle keeps the focus on process safety metrics.

2.2.1 Collecting quality data

At the end of the month the site process safety lead requests metric data and commentary from each metric reporter. Once received, the process safety lead reviews the data with a focus on metric reporter consistency and commentary quality. If necessary, the process safety lead will work with the metric owner to better understand the details behind the data. Metric data and commentary are compiled in a presentation pack, where metrics performing outside of their tolerance, or trending in the wrong direction, are highlighted. The information is shared with the facility process safety element champion as pre-read with the expectation that they attend the monthly process safety stewardship meeting well prepared to explain the detail.

2.2.2 Analyzing data

The process safety stewardship meeting is a forum where process safety performance is analyzed and monitored by a cross-functional subset of the plant leadership team including production, maintenance, engineering and health, environment and safety. During the transition from construction and commissioning to operations, representation from the project and transition roles was required to provide visibility and discuss performance to support a condition-based start up.

The meeting is chaired by the plant manager and facilitated by the process safety lead with discussions prioritized based on metrics by exception and taking facility operational performance into consideration. Core members of the plant leadership team are allocated a process safety element to champion. Their role is to recognize and report on threats or potential vulnerabilities that may impact process safety performance or reflect failure of safeguards and controls. When a threat or potential vulnerability is identified, the team hold a robust conversation to determine what appropriate actions and additional resources are required to drive improvement. In addition to threats and vulnerabilities, the team also recognize and discuss positive metric trends.

2.2.3 Act on the data

When the plant leadership team agrees that intervention is required, SMART actions are prioritized, and resources allocated based on what the team believes will make the biggest difference in process safety performance. SMART actions are specific, measurable, actionable, relevant, and time-limited. These actions are owned by the relevant process safety element champion and are tracked to closure using an action tracking tool.

2.2.4 Monitoring data

Process safety metrics mentioned in this paper are primarily monitored at the facility level during the monthly process safety stewardship meetings. The status and outcomes of actions identified are also discussed during this meeting, while progress of action item closeout is tracked by the site process safety lead. The metric is closely monitored to assure the corrective action is effective. A subset of these metrics requiring increased oversight are discussed by exception in the daily tactical operations meeting on site. Corporate and local Business Unit metrics are also discussed in other forums involving higher levels of management within the Company.

2.3 What was learned from site implementation

2.3.1 Less is more

How many metrics are required to improve process safety performance? There is no clear answer, however what became apparent at this facility was that there were too many metrics and most required significant manual data manipulation by metric reporters at the end of each month.

Critical metrics were difficult to locate, and the health of facility safeguards and controls was not clear to facility leaders. Too many metrics resulted in a monthly "data-fest" for the site process safety lead during the "collect quality data" step, the output being a comprehensive presentation deck exceeding 100 slides. The facility was drowning in data and the plant leadership team became disengaged. To address this problem, the site process safety lead leveraged the experience of the plant leadership team to identify the high-value metrics and remove low-value metrics. The number of site stewardship metrics was optimized from 65 to 34 (Figure 2) by removing metrics reported in other forums (such as an operational discipline meeting) and applying a risk-based approach focusing on metrics specifically related to the presence and effectiveness of safeguards and controls. The presentation deck was simplified, and the plant leadership team was coached through the changes.

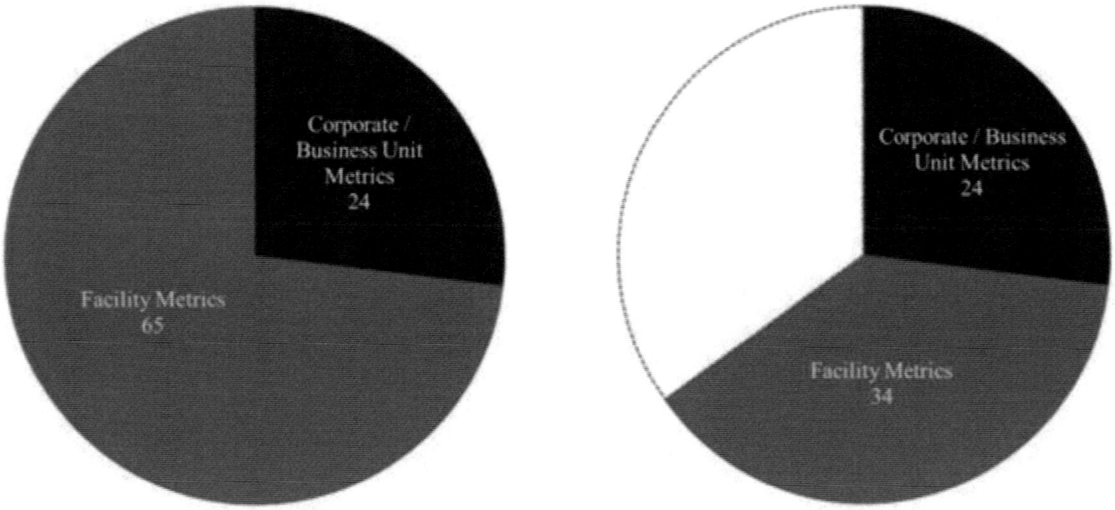

Figure 2. Removing duplicated and low-value metrics allowed the facility to optimize the number of metrics from 65 to 34.

The key message to takeaway is this: Don't drown in data. Simplify the number of metrics by focusing on the few that are meaningful for the facility and its stage of operational maturity.

2.3.2 Manage metrics by exception

Initially all metrics were discussed during the process safety stewardship meeting, which proved an impossible task due to the large number of metrics. Once the number of metrics was rationalized, the decision was made to talk to metrics by exception. That is, metrics outside of their tolerance, or metrics trending in the wrong direction were discussed. After gaining operational experience, the plant leadership team also discussed the need to modify previously set metric thresholds to ensure that resources were focused in the right place. Additionally, metrics previously identified as bad actors but now showing significant improvement were also discussed. A key message to takeaway is this: Be flexible. Don't be afraid to review metrics and adjust performance expectations as you learn more about how your facility operates.

3 Using Metrics to Drive Improvement

The aim of process safety metrics is "...to help prevent major incidents that are generally the result of multiple, simultaneous barrier failures" [IOGP, 4]. Building on this, sociologist Andrew Hopkins states "...in the quest for process safety, the important thing is to identify measures of how well the process safety controls are functioning" [7]. In the context of the short operational window for a new facility, the absence of major incidents cannot be correlated to an improvement in performance. There are not enough data points and major incidents are rare. Exploring this thought further, one could argue that a reduction in all process safety events, including minor incidents and near misses, may indicate an improvement in process safety performance. Todd Conklin challenges this argument by stating that, "...safety is not the absence of accidents, but the presence of safeguards." IOGP, Hopkins and Conklin's statements could be used as the basis for claiming improvement when metric trends measuring safeguard and control performance reflect either a reduction in safeguard and control failures or an improvement in their health.

Irrespective of how a facility claims improvement, a key enabler for improved process safety performance is a strong process safety culture. CCPS [8] lists a commitment to process safety culture and operational discipline as one of four critical success factors for a robust process safety management system. CCPS continue to identify twelve essential features of a good safety culture, all reliant on management commitment through both actions and words. More recently the CCPS developed Vision 20/20 [9] which identifies "committed culture" as one of five industry tenets that will drive process safety performance. Therefore, a reduction in process safety events and improvements in metric trends associated with safeguard and control health supported by a good process safety culture may indicate an improvement in process safety performance.

3.1 Metric trend improvements

3.1.1 Number of process safety events

The number of reported process safety events, which includes incidents and near misses, was selected as a site stewardship metric to provide an indication of incident investigation and reporting process safety element health. Over the course of a year the number of reported process safety events trended downwards (Figure 3).

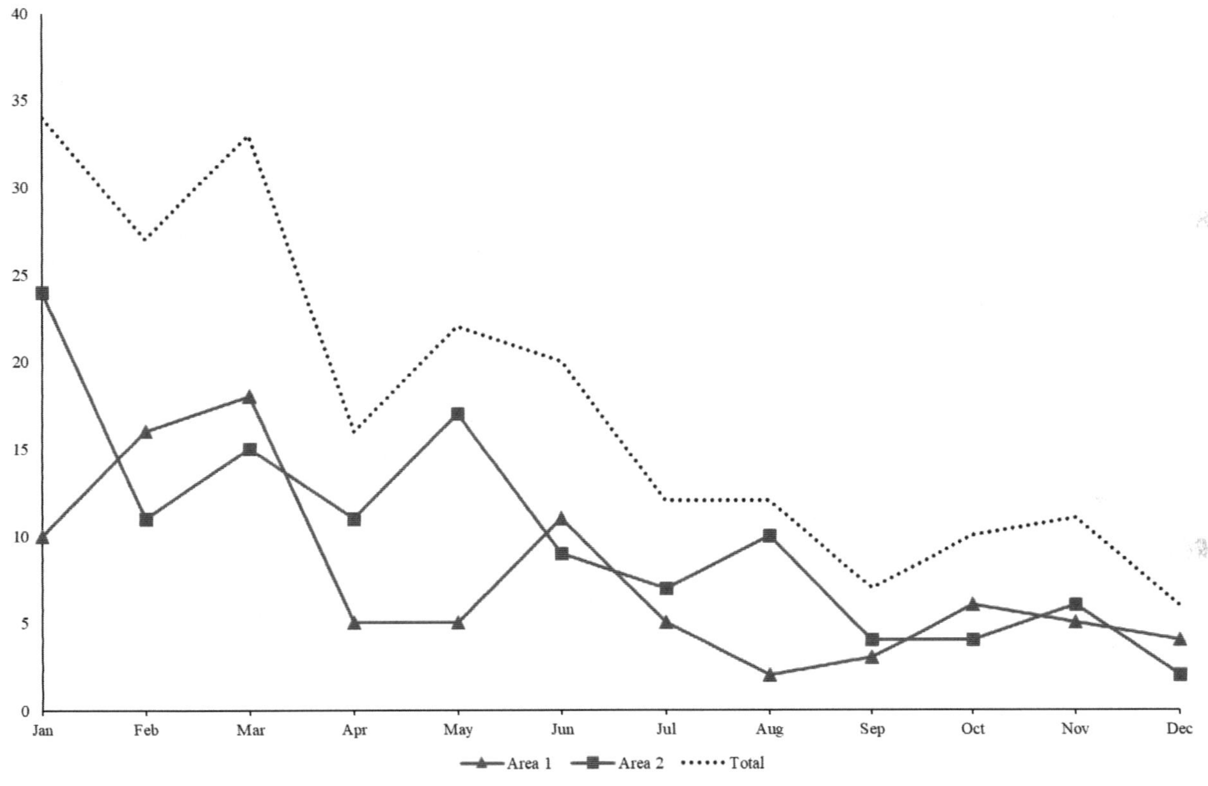

Figure 3. The number of reported process safety events for the entire facility decreased as a result of selecting the appropriate set of metrics and applying the metric review cycle.

During the "collecting quality data" step of the metric review process, all reported incidents and near misses were reviewed against a process safety event decision logic flowchart (Figure 4) by the site incident investigation and reporting coordinator. This decision logic allowed for consistency in identifying process safety events by multiple metric reporters (a requirement for a rotational work force) and directed questions to the site process safety lead for additional determination. A summary of these incidents was gathered each month for discussion in the site process safety stewardship meeting.

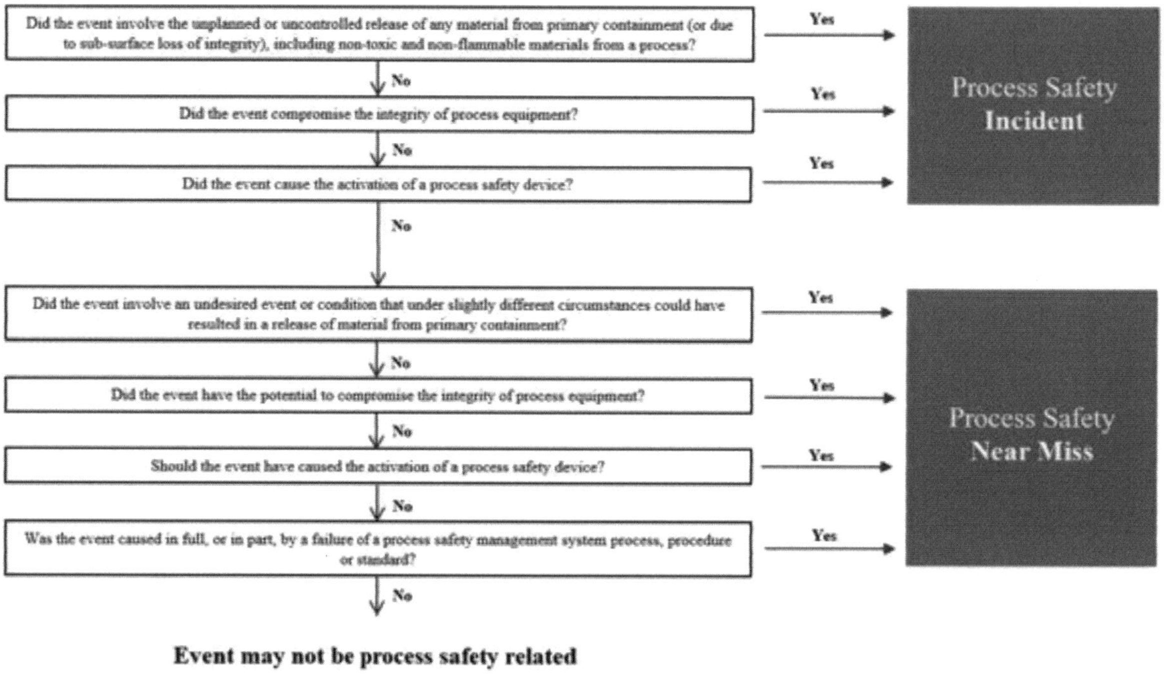

Figure 4. The decision logic flow chart enabled consistency within the workforce to identify when an event is a process safety incident or near miss.

During the "analyzing data" phase, the change in the number of process safety events from the previous month, the category and severity of events, and the status of investigation action items was discussed by the plant leadership team led by the site health, environment and safety superintendent. This holistic analysis of facility events by area allowed for identification of repeat events and potential systemic failures with other process safety elements. Resources were allocated, and performance monitored to ensure deteriorating trends were curbed. An example of an action arising from discussing this metric was the request for increased visibility over process safety events in the storage and loading facility. The data indicated an increased risk in this area and the plant leadership team sought more detail to determine what additional resources were needed to address the issue, and what specific leading metrics to focus on to prevent recurrence. As a result, a dashboard was developed to identify the relevant process safety events and the status of short-term and long-term mitigations to prevent recurrence. The plant leadership team regularly monitored the status of this dashboard, and prioritized resources until the number of process safety events in this area reduced to zero.

To combat under-reporting of process safety events, the plant leadership team at this facility worked hard to build and maintain a strong reporting culture. This was achieved through setting clear reporting expectations and reinforcing good behaviors through timely recognition. Although the number of reported process safety events decreased during the first year of operation, the workforce continued to report on events that fell outside of the definition of a process safety event.

Facility leadership is confident that a reduction in reported process safety events is reflective of improved process safety performance, supported through consistency, holistic analysis, resource allocation, performance monitoring, and the facility's strong reporting culture. The key message to takeaway is this: With quality data and a strong reporting culture, metrics can be a valuable way to monitor process safety performance.

3.1.2 Process automation metrics

During the first year of operation the number of console alarms, safety system forces (software bypasses), and safety shutdown system activations were measured. Being a brand-new complex process facility, the level of process automation was high and operational experience was low. The number of alarms, forces, and safety shutdown system activations was significantly higher than a mature facility, worsened by simultaneous operations between areas already handed over to operations and areas still under construction and commissioning control. To add to the noise, the systems, processes and performance expectations used to manage these metrics were immature and engineering resources were focused on commissioning and start up priorities.

Initially there was a strong focus on console alarm count, an indication of console operator workload, with a daily report out to production on the number of alarms and actions in progress to address bad actors. In contrast, production had little visibility of safety system forces and safety shutdown system activations. Facility leaders sought to better understand the detail behind the numbers so resources could be prioritized in accordance with the level of risk. Facility production and engineering leads reviewed the process automation metrics and established the following order of priority; alarm management, safety system force reduction, and safety shutdown system activation (trip) reduction.

Top priority was placed on daily review and reporting of frequent, flooding, and standing alarms. Bad actors were identified and addressed until console alarm counts remained close to the industry target of 288 per operator, per console, per day. Because of these efforts, the daily average console alarm count (averaged over the month) for two consoles reduced significantly during the first four months of operation, and then fluctuated close to the industry target for the remainder of the year (Figure 5).

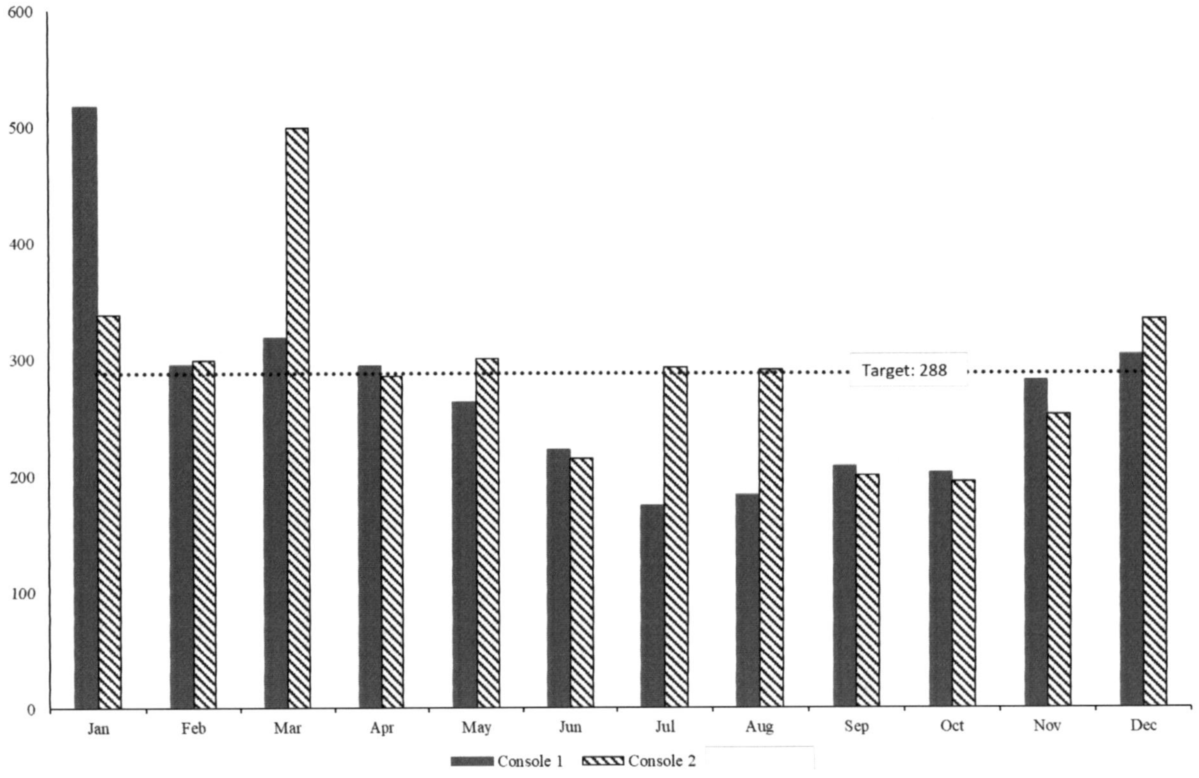

Figure 5. Prioritizing resources to focus on the console alarm count enabled the facility to eliminate bad actors. This approach reduced the alarm count average to industry accepted levels during the first four months of operation (Jan – Apr). The facility used the industry target of 288 during the first year of operation.

Second priority was given to reducing the number of active safety system forces, where a force is defined as a software bypass temporarily applied to force a device or system such that it will not perform its designed function. A weekly review was completed to identify and communicate the number of forces rendering a safety function unavailable. As the facility transitioned from commissioning to steady state operations, the number of active safety system forces steadily reduced. This can be seen with two decreasing trends occurring during Unit 1 start up (January-May) and Unit 2 start up (June-December) (Figure 6). This review provided an additional layer of protection to ensure that all forces applied by the project to support commissioning activities were either closed out or effectively handed over to production during facility start up.

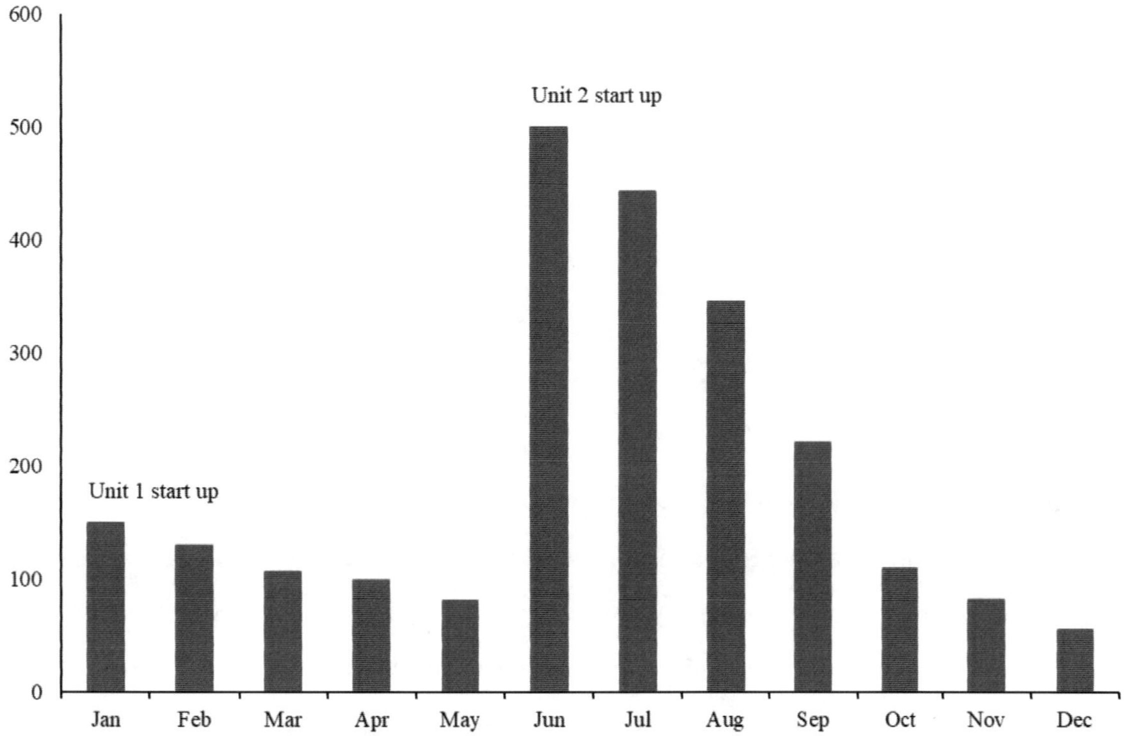

Figure 6. Complex process facilities are highly automated. A strong focus was applied to reducing the number of safety system bypasses during commissioning and start up of Unit 1 and 2. This approach assists with the safe transition to steady state operation.

Safety shutdown system activation (trip) reduction was prioritized third, where the monthly list of activations was used to identify and address safety systems with the highest activation count. These bad actors were reviewed, and resources were applied where necessary to ensure permanent changes were prioritized through the management of change process. The number of safety shutdown system activations generally trended downwards, with a spike occurring mid-year (Unit

2 start up) and in December (advanced process control commissioning during high ambient temperatures) (Figure 7).

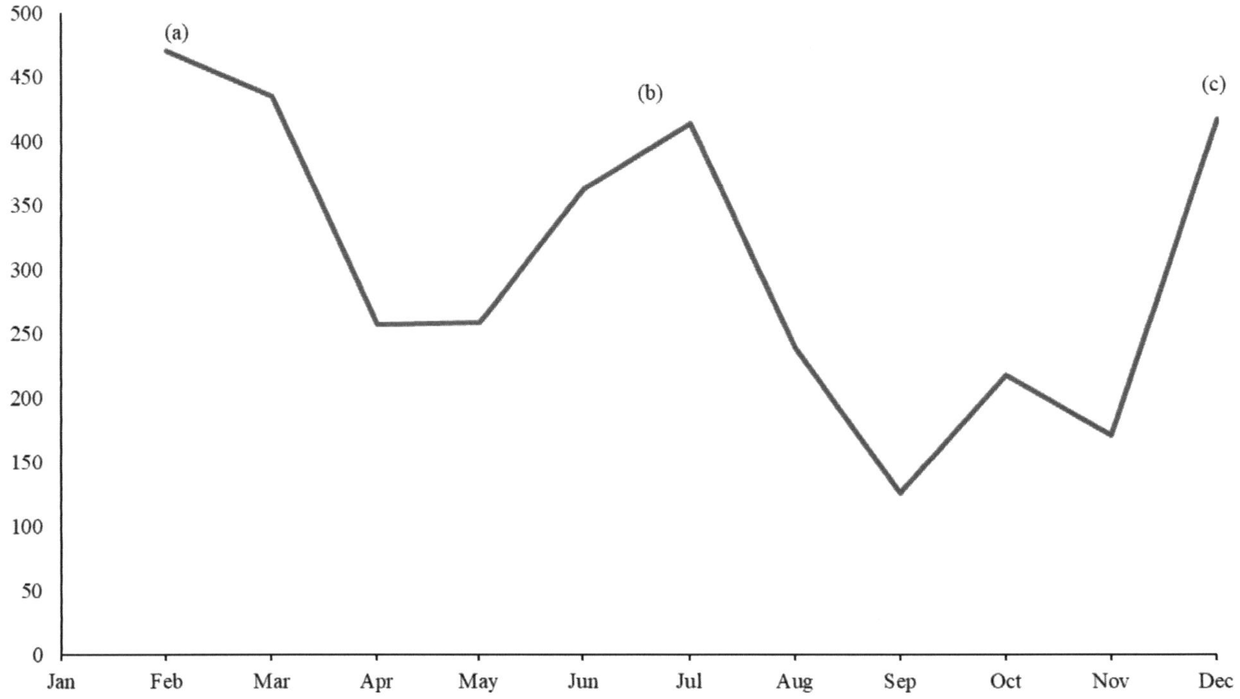

Figure 7. The monthly review of safety shutdown system trips identified corrective actions and prioritized permanent changes associated with bad actors. Addressing bad actors improves facility safety by preventing unnecessary stress on safeguards and controls. Unit 1 (a) and Unit 2 (b) start up reflect similar trends to Figure 6. The spike (c) in December is due to advanced process control commissioning to address high ambient temperatures at the facility location.

This facility's prioritization of process automation metrics is one example of how metrics can be used to manage facility risk and drive improvement in process safety performance. By prioritizing metrics and allocating resources to address bad actors, increased focus was brought to the things that matter the most while improving metric trends.

3.2 Process safety culture

A strong process safety culture at this facility fostered an improvement in process safety performance, evident by a reduction of process safety events and improvements in process automation metric trends previously discussed. Leadership ownership of metric selection and the monthly review process, along with their support to allocate resources to address deteriorating performance, demonstrates a strong process safety culture. Facility leaders used the daily console alarm count to support a condition-based approach to start up. When the number of console alarms on a panel exceeded the target line, the introduction of hydrocarbons was delayed until alarms reached a manageable level. This example of leaders using metrics to make condition-based decisions demonstrates their commitment to process safety culture.

Other areas thought to have had an impact on developing and sustaining a strong process safety culture includes:

- Major hazard facility training for all personnel working at the facility. The more people focused on process safety, the safer the facility and workforce will be;
- Making process safety performance visible through broad and effective communication to the workforce. Use of multiple communication vehicles to cascade key metrics, wins and improvement opportunities;
- Recognizing and rewarding good performance, particularly important to sustain a strong reporting culture; and
- Leadership engagement with the workforce during field process safety inspections, demonstrating management commitment to process safety.

4 Conclusion

Metrics identified and reviewed at this facility were used by leaders to make condition-based decisions during start up. By focusing resources on the right high-value metrics throughout the monthly review cycle the number of process safety events decreased and process automation metric trends improved. Therefore, the data would indicate that using leading and lagging metrics at this facility has led to improvements in process safety performance. Using a similar approach at other facilities fostered by a strong process safety culture may also result in improved process safety performance.

5 References

[1] The Report of the BP U.S Refineries Independent Safety Review Panel, available at http://sunnyday.mit.edu/Baker-panel-report.pdf, accessed on January 10, 2019.

[2] ANSI/API Recommended Practice 754. Process Safety Performance Indicators for the Refining and Petrochemical Industries. 2nd Edition. April 2016.

[3] CCPS. Guidelines for Process Safety Metrics. Center for Chemical Process Safety, American Institute of Chemical Engineers, New York, NY, 2009.

[4] International Oil and Gas Producers (IOGP) Report 456. Process Safety, Recommended Practice on Key Performance Indicators, available at https://www.iogp.org/bookstore/product/process-safety-recommended-practice-on-key-performance-indicators/, accessed on January 15, 2019.

[5] Energy Institute, High Level Framework for Process Safety Management, available at https://publishing.energyinst.org/topics/process-safety, accessed on January 10, 2019.

[6] United Kingdom Health and Safety Executive (UK HSE), Developing Process Safety Indicators: A step-by-step guide for chemical and major hazard facilities, available at http://www.hse.gov.uk/pubns/books/hsg254.htm, accessed on January 15, 2019.

[7] A. Hopkins, Thinking about Process Safety Indicators, Safety Science, 47, (2009), 460-465.

[8] CCPS. Guidelines for Implementing Process Safety Management. Center for Chemical Process Safety, American Institute of Chemical Engineers, New York, NY, 2016.

[9] The Center for Chemical Process Safety (CCPS). A Call to Action. Next Steps for Vision 20/20, available at https://www.aiche.org/ccps/resources/vision-2020, accessed January 30, 2019.

HOW TO IMPLEMENT A SYSTEM THAT EMPOWERS OPERATORS AND CREATES A CULTURE OF PERFORMANCE AND ACCOUNTABILITY IN PROCESS SAFETY, RELIABILITY AND PROFITABILITY

Claudio Schlosser
Petrobras America Inc.
Houston TX, USA
schlosser@petrobras.com.br

Luiz Eduardo Valente Moreira
Petroleo Brasileiro SA
Rio de Janeiro, Brazil
lvalente@petrobras.com.br

Socrates Fofano
Petroleo Brasileiro SA
Rio de Janeiro, Brazil
fofano@petrobras.com.br

Andre V. Barquette
Petroleum Refinery System Inc. -PRSI
Pasadena TX, USA
andre.vanzelote@petrobras.com.br

Prepared for Presentation at
American Institute of Chemical Engineers
2019 Spring Meeting and 15th Global Congress on Process Safety
New Orleans, LA
March 31 – April 3, 2019

How to implement a system that empowers operators and creates a culture of performance and accountability in process safety, reliability and profitability

Claudio Schlosser
Petrobras America Inc.
Houston TX, USA
schlosser@petrobras.com.br

Luiz Eduardo Valente Moreira
Petroleo Brasileiro SA
lvalente@petrobras.com.br

Socrates Fofano
Petroleo Brasileiro SA
fofano@petrobras.com.br

Andre V. Barquette
Petroleum Refinery System Inc. -PRSI
andre.vanzelote@petrobras.com.br

Keywords: Organizational culture, operational discipline

Abstract

In the petrochemical industry, chemical plants and refineries deal with very large volumes of flammable and toxic materials. The risk of fires, explosions or personnel exposure to toxic gases and chemicals is always present. In addition, as populations increase, the resulting growth in demand for land for residential and commercial purposes, causes refineries to face unrelenting pressure to seek permission to allow more and more people to live in closer proximity to processing and storage facilities.

Every workplace presents its own challenges and safety concerns. However, it is up to employers and employees alike to mitigate risks at work. This process starts at the top, provisioning of a safety management program that covers the risks associated with hazardous chemicals.

The management of operational risks in petroleum refineries is intended to guarantee the safe, reliable and profitable operation of production systems. This is accomplished by managing risks in the gap tolerance setting, considering the likelihood and severity of accidental scenarios and their consequences.

The main purpose of this presentation is to introduce an Operational Management System (OMS) that enhances stable operations in order to achieve environmental compliance, process safety and operational reliability, along with positive financial performance.

The system is based on seven key elements:

1. Empowering the operators
2. Culture of performance and accountability
3. Collaborative dynamic decision-making
4. Pinpointing individual contributions toward goal
5. Transparency and communications
6. Extensive use of analytics
7. Awareness & response

This system consists of real-time process data acquisition and analytics from the industrial process that the panel operators follow. Data analytics allows us to identify any deviation from a process variable (overtaking of defined limits). The data is collected, and the operator's performance is recorded. Score screening enables us to establish training programs. In a collaborative environment, operators and supervisors are able solve problems and improve the system. All of the data and performance assessments are available and allow for additional analysis. In order to identify improvements on process safety, reliability and economic results, the mechanical and chemical engineers can create a study case to identify and enhance areas based on operator and supervisor comments.

For the past six years this system has enabled users in all of Petrobras's 15 refineries and fertilizer facilities to create proactive indicators, which help standardize performance across all shifts. In addition, this tool defines a narrower operating range and identifies and fine-tunes training which reduces system failures and promotes stability control, safety, reliability and profitability.

1 Process System Based on Risks, and on Empowering operators, supervisors and engineers.

According to the definition adopted by CCPS [1,2], a management system consists of "*A formally established and documented set of activities designed to produce specific results in a consistent manner on a sustainable basis.*" These activities should be defined in detail for the workforce to reliably perform the required tasks.

Risk Based Process Safety Management (RBPS) is an evolution of strategic approaches that have been applied over the years in trying to answer the following question:

How can I better manage risks?

To manage risks, you must first understand the risk associated with an activity. This requires the following questions to be answered:

- What can go wrong?
- How bad could it be?
- How often might it happen?

Based on the level of understanding, an organization can decide what actions, if any, are needed to eliminate, reduce, or control existing risk.

1.1. The RBPS strategy approaching

In order to structure a robust risk based management system, the 20 elements need to be applied in an interrelated way. Regarding the OMS, let's focus on two of these elements. Although risk management systems differ, the standard approach is represented and recommended by Guidelines for Risk Based Process Safety, edited by CCPS [1].

In this approach, an RBPS management system incorporates four main pillars of accident prevention (fundamental blocks) as shown in figure 1:

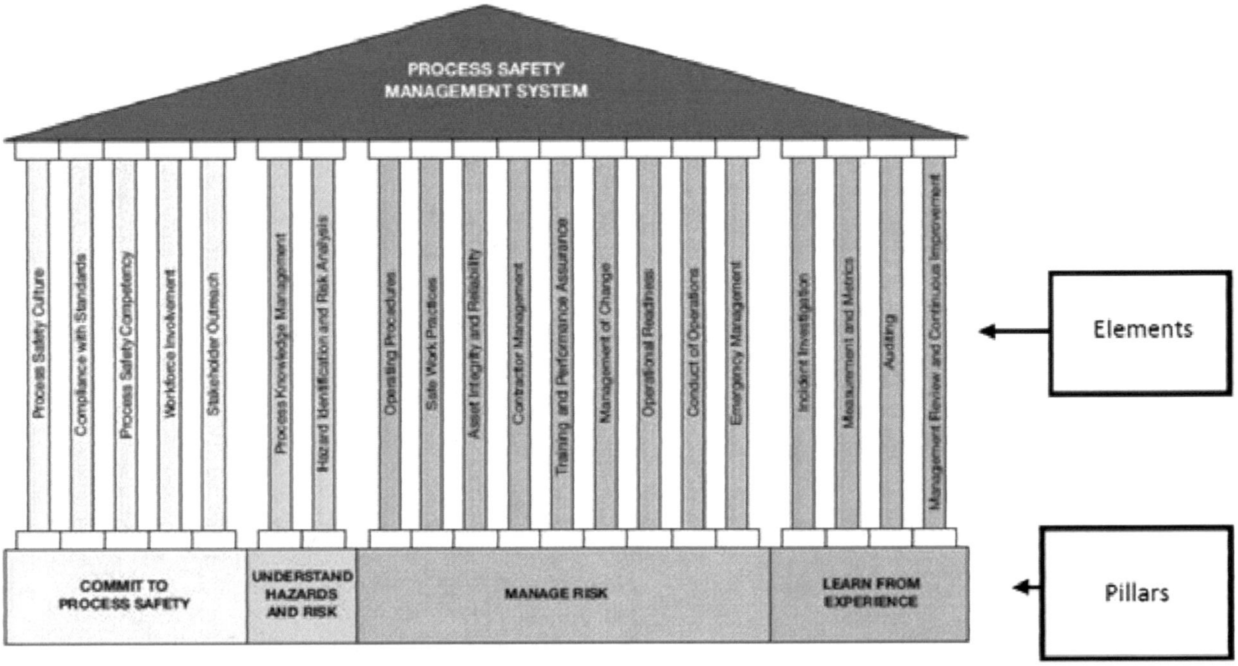

Source: CCPS, Guidelines for Risk Based Process Safety

Figure 1. Pillars (Foundation Blocks) and Elements of the RBPS

Regarding the OMS, these elements have focus on two, as follow.

- Active involvement of the workforce
- Conduct of Operations

Promoting the **active involvement of the workforce** at organizational levels is one of the five elements of the Pillar of Commitment to Process Safety. Everyone must have roles and responsibilities for improving and ensuring the safety of all operations. However, operators may not be aware of all their opportunities to contribute. Some organizations may also not take advantage of their employees' experience, or worse, may discourage workers from participating in problem solving. Therefore, involvement of the workforce enables employees and contractors to participate in the design, development, implementation and continuous improvement of the RBPS management system.

Developing and sustaining high standards is already one of the nine elements of the Risk Management Pillar. Conduct of operations consists of the execution of operational and managerial tasks in a structured way. Also called "operational discipline" or "formality of operations", it is closely linked to the culture of an organization. The conduct of operations institutionalizes the pursuit of excellence in the performance of every task . Workers are expected to perform their duties with full attention, knowledge, common sense and they will enjoy a greater sense of pride and responsibility.

In this approach, Petrobras has developed an Operational Management System (OMS), or *Sistema de Gerenciamento de Processos Operacionais* – GPO (in portuguese), that aims to stream line operations to ensure environmental compliance, process safety and reliability..

The system is based on seven key elements:

1. Empowering the operators
2. Culture of performance and accountability
3. Collaborative dynamic decision-making
4. Pinpointing individual contributions toward goal
5. Transparency and communications,
6. Extensive use of analytics
7. Awareness & response

This system consists of the acquisition and analysis of real-time data from the industrial process, which is monitored by the panel operator. Data analysis allows the identification of any process variable gaps (exceeding defined limits). Data is collected and operator performance is recorded. The score obtained can be used to design training programs. In a collaborative environment, operators and supervisors are able to provide solutions and improve the system. All data and performance appraisals are easily accessible for reference and further analysis. In order to identify improvements in the safety, reliability, and economic results from analysis the system can track, identify and enhance areas identified through feedback from operators and supervisors.

2 Operational activities and process safety

2.1. Areas of expertise of Petrobras

Petrobras is an integrated energy company focused on oil and gas. A Glance of operations is shown in figure 2:

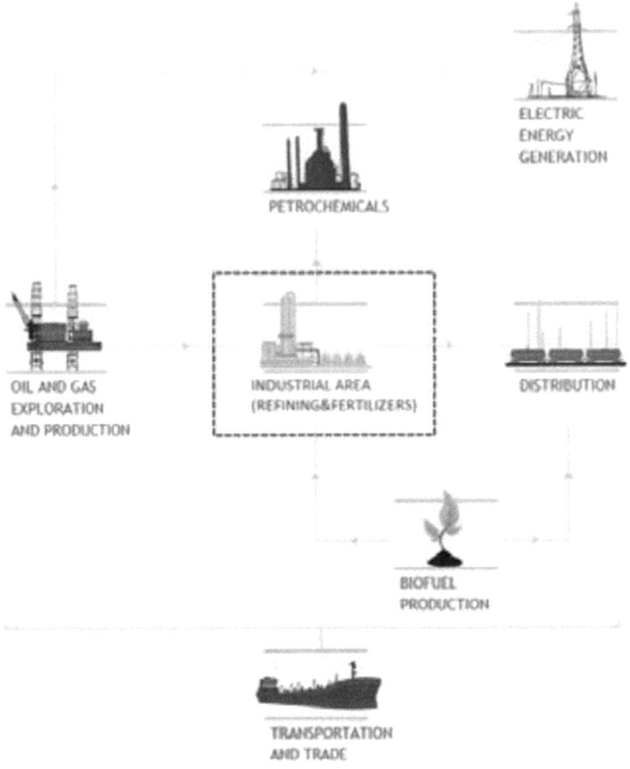

Source: www.petrobras.com.br [3]

Figure 2. Areas of Expertise

The industrial area (identified in figure 2) is composed of 15 operating units: 12 refineries, 2 fertilizer plants and a shale industrialization plant, distributed as shown in figure 3:

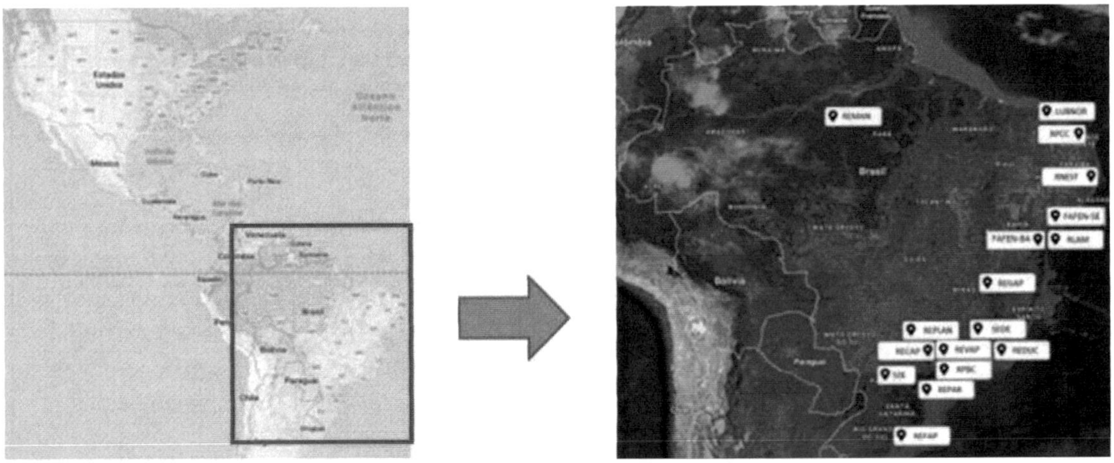

Source: www.petrobras.com.br [3]

Figure 3. Location of Petrobras' Industrial Areas and Facilities

For complex companies such as Petrobras, standarding safety management processes is crucial to achieving world-class results, because it is the most efficient way to ensure collective buy-in and teamwork.

2.2. Involving operators in facility risk management

Process safety performance depends on the organization's commitment to safety, as demonstrated through strong leadership. The workforce will follow suit and recognizing that the company values safety and promotes the slogan "do everything right, all the time" to support their effort. In an operation-oriented approach, it is critical that people working in these activities clearly understand the risks of their activities understand their responsibilities, be able to respond to emergency demands, and be able to incorporate and promote organizational learning as well. It is well known that the nature of the operating activities is critical, considering its performance characteristics and the human factors involved in the interface with technical systems of great complexity. For this, organizations must provide strategies and tools to properly engage the workforce and spur a sense of responsibility with clear expectations to reduce human error and improve performance.

3 Operational Management System (OMS) as an instrument for involving and empowering operators and achieving the best results in safety, reliability and efficiency

3.1. Operational control and its challenges

In general, the control of operations is accomplished either through automatic responses or based on control parameters established for a particular process plant. Figure 4 illustrates the layers of protection in which the operator effectively participates as agent in preventive management:

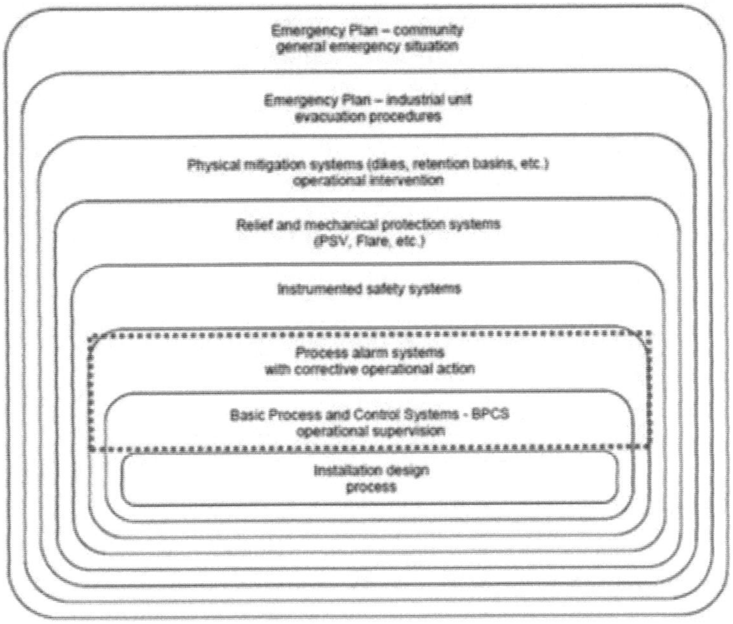

Figure 4. Preventive and Mitigating Layers of Protection for a Typical System (According to Petrobras Standard 2595) [4]

The response to demand depends on the operator's experience and the acceptable limits established for the plant's operation, which in general are defined exclusively by the safety requirements for the installation design.

This condition permits the system to operate within a wide range, not necessarily obtaining the best results in reliability and efficiency. In addition, each operator and shift supervisor can interpret and respond to fluctuations differently. Figure 5 illustrates a generic control example based solely on safety limits:

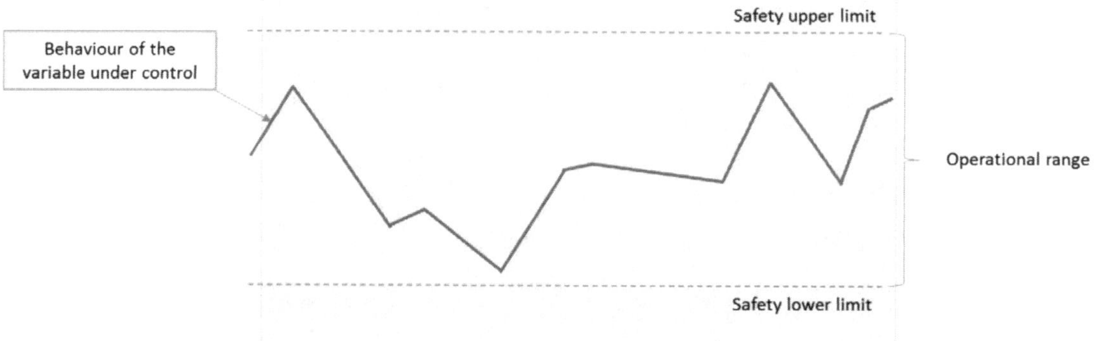

Figure 5. Behavior of a generic variable within the safety range

Although the control lies within the safety range, the system is not necessarily operating at best performance.

Further analysis will show differences in process control across shifts, which led to greater movement within the operational range.

In order to standardize and sustain the unit within a reliable, safe and optimal operating range, Petrobras' Industrial area set forth a continuous monitoring of critical operational variables kept within specified limits with a lower standard deviation.

The process is cyclical and relatively simple:

- The engineers select the operational variables to be monitored according to Safety, Reliability and Optimization.
- For each variable, maximum and minimum limits are established based on the unit's design.
- Engineers then choose the settings for each variable to achieve optimal efficiency,
- The system catches limit violations, known as "breaks", and each break must be justified in writing in the system itself.
- The system assigns a notation based on the number of breaks that occurred within a given period (usually one shift).

- Engineers, shift supervisors and operators meet regularly to discuss anomalies and work hard toward improving results.

The system thus provides greater transparency in managing the operational process, enabling better monitoring and control of actions during production. In addition, it affords greater involvement and alignment among the teams (Operation, Maintenance, Optimization, and Engineering), in order to achieve greater visibility of the operational restrictions and perform comparative team evaluations. Finally, the technical process standards (adequacy of results obtained using project boundaries) can be reviewed and employed to implement new training procedures.

3.2. How variables are organized in OMS

The following clusters have been established:

- Process safety: includes variables whose failure increases the risk of loss of containment, which would negatively impact the environment
- Reliability: includes variables whose breakdown increases the risk of loss of equipment or function.
- Optimization: measures the distance from the unit's current operational level to its optimum point. This dimension includes variables with an impact on economic performance or conditions that could result in a violation of best operating parameters. This includes, for example, product giveaway, energy efficiency, etc.

Figure 6 illustrates some typical variables by analysis groupings:

Dimension	*Description*
Safety	Furnace skin Temperature
	FCC - Catalistic Level
	Degasification Lean DEA
Reliability	Differential Pressure Feed Filter
	Velocity Steam - Decoking Furnace
	Chloride in Desalted crude oil
Optimization	Excess Air Furnace
	Flash Zone Pressure - atmospheric distillation
	Vaccum Furnace Temperature
	Make-up H2 - Diesel HDT

Figure 6. Examples of variables and their grouping

Figure 7 presents a generic example using the same information as in figure 5, but with specified limits for reliability and optimization.

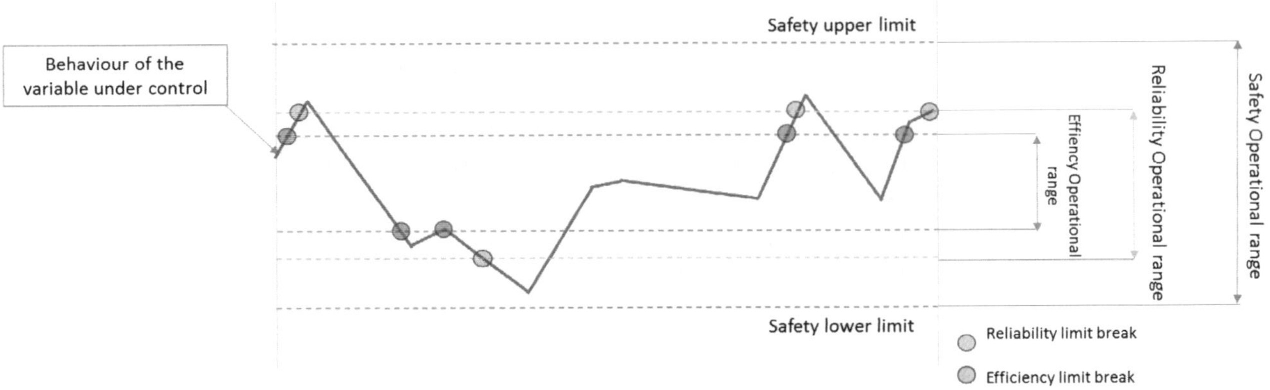

Figure 7. Behavior of the same generic variable shown in Figure 5, now indicating the limits defined for reliability and optimization

It was observed that, although no violations or breaks occurred in relation to the safety limits, 4 breaks are recorded on the reliability band and 5 for efficiency. Therefore, this behavior analysis indicates inefficiency, and the causes must be understood and treated.

3.3. The IT solution flowchart for data/information-sharing

Information for OMS is extracted from the Plant Information API (PI) system as shown in Figure 8:

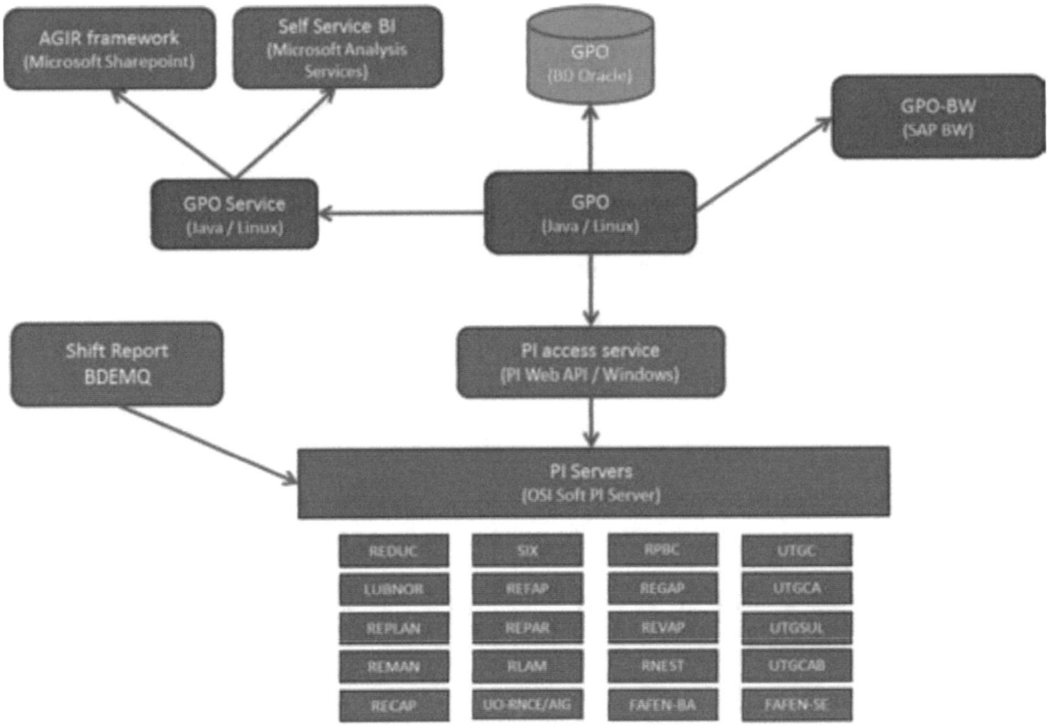

Figure 8. Solution Flowchart

The OMS provides a user-friendly interface that allows users to query real-time information on the status of critical variables for all industrial plants.

Figure 9 shows OMS screens with filters that can be used for performance analysis:

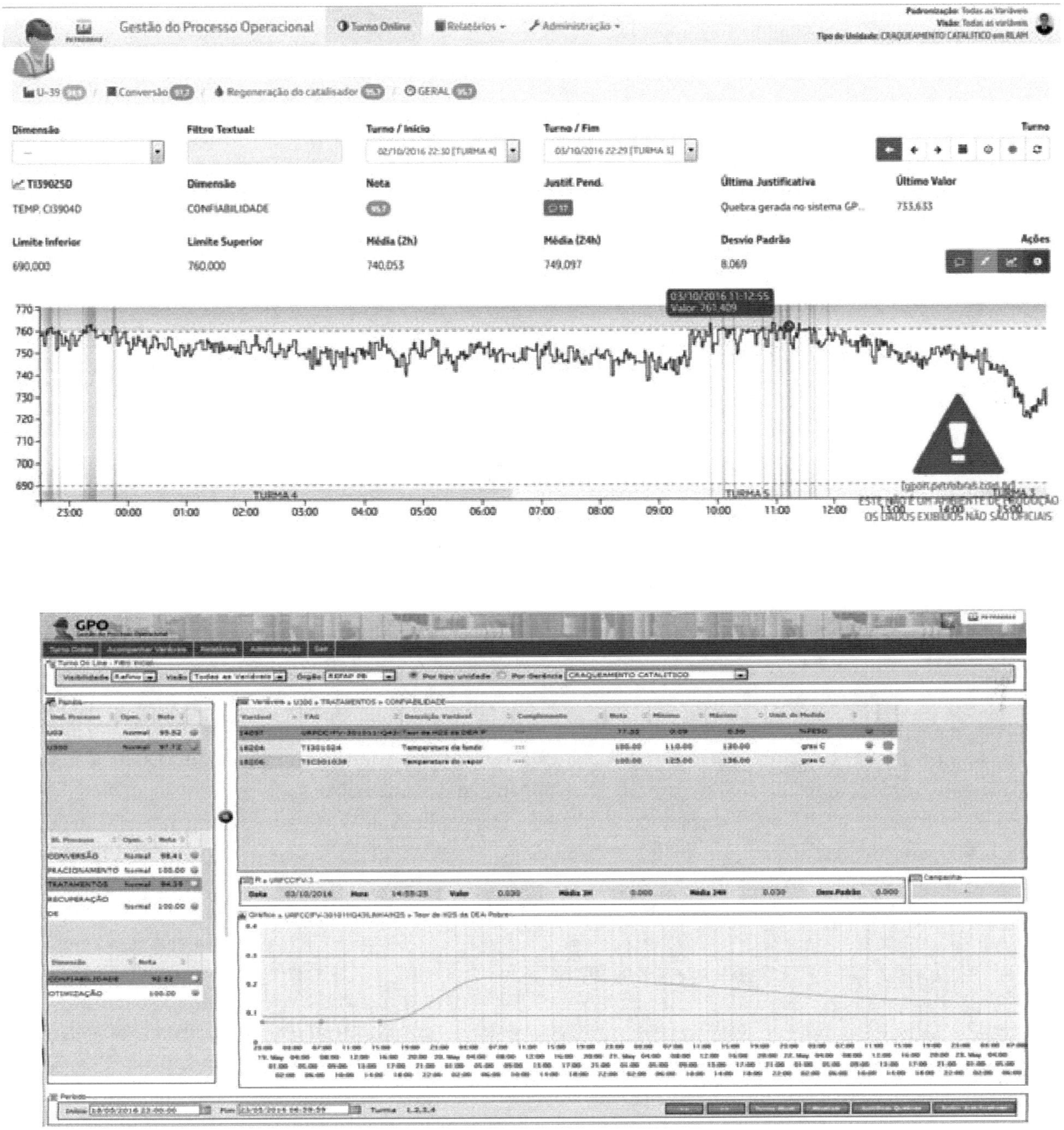

Figure 9. Screen Shots of the OMS Online

The OMS is available in a corporate network and allows consultation across all operating units among all the industrial plants. Both the justifications of breaks and performance analyses are displayed in the system allowing for timely feedback.

3.4. Workflow and decisions involved in OMS

The OMS workflow chart involves several key positions within the company. See Figure 10 below:

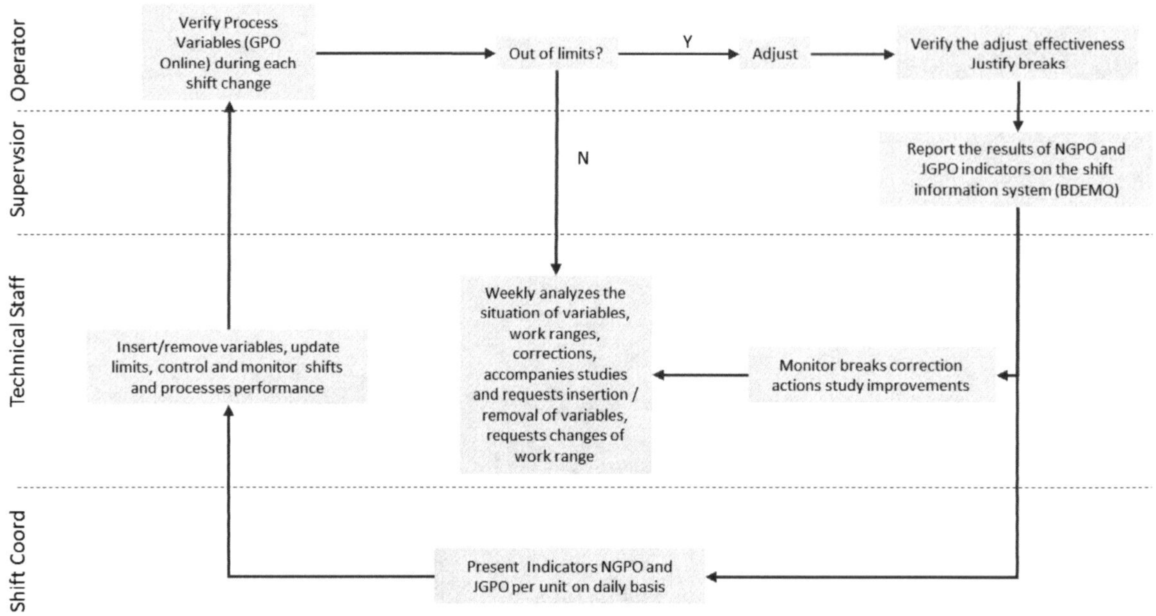

Figure 10. Flow of OMS Processes

The evaluation by the operators and supervisors is performed every shift. Two objective indicators were designed to follow up on this process, the NGPO, or management score, and the JGPO, the justification for failures.

On a weekly basis, technical support compares the shift performance against the critical variables in Safety, Reliability and Optimization and verifies the reasons for the breaks in order to determine whether they are systemic, or related to the quality of monitoring and process control by the operators. If systemic, the breaks may be related to the inadequacy of the defined ranges or control procedures, process sensitivity, or faulty instruments, adjustments or even design. In the latter case, the conditions that led to deviations are considered and retraining or procedural revisions are proposed to reduce misinterpretation or error.

3.5. NGPO and JGPO Performance Indicators

The results obtained by the shift are evaluated, and an operational process management score (NGPO) is generated that evaluates the quality of operation within the defined limits. The NGPO objectively measures the length of time the variable remained outside the range in relation to the total time of the analysis. Figure 11 shows the form of measurement in a hypothetical example:

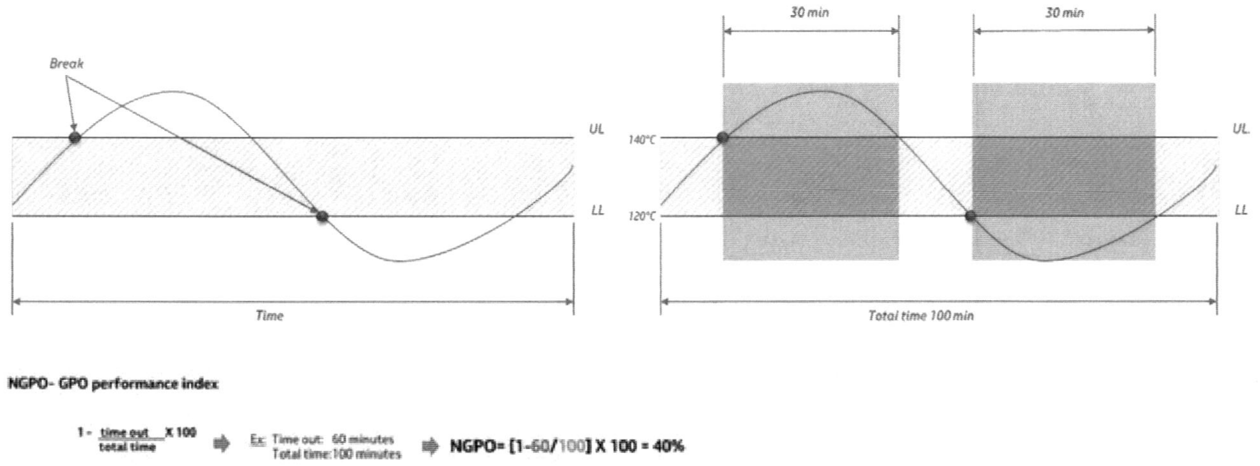

Figure 11. NGPO Calculation Example

In order to ensure that operators perform the break analysis, another indicator called JGPO will measure the percentage of justifications associated with the breaks, as shown in figure 12:

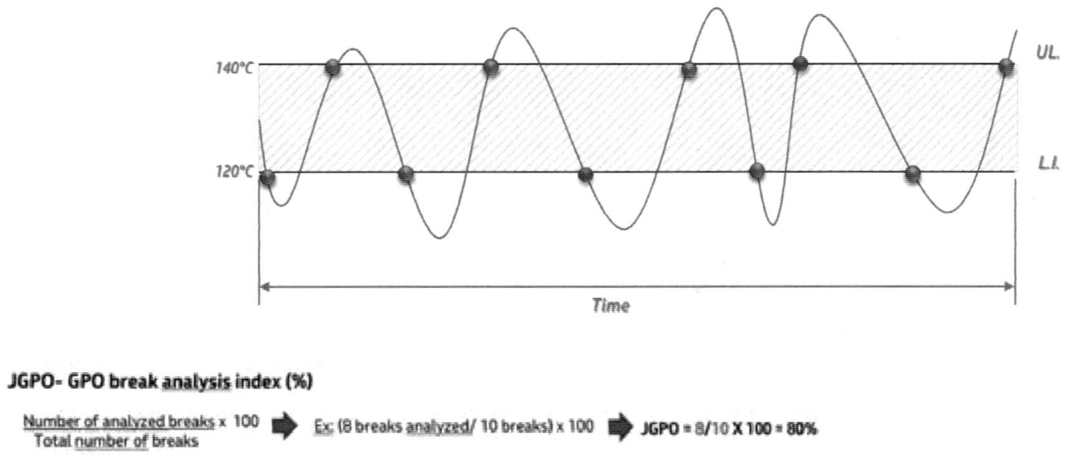

Figure 12. JGPO Calculation Example

The JGPO indicator promotes operational discipline by revealing the treatment of deviations clearly and comparing the results from shift-to-shift. This signals to the operator the importance of control within defined ranges to eliminate justifications altogether.

4 Results

This section describes results that the OMS has achieved in managing the processes at Petrobras Industrial plants:

a. ***Standardization of critical variables*** - The OMS was designed to define typical standard critical variables, applicable to all units of a given process, such as those at our Catalytic Cracking units:

Petrobras owns 12 of these plants, all controlled using the same set of critical variables. This standardization, in addition to allowing the comparative performance evaluation, provides digital transformation solutions, the use of expert systems, and artificial intelligence for evaluation and data analysis.

b. Standardization of the modus operandi - As different shifts perform the control process, the system compares performance variations among these groups over time. From this analysis, it is possible to verify whether the deviations are due to personal difficulties, the groups operating characteristics, system design, or even process restrictions.

Figure 13 illustrates the improvement in quality control that was achieved through standardization, monitoring and analysis. This figure shows the behavior observed in two different times with the same shift controlling the same variable, before and after the use of OMS. Reduced dispersal is evident.

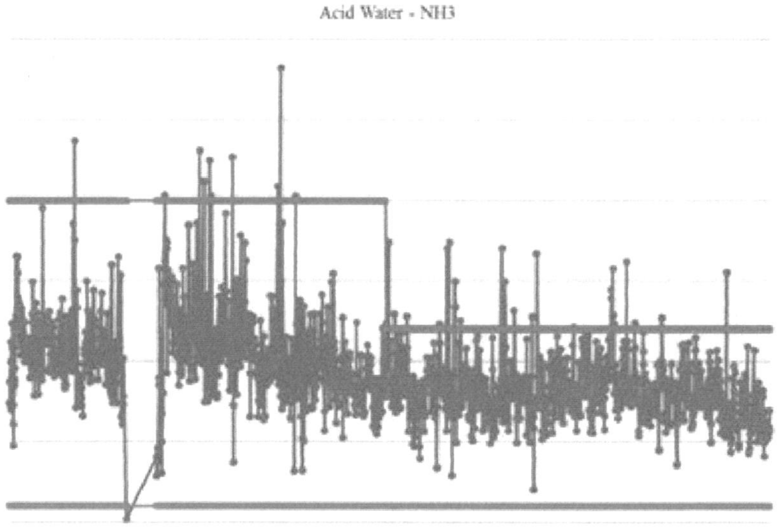

Figure 13. Performance observed with the same groups before and after OMS

In addition, with OMS, standardizing operational management processes has increased the refineries' operational performance and reduced differences among them. Figure 14 illustrates this improvement.

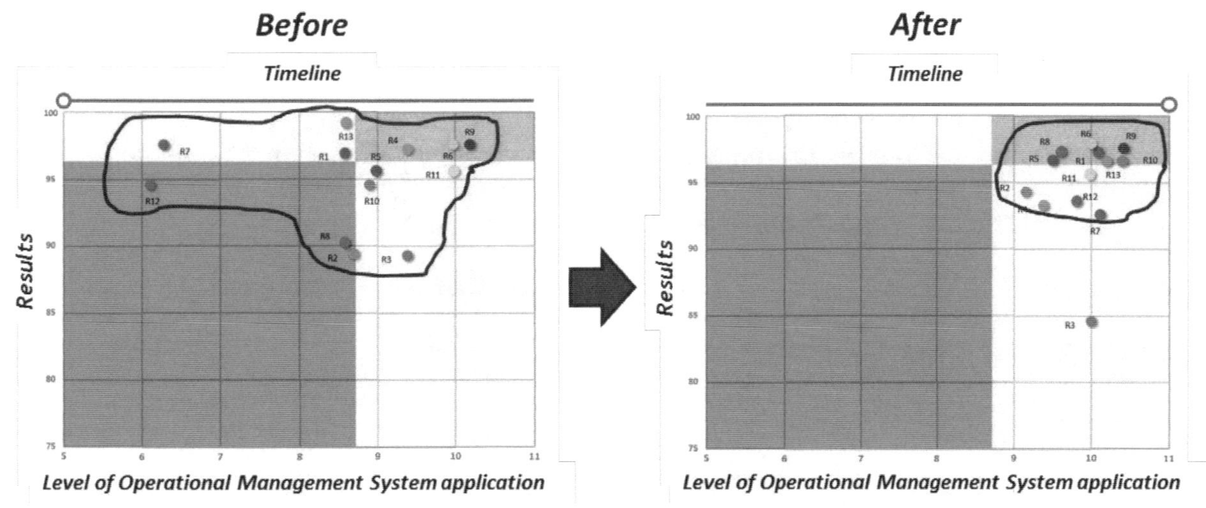

Figure 14. Refineries' performance

c. ***Quality Improvement of technical staff guidelines*** - Critical variables defined by engineering, whether optimization or reliability areas, may present deviations from the expected result. In these cases, the technical area is responsible for identifying the causes. OMS provides a way to calibrate optimization models and improve parameters for efficient operation management.

Here is an example of quality improvement after a review narrowed the operating range for better control:

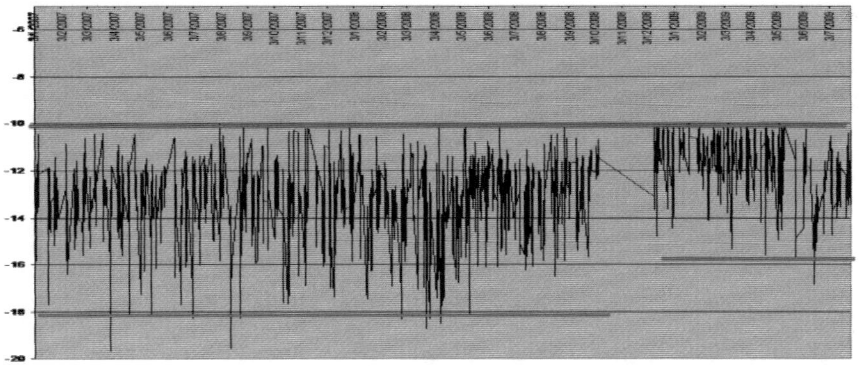

Figure 15. Changes in the control ranges resulting from OMS

Additionally, the analysis can lead, for example, to changing the parameters of process controllers.

d. ***Accountability by the panel operator*** - Clear accountability exists for process control. The system identifies the training needs and profiles of the operators, preparing them for every scenario.

e. ***Operational availability*** - Focusing on reliability reduces the risk of compromised equipment, and enables operation under control in its design ranges.

The OMS can also identify equipment or variables that are bad-actors in the system, allowing specific analyses for improving their performance and then correlating these results with their availability. Figure 16 gives an example of bad-actors identified in an industrial unit:

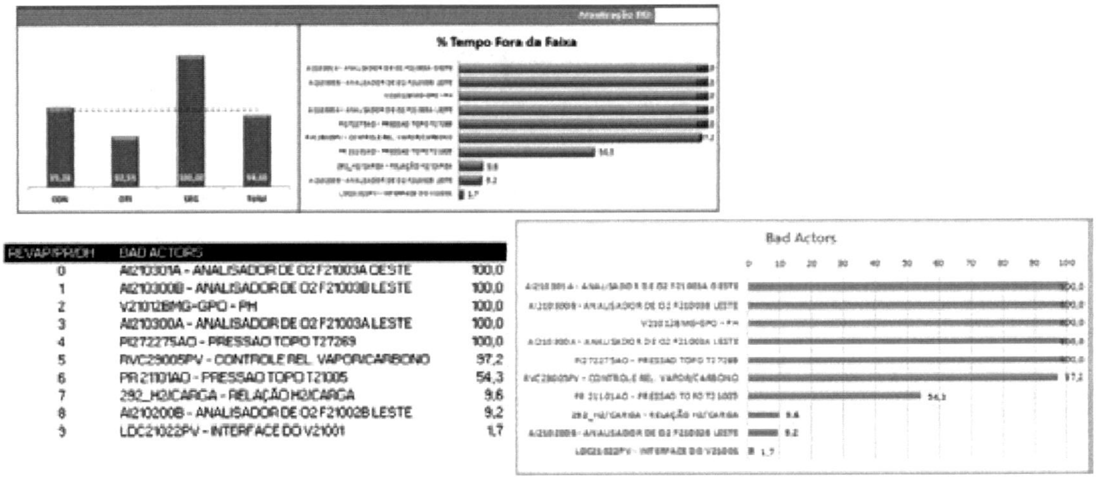

Figure 16. Bad-actors identified by the OMS

In addition, when the OMS was implemented, a significant reduction in unavailability was observed in unavailability due to causes associated with operation control. Figure 17 shows the availability loss between 2015 and 2018.

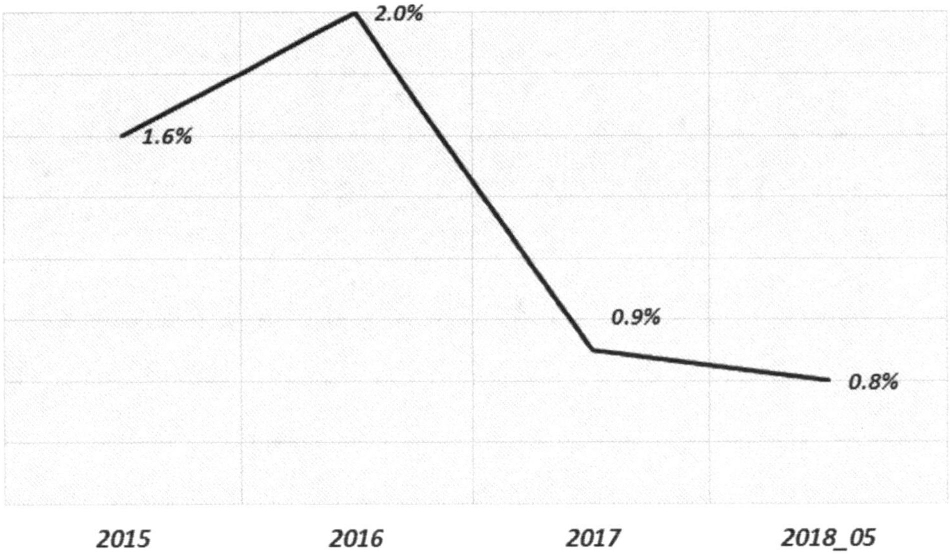

Figure 17. Availability loss due to operation (% of reference capacity)

f. ***NGPO and JGPO results*** - The improvement in process performance is evident when comparing the results obtained in the last 4 years for both NGPO and JGPO. See in figure 18:

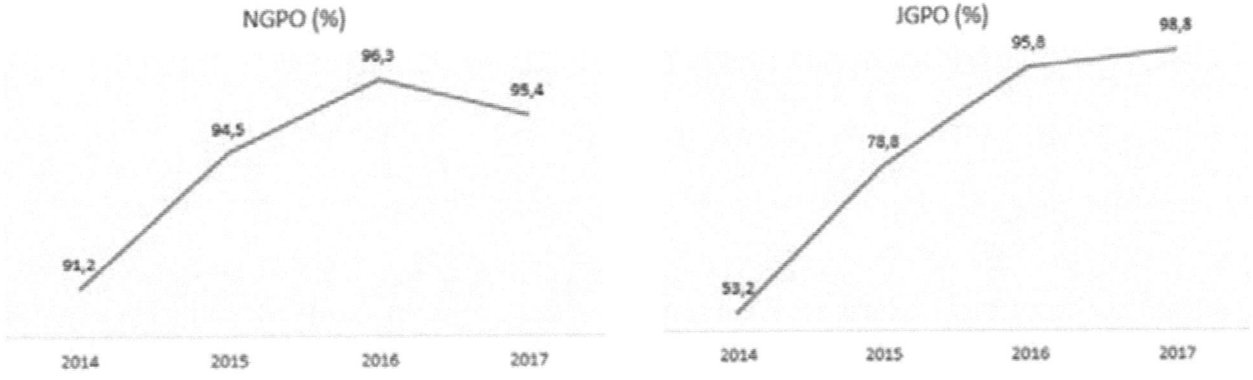

Figure 18. NGPO and JGPO Results for the Last 4 years

The reduction from 96.3% to 95.4% in NGPO between 2016 and 2017 is not due to a reduction in the quality of follow-up. In 2017, variables and workloads were reassessed and made more restrictive, to aim for better process performance.

Working in a collaborative environment allowed operators and supervisors to provide solutions to problems and improve the system. The JGPO results show the clear understanding of the relevance of monitoring and collaboration by operators. The jump in the percentage of justification of breaks has produced better analysis and reevaluation of the optimal bands, and the defining of critical control variables has positively influenced the units' bottom line.

That aside, it is possible to establish *other proactive* indicators for risk management for adding additional layers of protection.

5 Conclusions

The OMS is fully compliant with Risk-Based Process Safety concepts. The implementation of a reliable practice established a high performance standard that encourages a sense of ownership, reduces vulnerability and ensures empowerment of employees in successfully fulfilling their responsibilities. This is intended to open communication in an effective and collaborative manner in order to promote trust and timely feedback. This type of management system proposed and deployed using the OMS meets these requirements.

Monitoring performance and ensuring the quality of operations conduct are determining factors in the pursuit of excellent operations. Therefore, the clarity of the evaluations and follow-up defined by the performance indicators (NGPO and JGPO) allow comparability and stimulate the best results among shift.

Additionally, in an organizational view, the OMS provided standardization of the systematic analysis and definition of critical variables for similar production processes. This approach also allowed us to evaluate the performance of plants in relation to operating practices, control systems and operational arrangements.

The transparency of the results adds meaning to each activity and creates clarity of purpose when performance expectations are no longer organizational, but individualized.

The performance information also recognizes individual contributions, such as that of operators, toward achieving company goals, which creates a performance-based culture that values business profitability.

Regarding reliability, the system has allowed the units to operate within more suitable ranges for integrity management, reducing the damage rate and providing more reliable campaigns with fewer losses due to mechanical failure. Already, operating within more optimal bands has produced results that are more profitable. Implementing the OMS has been worth the effort.

6 References

[1] CCPS. *Guidelines for Risk Based Process Safety.* Center for Chemical Process Safety, American Institute of Chemical Engineers, New York, NY, 2007.

[2] Risk Based Safety Process Overview. Available at https://www.aiche.org/ccps/resources/publications/books/guidelines. Accessed on January, 2, 2019.

[3] Nossas-atividades/areas-de-atuação. Available at www.petrobras.com.br. Accessed on December 28, 2018.

[4] Petrobras Internal Standard N-2595: Criteria for Design, Operation and Maintenance of Safety Instrumented Systems in Industrial Units. Petrobras, 2016.

The Effects of Oxygen Enrichment

on the

Explosibility Properties of Flammable Gases and Vapors

Richard W. Prugh
DEKRA Process Safety
113 Campus Drive
Princeton, NJ 08540
www.dekra-process-safety.com
prughrw@aol.com

Prepared for Presentation at
American Institute of Chemical Engineers
2019 Spring Meeting and 15th Global Congress on Process Safety
New Orleans, LA
March 31 – April 3, 2019

The Effects of Oxygen Enrichment on the Explosibility Properties of Flammable Gases and Vapors

Richard W. Prugh
DEKRA Process Safety
113 Campus Drive
Princeton, NJ 08540

Keywords: Flammables, Oxygen, Explosion

Abstract

In the chemical industry, oxygen-containing chemicals are used in the production of partial-oxidation products, including many in the pharmaceutical industry. The vapor spaces in oxidation reactors can contain high concentrations of oxygen gas, in combination with flammable vapors. Thus, it is important to determine the explosibility properties of organic-compound vapors in atmospheres containing high oxygen concentrations.

1. Introduction

In the chemical industry, "nascent" oxygen is involved in the production of partial-oxidation products, including many in pharmaceutical operations. The oxygen is initially in the form of peroxides, such as hydrogen and sodium peroxides, and also as chlorates and perchlorates, periodates, nitrates, and dichromates [1]. The vapor space in oxidation reactors can contain high concentrations of oxygen gas, in combination with the flammable vapors of organic compounds. To evaluate the possible explosion hazards of such mixtures of fuel and oxidizer, it is important to determine the explosibility properties of organic compounds in atmospheres containing high oxygen concentrations.

The basic flammability/explosibility data for this study are from the following references:

Table 1. References for Flammability Data

REFERENCE	FLAMMABILITY/EXPLOSIBILITY PROPERTY					
	Flammable Limits	Auto-Ignition Temperature	Minimum Ignition Energy	Maximum Explosion Pressure	Deflagration Index Kg	Oxygen Conc. Limit
NFPA 49; NFPA 325 [2]	√	√				
NFPA 68 [3]				√	√	
NFPA 69 [4]						√
NFPA 77 [5]	√ *		√			
NFPA 497 [6]	√	√	√			
USBM 503 [7]	√ *					√
USBM 627 [8]	√ *	√ *				
Babrauskas [9]	√	√ *	√ *			
Britton [10]	√ *		√ *			√
Crowl and Louvar [11]	√ *	√	√	√	√	√
Yaws [12]	√	√				

* Limited data for oxygen

Several methods are presented in this paper for estimating explosibility properties for preliminary process design. However, it is prudent to confirm the fuel/oxygen relationships with tests where process safety depends on the reliable control of oxygen concentration.

2. Flammability/Explosibility Properties

The explosibility properties of combustible materials can be divided into Ignition Sensitivity and Explosion Severity [13, 14]. The Lower Flammable Limits, the Autoignition Temperatures, and the Minimum Ignition Energies would be important in evaluating the Ignition Sensitivity. The Maximum Explosion Pressure and the Deflagration Index – a measure of burning rate – would be important in evaluating the Explosion Severity. The following sections describe the relationships of these properties to oxygen concentration.

2.1 Lower Flammable Limit

Data from many sources [15, 16, 17, 18] indicate that there is no significant effect of oxygen concentration on the Lower Flammable Limit [LFL] from the 21 volume-percent oxygen in air to 100% oxygen. At oxygen concentrations greater than the LFL in air, the specific heat of oxygen is about the same as the specific heat of nitrogen; that is, excess oxygen behaves as a diluent at the LFL.

For comparison with the Minimum Explosible Concentrations of combustible dusts and for evaluating the hazards of mixtures of flammable gases/vapors of combustible dusts, the values of LFL can be converted to mass densities by:

$$C = [\,LFL / (100 - LFL)\,]\,[\,MW / 0.024\,] \quad \text{grams per cubic meter}$$

where 0.024 is the number of cubic meters per gram-molecular weight [at 25°C and one atmosphere pressure (14.7 psia)].

For example, the LFL for propane in air is 2.1 volume-percent and, for the molecular weight of 44.1, the Minimum Explosible Concentration would be given by:

$$C = [\,2.1 / (100 - 2.1)\,]\,[\,44.1 / 0.024\,] = 39 \text{ grams per cubic meter}$$

Since the LFL is not significantly affected by oxygen concentration, oxygen enrichment would not affect the ventilation and dilution or inerting precautions that should normally be taken for a flammable vapor or gas.

2.2 Autoignition Temperature

The Autoignition Temperature [AIT] of a fuel is a kinetic phenomenon; that is, the AIT is low for very-long exposures to a given temperature, and it is quite high for short exposures to a given temperature. This relationship can be shown to follow the Arrhenius "law" [19], as:

$$r = A\,e^{-Ea/RT} \quad \text{fraction per second}$$

where A is the pre-exponential factor (in "per second");
Ea is the activation energy (in calories per gram-mole);
R is a constant (1.987 calories per gram-mole per °K); and
T is the temperature (in °K).

If equal reaction rates are assumed for combustion in air and oxygen, at two different temperatures, a value for the activation energy can be obtained, from:

$$Ea = 1.987\,[\,\ln(100/21)\,] / [\,(1/T_{AIR}) - (1/T_{O2})\,] = 3.1 / [\,(1/T_{AIR}) - (1/T_{O2})\,] \quad \text{cal./gmol}$$

The Autoignition Temperatures for 90 chemical compounds in air and in 100% oxygen are available [20]. The average AIT in air was calculated to be about 394°C [about 667°K], and the average AIT in oxygen was about 361°C [about 634°K]. Thus, the differences in AIT for air and oxygen is quite small, at about 5% lower.

The calculated activation energy Ea, as based on the referenced data, would be about 40,000 calories per gram mole, which is a “reasonable” value for an oxidation/ combustion reaction [21].

Thus, the AIT for a given oxygen concentration can be estimated from:

$$AIT_{O2} = AIT_{AIR} / [1 + (1.987\ AIT_{AIR} / 40{,}000) (\ln (C_{O2}/21)] - 273 \qquad {}^{o}C$$

Oxygen enrichment would not affect the precautions that should normally be taken for a flammable or combustible vapor, such as avoiding exposure to hot surfaces in ovens, in dryers, and on electrical equipment such as motors.

2.3 Minimum Ignition Energy

Previous investigators had found that the Minimum Ignition Energy [MIE] was inversely related to oxygen concentration [22]. The Minimum Ignition Energies for 13 chemical compounds in air and in 100% oxygen are available [23]. The average MIE in air [excluding hydrogen (very low) and trichloroethylene (very high)] was calculated to be about 0.30 millijoule, and the average MIE in oxygen was about 0.0027 millijoule. Thus, oxygen concentration has a very strong effect on MIE, with about a factor of 100 reduction for 100% oxygen. The exponent for oxygen concentration for the 13 chemical compounds [including hydrogen and trichloroethylene] is about 2.9.

Thus, the MIE for a given oxygen concentration can be estimated [+/- 50%] from:

$$MIE_{O2} = MIE_{AIR} / [(C_{O2} / 21)^{2.9} = 6{,}200\ MIE_{AIR} / C_{O2}^{\ 2.9} \qquad \text{millijoule}$$

As indicated above, oxygen enrichment would have a significant effect on the precautions [such as inerting] that should be taken for flammable gases and liquid against very weak electrostatic [and other] ignition sources, including corona discharges [24].

2.4 Maximum Explosion Pressure

The Maximum Explosion Pressure [Pmax] – for combustion in air or in an oxygen-enriched atmosphere – can be calculated rather accurately if the heat of formation of the fuel is known and the degree of dissociation of the carbon dioxide and steam combustion products can be determined. For combustion in an oxygen-enriched atmosphere, the degrees of dissociation are very high; thus, the Pmax in 100% oxygen could be expected to show a rather small increase in pressure as compared to combustion in air.

There is a direct relationship between flame temperature and explosion pressure, as governed by the Ideal Gas Law, as:

$$P_{Max} = 14.7\ (\ T_{FLAME} + 273\ )\ /\ (\ 25 + 273\ ) - 14.7 \qquad \text{pounds/sq.in. [psig]}$$

The available flame-temperature explosion-pressure data are as follows [25]:

Table 2. Flame Temperatures and Explosion Pressures for Fuel/Oxygen Mixtures

FUEL	FLAME TEMPERATURES (°C)			MAX. PRESSURES (barg)		RATIO (bara/ bara)	IDEAL-GAS-LAW RATIOS (°K/bara)	
	In Air	In Oxygen	Ratio [°K/°K]	In Air	In Oxygen		In Air	In Oxygen
Methane	1,950	2,705	1.37	7.1	10.4	1.41	274	261
Propane	1,988	2,822	1.37	7.9	10.2	1.26	254	276
Ethylene	2,088	2,902	1.34	8.0	10.7	1.30	262	271
Acetylene	2,262	3,069	1.32	10.6	11.2	1.05	219	274

Thus, approximate values for Pmax in mixtures of oxygen and in air would be given by:

$$P_{Max,O2} = [\ (\ T_{FLAME} + 273\ )\ /\ 261\ ] - 1 \qquad \text{barg}$$

Thus, if the flame temperature can be calculated, from heats of formation and specific heats – with consideration of dissociations – the maximum explosion pressure for a hydrocarbon can then be calculated. It is likely that this relationship would also apply to other hydrocarbons and carbohydrates. Also, as indicated in the preceding Table 2, the values of P_{MAX} in 100% oxygen are about 30% greater than the value in air [excluding acetylene].

The data in Table 2 above can be used to estimate the Maximum Explosion Pressure [P_{MAX}] for a given concentration of oxygen as:

$$P_{MAX,C} = 0.035\ C_{O2} + 0.92\ P_{MAX,AIR} \qquad \text{bara}$$

The following Table presents the results of calculations, based on the above relationship, for two typical values for P_{MAX} in air [26]:

Table 3. Maximum Explosion Pressures for Fuel/Oxygen Mixtures

$P_{MAX,AIR}$	MAXIMUM EXPLOSION PRESSURES IN FUEL/OXYGEN MIXTURES (barg)							
	30 $\%O_2$	40 $\%O_2$	50 $\%O_2$	60 $\%O_2$	70 $\%O_2$	80 $\%O_2$	90 $\%O_2$	100 $\%O_2$
7 barg	7.4	7.7	8.1	8.4	8.8	9.1	9.5	9.8
8 barg	8.3	8.6	9.0	9.3	9.7	10.0	10.4	10.7
9 barg	9.2	9.6	9.9	10.3	10.6	11.0	11.3	11.7
10 barg	10.1	10.5	10.8	11.2	11.5	11.9	12.2	12.6

Thus, oxygen enrichment would have a significant effect on the extent of damage that could result from ignition of confined flammable gases and vapors, and could affect the design of explosion-protection systems and devices, such as explosion vents.

2.5 Deflagration Index, Kg

The Deflagration Index [Kg for flammable gases and vapors] is a measure of the speed of combustion and, thus, is directly related to the speed of flame through a mixture of fuel and oxidant. In the absence of strong turbulence, the flame speed is related to the fundamental burning velocity [FBV] by the ratio of the densities of the unburned mixture and the density of the combustion gases, by [27]:

$$S_{FLAME} = FBV\,(d_{UNBURNED}/d_{BURNED}) \quad \text{centimeters per second}$$

Typically, the ratio of the flame speed to FBV is about 6.

Previous investigators had observed that there is a relationship between the FBV and the Minimum Ignition Energy [MIE], as given by [28]:

$$FBV = K/MIE^{0.8} \text{ centimeters per second}$$

Some data are available for both the FBV and the MIE, as indicated in the following Table [29] :

Table 4. The Relationship Between FBV And Oxygen Concentration

GAS/VAPOR	FBV, Air (cm./sec.)	Kg, Air (bar-m/s)	MIE, Air (millijoule)	FBV, O_2 (cm./sec.)	MIE, O_2 (millijoule)	Kg, O_2 (bar-m/s)	Kg Ratio O_2/Air
Hydrogen	312	638	0.016	1400	0.0012	3,000	4.5
Methane	45	55	0.21	450	0.0027	450	4.6
Ethane	47.5	103	0.24	375	0.0019	670	6.5
Propane	46	100	0.25	370	0.0021	720	7.2
Butane	-	92	0.25	-	0.009	-	-
Hexane	-	-	0.24	-	0.006	-	-
Heptane	42.5	-	0.24	-	-	-	-
Averages	-	-	-	-	-	-	5.7

As indicated in the above Table, the Deflagration Index – and the speed of combustion – in oxygen is almost six times greater in 100% oxygen than it is in air [at 21% oxygen]. The data in the preceding Table can be combined to yield two approximate expressions for the Deflagration Index in given concentrations in oxygen, as:

$$Kg_{,O2} = 0.04\ (K_{G,AIR})\ (C^{1.2}) \qquad \text{bar-meters/second}$$

$$Kg_{,O2} = 1.85\ C^{1.1} / MIE_{AIR}^{\ 0.46} \qquad \text{bar-meters/second [excluding } H_2]$$

Thus, if the value of the Deflagration Index in air is known, or if the Minimum Ignition Energy in air is known, an approximate value for the Deflagration Index at a given oxygen concentration can be determined.

3. Combined Relationships

The explosibility/flammability properties of a given flammable gas or vapor can be compared with a standard or typical flammable gas, through the use of Ignition Sensitivity and Explosion Severity. These property combinations were first developed and utilized for the comparisons of the hazards of coal mines, starting in 1940, where combustible dusts presented an explosion hazard [30].

3.1 Ignition Sensitivity

The product of LFL [in terms of grams per cubic meter], AIT, and MIE form the basis for a calculation of Ignition Sensitivity. A high value of this product indicates a low likelihood of ignition by relatively-weak ignition sources, such as electrostatic discharges, friction, impact, and hot surfaces. For propane in a mixture with 40% oxygen, the product is only 39 x 445 x 0.035 = 600. For comparison, the similar product for typical combustible dusts is about 45,000.

To determine the Ignition Sensitivity of a combustible dust, the LFLxAITxMIE product would be divided into the corresponding product for Pittsburgh [bituminous] Coal Dust, at 2,000,000. Thus, the Ignition Sensitivity for typical combustible dusts is about 2,000,000/45,000 = 45. By comparison, the Ignition Sensitivity for propane in a 40% oxygen mixture would be about 2,000,000/600 = 3,300, indicating the much greater sensitivity of this propane mixture to ignition sources.

As another indication of the effects of oxygen enrichment, the Ignition Sensitivity for a stoichiometric [worst-case] mixture of propane and air would be about 450. Thus, the 40% oxygen mixture would be about 7 times more sensitive to ignition [3,300/450].

3.1 Explosion Severity

The product of Deflagration Index Kg and the Maximum Explosion Pressure Pmax [in terms of barg] forms the basis for a calculation of Explosion Severity. A low value of this product indicates a low likelihood of severe damage from explosion. For propane in a mixture with 40% oxygen, the product is about 275 x 9 = 2,500. For comparison, the similar product for typical combustible dusts is about 1,500.

For combustible dusts, the product is divided by the corresponding product for Pittsburgh [bituminous] Coal Dust, at 840. Thus, the Explosion Severity for typical combustible dusts is about 1,500/840 = 1.8. By comparison, the Ignition Sensitivity for propane in a 40% oxygen mixture would be about 2,500/840 = 3.0, indicating the greater likelihood of damage if a confined mixture of propane and 40% oxygen were to be ignited.

The effect of oxygen enrichment can be shown by the Explosion Severity for a stoichiometric [worst-case] mixture of propane and air, at about 0.9. Thus, the 40% oxygen mixture could be about 3 times more damaging than an air mixture [3/0.9].

3.1 Combination of Ignition Sensitivity and Explosion Severity

If the Ignition Sensitivities of fuel/oxidant mixtures are less than 0.2 AND the Explosion Severities are less than 0.5, then such mixtures "are generally not considered to be significant explosion hazards" [31].

For the mixture of propane and 40% oxygen – for which the properties are shown above – the combination of Ignition Sensitivity and Explosion Severity is about 5 orders of Magnitude [a factor of almost 100,000] above the "no significant explosion hazard" criterion. Thus, it is unlikely that a mixture of air with any flammable or combustible vapor could be considered to present "no significant explosion hazard".

3.1 Combination of Ignition Sensitivity and Explosion Severity as Explosibility Index

The product of Ignition Sensitivity and Explosion Severity can be considered to be an Explosibility Index [32 to 37]. This Index has been found to be useful in comparing the hazards of combustible dusts, but had not yet been applied to flammable gases and vapors.

4. Graphical Presentation of Flammability/Explosibility Properties

Figure 1 presents the flammability/explosibility properties of a mixture of flammable fuel, oxygen, and nitrogen, in terms of Ignition Sensitivity [IS], Explosion Severity [ES], and Explosibility Index [EI] at atmospheric pressure and at an initial temperature of 25°C, as compared to the similar explosibility properties of propane at room temperature and atmospheric pressure. The ratios are based on the available data and the results of the calculations that are described above. As could be expected, the explosion hazards increase as the stoichiometric line – where all carbon is oxidized to carbon dioxide, and all hydrogen is oxidized to steam – is approached.

5. Conclusions.

To evaluate the explosion hazards of mixtures of fuels with oxygen-enriched atmospheres, the available data were utilized to obtain methods for estimating the values of pertinent explosibility properties. The relationships are as follows:

Lower Flammable Limit	Equal to that of fuel/air mixtures	
Autoignition Temperature	$AIT_{O2}=AIT_{AIR}/[1+(2.0AIT_{AIR}/40,000)(\ln(C_{O2}/21)]-273$	°C
Minimum Ignition Energy	$MIE_{O2} = 6,200\ MIE_{AIR} / C_{O2}^{\ 2.9}$	millijoule
Max. Explosion Pressure	$P_{Max,O2} = [[0.035\ C_{O2}) + (0.92\ (P_{Max,AIR} + 1))] - 1$	barg
Deflagration Index	$Kg_{,O2} = 0.04\ (K_{G,AIR})\ (C^{\ 1.2})$	bar-meters/second
	$Kg_{,O2} = 1.85\ C^{\ 1.1} / MIE_{AIR}^{\ 0.46}$	bar-meters/second

In addition, a method is presented for evaluating the Ignition Sensitivity and the Explosion Severity of fuel/oxygen mixtures, using the above relationships, together with a method for graphically showing the relative hazards of such mixtures.

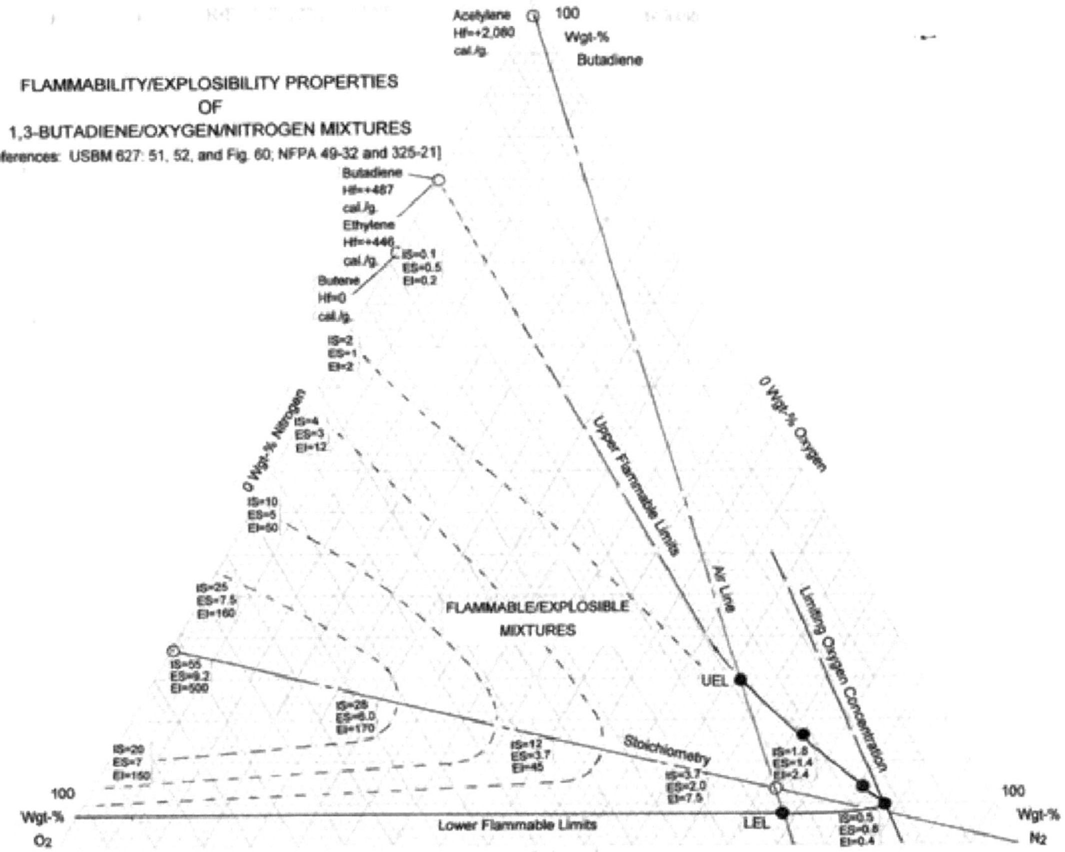

Figure 1 – Weight Concentrations of Gas Mixtures

with

Ignition Sensitivities, Explosion Severities, and Explosion Index

as compared to the properties of Propane

6. References.

1. Urben, P. G. [editor], "Bretherick's Handbook of Reactive Chemical Hazards", 6th Ed., Vol. 2, Page 287 (1999).

2. National Fire Protection Association, "Fire Protection Guide to Hazardous Materials", 13th Ed., including NFPA 49 and NFPA 325 (2002).

3. National Fire Protection Association, "Guide for Venting of Deflagrations", NFPA 68, Table D.1 (2002).

4. National Fire Protection Association, "Standard on Explosion Prevention Systems", NFPA 69, Table C.1(a) (2014).

5. National Fire Protection Association, "Recommended Practice on Static Electricity", NFPA 77, Table B.1 (2014).

6. National Fire Protection Association, "Recommended Practice for the Classification of Flammable Liquids, Gases, or Vapors and of Hazardous (Classified) Locations for Electrical Installations in Chemical Process Areas", NFPA 497, Table 4.4.2 (2017).

7. Coward, H. F., and G. W. Jones [U. S. Bureau of Mines], "Limits of Flammability of Gases and Vapors", Bulletin 503, Table 44 [and many "nose" Figures] (1952).

8. Zabetakis, M. G. [U. S. Bureau of Mines], "Flammability Characteristics of Combustible Gases and Vapors", Bulletin 627, Table A-1 [and many "nose" Figures] (1965).

9. Babrauskas, V., "Ignition Handbook", Table 1A (1992).

10. Britton, L. G., "Avoiding Static Ignition Hazards in Chemical Operations", Page 223 (1999).

11. Crowl, D. A., and J. F. Louvar, "Chemical Process Safety", 3rd Ed., Pages 251, 259, 261, 287, and 654 (2011).

12. Yaws, C. L., "Chemical Properties Handbook", Table 25-1 (1999).

13. National Fire Protection Association, "Recommended Practice for the Classification of Combustible Dusts and of Hazardous (Classified) Locations for Electrical Installations in Chemical Process Areas", NFPA 499, Paragraph A.5.2.3 (2017).

14. Frank, W. L., and S. A. Rodgers, "NFPA Guide to Combustible Dusts", Page 92 (2012).

15. Britton, L. G., "Avoiding Static Ignition Hazards in Chemical Operations", Page 223 (1999).

16. Babrauskas, V., "Ignition Handbook", Page 1023 (1992).

17. Zabetakis, M. G. [U. S. Bureau of Mines], "Flammability Characteristics of Combustible Gases and Vapors", Bulletin 627, Page 9 (1965).

18. Crowl, D. A., and J. F. Louvar, "Chemical Process Safety", 3rd Ed., Pages 263, 265 to 267, and 269 (2011).

19. Crowl, D. A., and J. F. Louvar, "Chemical Process Safety", 3rd Ed., Pages 398 and 400 (2011).

20. Babrauskas, V., "Ignition Handbook", Pages 1023 and 1024 to 1058 (1992).

21. Green, D. W., and M. Z. Southard, "Perry's Chemical Engineers' Handbook", 9th Ed., Page 19-24 (2019).

22. Babrauskas, V., "Ignition Handbook", Pages 75, 76, 203, and 625 (1992).

23. Babrauskas, V., "Ignition Handbook", Pages 1024 to 1055 (1992).

24. Babrauskas, V., "Ignition Handbook", Page 625 (1992).

25. Lewis, B., and G. von Elbe, "Combustion, Flames, and Explosions of Gases", 3rd Ed., Pages 717 and 719 (1987).

26. National Fire Protection Association, "Guide for Venting of Deflagrations", NFPA 68, Table D.1 (2002).

27. Babrauskas, V., “Ignition Handbook”, Page 35 (1992).

28. Babrauskas, V., “Ignition Handbook”, Page 74 (1992).

29. Zabetakis, M. G. [U. S. Bureau of Mines], “Flammability Characteristics of Combustible Gases and Vapors”, Bulletin 627, Page 43 and Figures 44 and 45 (1965).

30. Nagy, J., et al. [U. S. Bureau of Mines], “Explosibility of Carbonaceous Dusts”, Report of Investigations 6597, Page 2 (1965).

31. National Fire Protection Association, “Recommended Practice for the Classification of Combustible Dusts and of Hazardous (Classified) Locations for Electrical Installations in Chemical Process Areas”, NFPA 499, Paragraph A.5.2.3 (2017).

32. Dorsett, H. G., et al. [U. S. Bureau of Mines], Laboratory Equipment and Test Procedures for Evaluating Explosibility of Dusts”, Report of Investigations 5624, Page 2 (1960).

33. Jacobson, M., et al. [U. S. Bureau of Mines], “Explosibility of Dusts used in the Plastics Industry”, Report of Investigations 5971, Tables A-1, A-2, and A-3 (1962).

34. Jacobson, M., et al. [U. S. Bureau of Mines], “Explosibility of Metal Powders”, Report of Investigations 6516, Tables A-1, A-2, and A-3 (1964).

35. Nagy, J., et al. [U. S. Bureau of Mines], “Explosibility of Carbonaceous Dusts”, Report of Investigations 6597, Page 2 (1965).

36. Dorsett, H. G., and J. Nagy [U. S. Bureau of Mines], “Dust Explosibility of Chemicals, Drugs, Dyes, and Pesticides”, Report of Investigations 7132, Tables A-1, A-2, and A-3 (1968).

37. Grossel, S. S., “Safety Considerations in Conveying of Bulk Solids and Powders”, Page 62 in the “Journal of Loss Prevention in the Process Industries”, Butterworths, Vol. 1, No. 2 (April 1988).

7. Legal Disclaimer and Liability

(a) Limitation of Liability. The consulting services conducted by DEKRA Process Safety (the "Company") were performed using generally accepted guidelines, standards, and/or practices, which the Company considers reliable. Although the Company performed its consulting services pursuant to reliable and generally accepted practices in the industry, the Company does not guarantee or provide any representations or warranties with respect to Client's use, interpretation or application of the findings, conclusions, and/or suggestions of the consulting services provided by the Company. Moreover, the findings, conclusions, and the suggestions resulting from the consulting service are based upon certain assumptions, information, documents, and procedures provided by the Customer. AS SUCH, IN NO EVENT AND UNDER NO CIRCUMSTANCE SHALL THE COMPANY BE LIABLE FOR SPECIAL, INDIRECT, PUNITIVE OR CONSEQUENTIAL DAMAGES OF ANY NATURE WHATSOEVER, INCLUDING WITHOUT LIMITATION, ANY LOST REVENUE OR PROFITS OF THE CUSTOMER OR ITS CUSTOMERS, AGENTS AND DISTRIBUTORS, RESULTING FROM, ARISING OUT OF OR IN CONNECTION WITH, THE SERVICES PROVIDED BY THE COMPANY. The Customer agrees that the Company shall have no liability for damages, which may result from Client's use, interpretation or application of the consulting services provided by the Company.

(b) The Company's pricing of the consulting services provided does not contemplate that the Company shall have any liability resulting from its performance of the consulting services, except as otherwise set forth in the Quotation from the Company. Accordingly, the Customer shall indemnify and hold harmless the Company, its shareholders, directors, officers, employees and agents (the "Indemnified Parties") from and against any and all loss, cost, liability and expense, including reasonable attorney's fees and costs, which any of the Indemnified Parties may incur, sustain or be subject to, as a result of any claim, demand, action, investigation or proceeding arising out of or relating to either: (a) the consulting services provided by the Company; or (b) any material, equipment, specifications or safety information (or lack thereof) supplied to the Company (or which should have been supplied to the Company) by Customer and/or any failure of such materials, equipment, specifications and safety information to comply with any federal, state or local law or safety standard.

(c) For additional terms and conditions, which apply with respect to the provision of this report, see the Quotation provided by the Company and executed by Customer. If any terms set forth in the Quotation conflict with the terms set forth herein, the terms set forth herein shall apply.

Bridging Chemical Engineering (Risk Based Process Safety) to the Humanities – 7 Mega Elements' Approach (Protocol) of Accident Investigation in Light of Ethical Dimension, Architecture & Epistemology

Long(Steven) Zhang
No affiliation since Jul 2014 off-hiring from COP
11455 Bissonnet St Apt #514, Houston TX 77099
longz.houston@gmail.com

Prepared for Presentation at
American Institute of Chemical Engineers
2019 Spring Meeting and 15th Global Congress on Process Safety
New Orleans, LA
March 31 – April 3, 2019

Bridging Chemical Engineering (Risk Based Process safety) to the Humanities – 7 Mega Elements' Approach (Protocol) of Accident Investigation in Light of Ethical Dimension, Architecture & Epistemology

Long(Steven) Zhang
No affiliation since Jul 2014 off-hiring from COP
11455 Bissonnet St Apt #514, Houston TX 77099
longz.houston@gmail.com

Keywords: Investigation, Diagnosis, Humanities, *the Art of War, Narrative, Epistemology*

Abstract

In this paper, facing the challenges of Incident Investigation, actually whole CCPS 4th Pillar "Learn from Experience", chemical engineers' (process safety practitioners') brand new vocabulary (7 Mega Elements' narrative on failure diagnosis) would be presented in context of bridging chemical engineering (RBPS) to the Humanities, of which 3 core topics i.e. ethical dimension, architecture and epistemology would be explored and reflected upon so that a new era of RBPS diagnosis (Learn from Experience) predicted.

1 Introduction

In 2007 Center for Chemical Process Safety (CCPS) unprecedentedly presented to process industry Risk Based Process Safety (RBPS) 4 Pillars (and relevant 20 Elements), among which 4th Pillar "Learn from Experience" consists of 4 Elements: Incident Investigation, Measurement & Metrics, Auditing, and Management Review & Continuous Improvement[1].

In latest decades chemical engineers (process safety practitioners) have been facing some big challenges from accident/incident investigations, actually from whole CCPS 4th Pillar (Learn from Experience), for which there might have been many efforts, discussions, and even some 'solutions'. The author argues, however, the challenges comes not only from the needs of high quality and consistency of accident investigation, but, more fundamentally, from building up a brand new vocabulary (protocol) for chemical engineers to competently communicate of, and negotiate on, accident investigations or "why they occur and how ultimately to prevent".

Toward building a new vocabulary/protocol/narrative on accident investigation or, more broadly, on how Experience could be truly Learned (CCPS 4th Pillar), since 9th Global Congress, the author has had a series of efforts for exploration and reflections on the possible way forward, e.g. RBPS Core value (landlord for temple), Process Safety Cultural Communication (landlord's livelihood) & Egg model[2]; 4th Pillar as "espionage" not to enemies but ourselves[3][4]; 7 Mega Elements & 28 infra cultural activities (practices); RBPS Socio-Technical systems[5]; an ethical dimension of Vision 20/20[6] etc.. And in broader context of bridging chemical engineering (Risk Based Process Safety) to the Humanities, the author finally arrived at 7 Mega Elements' vocabulary/ protocol/ narrative/ "cosmology" on RBPS diagnosis for chemical engineers (process safety practitioners)[7].

2 7 Mega Elements' Protocol – Chemical Engineers' Vocabulary

Louis L. Bucciarelli claimed that he had taken seriously the "two cultures" challenge set by C. P. Snow and been working at bridging engineering to the Humanities for 35 years[8]; instrumental rationality does not suffice itself in engineering design, diagnosis and education[9]. And the author's experiences during the humble quest (through relevance of anthropology, intercultural communication, *the Art of War* etc.) for an ultimate solution to catastrophic (process safety) accidents (chemical engineering failure) seems not only resonating with Bucciarelli's bridging journey but a little bit unique, not only toward chemical engineering (Risk Based Process Safety) but also to engineering discipline at large, i.e., through benchmarking RBPS approach against Sun Tzu's *the Art of War,* 7 Mega Elements (vocabulary/ protocol/ narrative) not only were discovered for (chemical) engineers but function as 7 fundamental (chemical) engineering thoughts, which eventually sheds light on those entailed 28 practices of (chemical) engineering discipline.

Unlike Bucciarelli who tried to bridge philosophy and engineering, the author proceeded another way, i.e. first making relevance of anthropology & intercultural communication to chemical engineering (Risk Based Process Safety), then benchmarking Egg model against *the Art of War*, what the author calls 1st main ancient source for bridging (chemical) engineering to the Humanities, compared to 1st main modern one - Bucciarelli's *Engineering Philosophy*.

When guiding all visitors to some biggest lessons ("more original" two cultures) of ancient cultural traditions or civilizations of east and west, the author claimed that there should be an landlord (core value) to live within the temple building (CCPS 4 Pillars) and, furthermore, the landlord should have livelihood (Process Safety Cultural Communication (PSCC)) as well, and that should the landlord died the building seems still beautiful outwardly, however, with a purpose of tomb. In another words, when (chemical) engineers' vocabulary/ protocol/ narrative becomes deficient, or landlord and his livelihood are absent in the vocabulary, catastrophic (process safety) accidents are bound to happen, no matter how technically over competent relevant chemical engineers (process safety practitioners) might be, e.g., the lesson from the case of Bian Que–Duke Cai.

Through anthropology and intercultural communication, **RBPS Core Value** was defined in two ways, an unfolded one: *RBPS Core Value functions as the root of the tree of Process Safety*

Culture. Deficiencies are symptoms or signals for a dying or dead state of a core value, which means that the organization has strategically failed at the highest levels in managing process safety through a risk based approach – even if there has been no occurrence of process safety accidents till that moment. In another words, the RBPS temple is bound to collapse, sooner or later, due to the change or drift from the original design intent of the temple. And RBPS Core Value's simpler definition, through the concept of Power Distance in value studies: *there is no power, authority or governance on process safety except on behalf of process safety expertise (PSE).* (PSE means the divine essence of the highest strategic insights and vision of process safety management; the Tao for process safety culture; and the soul of *The Art of War* against catastrophic process safety accidents)

And process safety culture does not exist except through ***PSCC****, whose ultimate purpose is to convey the message of RBPS Core Value. Without nourishment from RBPS Core Value, the tree of Process Safety Culture or PSCC would get dry and dead.* So the temple building was eventually upgraded to Egg model, as showed within the scheme 1st. And these 2 new universes (or 2 Mega Elements) might be deemed what Bucciarelli called the rhetoric of engineering as the parcel of instrumental explanation[10], and probably 2 most fundamental (chemical) engineering thoughts as well.

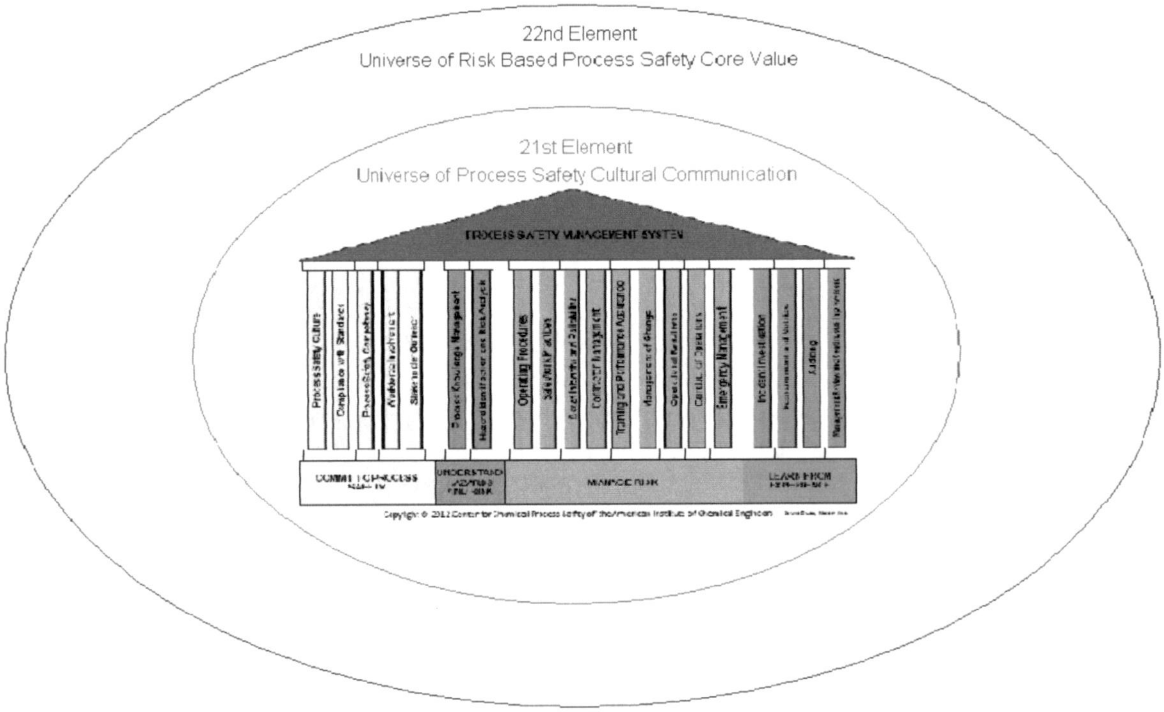

Egg Model (CCPS Temple + 2 universes) - Scheme 1st

The Temple (4 pillars) Figure Source: CCPS, Process Safety Boot Camp, contributed by David Guss, Nexen Inc. Copyright 2010 AIChE and used with permission

Then, in spite of the formidable strengths of most, actually 17, elements of CCPS 4 Pillars, except Process Safety Culture, Process Safety competency & Stakeholder Outreach, the outcome of benchmarking Egg model (4 Pillars + 2 Universes) against *The Art of War,* what the author calls a matured ancient vocabulary on engineering (military) diagnosis, not only shows that those 4 Pillars' (20 elements') narrative, even together with the new 2 Universes, would probably be not so architecturally organized at negotiation and communication on why process safety (chemical engineering) accidents (failure) occur & how to prevent, but makes the author further believe that Sun Tzu's vocabulary (6 Mega Elements, refer to Table 1st) might potentially become a narrative/ protocol paradigm for future (chemical) engineering diagnosis and *The Art of War* an main ancient source of bridging engineering to the Humanities.

For the fact that, however, we could not neglect the historical context of narrative/ protocol of *The Art of War*, the author, in light of modern globalization and global operations, fundamentally redefined Process safety Culture into Process Safety Cultural Communication (PSCC) as the 7th Mega Element, and upgraded another 3 ones i.e. Leadership, Weather/Climate & Espionage as below:

Leadership: a new chain of command, inherently from RBPS Core Value, rather than blindly from superiors;

Weather/Climate: RBPS Core Value enlightened outreach to the stakeholders & community, rather than only tactically maintaining seemingly good relationship with community; &

RBPS Egg		The Art of War	Philosophy of Process Safety	
RBPS Core Value - Tao - **RBPS Worldview**				Worldview
Commit to Process Safety	Process Safety Competency	Leadership	People	Process Safety Cultural Communication (Process Safety Culture) **RBPS Methodology**
	Workforce Involvement			
	Compliance with Standards	Weather/Climate	Plant	
	Stakeholder Outreach			
Understand Hazards & Risk		Terrain		
Manage Risk		Discipline	Procedure	
Learn from Experience		Espionage	Engineering & Humanities	

Table 1st

Espionage: 4th Pillar *Learn from Experience* (4 elements: *investigation, metrics, audit & review*) plays role "espionage", not to the enemies but to ourselves our organizational "national security"

And 7 Mega Elements might also be deemed as 7 (chemical) engineering thoughts.

3 Ethical Dimension (2 Universes) - What Chemical Engineers Count & How

While declared *there is no power, authority or governance on process safety except on behalf of Process Safety Expertise (PSE)* the author claimed that PSE shall eternally and exclusively chair at the highest strategic levels throughout all the processes of decision making and behavior determining, i.e. chemical engineers (process safety practitioners) need getting inspired from RBPS Core Value and persistently conveying the message of RBPS Core Value, both of which make up the ethical dimension (of 7 Mega elements' vocabulary/ protocol/ narrative), probably what Bucciarelli calls "best left sacred"[11], which ensures, or anthropologically designs, all relevant decision making and behavior determining correct. In this sense, at highest strategical levels, chemical engineers do not design at all but RBPS Core Value does.

From philosophical perspective, Bucciarelli claimed that the tensions at the interfaces between different design participants could not be solved by instrumental means alone [12]. And imitating American anthropologist Edward T. Hall's saying the author ever described process safety culture as a prison, unless we have RBPS Core Value as the key to unlock it and PSCC the brightness of the key. And Bucciarelli pointed out that *designing is a social process of negotiation, of iteration, of rectifying mis-steps, even misconceptions – a process rich in ambiguity and uncertainty* [13]. Similarly, the author in metaphor described PSCC as the photosynthesis of tree of process safety culture, the impulses from the heart (RBPS Core Value) and nourishment (highest designing principles) transportation from the root (RBPS Core Value). What does Bucciarelli mean by that 'social process'? What does he mean by "contribute to a sense of community among the citizenry" [14]? And what does the author mean by "photosynthesis" and "nourishment"? With hesitation & obligations, the author claims that the said social process, what the author puts PSCC, traditionally should be called Constitutional Democracy; and RBPS Core Value a nickname for Universal Value in chemical engineering.

As Universal Value & Constitutional Democracy in (chemical) engineering seem more subtle than those traditional ones the author here might call them Universal Value & Constitutional Democracy 2.0, what Bucciarelli described "old themes but in a new light" [15]. And this 2.0 might be acknowledged as two fundamental (chemical) engineering thoughts and the ethical dimension of (chemical) engineers' vocabulary/ protocol/ narrative on accident investigation and whole (chemical) engineering diagnosis (4th Pillars) as well. And probably only in this way could (chemical) engineers truly acquire what Bucciarelli called "new ways of perceiving and reading the world" [16], eventually leading the industries and communities into a new enlightened era of accident investigation and (chemical) engineering diagnosis. And the author might further states that, as the ethical dimension of (chemical) engineers' vocabulary (protocol), Universal Value & Constitutional Democracy 2.0 should function as the highest communication (designing) principles on accident investigation and (chemical) engineering diagnosis, and on accident prevention (foreknowledge) as well.

4 7 Mega Elements' Architecture – 28 Practices' Diagnosis Map

Inspired from Edward T. Hall's map of culture, the author developed a map of 28 infra cultural activities of Process Safety Culture (refer to 7 ones in the diagonal from upper left to lower right and 21 ones above), what Bucciarelli might like to call another, slightly different, "more complex vision of engineering thoughts and practices" [17]. And each infra cultural activity was defined as the critical statements & behaviors related to the interaction between Mega Element A and Mega Element B, e.g. infra activity 04 means: verbally & behaviorally communicate that the actual death of RBPS Core Value might be one of the root causes of accidents. And those 7 ones, extremely, are in fact of the critical verbal statements & behaviors related to the definition of the Mega Element (or engineering thought) concerned, e.g. infra activity 22: verbally and behaviorally embrace and reflect on Leadership - the engineering thought of new chain of command inherently from RBPS Core Value rather than blindly from superiors.

This matching between verbal statements & behaviors, or the inherent relationship between the message of RBPS Core Value and all relevant decision making & behavior determining, seems some resonance with Wittgenstein's recognition that rules and meanings (what the author calls ***tone***) in talk and texts (what the author calls vocabulary/ protocol/ narrative) were mixed with practice [18], or with what Bucciarelli put another way "the meaning of statements is embedded in practice" [19]. So these 28 infra cultural activities' map, an unfolded version of what & how chemical engineers (process safety practitioners) count, might accommodate as highest strategic encompassing diagnosis protocol (map).

	RBPS C.V. 0	P.S.C.C. 1	Leadership 2	Weather/C. 3	**Espionage (Diagnosis) 4**	**Terrain 5**	**Discipline 6**
RBPS Core Value 0	00	01	02	03	04	05	06
Process Safety Cultural Communication (PSCC) 1	10	11	12	13	14	15	16
Leadership 2	20	21	22 4 dimensions	23	24	25	26
Weather/ Climate 3	30	31	32	33	34	35	36
Espionage (Diagnosis) 4	40	41	42	43	44	45	46
Terrain 5	50	51	52	53	54	55	56
Discipline 6	60	61	62	63	64	65	66

28 Infra Cultural Activities' Diagnosis Map (Table 2nd)

As the Humanities e.g. philosophy, anthropology & intercultural communication, Sun Tzu's *The Art of War* etc. sheds more lights on the old themes (i.e. RBPS Core value, PSCC, Leadership and Weather/Climate), upon this Diagnosis Map, "another" complex vision of engineering thoughts and practices, the author claims that *the yellow grounded, or more exactly red number lettered area, are all about organizational "national security", and that the infra activity 44 (Espionage)'s 2nd rationality, next to the instrumental one, whose protocol area is painted green, SHALL be reactivated so that "espionage" to our organization would shift to a more proper name - "counter-intelligence", on accident investigation & prevention and on engineering diagnosis at large.*

Through maintaining the architecture/protocol of 7 Mega Elements could all these 22.5 infra activities of "national security" area be reliably visible, open and "problematic" over there, together with "traditional" green area (5.5 infra activities), to critique and diagnosis among board of directors, executive management, operation managers, strategy, legal, HR, chemical engineers (process safety practitioners), and even beyond. Pragmatically, from designing perspective, chemical engineers should pursue this more matured architecture/protocol/vocabulary, getting inspired from RBPS Core Value, persistently conveying its message and maintaining alive this complex vision of engineering thoughts & practices i.e. this 28 Practices (infra activities)' Diagnosis Map, otherwise the insightful topic "social construction of failure/ accident" [20] would turn out to be a truism. In another words, either RBPS Core Value highest strategically designs safety (and tranquility) through PSCC, or otherwise (chemical engineering/ process safety) failures/ accidents/ catastrophes are socially constructed.

5 Epistemology – Future (Chemical) Engineers' Knowledge Hierarchy

Walter Vincenti's understanding of *engineering knowledge,* as J. C. Pitt ever explained, *is task-specific and aims at the production of an artifact to serve a predetermined purpose* [21], of which, the author believes, CCPS 4 Pillars (20 Elements), as of (chemical) engineering/RBPS knowledge, might have been satisfactorily reflecting the main characters. Bucciarelli discovered, however, that engineering knowledge, of which he described *what engineers do, the reasons they do what they do, and their justification for believing the latter are good and true,* is *historically and culturally contingent, not fixed by the dictates of science and empirical fact alone.* [22]. The author might explain here, echoing the former statement "the landlord ... even to buy a new house with new style of design to live within", that the (chemical) engineering knowledge we have here with CCPS 4 Pillars might not be the single unique design/house for the landlord [23], but the landscape (landlord, livelihood & house) of the contingencies, what Bucciarelli calls "a picture of both possibilities and limitations of engineering thinking" [24], is unique. In another words, both the landlord (RBPS Core Value), deeply rooted in most cultural traditions and civilizations, and his livelihood (PSCC), what the author likes to call Constitutional Democracy, should preside the landscape of the contingencies and chair and guide at the highest strategic levels the whole engineering design process, for surety including on accident investigation and diagnosis, so that the objects of design, which "traditionally" have been confined to artifacts e.g. a mega petrochemical project/ operations, might also need to include all processes of decision making and behavior determining (refer to Table 2nd – 28 Infra Activities' Diagnosis Map).

The author might state that *a complete (chemical) engineering design process, including on accident investigation, diagnosis & prevention, shall consist of 3 components: RBPS Core Value (Universal Value), PSCC (Constitutional Democracy) and the application of (chemical) engineering knowledge (e.g. one with CCPS 4 Pillars).* And at same time, we recognize that *future (chemical) engineers need a new epistemology - a new hierarchy of knowledge which consists of Universal Value Knowledge, Constitutional Democratic Knowledge & the relevant (chemical) engineering knowledge.* Edward T. Hall had an earth-breaking saying: *culture is communication and communication is culture* [25], from which the author got inspired and developed the concept of PSCC. When (chemical) engineers verbally (of text) and behaviorally (of practice) abandon the ethical dimension of 7 Mega Elements' vocabulary/ protocol/ narrative, they lose the whole process safety culture or, we could say, both their knowledge hierarchy is deficient, akin to a genocide, and their engineering design process seriously flawed, like building concentration camps.

6 Conclusion

In context of bridging chemical engineering to the Humanities, on accident investigation, diagnosis and prevention, and on engineering design process at large, the author presents 7 Mega Elements' Vocabulary/ Protocol/ Narrative, 28 Infra Cultural Activities' Diagnosis Map and a newly galvanized old Hierarchy of Knowledge, so that some outlooks might be made: a) as one of the largest communities of engineering professions, AIChE would pursue leading roles at dialogues between engineering & the Humanities; b) some small groups of chemical philosopher/anthropologist-engineers would emerge; c) an enlightened era of (chemical/process safety) accident investigation, diagnosis & learning, and engineering design process at large, would arrive; & d) engineering education's future would get brighter than ever.

7 References

[1] Center for Chemical Process Safety, *Guidelines for Risk Based Process Safety*, John Wiley & Sons, Inc., 2007

[2] Zhang, L., An Upgraded Definition of Risk Based Process Safety Core Value and Its Impact in Light of Anthropology & Intercultural Communication on the Future Evolution of Process Safety Culture & Management, 2013 AIChE Spring Meeting and 9th Global Congress on Process Safety, San Antonio, April 2013

[3] Sun-Tzu and Ames, Roger T., Sun-tzu: *The Art of Warfare*, Ballatime Books, 1993

[4] Zhang, L., Benchmarking Risk Based Process Safety Egg (RBPS Temple+2 Universes) against The Art of War, 1st CCPS Asia-Pacific Conference on Process Safety, Qingdao, Sept 2013

[5] Zhang, L., People Means Leadership, NOT Simply mean People – 4 New Dimensions of Process Safety Competency, 2014 AIChE Spring Meeting and 10th Global Congress on Process Safety, New Orleans, April 2014

[6] Zhang, L., How Committed to Process Safety is Committed Enough? – Doing Process Safety Cultural Communication (PSCC) as the Inevitable Ethical Dimension for Vision 20/20, 2016 AIChE Spring Meeting and 12th Global Congress on Process Safety, Houston, April 2016

[7] Zhang, L. and Wasel, W., Corporate Memory Loss Ongoing – Except CCPS 4th Pillar (Learn from Experience) as Espionage, 2018 AIChE Spring Meeting and 14th Global Congress on Process Safety, Orlando, April 2018

[8] Bucciarelli, L. L., *Engineering Philosophy*, Delft: DUP Satellite, 2003, p.2

[9] Ibid. p.99

[10] Ibid. p.6

[11] Ibid. p.4

[12] Ibid. p.4-5

[13] Ibid. p.7

[14] Ibid. p.5

[15] Ibid. p.2

[16] Ibid. p.7

[17] Ibid. p.101

[18] Ibid. p.6

[19] Ibid. p.6

[20] Ibid. p. 101

[21] Pitt, J. C., "What Engineers Know", *Techne* 5:3, Spring, 2001, p.25

[22] Bucciarelli, *Engineering Philosophy*, p.99

[23] Pitt, "What Engineers Know", p.26

[24] Bucciarelli, *Engineering Philosophy*, p.99

[25] Hall, E. T., *The Silent Language*, Anchor Books, 1990, p.186

548354 Applying PSM in Vaca Muerta Shale Oil & Gas Operations: The Experience

How to develop, implement and sustain Process Safety in greenfield shale oil and gas projects? This has been a challenge to many new ventures.

Although conventional oil operations have been in place for decades, the exploration, drilling and fracking operations in unconventional shale fields are relatively new in the World. Most of the projects have kicked off less than eight years ago and there are some aspects of the operations that are unique to unconventional fields.

The objective of this paper is to share with the Process Safety global community, how a fit for purpose Process Safety program was developed and is being implemented in the Upstream Unconventional shale oil and gas YPF & Chevron joint venture in Vaca Muerta / Argentina. The project started in 2013, has currently 14 D&C rigs, 5 fracking sets, several separation and treatment production facilities, 3,500 employees and is expected to grow even more in the next 5 years. The paper describes how the project team developed the idea, planned the steps, engaged all stakeholders and selected the key elements of the Process Safety Management System that would add more value in each stage of the project, including drilling, fracking and oil and gas production. The paper also explains the key challenges, such as the difficulties faced to instill Operational Discipline and other key cultural and technical aspects related to Contractor field Operation crews. The journey to implement Process Safety in the Unconventional shale oil & gas operations is very dynamic and is in progress, however, there are already several positive aspects as well as pitfalls to take away as a lesson learned.

548596 A Survey of Process Safety Workers' Attitudes, Behaviors, and Outcomes: Moving Beyond Human Error and Toward a Systems Model

In an effort to move beyond the "human error" explanation for safety incidents, we explored issues with procedures from the worker's point of view by using an anonymous, holistic survey developed from interviews with then currently employed operators. The current sample ($N = 174$) included individuals employed in the process safety industry (MKO/Advanced Procedures Initiative) and were primarily from the Oil & Gas and Chemical industry. The survey was deployed over the course of approximately four weeks. Twelve distinct constructs emerged from the survey (e.g., perceptions of procedure quality, procedure deviation, attitudes toward the procedure change process, etc.). Results indicated that perceptions of procedure quality was the focal variable in all analyses including positive relationships with attitudes toward the procedure change process and negative relationships with procedure deviations, and safety incidents and near-misses. We conclude that these elements are important factors to consider when evaluating and developing procedure systems and provide additional information beyond what the industry considered paramount.

546680 A Method of the Evacuation Pathway Optimization Under Toxic Gas Releases in Chemical Plants

Toxic gases release in chemical plant is a key factor contributing to major accidents and may cause tremendous human life and economic losses. A scientific planning evacuation pathway is an important guarantee of reducing the losses of accidents. Many studies regarding evacuation planning in chemical plant have been carried out by simulating the accident scenarios. However, at present, a systemic method of evacuation pathway optimization has not been developed, and the effectiveness of emergency preparedness has been compromised. To address this problem, an approach based on the CFD modelling of hazardous gas release consequences to plan evacuation pathways and appropriate shelters is proposed in this paper. The methods stochastic programming and dose-response model are mainly employed in this paper. The toxic gas concentration can be converted into poisoning fatality probabilities by dose-response model. Then, stochastic programming method is used to select all the optimal pathways from the evacuation point to the safe area. Setting the minimal fatality probability as the optimization goal, the optimized evacuation pathways under all the leakage scenarios are finally determined through solving the stochastic programming model. As for places with limited evacuation opportunities, an approach of setting up a safe room is proposed in this paper and relevant design guidance are put forward in this paper considering the detailed situation of the chemical plant. A case study of the optimization of evacuation pathways and safe room setup in a salt chemical plant that contains 88 leakage scenarios of chlorine leakage is investigated. With the proposed methodology, the optimized evacuation pathway and the appropriate safe room can provide scientific references for decision makers and evacuees on reducing losses when toxic gases release in chemical plants.

Four Major Gaps That Are Preventing Most Companies Worldwide from Achieving Excellent Process Safety Performance

William G. Bridges
Process Improvement Institute Inc. (PII)
1321 Waterside Lane, Knoxville, TN 37922 USA
wbridges@piii.com

Jeff Thomas
Process Improvement Institute Inc. (PII)
jthomas@piii.com

Prepared for Presentation at
American Institute of Chemical Engineers
2019 Spring Meeting and 15th Global Congress on Process Safety
New Orleans, LA
March 31 – April 3, 2019

FOUR MAJOR GAPS THAT ARE PREVENTING MOST COMPANIES WORLDWIDE FROM ACHIEVING EXCELLENT PROCESS SAFETY PERFORMANCE

William G. Bridges
Process Improvement Institute Inc. (PII)
wbridges@piii.com

Jeff Thomas
Process Improvement Institute Inc. (PII)
jthomas@piii.com

Keywords: process safety, PSM, human factors, near misses, PHA, HAZOP, procedures

Abstract

Process safety requires implementing many management systems, specific engineered features, and operating and maintenance practices effectively. Most companies believe they have done just that, and yet major accidents continue to occur. Why is that? What is missing?

This paper looks at the statistics of major accidents, combined with results from audits and assessments from more than 50 chemical, petrochemical, oil/gas, and related processing companies world-wide. The paper illustrates the four major gaps that are common to the companies/sites that keep having major accidents, compared to those companies/sites that do not have such accidents:

- Accurate and clear operating and maintenance work instructions (procedures)
- PHA of all modes of operation, especially startup, shutdown, abnormal operations, and online maintenance modes
- Getting Near Misses reported and investigated
- Addressing human factors that are missing from most management systems

The paper also provides guidance for closing each of these gaps.

Introduction

"Process safety management (PSM) is the application of management principles to the identification, understanding, and control of process hazards to prevent process-related incidents"[1]. PSM entails development and implementation of programs or systems to ensure that the practices and equipment used in hazardous processes are adequate and are maintained appropriately. The primary categories of programs or systems have come to be called elements of PSM. However, the basic elements of PSM have been defined by many groups in a number of ways. Table 1 lists the elements of PSM systems from various industry and government groups. Many of the elements with different names have essentially the same meaning. For instance, "maintenance and inspection of facilities," together with some aspects of "personnel" practices, both under American Chemistry Council's (ACC's, formerly CMA's) Process Safety Code of Responsible Care™[2] are essentially the same as the single element, "mechanical integrity," under 29 CFR 1910.1190)[3].

Table 1. Comparison of PSM systems

OSHA 29 CFR 1910.119 EPA 40 CFR 68	AIChE/CCPS Risk-Based Process Safety (RBPS) Standard	Responsible Care© Process Safety Code
Management System	**Commitment to Process Safety**	**Management Leadership**
Employee Participation	Process Safety Culture*	Commitment*
Process Safety Information	Compliance with Standards	Accountability*
Process Hazard Analysis*	Process Safety Competency*	Performance Measurement
Operating Procedures*	Workforce Involvement*	Incident Investigation
Training*	Stakeholder Outreach	Information Sharing
Contractors	**Understand Hazards and Evaluate Risk**	CAER Integration
Pre-Startup Safety Review	Process Knowledge Management	**Technology**
Mechanical Integrity	Hazard Identification and Risk Analysis	Design Documentation
Hot Work Permit	**Manage Risk**	Process Hazard Information
Management of Change	Operating Procedures*	Process Hazard Analysis
Incident Investigation	Training and Performance*	Management of Change
Emergency Planning and Response	Safe Work Practices *	Facilities
Compliance Audits	Asset Integrity and Reliability	Siting
Trade Secrets	Contractor Management	**Codes and Standards**
	Management of Change	Safety Reviews
	Operational Readiness*	Maintenance and Inspection
	Conduct of Operations*	Multiple Safeguards*
	Emergency Management	Emergency Management
	Learn from Experience	**Personnel**
	Incident Investigation	Job Skills*
	Measures and Metrics	Safe Work Practices*
	Auditing	Initial Training*
	Management Review and Continuous Improvement	Employee Proficiency*
		Fitness for Duty*
		Contractors

* Contains some of the 6 elements of a Human Factors element, but all missing elements not covered completely. Mostly covered in CoC/OD element

Note that the newest definition of process safety is CCPS's *Risk Based Process Safety (RBPS)*[4] replaces their earlier process safety definition. In the older definition from CCPS [1], there was an element on Human Factors, which brought strong focus to this necessary element; in RBPS, the human factors subelements are now spread across 6 different elements, though some human factors were inadvertently weakened in the transition to *RBPS*.

Although nearly the entire industry agrees that implementing PSM is the right thing to do, interpreting and converting the PSM requirements into practices is unique to each company, and even unique to each plant site. Not only can the requirements be interpreted differently for each site based on local needs, but each site also starts from a different point when they begin to implement a system that is consistent with this regulation.

This paper looks at the statistics of major accidents, combined with results from audits and assessments from more than 50 chemical, petrochemical, oil/gas, and related processing companies world-wide. The paper illustrates the four major gaps that are common to the companies/sites that keep having major accidents, compared to those companies/sites that do not have such accidents:

- Accurate and clear operating and maintenance work instructions (procedures)
- PHA of all modes of operation, especially startup, shutdown, abnormal operations, and online maintenance modes
- Getting Near Misses reported and investigated
- Addressing human factors that are missing from most management systems

The paper also provides guidance for closing each of these gaps.

1. Procedures Need to be More Accurate and Clear

Operating and maintenance procedures have always been crucial to the safety, quality, and productivity of process systems. With the advent of new safety and quality standards such as OSHA's process safety management (PSM) regulation, OSHA's personal protective equipment (PPE) regulation, EPA's risk management program (RMP) regulation, and the ISO 9000 quality standard, many companies are facing the daunting task of developing or upgrading their procedures to satisfy varied and sometimes complex and conflicting requirements.

Although there are overlapping characteristics that the various regulations and standards share, each approach the procedure-writing process from a somewhat different perspective (e.g., quality or safety, protection of workers or protection of the public/environment), and the required level of detail for documenting the procedures differs greatly between these regulations and standards. For instance, the ISO 9000 quality standard advocates documenting procedures and work instructions that impact quality and suggests keeping the procedures as simple as possible, while the OSHA PSM regulation requires detailed procedures that address: all operating modes (startup, shutdown, etc.), operating limits, consequences of deviations, means to avoid hazards, safety and health considerations, and safety systems and their functions. How can companies develop and maintain procedures that ensure productivity and simultaneously satisfy the different regulations and standards? The key is to remember the ultimate goal of the regulations and standards: to reduce human errors that can impact quality, productivity, and/or safety.

Procedure-related errors are errors that occur because some characteristic of the procedure caused task performance to fail. **This is currently the most critical human factor at most sites since 90% of accidents have at least one root cause related to mistakes within procedures**. Reducing these procedure deficiencies can reduce human error rates by a factor of 2 to 20 [5].

1.1. Identified problems

Deficient Procedures are the most prevalent problem in process industries since procedures have not traditionally been developed from the perspective of optimizing human factors; instead procedures have been traditionally developed to meet a compliance requirement. Examples of procedure deficiencies (inaccuracies) include:

- Incorrect/incomplete/nonexistent (most procedures we have audited have been only 70-75% accurate – the inaccuracies include missing critical steps, steps as written are not what needs to be done, or the steps are out of sequence) [5]
- No/misplaced/incorrect information in warnings (for example, a warning should **never contain** the action to take; it should instead **emphasize** the action to take) and warnings should be located BEFORE the step applicable to the warning
- Poor format and presentation rules

1.2. Objective

The flowchart (Figure 1) nd the procedure quality checklist (Table 2) describes the basic sequence of steps for writing effective procedures. Note that some basic rules are contained in the flowchart, such as "operators write operating procedures," "maintenance craft-persons write maintenance procedures," and "Lab technicians write lab procedures." Procedures are NOT written by engineers or department superintendents. This is a critical to ensure procedures are written in the user's common language and are written at the right level of detail.

Figure 1. Best Approach to developing written procedures, once the unit has decided which procedures are needed (Extracted from PII Course, Writing Effective Operating and Maintenance Procedures © PII 2003-2017) [6]

The procedure content controls the human error related to "procedure-based" error – if content is wrong, the trainee is very likely to learn the step wrong (trainer uses the procedure as a basis for training or the trainee refers to the procedure step that is wrong when executing the procedure).

Even if the content is correct, not following page formatting best practices can still increase human error rates. If the content of the step is wrong it does not matter if the presentation is clear. If the content is perfect, you can further reduce error rates by 50% to 70% by presenting the information properly.

Rules for developing content[7]

- The procedures should be written at the level of someone who has just completed the basic training for that task. Do not write the procedure for someone just hired or for the 10+ veteran.
- Clearly identify ahead of time what activities need procedures to help ensure error rates are controlled low enough, and identify which activities are considered common "skills" of all of the staff.

  Example: Starting a pump is typically considered a "skill." This means a procedure step can simply be "Start the benzene recycle pump (P-119)." The procedure does not need to explain how to start the pump. But, starting a pump is a skill that nonetheless must be learned; it requires doing several sub-steps related to positioning of intake and discharge valves, how the pump is throttled at startup, checking local pressure gauge, etc. So, a training module is needed for starting pumps of various types, and the operator needs to learn this necessary skill. Once the skill is learned, it will be applied to so many various cases and so often, that a procedure is no longer needed as a "refresher" on how to "start the pump."

- The first draft of the procedure must be walked-down in the field by another user. Simply reading through the procedure does not catch enough of the mistakes.
- The revised draft should be reviewed in the field by a technical staff person, such as a process or operations engineer (for operating procedures).
- The procedure needs to be checked to ensure it follow the page format and step writing rules described in the next section. This ensures the accurate steps are clearly presented.

The target is to reach an accuracy of 95% or better[7] (so, no more than one wrong or missing step out of 20 steps). The following have been observed in the field by PII staff by direct data collection at more than 110 sites/plants around the world, based on walk-down review of several to dozens of procedures at each unit/plant:

- At 95% accuracy or better and when the same procedures follow 80% of the rules for procedure clarity (presented next), then most users will follow the written procedures and will try to keep the procedures up-to-date
- At about 85% accuracy or less, about half of the users stop using the procedures
- At about 75% accuracy or less, less than 10% of the users will refer to the procedure or will try to keep it up-to-date. **So, the written procedures are not very effective.**
- Unfortunately, the typical operating procedure walked down by PII staff (accompanied by senior operators, a process engineer, and a shift supervisor) is about 75% accurate (so one step in four is missing or wrong). Usually these inaccurate procedures also follow less than half of the best practices for procedure format, presented next [7].

Procedure Clarity

The importance of the accuracy of the procedure steps were discussed above. The clarity of the procedure steps (how they are written and how the page is formatted) is also important. Following best practices for step and page format reduces human errors by a multiplying factor of 3 to 5. [5], [7]

The best practice rules for writing and validating procedures have been published for many years *(see Bridges & Williams, 1997; Madden & Bridges, 2016, 2017)* [7],[8],[9]. These rules include best practices for formatting of the pages and steps. These have been gradually improved over the past decades and now are incorporated into PII's training materials.[6]

Table 2. Procedure Quality Checklist (courtesy PII, copyright 2003-2017)[6]

#	Issue	Response
	Procedure Content Checklist	
1	Is the procedure drafted by a future user of the written procedure? (Engineers should not author procedures to be used by operators or maintenance staff.)	
2	Is the procedure validated by a walk-down in the field by another future user of the procedures?	
3	Is the procedure reviewed and commented on by technical staff (engineers or vendors)?	
4	Is the procedure checked versus the Page and Step format rules below?	
5	Is a hazard review of step-by-step procedures performed to make sure there are sufficient safeguards (IPLs) against the errors that will occur eventually (when a step is skipped or performed wrong)?	
6	Is the content measured using "newly trained operators" to judge the % of steps that are missing, steps that are confusing or wrong, and steps that are out-of-sequence? (A score of 95% accuracy of content is good; 98% should be the targeted average.)	
	Page Format Checklist	
1	Is the title of the procedure the largest item on the page?	
2	Is the procedure title clear and consistent with other titles, and does it uniquely describes the topic?	
3	Are the document control features the smallest items on the page?	
4	Are temporary procedures clearly identified?	
5	Is white space used effectively? • Is there one blank line between each step? • Does indentation help the user keep their place? • Are the margins large enough to reduce page congestion?	
6	Is type size is 12 pt font or larger?	
7	Is mixed case used for words of steps, with ALL CAPS used only for special cases (such as IF, THEN, AUTO, and WARNING)?	
8	Is the step number very simple (single level of number used)? Only an integer?	
9	Have sections or information not necessary to performing the steps been moved to the back or to another part of the manual or training guide?	
10	Are section titles bold or larger than the text font? Do sections have clear endings?	
11	Is the decision on electronic presentation versus hard copy correct? Are electronic linkages to procedures clear and accurate and easy to use? If paper is chosen, is the color of the paper appropriate?	
12	Is the overall page format (such as Outline format or Two-Column [T-Bar] format) appropriate to the use of the procedure?	
13	Are play script features added for tasks that must be coordinated between two or more users? • Play script is normally used when there are two or more hand-offs of responsibility for steps.	
14	Are rules followed for formatting of Warnings, Cautions, and Notes? (See annotated rules, such as Warnings are for worker safety and Warnings must clearly standout from rest of page.)	
	Step Writing Checklist	
1	Is each step written as a command?	
2	Is the proper level of detail used throughout? This is judged based on: • *Who* will use the procedures • Same level of detail used in similar procedure steps	
3	On average, is there only one implied action per instruction? Best practice is to average 1.2.	
4	Does the procedure indicate when sequence is important? • If sequence matters, each step should be numbered (with an integer or letter)	

	• If sequence does not matter, bullet lists should be used	
5	Are only common words used? Apply "education" level test (5 grade reading level is best)	
6	Do all acronyms, abbreviations, and jargon aid understanding? • Develop a list of such terms for use in procedures *and* communication. • Use terms that users use (within reason)	
7	Is each step specific enough? No room left to guess/interpret: • The meaning of a word or phrase (Check vs. Make sure) • The intent of a step or series of steps • A desired quantity or value • To what equipment the step applies	
8	Is the procedure free of steps that require in-your-head calculations? • Values expressed as ranges rather than targets with error bands • Conversion tables, worksheets, or graphs provided where needed	
9	Are graphics to the user's advantage? • No explanatory paragraphs or lengthy instructions that could be replaced by a picture • No impressive graphics that provide no real advantage	
10	Are references to the user's advantage? • No lengthy explanations or instructions that could be replaced by branching to a reference • No references to a procedure that references still another • No gaps or overlaps between this procedure and a referenced document • If branching, must branch to a procedure, not to a specific step in a procedure	
11	Are rules followed for writing warnings, cautions, and conditional steps? • No more than 2 per page • No actions within a warning or caution (actions must always be numbered steps) • Warnings and Cautions contain descriptions of potential consequences	

For a complete description of best practices, refer to PII's paper *"Best practices for writing operating procedures and trouble-shooting guides"*. [7]

2. PHAs (Process Hazards Analyses) Need to Close Two Major Gaps

Process Hazards is a broad term. The relationship between many components define process hazards. For a PHA team, Process Safety Information (information on the chemicals, technology and equipment) and operating procedures are necessary to identify specific components of and analyze a hazard scenario. These components include: process deviations (parametric and procedural), their causes, the process safety consequence and various safeguards for detecting deviations and causes, preventing causes, detecting and mitigating consequences and intervening safeguards. A thorough review of incident investigations also offers an opportunity to further analyze process hazards.

2.1. Identified problems

Over the last twenty years the CSB has brought needed attention, through its investigations of industry incidences, to two significant gaps in many current PHAs; where the teams did not identify and analyze:

- Process hazards unique to non-routine modes of operations resulting in catastrophic consequences, and
- Process hazards related to Damage Mechanisms.

Though the intent of the US OSHA PSM regulation has always been for PHA teams to analyze the hazards for all modes of operation, OSHA did not give specific requirements for addressing damage mechanisms and only recently have they increased their enforcement attention on PHA of non-routine modes of operation.

Industry has responded by modifying the *Guidelines for Hazard Evaluation Procedures*, 3rd Edition, 2008, CCPS/AIChE[10] to improve the coverage of these two main issues. API in turn issues guidance on addressing damage mechanisms[11]. Industry best practices now exist for addressing both of these weaknesses. Now, with proper approaches, a team with process safety competencies and process knowledge can thoroughly address the issues during a PHA.

Another less significant gap noted by regulators and by the US CSB has been an incomplete consideration, within the PHA, of previous incidents.

This section of the paper lays out the case from the US CSB and US OSHA on these PHA gaps:

- Poor or no coverage of hazards during non-routine modes of operations
- Poor or no coverage of damage mechanisms

2.2. Objectives

2.2.1. Hazard Evaluation of Non-Routine Modes of Operation

Non Routine modes of operation include start-up, shut-down, online maintenance and other abnormal operations. The hazard evaluation of non-routine modes of operation involves reviewing procedures using a HAZOP, simplified HAZOP, or What-if analysis to uncover potential accident scenarios associated with non-routine operations, for continuous or batch operations. Human error is more likely and more critical during non-routine operations. By analyzing procedural steps where human error is more likely, and where human error or component failure could lead to a consequence of interest, risk can be reduced. The hazard evaluation team's objective is to evaluate the risk associated with skipping steps and performing steps wrong.

The purpose of a hazard evaluation of non-routine modes of operation (governed by written procedures) is to make sure an organization has enough safeguards for the inevitable instance when a step is either performed wrong or skipped (inadvertently or due to shortcutting or other reasons).

Industry has found that a HAZOP or what-if analysis, structured to address procedures, can be used effectively for finding the great majority of accident scenarios that can occur during non-routine modes of operation[10],[12],[13],[14]. Experience shows that reviews of non-routine procedures have revealed many more hazards than merely trying to address these modes of operation during the P&ID driven hazard evaluations.

General guidelines for analyzing non-routine modes of operation or batch processes

- Define the assumptions about the system's initial status. "What is assumed to be the starting conditions when the user of the procedure begins with Step 1?"
- Define the complete design intention for each step. "Is the step actually 3 or 5 actions instead of one action? If so, what are the individual actions to accomplish this task?"
- Don't analyze safeguard steps that start with ensure, check, verify, inspect, etc., or where the consequence of skip is "loss of one level of safeguard/protection against…" There is no reason to analyze these steps since they will show up as safeguards of deviations of other steps. This approach is similar to not analyzing a PSV during a HAZOP of continuous mode (i.e., during a parametric deviation analysis); instead the PSV is shown as a safeguard against loss of containment.
- Together with an operator before the review, identify the sections of the procedures that warrant use of:
  - 7-8 Guide Words (extremely large consequences can happen if deviations occur)
  - 2 Guide Words (the system is complex, mistakes are costly, or several consequences could occur)

  - On others, use What-If (no guide words or guide phrases; for use on simpler or lower hazard systems)
- Decompose each written step into a sequence of actions (verbs)
- Apply guide words directly to the intentions of each action

The following preparation steps may also be needed prior to the procedure PHA:
- Walk through procedure in the plant with one or more operators to see the work situation and verify the accuracy of the written procedure.
- Determine if the procedure follows the best practices for "presentation" of the content; the best practices will limit the probability of human error
- Discuss generic issues related to operating procedures, such as:
  - staffing (normal and temporary)
  - human-machine interface
  - worker training, certification, etc.
  - management of change
  - policy enforcement
- Review other related procedures such as lock out/tag out and hot work
- **If the procedures are NOT >90% accurate, then redo procedures before the PHA.**

2.2.2. Damage mechanisms

Recently, more and more regulators, including the US CSB, are making analysis of Damage Mechanisms (DMs) a PHA requirement. This requires that potential process damage mechanisms and their potential consequences of interest be identified. Damage mechanisms can be broken down into three main categories. These categories and examples of each are shown in the following table (Table 3).

Table 3. Types of Damage Mechanisms [15]

Category	Examples
Mechanical	• Mechanical loading failure • Mechanical fatigue • Buckling • Cracking • Embrittlement • Ductile fracture • Brittle fracture

Chemical	• Corrosion
	  - Uniform
	  - Localized
	  - Pitting

Physical	• Thermal failures
	  - Creep
	  - Thermal fatigue
	  - Transformation

The team may use aids such as piping specs, established literature and standards, any applicable MOC documents, etc. to help identify DMs. The PHA leader has several options of how to address damage mechanisms, but PII preferred approach is to discuss during HAZOP under "Loss of Containment" deviation in each node/section.

Damage mechanisms can (and sometimes ***should***) be reviewed prior to the PHA. For example, all MOCs should consider damage mechanisms and their potential consequences of interest before the change is approved. The MOC should also require updates to process safety information which might be relevant to protect against DMs.

Coverage of Damage Mechanism (DM) for each major section [14]

As mentioned above, this is PII's preferred method for thoroughly covering all DM within a PHA/HAZOP of an entire unit. The benefits are that the team can more easily catch changes in DM from section to section and they can more easily identify when unique safeguards, such as Remote Isolation Valves, segregated containments (dike), and unique materials of construction are needed. To facilitate this, a list of generic causes of loss of containment and a generic list of safeguards against loss of containment are covered in the Loss of Containment (LOC) deviation of each section.

For each node (each line, each vessel, each column, etc.), the PHA team should discuss and document each damage mechanism listed in Table 3 (as a cause of loss of containment), consequence of the failure if the damage mechanism occurs, and the safeguards in place to prevent the damage mechanism, detect the mechanism before failure, prevent the release, detect and response to the release, and mitigate or contain the release.

The documentation style varies between PHAs. There are two different styles that have been acceptable in the past.

- One style (Example A, Table 4) uses a reference to a summary table of typical causes (this summary table is not shown but is similar to Table 5 of this paper) instead of listing each individual damage mechanism. In Example A, the same approach is used for safeguards. This approach saves redundant text and some time, but it requires the leader and scribe to be diligent to cover everything in the Typical tables that were referenced in the Cause and Safeguard columns. A modified approach is to relegate some generic causes and safeguards to Typical tables for reference and then to carefully list the specific cause and/or safeguards of interest for LOC for each specific node.

Table 4. Example A – Referencing Generic Tables for Typical Causes and Typical Safeguards [14]

Dev	Deviation	Causes	Consequences	Existing Safeguards	Recommendations
9.10	Loss of containment	Accelerate corrosion External fire High pressure (linked from 9.7) Typical causes of loss of containment (see Table A.1)	Release to atmosphere leading to potential injury of workers and/or community	Pressure Relief Valve's 231A, B, C on Reactor and no valves in line to reactor Most lines / connections are welded construction; only a few flanges Drills conducted each year on evacuation, rescue and isolation Emergency Response personnel are trained at SABIC FTC Generic safeguards protecting against or mitigating process material releases (see Table A.2)	Safety 7. Consider changing the ITPM schedule for managing most PSVs whose inspections are too infrequent based on industry standards and best practice. For instance, the current inspection frequency for PSV-8220 on the Ammonia Receiver (I-2005G) is 9 years, whereas, consensus codes typically recommend testing/inspection every 1-4 years for PSVs in highly toxic services.

- Example B (**Table 5**) does not use or reference a generic or Typical list of causes (damage mechanism) or Typical safeguards, but instead develops a specific listing for the LOC deviation of each node. This style has proven easier to justify to regulators and other outside reviewers but takes more documentation effort.

Table 5. Example B – Specific Listing of Causes and Safeguards [14]

Dev	Deviation	Causes	Consequences	Existing Safeguards	Recommendations
1.9	Loss of containment	Corrosion/erosion External fire and/or flame impingement Gasket, packing, or seal failure Improper maintenance Material defect Operator failing to close or inadvertently opening a valve to the atmosphere (e.g., a valve at a hose connection) Railcar inadvertently derailed Valve leaking to the atmosphere High pressure (linked from 1.5) Acid corrosion caused by high concentration of water (linked from 1.8) High ambient temperature External impact (such as from a mini-engine or another railcar)	Catastrophic release of chlorine from a ruptured railcar Steady release of chlorine from a ruptured connection Steady release of chlorine from a leaky connection High pressure caused by thermal expansion of liquid chlorine if railcar is also over-full	Chlorine repair kit Derailer and warning flag to prevent impact by a mini-engine or another railcar Limited vehicular access to area Maintenance/operator response as required, including isolation if needed Operator periodically monitoring the railcar valves while unloading Personal protective equipment in the area Plugs installed in all chlorine valves to the atmosphere when the valves are not in use Relief valve on each railcar for mitigating releases caused by overpressure Supplier maintenance of railcars (per strictly enforced US DOT requirements) Video monitoring of the unloading area Concrete crossties on rail spur Dike preventing any combustibles spilled nearby from reaching the unloading rack area Concrete railroad ties in chlorine unloading area to prevent fires near railcar	10. Consider installing a chlorine detection system in the unloading and vaporizing area to help detect chlorine releases (especially at likely release points) 11. Verify that periodic maintenance and inspections are being performed in accordance with Chlorine Institute recommendations 12. Review the drainage system for the unloading are, and identify the areas that may be affected by a large chlorine release 13. Consider prohibiting the use of heavy equipment (e.g., cranes) in the unloading and vaporizing area unless special precautions to prevent equipment damage are enacted 33. Consider providing a high pressure alarm for each vaporizer 40. Consider providing a water deluge system in the unloading area to help mitigate chlorine releases from the railcar

Regardless of method, the team leader and scribe must ensure the team rigorously discusses all of the damage mechanisms in Table 5 and adequately documents the results of the damage mechanism review in the PHA/HAZOP analysis tables.

For a complete description of PHA best practices, refer to PII's paper *"Recipe for a Complete Process Hazard Analysis".* [14]

3. Near Misses Need to be Reported and Enough of Them Investigated

We must learn from accidents and near misses to prevent recurrence. The first step in the learning process is investigation to determine the causes and underlying reasons why accidents and near misses occur. A thorough investigation of root causes will identify the management system weaknesses. Learning which management system weaknesses are leading to near misses and accidents is one of the highest value activities in which a company can invest, and learning from near misses is much cheaper than learning from accidents. Many chemical companies have implemented process safety management systems, and now they are beginning to focus on getting near misses reported and on root cause analysis. This is a very exciting trend. Unfortunately, the chemical industry gets very few near misses reported (the chemical industry is certainly not the only industry with this problem).

To understand more about near misses and getting them reported, it is best to first review the basic definitions.

An incident is either an Accident or a Near Miss.

An Accident is a sequence of unplanned events and conditions that result in harm to people, environment, process, product or image.

A Near Miss is an unplanned sequence of events that could have caused harm if conditions were different or is allowed to progress, but did not in this instance.

Given a consistent understanding of the definition of a near miss, it is possible to estimate how many near misses should be reported for every accident. Studies in several industries indicate that there are between 50 and 100 near misses for every accident. Also, data indicates that there are perhaps 100 erroneous acts or conditions for every near miss. This gives a total population of roughly 10,000 errors for every accident. **Figure 2** illustrates the relationships between accidents, near misses and non-incidents.

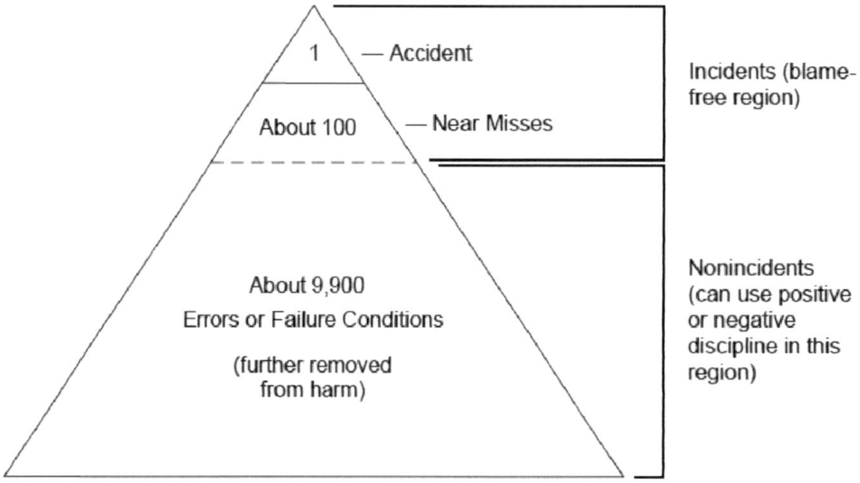

Figure 2. Relationship between Errors and Potential or Actual Impact [16]

Investigating near misses is critical to preventing accidents, because near misses share the causes and root causes of accidents; they are one or two barriers away from the loss/accident. We are very likely preventing many apparently unrelated accidents when we prevent the ones that are obviously related.

3.1. Identified problems

During the past 15 years, we have asked more than 5,000 students of our process safety management (PSM) courses and 3,500 students from our investigator leadership training courses how many near misses they get reported for every accident. The students represented about 400 companies, predominantly in chemical-related process industries (chemical, polymer/plastic, petrochemical, refining, oil and gas exploration, pharmaceutical, pulp and paper, etc.). The answers are quite disturbing. More than 95% said that their ratio of near misses reported to accidents reported falls in the range of 0 to 20. Less than 5% of the individuals said the ratio was greater than 5, and less than 2% said the ratio was higher than 10. Students noted that fear of disciplinary action, lack of management commitment, and lack of understanding of the difference between a near miss and a non-incident were the main reasons why near misses do not get reported. [16]

In conducting about 150 PSM audits and 7000 process hazard analyses (PHAs) during the past 22 years, we have found ratios from 0 to 105. For the first half of the 1990s, more than 90% of the facilities we talked to had ratios in the range of 0 to 0.5, and more than 95% had ratios in the range of 0 to 1. In the last half of the 1990s, more than 90% of the facilities had ratios in the 0 to1 range, and more than 95% had ratios in the 0 to 2 range. In addition, only a few (less than 2% of the facilities) had a ratio higher than 5. By 2005, many facilities had ratios in the 20 to 100 range and each of these had seen great gains (more on this in future articles), but the average across all companies we talked to increased only slightly. By 2011, the ratio of near misses to accidents (any loss event) was about 2 for companies we deal with. But by the end of 2011, many major companies had ratios well above 50! [16]

Our auditors and PHA leaders commented that the primary reasons for lack of reporting were about the same as those found in the survey. The barriers to getting near misses reported, discussed in detail later, are:

- Fear of disciplinary action.
- Fear of teasing by peers (embarrassment).
- Lack of understanding of what constitutes a near miss versus a non-incident.
- Lack of management commitment and lack of follow-through once a near miss is reported.
- An apparently high level of effort is required to report and to investigate near misses compared to low return on this investment.
- There is No Way to investigate the thousands of near misses per month or year!
- Disincentives for reporting near misses (e.g., reporting near misses hurts the department's safety performance).
- Not knowing which accident investigation system to use (or confusing reporting system).
- Company discourages near-miss reporting due to fear of legal liability if these are misused by outsiders.

The good news is that near-miss reporting appears to be improving, and the industry appears to recognize most of the barriers to near-miss reporting. The bad news is that the ratio is still very poor and improvement appears slow!

The number of companies that participated in the informal surveys suggests that the results are a statistically significant representation of the chemical industry.

A formal (written) survey was developed and e-mailed, faxed, and/or mailed to more than 100 companies. Of these, more than 12 replied originally in 1997. Since then, we have reviewed data in detail from more than a dozen major companies. These 25+ companies or affiliates provided data from more than 400 facilities, including more than 150,000 employees in manufacturing and operations. The data were from the chemical industry, polymer industry, refineries, drug/pharmaceutical companies, pulp and paper mills, petrochemical companies, and oil exploration/production.

Table 6 summarizes the results of both the informal and formal surveys.

Table 6. Summary of survey results [16]

Survey Items	Informal surveys		Formal Survey (or actual organization data)
	Classroom polls	Found during Audits and PHAs	
Current Near-miss Reporting Ratio Range (0-5 years prior)	1 to 105	2 to 20	0 to 105
Previous Near-miss Reporting Ratio Range (5-10 years prior)		1 to 5	NA
Current Near-miss Reporting Ratio Average (0-5 years prior)	2-3	3	5
Previous Near-miss Reporting Ratio Average (5-10 years prior)		1	NA
Goal for Near-miss Reporting Ratio (average)	20+	NA	20
Theoretical Upper Limit on Near-miss Reporting Ratio (average)	About 100	NA	100
Number of Participating Companies	400	300+	25+
Number of Participating Facilities	8500+	500+	400+

3.2. Objective

A company should strive to reach a ratio of 50-100 and investigate about 20 near misses per accident. This will provide a statistically significant sample of all incidents (and all important errors) and provide a company with sufficient feedback on which management system weaknesses are causing the errors and component failures. Various companies with different cultures have achieved high ratios with great return on investment.

Table 7 shows a brief description of possible solutions to overcome each one of the barriers identified in the survey.

Table 7. Summary of Survey Results[16]

Barriers	Solutions
Fear of disciplinary action	Implement a policy to NOT punish individuals when their errors lead to accidents and Near Misses.
Fear of teasing by peers (embarrassment)	Ensure that all employees understand the importance of near-miss reporting; demonstrate, through feedback of lessons learned.
Lack of understanding: Near miss vs Non-incident	Develop a list of "in-context" examples that illustrate what you consider to be Near Misses and what you consider to be non-incident
Lack of management commitment and lack of follow-through on reported near misses	Hold management accountable for achieving a Near-Miss reporting ratio
Apparently high level of effort is required to report/investigate Near Misses	Ensure that the data are entered in a database and queried regularly. Share the results with employees so they can see the value of the reported near misses
There is no way to investigate the thousands of Near Misses per month	Let front-line foremen or supervisors decide if a Near Miss or accident needs to be investigated
Disincentives for reporting Near Misses	Ensure that goals and incentives are not tied to lower incident rates (since this discourages reporting), but instead provide incentives for high Near-Miss reporting ratios
Not knowing which accident investigation system to use	Have ONE incident reporting system with ONE approach
Company discourages Near Miss reporting due to fear of legal liability	Involve legal on major Near Misses and accidents to ensure the results are protected as much as possible under attorney/client privilege.

For a complete description of the possible solutions for each barrier, refer to PII's paper *"Gains from Getting Near Misses Reported"*. [16]

4. Key Human Factors are Not being addressed

All accidents (or nearly all, if you consider that there are some natural phenomena that we either cannot guard against or choose not to guard against) **result from human error**. This is because humans govern and accomplish all of the activities necessary to control the risk of accidents. Humans influence other humans in the process – not only do humans cause accidents (unintentionally) by making errors directly related to the process itself, but they also cause errors by creating deficiencies in the design and the implementation of management systems (i.e., we make errors in authorities, accountabilities, procedures, feedback, proof documents and continual improvement provisions). Ultimately, these management systems govern the human error rate by directly causing or indirectly influencing the process. The process-related activities where errors have the most influence include:

- Designing a process
- Engineering a process
- Specifying components

- Receiving and installing equipment
- Commissioning
- Operating a process
- Maintaining, inspecting and repairing a process
- Troubleshooting and shutting down the process
- Managing process, procedure, materials, facility and personnel changes

The 10 human factors categories to be controlled by companies are:

- Available Time (includes staffing Issues) – for responses only
- Stress/Stressors (includes staffing issues)
- Complexity & task design
- Experience/Training
- Procedures
- Human-Machine Interface (includes tools)
- Fitness for Duty
- Work processes & supervision
- Work Environment
- Communication

4.1. Identified problems

Recent major accidents have highlighted the need for increased focus on human factors. The U.S. Chemical Safety Board (CSB) cited (US CSB Video, 2006) human factor deficiencies as one of the main contributors of the catastrophic accident at the BP Texas City Refinery in March 2005[17]. The human factor deficiencies included lack of control of worker fatigue, poor human-system-interface design, poor communication by radio/phone, out-of-date and inaccurate operating procedures, and poor (no) communication between workers at shift handover. The CSB cited similar issues from many other accidents[18] and has urged industry and the U.S. OSHA (the regulator) to pay much more attention to human factors. As a result, the recent U.S. OSHA National Emphasis Program for Refineries included human factors as one of the 12 core elements.

Implementing human factor engineering and policies to prevent accidents is not a new concept. Nearly all (or all, from a more complete perspective) of the causes and root causes of major accidents in the past 30 years have been the result of poor control of human factors. This has been cited in many root causes analysis reports and papers concerning these major accidents.

Process Safety Management (PSM) systems based on OSHA's PSM standard are likely lacking the fundamental human factor elements and implementation guides that, if applied across the applicable PSM elements, would work together to reduce human error. The *Risk Based Process Safety* guideline from the CCPS/AIChE (2007) does contain these human factor requirements.

Many site and company staff do not know there are standards for the control of human factors. However, at last count there were more than 300 non-governmental standards (such as from ANSI, ISO, IEEE, etc.) for the control of human factors, and there are more than 100 government regulations and standards and guidelines. Some of the better government regulations cover aviation, marine operations (shipping) and aerospace. An organization should devote resources to finding and implementing such best practices.

The following is a description of each of the human factors mentioned above and specific ways to control each. Our data from review of more than 2000 incidents reveal that:

- 90% of accidents had at least one root cause related to deficiencies in written work instructions.
- 70% of accidents had at least one root cause related to miscommunication between workers or between workers and supervisors.

- 40% of accidents had at least one root cause related to excessive fatigue of the worker. [5]

4.2. Objective

Not all errors can be prevented. Since the beginning of time, humans have tried to control error rate with more or less success. Human errors have been measured for hundreds of years. Psychologists have studied why humans make mistakes and have gradually put a science around human error probability. Today, the best models for control of human error in the workplace are generally agreed to be related to control of human factors; in turn, these have grown out of what was previously denoted performance shaping factors. What can be done to control each Human factor is discussed in greater detail next.

For a complete description of each human factor, its importance and how to control them; refer to PII's paper *"Human factors and their optimization"* [5].

Time available for the task

Task design is directly linked with this category. Process design, procedure development and worker scheduling need to take into consideration not only the time it takes to complete each task but also the time it takes to manage usual as well as potentially unusual or unexpected situations. True; we do not create norms for exceptions, but exceptions sometimes are more usual than realized.

Stress/Stressors

A well-designed, organized and managed work process helps maintain and promote individual health and well-being, but stresses will arise through no fault of the organization. Supervisors should be trained to recognize stress in workers and, in severe cases, reassign them and help them get assistance to deal with the stress and in other cases, reduce their job pressures for a day by shifting assignments. A wellness program may also be necessary to help cope with stress. Simple approaches, such as training workers to recognize and deal with stress, can help most individuals. Approaches to deal with stress include positive self-talk and practicing relaxation techniques (i.e., breathing control).

Complexity & task design

If roles and responsibilities for accomplishing tasks are not clearly defined, then there will be a risk of serious errors. It is also important that employees and engineers/designers understand the characteristics of the work elements involved, how each element passes information to the other, how each person involved communicates, and that each person has guidelines to adhere to during task design and procedure development.

For optimal performance, task designers often must integrate human and automated equipment in their task analysis.

It is crucial for task designers also to have an understanding of the appropriate allocation of roles and responsibilities to the various participants in a task. Understanding the characteristics and limitations of all of the humans and automated systems involved in the task (particularly critical tasks) will allow for establishing additional controls where needed.

The task analysis process includes *Task definition* which is the initial operational description of the operator tasks required for execution of a given system function; *Performance analysis* of all tasks with respect to human performance requirements; *Error and workload analysis* to be included in the design process, particularly when the likelihood of unacceptably operator error is high; and *Evaluation* since evaluating the design of a new or complex task is important to ensure that the task can be effectively performed as designed. For very high consequence scenarios involved complex human interaction, it is likely prudent to invest in a human reliability analysis (HRA).

Experience/Training

Training to master the Knowledge, Skills, and Abilities (KSAs) required for a job may be obtained from a variety of sources. Companies provide the training or it is obtained from other sources (e.g., trade schools, universities, contractor organizations, etc.) to reach the skills and knowledge established in job descriptions. There are several factors that may result in personnel not mastering the required KSAs for a job. These include course design and delivery methods, course completion, practical skills demonstrations or simulation, and training frequency. Course design begins when the learning objectives have been identified. The design process consists of determining the delivery methods (simulator, on-the-job training, etc.), number of hours required to cover the materials, instructor qualifications, etc. Although some methods, materials and instructors may be more effective or efficient than others, the important issue is that the course content is complete and addresses all of the relevant KSAs, so that the learning objectives are met and the KSAs mastered.

Another determinant of KSA mastery is course completion. Although this factor appears obvious, there are often competing demands on personnel that may pull them out of training at times. As a result, they may miss the instruction related to specific KSAs. Testing may not identify the KSA deficiency because it is impossible to test mastery of all KSAs. Sampling techniques are used to generate examinations. If attendance and participation are not controlled, some personnel may miss training on specific KSAs and testing may not identify the deficiencies before an error is committed.

Another factor affecting KSA mastery is forgetting. An individual's ability to perform a task will degrade over time unless the relevant KSAs are refreshed. Proficiency training (i.e., refresher) will be required for some tasks to maintain the level of mastery that was demonstrated following initial training. One function of training programs is to identify those tasks that require proficiency training.

If certain tasks are performed frequently, proficiency training may be unnecessary. By performing a task, personnel practice the task and obtain feedback on where they have weaknesses. Task performance refreshes the KSAs and successful task performance verifies that proficiency has been maintained. Furthermore, since humans learn by comparison to similar activities, as a worker practices one task, they are learning about similar tasks. Care must be taken here to avoid the practice drifting away from what the procedure requires. Also, if new operators are trained by experienced operators, you must ensure that the new employee is trained according to the procedure, not whatever practice is actually being used. (Of course, ideally, the practice and procedure will be identical.)

Operating and maintenance procedures

Many procedures do not follow best practices for controlling human error, and so the written procedure actually "contributes" to increased error rates. Additionally, many organizations do not have guides on how to troubleshoot (what to do when process deviations occur). The best practice rules for writing and validating procedures have been published for many years. [6], [7], [8], [9]

Procedures have not traditionally been developed from the perspective of optimizing human factors; instead, procedures have been traditionally developed to meet a compliance requirement to have written procedures. Examples of procedure deficiencies (inaccuracies) include:

- Incorrect/incomplete/nonexistent (most procedures we have audited have been only 70-85% accurate – the inaccuracies include missing critical steps, steps as written are not what needs to be done, or the steps are out of sequence)
- No/misplaced warnings or misuse of warnings (for example, a warning should never **state** the action to take; it should instead **emphasize** the action to take an the statement of the action should be in a numbered step)
- Poor format and presentation rules

Human-machine interface

Design for operability refers to designing the Human-System Interface (HSI) to be consistent with the abilities and limitations of the personnel who will be operating it. Weaknesses in the design processes can result in an HSI that is not well suited to the tasks that personnel must perform to ensure plant safety, resulting in increased workload, decreased performance by personnel, and an increased likelihood of errors.

Design for maintainability refers to designing the HSI and associated plant equipment to ensure that personnel are able to perform necessary maintenance activities efficiently. Weaknesses in the design process can result in systems that impose excessive demands on personnel for maintenance and, therefore, are prone to maintenance errors or problems with reliability and availability. *If it is hard to reach, workers will make more errors, including the error of deciding it is not worth it.*

Design for flexibility refers to the way that changes, such as upgrades to the HSI, are planned and put into use. A new HSI component may require the user to perform functions and tasks in new ways. Skills that the user developed for managing workload when using the former design, such as ways for scanning information or executing control actions may no longer be compatible with the new design. The new HSIs must also be compatible with the remaining HSIs so that operators can use them together with limited possibilities for human error. Also, HSI modifications may not be installed or put into service all at one time, causing the user to adapt to temporary configurations that are different from both the original and final configurations. Weaknesses in HSI implementation can increase operator workload and the likelihood of errors.

Fitness for duty

Implement a company fitness-for-duty program with a primary responsibility for detecting and preventing impaired personnel from performing tasks that may affect productivity and safety. Medical evaluations of personnel, behavioral observation programs, employee assistance programs, and drug and alcohol testing are used to detect impairment. Weaknesses in this program may allow impaired personnel to have access to vital areas in a plant where they could commit errors.

Overtime Policies and Practices – Most companies establish limits for work hours to reduce on-the-job fatigue and the potential consequences for poor task performance. Routine authorization for work hours in excess of those recommended may result in fatigued workers. Furthermore, a practice of excluding training or meetings that occur outside of an individual's normal work schedule from work-hour limitations will also contribute to fatigue. But in many cases these have weak enforcement, especially during shutdowns when the company is under stress to re-start soonest.

Shift Scheduling – Shift scheduling may also affect the likelihood that personnel will show performance decrements due to fatigue. A change in the assigned shift or a rotating shift schedule will disrupt circadian rhythms and may increase the likelihood of errors. So, a company must choose the proper shift rotation to allow adjustments to sleep patterns.

Work processes & Supervision

Written plant policies and procedures are meaningful only when they are enforced; otherwise, worker practices are the policy. Once discrepancies are tolerated, individual workers have to use their own judgment to decide what tasks are necessary and/or acceptable. Supervisors and managers should vigorously enforce plant policies/procedures and ensure that plant policies/procedures and practices are revised as necessary to be consistent. To enforce well, a supervisor must (1) lead by example, (2) watch workers often to ensure they are following company policies and procedures, and (3) determine the correct course of action (including recommending changes to policy and procedures) when workers are found deviating from procedures.

Work environment

Programmatic causes of task environment errors are typically found in the company's processes for designing human-system interfaces or in managing maintenance activities. Other programs may also be implicated. Common programmatic causes of task environment errors include:

Industrial Hygiene and Radiation Protection – These programs are responsible for ensuring that task environments have been evaluated to identify hazards and that needed controls are implemented to minimize exposures. Weaknesses in these programs may result in personnel working in task environments that are conducive to errors.

Work Planning and Control – Weaknesses in the work planning and control system may allow work to be planned without consideration of adverse environmental conditions and performed without the necessary compensatory measures. For example, communication devices may not be provided in noisy environments to support task performance. For tasks that involve unusual physical positions or cramped workspace, additional time to complete the task may not be scheduled. Rest breaks for hot and cold environments may not be planned into the work, or additional temporary lighting may not be provided if the work site is not adequately lighted.

Procedures – Weaknesses in the company's procedure development process may result in the design of procedures that are inappropriate for the conditions in which they will be used. For example, procedures that may be used at night, outside and in the rain should be laminated and the type size should be larger to ensure the procedure can be read. Procedures that will be used in vibration conditions may also require larger type size than procedures read in the stationary environment of the control room.

Human Factors Engineering – Weaknesses in the human factors engineering program may result in the installation of new equipment or systems without consideration of task environment characteristics. For example, the impact of control room lighting on the visibility of digital displays or effects of vibration on the legibility of dials or gauges at local control stations should be considered before installation.

Communication

Successful communication requires several steps. The sender first develops the intention to communicate either verbally or in writing. The sender then composes a message that presents the meaning as clearly as possible. The receiver must pay attention to the message and then interpret its meaning. If the communication is successful, the receiver interprets the message consistently with the sender's intended meaning. Table 8 summarizes the rules for successful verbal communication.

The similarity of the meanings given to the message by the sender and receiver can be verified through feedback. An example of feedback verification in verbal communication is when the receiver "repeats back" the message and the sender either agrees with the receiver's repeat back or corrects it. Verification feedback serves an important error-checking function in the communication process. It also allows supervisory oversight of communications to catch errors before they have consequences.

A sender and receiver must both be active for communication to be effective. The sender and receiver share responsibility for ensuring successful communication. However, when companies analyze the causes of events, errors in sending messages are more often identified than errors in receiving. The reasons for the difference are unclear. A company's investigation should consider sending and receiving errors and corrective actions should address both to be effective.

Shift handover is another critical aspect of communication. The status of equipment, repairs that occurred, problems faced, upcoming chores, etc., are all key information the oncoming worker will need to know. One good guide on shift turnover is U.S. DOE 1038-93. [19]

Table 8. Rules for effective verbal (face-to-face, radio, phone) Communication (© PII, 2012)

#	Communication rules
1	Senders MUST require repeat back by the receiver; ask questions to make sure of understanding.
2	Receivers of messages MUST repeat a message they receive.
3	Communicate one task at a time.
4	Sender & receiver must use the standard name and component number (and building name and number) and/or for related reference material (drawing, procedure name/number) in each message.
5	Sender must require receiver to give feedback of foreseeable conflicts. Sender (if message is to subordinate) should follow-up in the field if possible, to make sure message was understood.
6	Receiver must request clarification anytime they believe the message is confusing.
7	Do NOT give verbal instructions to workers who do not have the demonstrated skills to correctly understand and perform the task assignment (unless you plan to supervise/coach the task yourself).
8	Communicate to the most senior member on the work crew.
9	Know your audience & change message accordingly to compensate for lack of knowledge (when appropriate; see other related rules).
10	Use units (and use approved units, such as SI) for process parameters.
11	Always count from left to right when giving instructions related to multiple choices of equipment.
12	Use approved jargon only.
13	Use the words **check**, **make sure**, and **actions** consistently (see procedure writing rules).
14	When possible, write out the task (use a special procedure or work order) rather than communicating verbally. Use formalized templates for consistency and make sure the receiver is provided all information (such as work orders and drawings) necessary for the task.
15	Perform pre-job meeting with work crew (including with contractors or construction company).
16	Perform walkthrough at the location where the work will be completed.
17	Have a backup communication method. Don't rely on one mode of communication. In emergency situations have hand signals as backup to loss of verbal communication.
18	Talk to the operator or technician who is doing the work (no delegation of work); but see Rule 8.
19	Receiver (worker or group) must report back when work is complete.
20	If confusion exists in implementing a task, the shift supervisor (and higher, if supervisor is unavailable) must be contacted to make sure he/she understands the problem.
21	Do NOT use a PA (public address system) for process instructions, since this method of communication does not allow repeat-back.
22	Use Open questions and non-confrontational questioning methods when requesting clarification.
23	When communicating remotely (by phone or radio), if the message is not understood on the second attempt at clarification the Sender must find the Receiver and communicate face-to-face.
24	For large or complicated jobs, in addition to repeat back, the sender must ask the receiver for an assessment of the pre-job briefing to ensure the workers are not confused.
25	Workers and supervisors must keep a shift log to aid in turnover between shifts. Workers and supervisors and support staff (if necessary) must have a minimum or 15 minutes overlap with a relieving shift. *See shift turnover standard for more details.*

Conclusion

Closing the gaps in these 4 areas has a greater than 100:1 payback!

Acronyms and Abbreviations

ACC - American Chemistry Council
AIChE – American Institute of Chemical Engineers
API – American Petroleum Institute
CCPS – Center for Chemical Process Safety (of AIChE)
CMA - Chemical Manufacturers Association
CSB – Chemical Safety Board
DM – Damage Mechanisms
FMEA – Failure Mode and Effects Analysis
GCPS -- Global Congress of Process safety
HAZOP – Hazard and Operability; as in HAZOP Analysis or HAZOP Study
HRA – Human Reliability Analysis
HSI – Human-System Interface
KSA – Knowledge, skills and abilities
LOC – Loss of containment
LOPA – Layer of Protection Analysis
MOC – Management of Change
OSHA - US Occupational and Health Administration
PHA – Process Hazard Analysis
PSM – Process Safety Management
RBPS – Risk Based Process Safety
SOP – Standard Operating Procedure

References

[1] *Guidelines for Technical Management of Chemical Process Safety,* CCPS/AIChE, 1985.

[2] "The Cost and Benefits of Process Safety Management: Industry Survey Results," W. Bridges, *Process Safety Progress*, AIChE, January 1994.

[3] "Necessity of Performing Hazard Evaluations (PHAs) of Non-normal Modes of Operation (Startup, Shutdown, & Online Maintenance)", W. Bridges and Mike Marshall (US OSHA), *18th Annual International Symposium, Mary Kay-O-Connor Process Safety Center, College Station, TX, October 2015.*

[4] *Guidelines for Risk Based Process Safety*, CCPS/AICHE, 2007.

[5] "Human factors and their optimization" W. Bridges and G. Collazo-Ramos, *8th Global Congress on Process Safety*, Houston, AIChE, April 2012.

[6] PII Course 5, *Writing effective operating and maintenance procedures*, © 2003-current

[7] "Best practices for writing operating procedures and trouble-shooting guides", W. Bridges and R. Tew, *13th Global Congress on Process Safety*, San Antonio, AIChE, March 2017.

[8] "Effective Procedure Writing – A Two-Part Series (Part 1: Reducing Risk of Worker Errors; Part 2: Addressing Process Safety Management Requirements for Operating Procedures)," *Chemical Engineering Progress*, 1997, W. Bridges with T. R. Williams (both of JBFA).

[9] "Best Practices for Writing Operating Procedures and Trouble-Shooting Guides", L. Madden and W. Bridges, *12th Global Congress on Process Safety*, Houston, TX, AIChE, March 2016.

[10] *Guidelines for Hazard Evaluation Procedures*, 3rd Edition, 2008, CCPS/AIChE.

[11] "API RP 571 - Damage Mechanisms Affecting Fixed Equipment in the Refining Industry", API, 2nd edition, 2011.

[12] Bridges, W.G., et. al., "Addressing Human Error During Process Hazard Analyses," *Chemical Engineering Progress*, May 1994.

[13] "Necessity of Performing Hazard Evaluations (PHAs) of Non-normal Modes of Operation (Startup, Shutdown, & Online Maintenance)", W. Bridges and Mike Marshall (US OSHA), *18th Annual International Symposium, Mary Kay-O-Connor Process Safety Center, College Station, TX, October 2015 and 12th Global Congress on Process Safety, April, 2016.*

[14] "Recipe for a Complete Process Hazard Analysis – Especially Addressing the Key Demands from US CSB", R. Tew and W. Bridges, *13th Global Congress on Process Safety*, San Antonio, AIChE, March 2017.

[15] PII Course 8, *PHA/HAZOP Leadership training*, © 2003-current.

[16] "Gains from Getting Near Misses Reported", W. Bridges, *8th Global Congress on Process Safety*, Houston, AIChE, April 2012.

[17] US Chemical Safety and Hazard Investigation Board (CSB), *Anatomy of a Disaster: Explosion at BP Texas City Refinery,* published 2006.

[18] US Chemical Safety and Hazard Investigation Board (CSB), *"Pesticide Chemical Runaway Reaction Pressure Vessel Explosion – Bayer CropScience, LP",* published 2008.

[19] "DOE-STD-1038: Guide to Good Practices for Operations Turnover", US DOE, 1998.

546668 Recycle Line from NGL Spheres to Residual Gas Header

Abstract:

HNGL plant is received 1.6 BSCFD of Haradh gas flows into the DGA Gas treating units where CO2 is removed and the produced sweet gas is directed to the NGL recovery units .Hawiyah sales gas is pretreated at HGP for CO2 removal ; hence the produced 2.4 BSCFD flows directly to the NGL units where it is mixed with the Haradh gas. The mixed gas then is processed in the NGL units producing C2+ NGL and sales gas stream. The NGL is pumped to storage spheres and subsequently shipped to NGL piping network through Shedgum Gas Plant. The Residue Gas flows to the sales gas compression area which increase the pressure allowing routing back the gas to the sales gas grid.

The purpose of this project is to install vent line from the two NGL Spheres to Sales Gas (Residue Gas) common header at Hawiyah NGL Plant. The aim of this project is to minimize the continues flaring from the NGL spheres , safe plant resources and protect the environment

Sphere PVs were venting the NGL vapor to the cryogenic burn pit when the pressure of the sphere exceeds 350 psig. In order to lower sphere pressure without losing NGL production, newly installed PVs are set at lower pressure set point of 305 PSIG to recycle the spheres vapors to NGL trains residue gas line, downstream of the brake-compressors..

During summer months July and August (if associated with maximum plant production), spheres are venting about 4.0 MMSCFD of Methane. Based on Methane price of 4.4 $/MMBTU then total saving is 1.05 MM $/year.

547776 Practical Notes on Ragagep Compliance

Process Safety Management (PSM) and Risk Management Plan (RMP) regulations have long required facilities to ensure that they comply with Recognized and Generally Accepted Good Engineering Practice (RAGAGEP). In recent years, authorities having jurisdiction have placed increased emphasis on RAGAGEP compliance. Specifically, emphasis has been placed on documentation and compliance with the industry codes and standards to which a facility is designed and maintained. This poses a unique challenge to older facilities, which have been designed to codes and standards long out of date and especially to owners and operators that do not have access to engineering documents used in the original design of their facility. For facilities using outdated codes, OSHA requires that the operating company document that "equipment is designed, maintained, inspected, tested, and operating in a safe manner."

Over time, the gradual evolution of industry codes and standards may cause facilities to fall out of compliance with the RAGAGEP requirements of the regulations. This paper provides example cases based on authors' experience with various owners and operators of older facilities. Practical methods are discussed for achieving and maintaining RAGAGEP compliance. It includes examples of comparing systems designed to outdated codes and standards against the latest codes, to ensure that they are able to maintain compliance with RAGAGEP.

548482 Propane Safety: Investigation Findings and Lessons Learned in the 2014 Philidelphia FOOD Truck Explosion

On July 1, 2014, a propane cylinder catastrophically failed and exploded on the back of a food truck in Philadelphia, PA. The explosion led to a sudden release of pressurized propane vapor and superheated propane liquid, which rapidly evaporated resulting in a large white cloud that entered and engulfed the rear of the food truck. This cloud ultimately found an ignition source within the food truck and resulted in a significant fireball burning both inside and outside of the food truck. The explosion and ensuing fire fatally injured the food truck owner and her daughter, and also caused injuries to 11 other individuals. This paper will first present the results of our investigation into the cause and origin of the catastrophic failure and explosion event, which included: (1) analyzing video footage of the explosion and ensuing fire; (2) applying propane phase diagrams to determine the expansion and resulting pressures within the cylinder prior to failure; and (3) applying blast techniques to determine the post catastrophic blast and ignition during the event. These techniques included advanced computational fluid dynamic modeling.

The paper will also discuss safety critical design features and processes in the propane industry that must be in place to ensure the safe use and filling of propane cylinders, which include: cylinder requalification; having a fixed maximum liquid level gauge; having a pressure relief valve; and filling cylinders by weight or volume.

548391 Using the Risk Analysis Screening Tool (RAST) in Process Design to Manage Risk

The Risk Analysis Screening Tool (RAST), available through the Center for Chemical Process Safety (CCPS), courtesy of the Dow Chemical Company can enable users to identify hazards and risk scenarios to better manage process safety risks. RAST is a spreadsheet based tool that uses simplified algorithms to identify and evaluate risk scenarios of a unit operation. The risk scenario identification is based on generic HAZOP deviations applied to the unit operation. An engineer can use RAST to identify risks to proposed process designs or plant modifications in the design phase of a project, before P&IDs are developed and a full process hazard analysis study is done. This can save time in process design by weeding out higher risk process alternatives as the process flow diagram is being developed.

The utility of RAST is demonstrated by application to actual case studies. RAST is applied to the runaway reaction at T2 Laboratories in 2007 and to the chlorine release at DPC Enterprises in 2003 showing how the hazardous events and consequences could have been analyzed during the design phase of the projects.

Drug Product vs. Process Safety: Integrating Both into the Design Process from Concept Development to Start-up

Jeffrey Heil, P.E.
Lead Process Engineer
DPS Group Global
959 Concord Street, Suite 100
Framingham, MA 01701
Jeffrey.Heil@dpsgroupglobal.com

Prepared for Presentation at
American Institute of Chemical Engineers
2019 Spring Meeting and 15th Global Congress on Process Safety
New Orleans, LA
March 31 – April 3, 2019

Drug Product vs. Process Safety: Integrating Both into the Design Process from Concept Development to Start-up

Jeffrey Heil, P.E.
Lead Process Engineer
DPS Group Global
959 Concord Street, Suite 100
Framingham, MA 01701
Jeffrey.Heil@dpsgroupglobal.com

Keywords: Process Safety, Process Design, Biotechnology, Inherently Safe Design

Abstract

Process safety has been around for over 350 years but may easily be overlooked or treated as an afterthought in the design process, often leaving hazard reviews and safety valve considerations to the end of detailed design. In the chemical industry, emphasis on safety is generally in respect to process safety, but in the growing biotechnology industry the emphasis is often shifted to product and patient safety. Due to this subtle yet significant difference in focus, it is often challenging for the design team on these projects to secure the necessary project resources to ensure both robust product and process safety in the facility being designed, often on an accelerated schedule. Leveraging experience from working in the oil and petrochemical industry for over 10 years, an in-depth understanding of the project design cycle, and by asking some simple questions along the way; process safety in design can be assured for even the most challenging projects, without a significant impact on project resources.

1 Introduction

1.1 Brief History

The first safety valve, developed in 1681 by Frenchmen Denis Papin, protected the first ever pressure cooker, one of which was used to make meat jelly for England's King Charles II. [1] Papin first invented the steam pressure cooker in 1676 to make the jelly and other food products but then realized no one would be willing to operate the cooker given the risks. Having identified the hazard, he set out to invent a way to mitigate that hazard while ensuring the King got his jellied meat pies.

Following development of the first weighted lever style safety valve invented by Papin, little progress was made until the early 1800's when locomotive steam engines gained notoriety both as a superior means of transportation and a hazardous machine often operated beyond its limits in the name of more power and speed. Safety valves were revised throughout the early 1800's to make these locomotives safer and formally modernized in 1856 with the invention of the first spring actuated pressure relief device that could not be modified to a different pressure, while in service [2].

Modern process safety, as we can still see it in existence today, dates to the early 1900's with the development of modern oil refineries and all the hazards that they include. With these refineries and their complex chemical processes came a wealth of experimentation, safety system design, and industry supported safety initiatives. The goal was the same as it is today, ensuring safety to the employees, the environment, and the facility. The problem with many of these early safety initiatives is that different companies had different ideas of what safe meant and relied on historical data to predict the future. The notion that "it has been running like this for 20 years and will continue for another 20 years" lead to a raising number of deadly process safety incidents, particularly throughout the 1970's to 1980's (i.e. Bhopal, India Methyl Isocyanate release and Piper Alpha explosion in the North Sea.). Plants were operating to the status quo, not improving their practices new technology became available. In parallel, urban sprawl forced industrial plant operation closer to populated areas, increasing risk to the community. The OSHA process safety management (PSM) initiative was introduced into law in the United States to address this very problem [3]. The PSM standard was performance-based meaning that as new technology was developed and safer systems could be designed, those safer designs had to be implemented; i.e., the status quo was no longer good enough. The PSM standard was ratified in 1992: hence, even though the oil and petrochemical industries in the United States have been around for about 100 years, they have only been operating under this standard for 26 years.

Contrast this with today's biotechnology companies. Gentech, one of the earliest, received FDA approval for the insulin drug Humulun in 1982. Like any new technology, investment and growth in biotechnology was slow to start, not taking off until early 2000's. There are many metrics for tracking growth of the Biotechnology industry such as venture capitalist investment and the number of companies in operation, but the high number of mergers and acquisitions over the past 30 years makes such comparisons incomplete. The Nasdaq Biotechnology Index (NBI) was established in 1993 with a baseline index value of 200 and provides a good metric for tracking industry growth over time. After holding steady in the low 200-300s throughout the 1990's, the NBI surged to 1,300 in 2000 as the Biotechnology bubble was forming. The index has steadily risen to a peak of 4,003 on 01-July-2015 and is hovering currently around 3,500 (Q1 2019), exemplifying the continued growth of the industry over the past 20 years. [4]

Pharmaceutical companies (e.g., Aspirin, invented by Bayer in 1899) have been around almost as long the oil companies have and are sometimes confused with biotech. Biotechnology companies are adolescent at best, having typical company life of less than 20 years with little process safety oversight beyond building code requirements as they do not meet the hazard requirements of the OSHA PSM Standard. Not meeting the hazard requirements of the PSM standard does not exempt these companies from a responsibility to ensure a safe working environment. If anything, their proximity to high density population areas and minimal operating experience places them at a higher risk with respect to process safety, but in many cases the operators and designers have only worked in these types of facilities and may not fully appreciate the safety concerns present. Coming from a background where process safety is paramount and the product quality, while important, takes second place, the author has made it his personal mission to increase awareness throughout the Biotechnology industry one project at a time.

1.2 Process vs. Product Safety

In the chemical industry, emphasis on safety is always in respect to process safety; ensuring that the public, employees, environment, and facility (in that order) are sufficiently protected from inherent hazards of a process. "Safety" refers to containment of the process and protection of personnel from harm. If a hazardous material cannot escape the process; it cannot be breathed in, released to the environment, set on fire, or result in damage to equipment that may cause injury to nearby personnel. The previously described process safety management standard formalized the set of tools used by these industries to ensure current best practices are used when it comes to process safety in hazardous environments. The goal of the PSM standard and implications of it being "performance based" is summarized as follows:

> *"... Recognizing the dangers associated with highly hazardous chemicals, Congress mandated (by the Clean Air Act Amendments of 1990) that OSHA set requirements for the management of highly hazardous substances to prevent and mitigate hazards associated with catastrophic releases of flammable, explosive, reactive, and toxic chemicals that may endanger workers. As a result, OSHA developed the Process Safety Management (PSM) standard (issued in 1992), which covers the manufacturing of explosives and processes involving threshold quantities of flammable liquids and flammable gasses (10,000 lbs.), as well as 137 listed highly hazardous chemicals. The PSM standard (29 CFR 1910.119) requires employers to implement safety programs that identify, evaluate, and control these hazards. Unlike some of OSHA's standards, which prescribe precisely what employers must do to comply, PSM is a "performance-based" standard-it outlines key features of safety programs for controlling highly hazardous chemicals, and employers have the flexibility to tailor their safety programs to the unique conditions at their facilities..." [4]*

The biotechnology industry on the other hand often has a more focused emphasis on product safety. Product safety ensures that all current good manufacturing practices (cGMP) are being followed with the goal of protecting the end-user or patient from hazards of a contaminated biological drug product. Product safety was important for the pharmaceutical industry because the medicines being produced were intended for use in the human body and would chemically alter the bodies function. Product cross contamination could result in serious health side effects, especially with highly potent or specialized substances. Biotechnology processes work by growing viruses and manipulating this viral material to deliver highly specialized medicines to the patient. Poor manufacturing conditions could allow unintended bacterial growth alongside the desired product, leaving the future patient with a potentially deadly virus or bacterial infection instead of a cure. The US Food and Drug Administration (FDA) passed the first cGMP regulations in 1969 [6], making them older and with many respects better understood in the industry than the PSM standard is in the oil and petrochemical industry. The practiced engineer will note several similarities in the quoted statements describing the PSM standard and the cGMP standards.

> *"The CGMP requirements were established to be flexible in order to allow each manufacturer to decide individually how to best implement the necessary controls by using scientifically sound design, processing methods, and testing procedures. The flexibility in these regulations allows companies to use modern technologies and innovative approaches to achieve higher quality through continual improvement. Accordingly, the "C" in CGMP stands for "current," requiring companies to use technologies and systems that are up-to-date in order to comply with the regulations. Systems and equipment that may have been "top-of-the-line" to prevent contamination, mix-ups, and errors 10 or 20 years ago may be less than adequate by today's standards." [7]*

Product safety requires process containment too, but in many cases the consequences are less likely to cause bodily harm so long as product quality control procedures ensure quarantine of the affected drug product. Note that this statement applies to the biotechnology industry, the consequences of a loss of containment in the pharmaceutical industry can be as deadly as in a chemical plant, depending on the substance released. Fifty years ago, when the cGMP's were adopted, pharmaceutical operators already knew about process safety and adopted it as readily as oil refineries and industrial chemical plants did. Twenty years ago, biotechnology was in its infancy and there were few large-scale biological processes in operation. Ten years ago, most biological processes ran at atmospheric pressure using water-based chemistry. The risk associated with potential process safety events was minimal in comparison. Today, many biotech processes use advanced solvent-based chemistries needing large volumes of hazardous flammable solvents, a key element in the severity of an incident when a loss of containment occurs. Add to this unique process chemistries such as dichloroacetic acid, methyl chloride, hydrochloric acid, and sodium hydroxide; all of which can easily corrode stainless steel, a common metal alloy used in biological processing, and unique hazards become a reality even at small scale.

2 The Design Process

Designing processes to produce goods takes time, time that the future producer of those goods doesn't always like to part with because "time is money". Each month spent designing or building the facility to manufacturer a product is a month of revenue missed or often in biotech, a month of trials that could be running. A typical design process includes several independent stages or phases. An ideal project would separate each of these phases with well-defined checks creating structure, an opportunity for detailed reviews, and adding time to the project. Initial concept development, feasibility, and design of large-scale chemical or industrial facilities often involves acres of real estate, spans years, and costs millions of dollars. By contrast small-scale bioprocessing facilities, often less than 40,000 ft^2, are fully designed in terms of weeks or months and leverage parallel work paths where possible.

The impact of such a short design schedule is that time for staggered activities, separated by well-defined stage-gates, is often sacrificed in the name of the project schedule. It is possible to crash a project schedule, completing the design early, but doing so limits the amount of design work that can be completed based on previously completed design elements. If a project schedule allots for 6-8 weeks between establishing the basis of design for a facility and documents that are to be issued for vendor bid, scheduling one week for a hazard analysis and another week to implement the resulting design changes (25% of the schedule) is a challenge. By the time the hazard analysis is complete, many aspects of the design have been progressed and resulting changes will result in unbudgeted rework. The same is true for the design of key safety systems such as relief devices or process interlocks that require a complete system design to be finalized. The engineer can anticipate credible overpressure scenarios and safe operating limits for a process but cannot complete these design elements until piping to and from the relief device is complete and the hazard analysis is approved, accepting the level of protection assigned to each process step. It is for these reasons that it is not uncommon to see relief devices for a process being sized and planned for as one of the final project tasks. Explaining to an architect that a storage tank must be moved to meet a flammable liquid set-back identified during a hazard review or that another roof penetration is required for a relief device tail pipe two weeks before the completion of a design are just two examples of real-world challenges that project teams may be faced with.

Understanding the different design phases or stage-gates, the tasks that occur during them, and their impact on the ultimate facility design is important if one is to create a plan for addressing process safety on such a short schedule. Table 1 summarizes one example of a design process, key stage-gates, and design elements to be considered in each phase. Note that jumping a phase and for example combining schematic design and design development (the most common violation of the design process) into one phase can confuse the separation of tasks between these phases and eliminate checks important to the design process.

Table 1 – Key Aspects of the Design Process

Phase	Process Design	Process Safety
Feasibility Study	• Identify product to be manufactured and major unit operations • Establish block flow diagram of process • Identify facility requirements • Confirm business case for proposed facility	• Identify hazardous materials used in the process • Identify hazardous operating conditions (i.e. high pressure) • Establish acceptable risk criteria for operation of the facility

Phase	Process Design	Process Safety
Concept Design	• Breakdown individual processing steps to the equipment level • Develop process flow diagrams for each major process step • Develop preliminary heat and material balance • Develop equipment list and identify long lead equipment (to prioritize design development tasks) • Detailed layout of process equipment • Identify utility requirements and size utility systems • Size and layout major process headers	• Develop a comprehensive hazardous and flammable materials inventory / determine if additional design precautions are required • Identify potential material compatibility concerns and material specifications for each process step • Complete a code review in collaboration with the full design team to ensure all building, fire, and design codes are being followed • Ensure facility citing requirements are met in the layout
Design Development	• Develop initial P&IDs for the process • Identify and document key control loops and automation requirements • Finalize design of all process equipment (size, pressure rating, operating conditions) • Complete detailed sizing and specification of major process headers and utility lines • Finalize process heat and material balance • Prepare equipment and piping specifications and datasheets	• Confirm inherently safe design where possible (pressure profile versus equipment design) • Identify safe operating limits for each unit operation, add control points as required • Review P&IDs for relief devices and confirm their discharge locations • Complete scenario analysis and preliminary sizing for all relief devices. Based on preliminary sizing, define relief device inlet/outlet piping • Confirm piping specifications versus services • Confirm vessel sizing versus hazardous material inventory and update as required • Complete a preliminary hazard analysis on the full process at the end of the design development stage

Phase	Process Design	Process Safety
Detail Design	• Incorporate all recommendations from the design development preliminary hazard analysis • Finalize all equipment and line sizing • Fully specify all process components including equipment, lines, valves and instruments	• Ensure that all actions from the preliminary hazard analysis are incorporated into the design • Review and finalize relief device sizing • Complete sizing and design of effluent handling system for relief device discharges • Double check the set points of key process variables versus the operation's safe operating limits, previously defined. • Review final equipment specifications against material compatibility matrix and equipment specifications

3 Safety Throughout the Design Process

Process safety should be considered in the earliest stages of design. One principle of process safety design is that inherently safe design, creating a system that physically cannot fail, is the only way to prevent an incident from occurring. The easiest example of an inherently safe system is designing a pressure vessel to withstand maximum system pressure, even if all process controls were to fail. While an easy goal on paper, inherently safe design is not always possible and often never fully achievable. A vessel can be designed to the proper pressure, but an unforeseen corrosion issue could reduce the design pressure making the vessel susceptible to premature failure. In anticipation of this and to protect against it, engineers design-in multiple layers of protection to a process starting with inherently safe design at the concept level and ending with proper emergency response procedures developed during commissioning and simulated during plant operation. Continuing with the vessel example, minimizing the quantity of material stored and having an inventory management plan to support smaller on-hand inventories is another way to reduce hazards associated with the process. Additional design principles or layers of protection used are summarized in Figure 1. Different layers of protection against a process safety event are established during concept development and built-up throughout the design process.

Figure 1. Typical Layers of Protection Applied to a Design

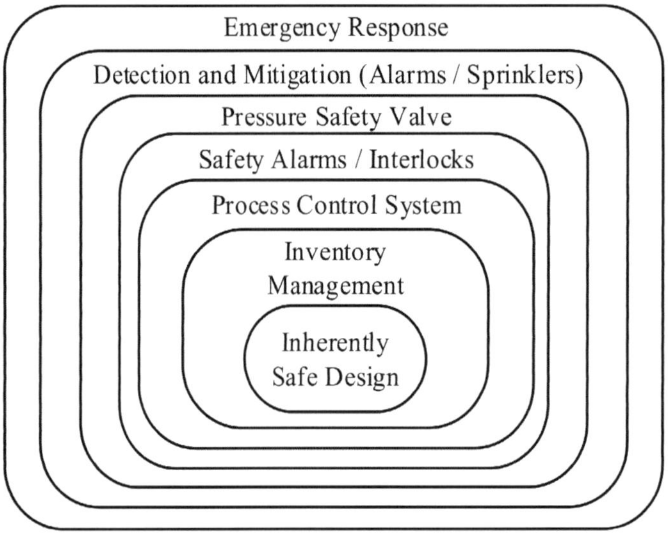

3.1 Concept Design

Concept design is one of the earliest stages of the design process. During concept design types of process equipment are first selected, their location mapped out, and consideration given to how materials will be heated up, cooled down, transferred, and stored when not in use. By asking a few questions as these fundamental design decisions are made, the design predisposes its self to being inherently safer.

- What is the worst-case operating condition, can the system be mechanically designed to handle those conditions?
- Where will hazardous materials be present and can that equipment be installed away from high concentrations of personnel?
- What hazards do the materials pose (toxic, flammable, corrosive)?
- How much of a potentially hazardous raw material must be on-hand? Weekly vs. monthly delivery may reduce stockpiles by 75%.
- Is there a safer way to produce the desired product?
- Is there a safer way to complete this unit operation?

3.2 Detail Design

Concept design develops enough detail to prove that the processing goal is obtainable and business case can be made. Concepts developed such as system location, connectivity, and required unit operations are further developed into a constructible design that can ultimately be operated to produce a finished product. In detailed design, detailed questions are needed as more layers of protection are added. These questions are to be asked informally throughout design and will be formalized in a process hazard analysis (PHA), completed during the detailed design phase.

- Where are the areas of high pressure, temperature, and hazardous material inventories; can they be reduced?
- Are there any specific code requirements or industry best practices for this system?
- What are the expected operating parameters, how can they be exceeded, identified ahead of time, and mitigated?
- Are all systems protected by a mechanical pressure relief device (if needed), do those devices discharge to safe locations?
- How and where could a loss of containment occur, how can it be quickly identified (leak or fire detection)?
- How will the consequences of an event be mitigated once started; how can the event be stopped (process shutdown or fire protection)?
- How can the public, employees, and environment be protected from the consequences of an event if it were to occur (emergency management)?

3.3 Start-Up

Following detailed design and construction, a process must be started, commissioned, qualified, and transitioned to manufacturing. Throughout this process, design assumptions must be confirmed, formal procedures verified, and changes to the design noted so that their impact on the process can be determined. Process safety considerations during start-up must include the full team from engineering to operations and maintenance.

- How does the as-built design compare to the final design, what are the impact of any changes?
- Were changes made to the operating envelope for any of the systems (pressure, temperature, level, concentration)?
- Has a pre-start-up safety review (PSSR) and as-built PHA been completed?
- Do operating procedures match the actual operation of the process?
- Have emergency drills been completed, are there lessons learned from these?

Addressing process safety throughout the design process, versus at the end of design when final design reviews are completed; minimizes late stage design changes, results in a better designed facility, and ensures safe facility operation for years to come.

4 Conclusion

Process safety is a well-established element in the design of oil and petrochemical facilities, taking a front row seat in the design process and being an area of focus once operations begin. Product safety holds a similar role and has been common place for a similar period in pharmaceutical and recently bioprocessing facilities. While the gap between these two elements of design has been closing in recent years as more pharmaceutical companies enter the bioprocessing arena and bioprocessing operations grow, a considerable gap remains. Based on the authors experience designing processes and facilities in both of areas, a few causes for the gaps are easily identifiable:

- Historically, bioprocessing facilities have not contained hazardous processes. Areas of pressure were limited to gas delivery systems (i.e. nitrogen and compressed air) and storage tanks contained aqueous based chemistries (i.e. bioreactors) minimizing the hazards with a loss of containment of these products.
- The shortened design timeline for bioprocess facilities challenges the ability to incorporate process safety into the design at distinct stage gates. The schedule does not allow for a discrete process safety design effort while other design work is on hold.
- There is a general lack of knowledge regarding safety system design in the biotechnology industry. This goes back to the opening statement that in many cases engineers responsible for these systems don't know what they don't know.

Product and patient safety are well defined requirements for biomanufacturing companies and are considered in the earliest stages of the drug discovery process. Applying similar principles to process safety throughout the design lifecycle, and not at distinct stage gates, will ensure a smooth transition from process development to design, validation, and ultimately commercial manufacturing.

5 References

[1] F.M. Editors of Encyclopedia Britannica. "Denis Papin." *Encyclopedia Britannica*, Encyclopedia Britannica, Inc., 18 Aug. 2018, www.britannica.com/biography/Denis-Papin.

[2] Ramsbottom, J. "On an Improved Safety-Valve." *Process Institute of Mechanical Engineers*, vol. 7, 1856, pp. 37–47.

[3] US Department of Labor Final Rule, "Final Rule on Process Safety Management of Highly Hazardous Chemicals; Explosives and Blasting Agents", 1910.109, *Fed Register #: 57.6356*, 24 Fen. 1992.

[4] NASDAQ Biotechnology Index, NBI, https://finance.yahoo.com/quote/%5Enbi/.

[5] *Process Safety Management Introduction*. Occupational Safety and Health Administration, www.osha.gov/dsg/psm/index.html.

[6] US Food and Drug Administration, *Good Manufacturing Practices (GMPs) for the 21st Century - Food Processing*, "GMPs Section One: Current Food Good Manufacturing Practices", 09 Aug. 2004, www.fda.gov/food/guidanceregulation/cgmp/ucm110907.htm.

[7] US Food and Drug Administration, "*Facts About the Current Good Manufacturing Practices (CGMPs)*", 25 Jun. 2018, www.fda.gov/food/guidanceregulation/cgmp/ucm110907.htm.

546706 LOPA and Sil Risk Analysis As Critical Tools for PSM Culture and Accidents Prevention

Industrial facilities that operate with dangerous products, whether chronic or acute, are increasingly common to be present and part of society at large, often allowing an aggregate of increasingly rich and economically beneficial professions and specialties to the community, be it directly or indirectly. Accidents in these types of units, especially those operating with hazardous products of acute consequence (those that have the capacity to cause fatalities when released to the environment), such as flammable, toxic, explosive products, due to the possibility of occurrence of undesirable events and the behavior of the effects resulting from releases of these types of product, could cause serious damage to the people and / or the installations that are in its vicinity. The various types of accident studies are known as EAR (Risk Analysis Studies), some of which are used more widely and others are specific to specific response needs.

According to BERNSTEIN (1997), the word "risk" derives from the ancient Italian risicare, which means "to dare". While DUARTE (2002) describes that the meaning of the word risk is an expression of thought that has a lot to do with the social, scientific and technological evolution of man. As BILLINTON and ALLAN (1992) note, society has great difficulty distinguishing between a hazard, which can be prioritized in terms of its severity, but does not take into account its probability, and risk, which considers not only the dangerous event, but also their probability of occurrence. KAPLAN (1997) explains that when a person asks "What is the risk?", 3 (three) questions are actually being questioned: What can happen? How can this happen? And if this does happen, what are the consequences? The answers define the scenario, frequency and consequence.

Within the agencies that tend to manage corporate risk, it is noticeable that the risk associated with a particular accident scenario is calculated according to its frequency of occurrence and its severity. Frequency is related to the temporal chance of occurrence of the initiating event, which leads to the release of dangerous product, such as LPP (Liquefied Petroleum Gas) emanation by ball rupture through overpressure of this equipment or, as a second example, the intrinsic rupture of a stretch of line. The consequence, also called severity, considering personal safety, is the capacity that the physical effects, due to the release of dangerous product, associated to the analyzed initiating event, can cause fatality (s), related to the potential of damage related to each effect analysis. The risk, qualitatively or quantitatively observed, and defined in a simplified way, is evaluated as the multiplication of frequency of occurrence and severity. The total social risk related to a particular facility, which can be defined as the risk that a community outside the facility is subject to by its presence in its vicinity, can be calculated as the sum of the risks of all the initiating events arising of the same. In order to do this, for each accident scenario, not only the generating causes, known as Initiating Events, but also the likelihood of ignition, the probability of affecting people, and several other parameters are considered, being one of them very important and permissible of analyzes, which are the failures

safeguards. These, if they act effectively when demanded, can prevent the accident from occurring, since the Initiating Event happens.

Hazardous product risk studies, such as qualitative (Hazard Operational Preliminary Analysis (APPP), Hazard Operability (HAZOP, among others) or those known as quantitative analyzes (Vulnerability Analysis and Quantitative Risk Analysis per se), in addition to being increasingly requested by regulatory agencies, have often been seen as an internal requirement of companies, representing one of the pillars of industrial safety management, in order to keep their risks at low or acceptable levels. However, these analyzes are usually performed in stages

until the final definition of the risk, comparison with acceptability criteria, and later mitigation, if necessary, each of these stages having a minimum time required for its execution, which may make the analysis complete rather time-consuming. Situations similar to these are detrimental to Risk Management when, in fact, what is desired is not a complete risk analysis, but only the achievement of a specific response, such as if my identified risk scenario is effective and properly protected.

The extent of possible damage related to each identified accident is delimited by the intensity of the physical effect causing the damage, and the relationship between the intensity of the physical effect and the corresponding damage is established through the vulnerability models.

On the other hand, semiquantitative risk analyzes such The extent of possible damage related to each identified accident is delimited by the intensity of the physical effect causing the damage, and the relationship between the intensity of the physical effect and the corresponding damage is established through the vulnerability models.

On the other hand, semiquantitative risk analyzes such as LOPA (Layer of Protection Analysis) and SIL (Safety Integrity Level), even though they are less demanded by chemical and petrochemical companies than qualitative or quantitative steps can have beneficial and effective results in terms of knowledge of all the characteristics of the analysis scenarios and of their protection barriers and, in a practical way, much faster to be developed. In addition, they can provide results that do not make up the analysis range of qualitative and / or quantitative studies in depth, such as knowledge and prior review of safeguards, many of which are known as Independent Protection Layers (CIP or IPL - Independent Protection Layers).

Therefore, with the anticipation of an increase in the number of future installations operating with hazardous products and the current installations that already do so, the semi-quantitative studies of LOPA represent an enormous value for the observance of the existing protections in each scenario of process accidents, also serving as the basis for the determination of new Independent Protection Layers or even the reduction of their number, tending to simplify the steps of determining the frequency of the scenario., even though they are less demanded by chemical and petrochemical companies than qualitative or quantitative steps can have beneficial and effective results in terms of knowledge of all the characteristics of the analysis scenarios and of their protection barriers and, in a practical way, much faster to be developed. In addition, they can provide results that do not make up the analysis range of qualitative and / or quantitative studies in depth, such as knowledge and prior review of safeguards, many of which are known as Independent Protection Layers (CIP or IPL - Independent Protection Layers).

Therefore, with the anticipation of an increase in the number of future installations operating with hazardous products and the current installations that already do so, the semi-quantitative studies of LOPA represent an enormous value for the observance of the existing protections in each scenario of process accidents, also serving as the basis for the determination of new Independent Protection Layers or even the reduction of their number, tending to simplify the steps of determining the frequency of the scenario.

546375 Controlling Static Discharge Using Antistatic Additives

Controlling Static Discharge using Antistatic Additives.

Abstract

The presentation will discuss controlling static discharge in non-fuel, non-conductive/low-flash point fluids. The scope of the presentation will concentrate on industrial producers, distributors, consumers of these fluids. Furthermore,furthermore, the presentation covers several industries where these fluids are most commonly found. Common Static-Accumulating Flammable Liquids That May Form Ignitable Vapor-Air Mixtures:

- VM&P naphtha
- Cyclohexane
- n-Heptane
- Benzene
- Toluene
- n-Hexane
- Xylene
- Ethyl benzene
- Styrene

Static discharge has caused catastrophic explosions and fires (U.S. Chemical Safety Board, Static Spark Ignites Explosion Inside Flammable Liquid Storage Tank No. 2007-06-I-KS).

Issues

- Nonconductive flammable liquids can accumulate static electricity during transfer and storage.
- Static sparks can readily ignite flammable vapor-air mixtures inside storage tanks.
- Material Safety Data Sheets (MSDSs) often do not adequately communicate hazard data and precautions.

Key findings:

- Safety publications omit antistat options. (see Process Safety Beacon, August, 2016)
- US affiliates of multinational corporations have different guidelines for antistatic use than their parent companies.
    - NFPA®77 and the American Coatings Association's, *Generation and Control of Static Electricity in Coatings Operation* has guidelines that are different than EU regulations ATEX- Appareils destinés à être utilisés en ATmosphères EXplosibles and ESIG's *Flamability: Safe Working With Industrial Solvents.*
    - The presentation will highlight these differences and how they contribute to the disparity in antistatic additive applications.
- Normal Bonding and Grounding May Not Be Enough! Companies that handle, transfer, and store flammable liquids should contact manufacturers to determine if these liquids can accumulate dangerous levels of static electricity, and if they can form explosive vapor-air mixtures inside storage tanks. If so, *extra precautions*—beyond normal bonding and grounding—may be necessary. Antistatic Additive is an extra precaution.

Recommendations: Knowledge members can use

- Who adds the Antistat?
- How much Antistat is added?
- How is the Antistat added?

546677 Led Your Process Safety Culture, Lead Beyond Hierarchies

The Oil & Gas industry is trying to advance at the next crossroads: "How to achieve that leaders without skills and without training in the management of culture, transform their own culture?"

Our research presents three key ideas that leaders from all over the world

can apply immediately to lead risk management with effectiveness:

frequency, consistency, meaning.

When specific leadership competencies to manage risk have not yet been developed,

Leaders can use this three concepts of easy adoption.

Increasing frequency of interactions with people will allow us to generate safer operations. In full attention economy and contexts of increasing complexity, for risk management, more communication is better communication. But it is not enough to improve the frequency of interactions. That frequency must be accompanied by consistency. The challenge is to achieve possible consistency between what we say and what We do. Leading by example is one of the central keys of risk management. The lack of consistency provides symbolic permissions to normalize deviations. We must maintain a high frequency of face-to-face interactions, but in addition, we must be consistent at every opportunity. And finally, to give meaning to our collaborators. It is necessary to ensure the understanding of risks. Each collaborator need to build his or her own sense of vulnerability and Operational Discipline.

We need workers who operate according to procedures and not adjusted to what the bosses say. This is what meaning allows.

548192 The Importance of Using Color Scale in Process Safety

Everyone knows the importance of color. From physics to fashion going through the automotive industry she will always be present. There is no world without color. Often, they come up on the scale form or not, but the variation in the scenarios are always important. It is so on the traffic sign and the ph tape. From the Diamond of Hommel to across the street. Colors, colors and colors. In addition to not necessarily needing ididomas, such as color of the sky and molten metal have a variety of shades. The main idea of the work is to choose a suitable color to be used in communication within a work area, choosing which and how many elements can evaluated suggest a color scale from the chosen one and present its contribution in the process safety.

550372 Flammable Gas Release Modeling for Real-Time Analysis Using Adversarial Variational Bayes

In this study, we have suggested a method to calculate high-accuracy damage range from flammable material in petrochemical complex. There are many flammable and toxic substances in chemical processes. Computational Fluid Dynamics(CFD) is one of the best way to calculate and simulate scenarios of release range of various fluids or gases. However, it is not used to calculate damage range, because of a lot of time to simulate one scenario. Thus, lots of researchers have enthusiastically studied the surrogate model or reduced order model to get the results of release range in real-time [1-3]. Recently, deep learning based generative model has receive attention because of its superior performance of non-linear manifold estimation [4]. Herein, we propose surrogate model based on adversarial varational Bayes (AVB) [5] for real-time analysis of probability of death. The proposed surrogate model can extract the latent space of high-dimensional flammable and toxic substances dispersion contours. Furthermore, it can generate the estimated contour image from variables of interest much faster and more precise than CFD data memorization with linear interpolation because it does not memorize the data but train the non-linear manifold regarding exact posterior distribution on given observation which basis is predefined in standard variational autoencoder (VAE) [4, 6].

The acrylonitrile was studied about risk assessment for health damage like cancer hazard.[7] But the component has flammable and toxicity, so we should calculate range of damage. The main scenario in this study is the acrylonitrile released from underground model, and the result is shown different damage following hole size, mass flow rate, wind velocity, and wind direction. The superior performance of the proposed model was exemplified by comparing with other surrogate models.

[1] Chen, T., Hadinoto, K., Yan, W., & Ma, Y. (2011). Efficient meta-modelling of complex process simulations with time–space-dependent outputs. *Computers & chemical engineering*, *35*(3), 502-509.

[2] Wang, K., Chen, T., Kwa, S. T., Ma, Y., & Lau, R. (2014). Meta-modelling for fast analysis of CFD-simulated vapour cloud dispersion processes. *Computers & Chemical Engineering*, *69*, 89-97.

[3] Long, K. J., Zajaczkowski, F. J., Haupt, S. E., & Peltier, L. J. (2009). Modeling a Hypothetical Chlorine Release on a College Campus. *JCP*, *4*(9), 881-890.

[4] Na, J., Jeon, K., & Lee, W. B. (2018). Toxic gas release modeling for real-time analysis using variational autoencoder with convolutional neural networks. *Chemical Engineering Science*, *181*, 68-78.

[5] Mescheder, L., Nowozin, S., & Geiger, A. (2017, August). Adversarial Variational Bayes: Unifying Variational Autoencoders and Generative Adversarial Networks. In *International Conference on Machine Learning (ICML)* (pp. 2391-2400). PMLR.

[6] Kingma, D. P., & Welling, M. (2013). Auto-encoding variational bayes. *arXiv preprint arXiv:1312.6114*.

[7] Haber, L. T., & Patterson, J. (2005). Report of an independent peer review of an acrylonitrile risk assessment. Hunam & Experimental Toxicology, 24, 487-527.

546615 Modeling and Analysis of Flammable Gas Dispersion and Deflagration from Offshore Platform Blowout

The operational safety of offshore platform is challenged by a series of adverse factors including harsh environmental conditions and process accidents. Blowout is a type of major accident that may result in catastrophic consequences. This paper uses a computational fluid dynamics (CFD) model to simulate the blowout accident on an offshore drilling platform. The simulation includes the dispersion of flammable gas and the resulting deflagration based on dispersion results. Through the modeling, the blowout gas dispersion processes under different wind speeds and directions are analyzed, and a deflagration scenario due to the ignition of flammable gas is predicted in which the impacts of jet fire, overpressure and heat radiation are assessed. The simulations indicate that wind speed and direction have a great impact on the dispersion behavior of blowout gas. During the deflagration, the maximum overpressure is 280 KPa, the jet fire mainly distributes in the derrick, and the maximum radiation flux of thermal radiation field is 2300 W/m^2. Essentially, the outcomes from simulations could help to conduct a prior risk assessment and make an emergency response plan. Additionally, it could give a reference for the inherent safety design of offshore platform to mitigate the unexpected consequences due to blowout accidents.

548682 Application of Inherently Safer Design Tool (*i*-SDT) for Hazards Assessment and Safer Operation

Inherently ***S***afer ***D***esign ***T***ool (*i*-SDT) is a newly developed semi-quantitative metrics for hazards identification and characterization[1]. It is built on property platform, probabilistic analysis of historical accidents information and characteristic equations of safety parameters (i.e., flammability, toxicity, explosiveness, reactivity, etc.) which are developed in-house to assess potential hazards under various operating conditions. Designers can identify dangerous components and vulnerable equipment of the investigated industrial process and can take necessary corrective actions by applying the inherent safety principles[2] (e.g., minimization, substitution, moderation and simplification). The final output of this *i*-SDT tool is a cluster safety parameter score (C_{SP}) which provides insights regarding the development of safe design and operational parameters to manage the hazards using a very limited amount of process information and safety incidents information.

In this work, the application of the proposed *i*-SDT tool is explained for assessing different alternative routes of an industrial process supply chain. The proposed *i*-SDT safety metric identifies the chemical components with highest risk and the major accident-prone operational units. The detail analysis using *i*-SDT tools shows how different operating conditions can affect the safety performance of the given process. The ultimate object is to find the safer route while considering the techno-economic and environmental performance. It provides Pareto curve to compare alternative routes and gives the freedom to the end-users choosing their safer operating route/technology and process conditions. Therefore, safety can be considered as an important selection criteria for choosing best alternative from the beginning.

Final cluster safety parameter scores (C_{SP}) obtained from the analysis enables designers to compare between available technologies and provides information on various hazardous situations. It can also provide in depth knowledge regarding the cause behind the hazardous behavior. The proposed *i*-SDT metrics allows designer to do the safety analysis simultaneously while doing the initial process synthesis which will reduce the future modification cost of any add-in safety features. This work shows how the unique features of *i*-SDT tool can be applied in a real life scenario while doing inherent safer design during early stage of design.

1. Kazi, M.–K, Eljack F. T., & AlNouss, A. Application of a property-based inherent safety quantification framework for integrating risk assessment into process safety life cycle. *AIChE Spring Meeting, Orlando, Florida, USA, 2018.* (156c), p. 2260.

2. Kletz, T. A., & Amyotte, P. (2010). Process Plants: A Handbook for Inherently Safer Design, Second Edition: Taylor & Francis.

546631 An Approach to Understanding Worker Interactions with Procedures Under Normal Working Conditions

Incidents in high-risk industrial settings are often attributed to the way in which procedures are used. Despite the identification of procedures misuse, little research has been conducted on how workers have interacted with procedures in normal conditions. To understand these interactions a behavior assessment technique has been developed which can describe how a worker interacts with a procedure to complete tasks in everyday work situations. interactions under normal conditions. Using the concept of work-as-imagined/work-as-done, the techniques describes the interactions at a macro whole-of-task level as well as an individual step level. At the macros level, assessment considers if the task was completed correctly and safely (Tier 0). At the micro step-level, assessment considers if the step was completed as it was imagined (Tier 1) or if the step was done in an alternate way (Tier 2). For the alternate approaches, assessment considers if the worker struggled or required assistance, completed the steps out of order or with iterations or if the step was not completed as intended.

Two studies, one involving industrial tasks Second Life® (Study 1), the other at Shells Robert, LA training facility (Study 2), were used to test and develop the behavior assessment technique. In Study 1, the behavior of 13 participants from offshore drilling conducting tasks in either day or night conditions was assessed. In Study 2 the behavior assessment technique was used to test if experience accounted for behavior difference in 25 participants from high-risk industries. Results demonstrate the success of applying the assessment techniques through identifying how the behavior of the participants changes according to their experience and familiarity with the tasks the were performing. Further analysis identified recommendations for design and work flow changes of both written procedures and specific equipment. In conclusion, the initial results suggest the assessment technique is an important contribution to the methods use for understanding early interventions to improve safety in the process industry.

PROCESS HAZARD ANALYSIS (PHA) PREPARATION: A ROLE BASED OPTIMIZATION

Susan S. Frizzell, PE
Adley Services, LLC
Lafayette, LA
Susan@AdleyServices.com

Biography:

Susan Frizzell, PE is the President of Adley Services. Shortly after high school, Susan began her career alongside her father on a drilling rig. Upon graduating from the University of Louisiana with a BS in Mechanical Engineering, she obtained her MBA and began working for a firm performing engineering and process safety reviews on refineries and other land facilities in the New Orleans area. After Susan relocated to Lafayette, LA, she began engineering offshore facilities, complimented with process hazard analysis facilitation. During her almost 20-year career, she has worked upstream, midstream, downstream, and all points in between. She is a licensed professional engineer and an executive member of various organizations.

Abstract

A PHA meeting consumes large resources and can place a strain on departmental manpower. Teams are often faced with the want to be more efficient in PHA conflicting with the need to be thorough and accurate. The key to solving this paradox is efficient preparation. By examining the lessons learned from past PHA's, a role by role analysis of best practices with regards to preparation will allow a PHA team to reduce the costly time spent in the meeting and yet still stay true to the methods and deliver a great, safe result.

When it comes to an optimal PHA, preparation is everything. For example, in a 6-member team PHA, 60 minutes of misused time equates to 6 hours of misused manhours. Whether it is managerial, operations, engineering, or even facilitation, there are lessons learned from previous PHA's that when shared, will not only make the PHA more efficient and the facility safer, but the industry as whole better. The goal being that those reading this be able to take back to their facility these practical preparation tips to share with their present and future PHA team members so that the resources of the facility can be better spent.

1. PHA Preparation Optimized

To fully optimize the Process Hazard Analysis (PHA) preparation, the team needs a clear vision of the end product: "To identify, evaluate, and control the hazards" [1, 2] of an operation or process to minimize the likelihood of occurrence and the severity of consequences of release of a dangerous substance [3]. With that in mind, the team can focus time and energy preparing for the PHA on factors within their control to meet that goal. An efficient, cost-effective PHA begins with including the different roles in the preparation. The PHA planning team should be comprised of management, engineering, operations, the scribe, and the facilitator.

An active commitment to reduce risk begins by assigning tasks to each role prior to the PHA. From the start, management should provide support and direction for the PHA team. They set the tone for obligation and importance. Second, engineering involvement in the preparation ensures accurate and complete process safety information (PSI) is available and accessible. Without correct information, the hazards may not be fully identified. Next, operations should walkdown the facility and identify areas of concern. They have the most intimate knowledge of the workings of the process, which is extremely valuable to the PHA when properly communicated to the team for hazard recognition. The scribe should have the technical skills to ensure equipment is functional and to properly and efficiently document the team's findings. Lastly, the facilitator needs to coach the team before and during the PHA to ensure the methodology is followed.

Successful PHA preparation can improve risk identification and limit cost through appropriate management direction, current and detailed documentation, and clear expectations of the timeline leading up to the PHA. We identify ways that companies can assist in preparation in the months leading up to a PHA and the benefits that preparation has on hazard recognition, team morale, and, most importantly, the bottom line.

2. Management – "*Set the Tone*"

It is important for management to set the tone of the PHA. More specifically, management needs to set the tone before, during, and after. In a PHA, the attitude, the posture, and the dedication of the team is usually a direct reflection of management's attitude, posture, and dedication. In our experience of both facilitating and participating in PHAs, we've seen attitudes of management in a variety of flavors. The following are observations we've made in the past regarding a successful PHA and one not as successful.

The corporate culture and support infrastructure set by management begins well before the first day of the PHA. A typical PHA is done every 5 years [1]. Given that schedule, it is imperative that management starts out on the right foot and begin planning well in advance. A management component that is very serious and very organized will typically induce an attitude of seriousness and organization to the team members. For example, a management team that waits until the last moment and rushes preparation will typically have PHA members that delay and do little preparation, if any at all, for the PHA. PHAs are very expensive and very time consuming. Lack of adequate and efficient preparation will be very damaging to an efficient use of PHA budget and resources. When management in advance sets the tone that a PHA needs to be done correctly and efficiently, the PHA team members tend to take preparation seriously. Just as it is well documented that an effective safety culture of a company starts at the top with management, effective preparation for a PHA also starts at the top with management [4]. The management teams we've seen in the past that do this most effectively set the tone of preparing and doing a PHA as part of your job, not as additional work on top of your daily duties. In other words, management sets the tone that PHAs are in your job description and that we are all PHA team members. For example, a PHA is not additional work given to engineers, it is in an engineer's job description.

The other things we have observed that effective management does by setting the tone is setting up a support team for the PHA team members. These PHA team members will have to transition out of their daily duties, be present at the PHA, and transition back into their regular duties. Often, their project managers and/or functional managers may give some resistance to that. It is imperative for upper management to set the tone with the managers of the team members that they need to have a plan for them to transition into the PHA and transition out of it with regards to their regular duties. Most of these facilities run around the clock all year long and the work never stops. Therefore, it is imperative for management to have a plan for these employees to be present at the PHA while it is going on. A PHA which is constantly stopped for employees dealing with issues outside of the PHA is an unacceptable waste of resources. For example, if a PHA consists of ten team members and one team member must stop the PHA to deal with outside problems unrelated to the PHA, the nine other team members are wasting company resources. When management sets this tone early on that we are dealing with real human lives and significant company resources, the employees and their managers tend to put greater weight on a PHA. And sometimes, the employees just need the help of upper management to persuade lower and mid-level management that their presence, uninterrupted, is necessary to conduct a PHA.

During the PHA, management's role is typically only at the very beginning. It is very effective during the opening remarks of a PHA for a member of management to also make an opening remark. During this address, management will set the tone on several things. It is not uncommon for management to begin opening remarks with a thank you for the team members for their hard work in preparing and committing of their time. Some members of management stress employees' safety; others may stress the importance of safety to the community or environment. Opening remarks is a great time for management to state known worst-case consequences initiated within the limits of the PHA scope or define the assumptions regarding severity. This guides the team towards consistent company-wide risk rankings and avoids the team paranoia over risk ranking too high or too low. It's not management's position to force upon the team a subjective belief; it is management's position to reaffirm the team uses best judgement and not be hesitant of ranking the risks as they believe they should be. Finally, management can give the PHA team members feedback from previous PHAs as to the wording, substance, and ranking of the recommendations to assist with review and completion post-PHA.

Once the PHA begins, management's role is very limited. We've seen effective managers from time to time pop-in, review recommendations, and discuss with team members their thoughts on such. Effective managers will pick a recommendation, look at the supporting data, and ask specific questions of how the team arrived at its conclusion. This serves the purpose of correcting any doubts that the PHA team may be having and serves the purpose of reaffirming to the PHA team what they are doing is correct. Management can also review recommendations with only the facilitator at specific progress intervals of the PHA. The facilitator can then bring suggestions back to the team without the direct watchful eye of management. This minimizes the intimidation factor but gives the team the opportunity to clarify wording or details.

We find in PHAs that the PHA team members want to be very cooperative with management. That is why it's imperative for management to not only set the expectations and tone, but also provide a supportive means and network for a high-quality PHA to be conducted. By management providing a support team and stressing the importance of a PHA during the preparation phase, the PHA team members are motivated to attend the PHA prepared with their full attention. It goes back to the old saying, garbage in, garbage out. Anyone who has sat in a PHA will attest to the opinion that PHAs are grueling and a marathon of being attentive. The PHA team members tend to be very technically sound and understand the PHA process and goal. The biggest hurdle is usually morale and minimizing outside distractions. A great opening statement by management and periodic check-ins to keep morale and encourage the team members goes a long way. Nothing lifts the moral of battle wary troops like a visit to the battlefield by the general. After the PHA is over, feedback and encouragement go a long way in instilling confidence for the PHA team members that will carry over to their next PHA. Although hardly any of this is required by the statutes, these are observations we have noticed with regards to effective management participation in PHAs.

3. Engineering – "The Devil is in the Details"

As an engineer, it is understood that your job is to ask the questions that have not been answered, review the material that has not been presented, and observe that which is not obvious. PHAs are no different. One of the duties of the engineer during a PHA is to gather all the process safety information (PSI). However, we've observed the more effective engineers will take this list, check how current it is, and review the design pressures and temperatures, such as pump discharge pressures. This review helps determine if the PSI is accurate. It also serves the dual purpose of observing any abnormalities that will need to be discussed during the PHA. Looking at this before a PHA is very effective when the engineer's eyes are fresh, and his mind is clear. Once the PHA is ongoing, this task becomes a little more complicated with lower morale and PHA fatigue. This holds true to the majority of the engineer's tasks.

Drawings, drawings, drawings. One of the largest time and money wasters we observe in PHAs is showing up to a PHA with either outdated drawings or needing drawings that are missing. It is imperative for the effective engineer to not only review the drawings to make sure they are up to date and accurately reflect what is present in the field, but it is equally important to review which drawings have been gathered and submitted to ensure that there are no surprises during the PHA. Anyone who has sat in a PHA and heard the words "these are not the most up to date drawings" or "where are all the other drawings at", will attest that it is going to be a long day. The PHA facilitator, the scribe, and pretty much everyone else in the room relies on the drawings being both accurate and complete in their preparation. If the drawings submitted for the PHA are either outdated, incomplete, or just wrong; it will have a ripple effect for the rest of the team both before and during the PHA. If the reader of this paper takes nothing away but one thing, please note that accurate and complete drawings are imperative for an effective PHA team.

PHA team members include those who are present in the room full time and other team members who may be called upon as needed. Effective engineers anticipate ahead of time what possible other resources will be needed during the PHA. For example, engineers for adjacent facilities need to be made aware of an upcoming PHA and given a heads-up as to be ready to answer questions regarding upset conditions and their effects. Likewise, engineers for skid packages or other equipment requiring expertise of vendor engineers need to also be made aware of an upcoming PHA and request their availability if called upon. Similarly, engineers of other disciplines may have better knowledge of failure mechanisms or a specific design review already conducted which is beneficial to the PHA team. It is not uncommon in a PHA to have a question that needs to be answered by a vendor engineer, subject matter experts, or an adjacent facility engineer. Examples of information needed include possible flow rates or pressure to/from receiving/feeding units, adequacy of safe guards in packages provided by vendors, or electrical area classifications. Oftentimes, a simple phone call to these engineers asking if they would be available by phone during the PHA is enough. Other times, effective engineers will require that other personnel be available to come in as requested. PHAs involve a lot of moving parts with each part having its own calendar or schedule. Giving the other possibly necessary personnel a heads up will ensure that should their presence become necessary, the PHA team will not have to recess until they are available.

Effective engineers will typically review MOCs and make phone calls or e-mails to update their current status. The adage that change is constant is very applicable to MOCs. MOCs may have been implemented, postponed, in progress, or completed. It is important to update the status of an MOC. One that has been canceled may require different attention than an MOC that is in progress or completed.

Effective engineers provide the technical expertise of the specific process under review [1]. Before a PHA, effective engineers will take a closer look to evaluate operating trends and upset conditions looking for a pattern that is not readably apparent. Engineers should discuss the facility and how it runs day-to-day with operations prior to the PHA. They should review procedures to ensure they are current and understood. It can be an awkward moment during a PHA when the engineer says, "we've never had a problem with that", and the operator rebuts with "we have a problem with that every day". Engineers know that engineered controls are the strongest defense against hazards. They look for ways to solve problems without relying on administrative controls.

Finally, effective engineers are present during a PHA. This does not mean they are sitting in a chair, it means they are focused, engaged, and completely dedicated to the PHA while they are in that room. Engineering is a life of dealing with problems. To be an effective PHA team member, it is important to make sure the engineer has support to deal with problems while he/she is in the PHA. Effective engineers can have a clear mind because they know that if normal, everyday problems occur, someone else is taking care of them. An engineer that is constantly worrying about what great surprises are awaiting them when they check their email at the end of the day, are not present during a PHA.

4. Operations – "The Ambassador of the field"

The operator is the ambassador of the conditions in the field. The operator, not engineering, management, or the facilitator, is the eyes and ears of the field. As an ambassador, it is important that the operator speak to his constituents (i.e. opposite crews), learn how to effectively communicate, and be present for the meeting of the PHA team. Effective operators speak to their coworkers and get a history of problems, concerns, or other issues they may be having that might not be well documented. Effective operators also walk around and observe conditions of the facility and document such. Documentation being key. Effective operators also review previous incidents not only in their facility, but other similar facilities or neighboring facilities to bring up any concerns that may not be readily apparent when reviewing documentation during a PHA. A great ambassador becomes the eyes and ears of his constituents by listening to what the other operators have to say and physically observing the condition of the facility.

Other things effective operators will do is review operating procedures and note deficiencies. Sometimes operating procedures are written in a way that no one understands them making them ineffective as a safeguard. Other times, there is no written procedure, forcing operations to react to conditions "the way we always do it." If operators are not following the procedures because they are vague, ambiguous, confusing, or out of date, it is up to the operator during the PHA to bring that to the attention of the team members. Frankly, effective operators understand that management and engineers assume that everything they write is perfect, clear, and easy to follow. It is up to the operator to explain why that may not always be the case. Some of the most effective operators we have encountered in a PHA not only describe what the problems are but are very specific about why these are problems.

The next characteristic of the operator as an ambassador for the field conditions is communication. For an operator to communicate with a PHA team, they must speak the same language. Effective operators know that to communicate with a PHA team, they must become a master at reading P&IDs. Being able to understand how the elements of a facility and its attributes are represented on a P&ID are extremely important for an operator. Effective operators can look at a P&ID with little or no guidance and interpret it to see if it is an accurate reflection of what is going on in the field. When the engineer asks a question regarding the P&ID, effective operators are on the same page. If the reader wanted to take away one important thing from this section on operators, please take away that an operator being a master of a P&ID is extremely important to be an effective operator in a PHA.

5. Scribe – "The Show Must Go On"

In American football, a great offensive lineman only receives attention when things do not go as planned. The same can be said about a scribe in a PHA. An effective scribe will give the outward appearance of very little responsibility or having very few complicated tasks. Nothing can be further from the truth. With the goal of quickly and accurately recording information [4], effective scribes understand that a lot of thought and preparation must go into making sure that nothing goes wrong, and if something does go wrong, the PHA suffers minimal interruption. Also, effective scribes are great at being very efficient in the tasks they complete.

At the end of the day, a scribe's responsibility is to ensure that the audio/video equipment works as planned. Although some facilities may be tasked with providing projectors and so forth, the scribe cannot scribe unless all this equipment works. Therefore, an effective scribe will oversee the equipment being used to make sure the parties who are responsible for providing it are communicating and that all the items are compatible. The scribe will ensure that both the scribe's computer and the display have the proper connectors and cables needed. An effective scribe not only brings the connectors and cables needed, but also bring the connectors and cables that were not anticipated as being necessary. An example is the conference room being used for the PHA has HDMI and ethernet and Wi-Fi…or at least that what was told to the PHA facilitator. In the past, we have shown up to conference rooms to find that the HDMI port is in fact a VGA port, or that the HDMI cable is three feet long and the scribe's position is ten feet away. Very effective scribes keep with them a toolbox of supplies such as VGA to HDMI converters, extra batteries for mice and Bluetooth keyboards, and extra cables. Sometimes, the right connectors and cables are present but are malfunctioning. The truly great, effective scribes seamlessly work through these problems and the show continues without interruption. Along those lines, a great practice for scribes to have is to not only back up the PHA file regularly, but also have a written procedure that has been reviewed regarding the backup. We always joke that we PHA the PHA. In other words, what are the risks for file corruption or data loss. Which risk are we willing to accept, and which risk are we going to mitigate? Knock on wood, so far, it's been very effective.

A lesser problem, but a problem none the less, is having to wait on a scribe to catch up. Although not as catastrophic as the PHA having to be stopped completely while technical issues are worked out, a scribe who is not able to keep up with the PHA team stifles the free flow of thought and can leave the door open for other interruptions such as taking phone calls and checking email. Very effective scribes will have experience and practice with the software, know the shortcut keys, and have a vocabulary list prepared for technical terms and names. A typical vocabulary list will be something as simple as a word document open where the scribe has already checked the spelling of names and technical equipment and as the scribe goes along, they can copy and paste the names and equipment labels very quickly without having to correct for spelling. A truly exceptional scribe will have already created templates with standard verbiage for safeguards, recommendations, and comments. For example, when the team decides to make a recommendation, the scribe can copy and paste it into the software and the team can edit the few words that need to be changed to modify the recommendation for the specific application. What this does is it allows the team to make a recommendation and move on to the next node without getting bogged down in the weeds of language choice. We find it is always easier to give a group of people a starting point for a recommendation as opposed to generating the language of a recommendation by consensus of the group. Along those lines, effective scribes review the drawings, MOCs, and other information to become familiar with the vocabulary in the drawings and have an inventory of what PSI they have and what node it corresponds to. Therefore, if a team member erroneously refers to a drawing that does not exist or refers to a drawing or MOC by incorrect name, the scribe will be quick to point out that the item does not exist in the inventory.

Effective scribes will also begin opening vital documents for a specific node as soon as the review of that node begins. Therefore, when the team would like to reference a specific document such as a drawing, MOC, or incident report, the document will be readily available. A common, effective means of doing this is having a second monitor only seen by the scribe, in addition to the screen shared with the projector. The scribe can work on his/her monitor opening documents or correcting language. As the team is ready to view specific information, he/she can drag the window over to the extended screen that is being displayed overhead. This allows the scribe to work without being micromanaged by the team and minimizes the distraction of the team from the scribe flipping through windows or searching for files. Again, the overall theme for an effective scribe is to keep the free flow of thoughts and the team moving. Anytime the team must pause or take a break because of the scribe performing its duties, the team's momentum and productivity can be lost.

Finally, truly effective scribes are great at covering the intangibles. This can include filling all the supplies or doing things to the environment in the room to lift morale. For example, effective scribes will typically have a box of random office supplies, whether it be pens, hi-liters, or post-it notes. At the beginning of each day and after each break, he/she should refill the table or the workspace of the PHA team with extras. So, if a team member would require a specific color pen, they would simply grab it and not have to pause the PHA. This gets back to the overall theme that an effective scribe keeps the evaluation moving forward without interruption. Another intangible that we often see effective scribes will do is they will place motivational posters around the room, such as large check-off list of nodes to visually see progress, to hang inspirational quotes, and so forth. PHAs are really an endurance test of morale and motivation and these little intangibles that the scribes do are of very limited cost, but of great benefit. In all honesty, we often do not realize how great and effective our scribe is until we do a PHA with an ineffective scribe. When you do a PHA with an ineffective scribe, you feel like you're stuck in the mud and going nowhere fast. Although realistically, the delay may add just a few minutes a day, with regards to morale and motivation, it feels like an eternity.

6. Facilitator – "What's Your Plan, Coach?"

Like a good coach, an effective facilitator enables great people to do great things. Effective facilitators setup the team members to be successful and, during the PHA, guide them to produce a high-quality product. Although a facilitator's duty is to ensure the methods are adhered to and the regulations concerning a PHA are complied with [1], a great PHA facilitator is so much more.

A great PHA facilitator, like a coach, ensures that the team is prepared come game day. Effective PHA facilitators will do such things as set deadlines for collection and development of information and suggest further actions to better prepare for game day. Not only do they set the deadlines, an effective facilitator will check in with the team members prior to those deadlines to ensure that progress is proceeding as planned and that the deadline will likely be met. Effective facilitators are great at helping team members overcome obstacles and hurdles in collecting this information. A team being well prepared on game day when they take their chairs and listen to management's opening remarks is no accident. Also, during the preparation, an effective PHA facilitator will ask the hard questions of the team members, such as: Do you have a transition plan for someone to fulfill your role during the PHA? Or, are there any allergies, physical, or dietary restrictions a team member may have so that these can be addressed well ahead of the PHA? To be an effective PHA facilitator, it really comes down to having the team prepare for the work they anticipate doing and addressing as many possible obstacles as one can prior to the PHA commencing. The litmus test for a PHA facilitator is simple - once the PHA commences, no surprises.

In addition to ensuring the rest of the team is preparing for the PHA, the effective facilitator puts in just as much effort as the rest of the team into preparing the PHA file, determining the node segregation, and sharpening his skills as a technical guide. On the day of the PHA, the PHA file should be prepopulated with node descriptions, drawings, common safeguards, etc. The facilitator should have reviewed the node list and noded drawings with other member(s) of the team to ensure the segregations make sense and all parts of the process are included. The effective facilitator should also review recent changes to regulations or process conditions that might affect the hazard identification. He should gain a basic understanding of the process under review along with the potential hazards specific to the facility. Effective facilitators have a list of questions to pose to the team during the appropriate node to ensure all hazards have been addressed.

Again, following the theme of the PHA facilitator being a coach, once the PHA commences, it is game day. It is the PHA facilitator's job to adhere to the methods and regulations regarding PHA facilitation. However, great, effective PHA facilitators understand that the true challenge in a PHA is morale, motivation, and focus. Effective PHA facilitators will typically start off the PHA with a "rules of the road" presentation. They do not assume that everyone in a room regularly participates in PHAs and is familiar with the process. They review the PHA process, the end goal of risk mitigation, the timeline, the daily schedule, acceptable and unacceptable behaviors, and expectations of participation. An effective PHA facilitator will go over these rules of the road and get each team member to commit to their role, to fully participate, and to strive towards the goal. It creates a sense of responsibility and obligation among the members. Once the PHA commences, the effective PHA facilitator will have a sense of when progress is slowing down and when to step in to move the team along. The effective PHA facilitator knows that the team must be given opportunity to discuss causes and consequences, but also has goals and objectives that need to be met within the time allotted.

As the team leader, the facilitator should plan on how to react to differences among team members. The great PHA facilitators have a way of getting conflicts resolved without upsetting or demeaning anyone in the room. Typically, in an argument, somebody must be wrong and somebody must be right. However, the great PHA facilitators have a way of allowing both parties to express their beliefs, guiding them to understand the beliefs of others, and concluding with a common solution. A great way to kill the free flow of ideas in a PHA and stifle productivity is to tell someone, "you're wrong, please be quiet". Great PHA facilitators know how to do that in a way that allows the person to maintain dignity and continue to perform in a PHA.

Another attribute of great PHA facilitators is that they are systematic and thorough. They have a plan of attack for each node. This usually includes the node description and limits review, a review of the appropriate drawings, written list of deviations to encourage the team, and review of MOCs, previous recommendations, and incidents related to the node. The effective facilitator also has a plan in place to limit idle time when critical information necessary to properly review a node is found to be missing. The action is documented as a follow-up for a specific team member to complete prior to the next day allowing the PHA to move forward. Any outstanding to-dos are documented and presented to the team members at the end of each day prior to dismissal. The PHA facilitator is responsible for meticulously documenting the team's findings, which cannot be done without adequate preparation of a game plan.

Finally, great PHA facilitators understand they have an obligation to future PHA facilitator to send their team off with a sense of accomplishment and pride in PHA participation. If the team members of the PHA leave with a bad taste in their mouth or feel discouraged, their participation in the next PHA may not be to the level it could be. Great PHA facilitators compliment their team members and make sure they leave the PHA on a positive note. Therefore, when the team members approach their next PHA, the new facilitator will not be dealing with baggage leftover from the previous PHA. It is a courtesy that great facilitators give to and expect from other facilitators. PHA facilitators are often asked by team members during a PHA, "how can you do this every day?" The response we often use is, "what can we do better to make you volunteer for the next PHA?" Soliciting feedback from team members besides a multiple-choice survey is imperative to be an effective facilitator. The feedback needs to be of substance that can be acted upon and improved upon. We truly love doing PHAs and at the end of every PHA, we express that to the team and solicit their feedback to what we can do better to make them truly love doing it too.

7. Summary

In conclusion, when it comes to an optimal PHA, preparation is everything. Each team member plays an important and specific role. Management sets the tone with their level of commitment to preparation, resources, and action. Engineering ensures the quality of information available to the team and contributes a deep understanding of the process design. Operations represents the field conditions, the procedures, and the most intimate knowledge of the process. The scribe provides accurate documentation through technical skill, proficiency with hardware and software, and team morale support. The facilitator leads the team to identify, evaluate, and control the hazards utilizing a systematic approach to preparation. It has been proven that a small amount of time spent preparing for each role will result in improved hazard identification, higher team morale, and reduced PHA cost.

8. References

1. Occupational Safety and Health Standards, Toxic and Hazardous Substances, Hazard Communication, Available at https://www.osha.gov/laws-regs/regulations/standardnumber/1910/1910.1200 Accessed on December 12, 2018.
2. Bureau of Safety and Environmental Enforcement, Department of the Interior, Offshore, Oil and Gas and Sulphur Operations in the Outer Continental Shelf, Available at https://www.ecfr.gov/cgi-bin/text-idx?tpl=/ecfrbrowse/Title30/30cfr250_main_02.tpl Accessed on December 12, 2018.
3. American Petroleum Institute Recommended Practice 750: Management of Process Hazards, First Edition, American Petroleum Institute, Washington D.C., 1995.
4. CCPS. *Guidelines for Hazard Evaluation Procedures*. Center for Chemical Process Safety, American Institute of Chemical Engineers, New York, NY, 2008.

551204 Building Process Safety Intelligence: What Should Every Graduating Chemical Engineer Know?

Process safety intelligence is derived from derived from a vast corpus of sources - including textbooks, handbooks, journals, conferences, incident reports, safety datasheets, and available materials and thermophysical property data. Dedicated decision-support tools enabling practitioners to harness this intelligence in their daily work, such as Knovel are an important component of chemical engineering practice today. However, many students don't realize the breadth and depth of technical knowledge needed to support process design and safety – they tend to overrely on non-specialist search, and miss pedigreed information available on specialized knowledge platforms, such as Knovel.

The undergraduate engineering space offers an excellent opportunity to plant the foundational information seeking behaviors in future process safety professionals and leaders. Knovel focuses on applied a preview of the types of insight they will be forced to leverage in industry careers, without a direct cost to students and faculty.

Common Process Safety Management Gaps Based on Occupational Safety and Health Administration Process Safety Management of Highly Hazardous Chemicals Violation Trends

Janet Persechino
Engineering Planning and Management, Inc.
959 Concord Street
Framingham, MA 01701
jap@epm-inc.com

Ryan Welch
Engineering Planning and Management, Inc.
959 Concord Street
Framingham, MA 01701
rdw@epm-inc.com

Prepared for Presentation at
American Institute of Chemical Engineers
2019 Spring Meeting and 15th Global Congress on Process Safety
New Orleans, LA
March 31-April 3, 2019

Janet Persechino
Engineering Planning and Management, Inc.
959 Concord Street
Framingham, MA 01701
jap@epm-inc.com
Keywords: OSHA, violations, regulations, process safety

Abstract

On July 17, 1990 the U.S. Occupational Safety and Health Administration (OSHA) published in the Federal Register the proposed standard *Process Safety Management (PSM) of Highly Hazardous Chemicals (HHCs)*. Approximately four months later it was enacted into law via Clean Air Act Amendments (CAAA). These amendments require that the U.S. Secretary of Labor, in coordination with the U.S. Administrator of the Environmental Protection Agency, promulgate, pursuant to the Occupational Safety and Health Act of 1970, a chemical process safety standard to prevent accidental releases of chemicals that could pose a threat to employees [1]. The full text of the final standard was issued in the Federal Register on February 24, 1992.

The OSHA standard is listed under Title 29 (Department of Labor) of the Code of Federal Regulations (CFR) as 1910.119. This standard leans towards a performance-based guidance document that outlines methods and programs to decrease the likelihood of flammable, toxic, reactive and explosive chemical releases. The intent is to reduce the risk to employees, visitors and contractors that may enter or work at chemical processing facilities.

Facilities are subject to the PSM regulation if they hold a threshold quantity of any HHC that is listed in Appendix A of the standard, with some exceptions. The violations that OSHA issues for the 1910.119 standard will cite the specific element and its relevant sub-section. Published data allows for review of the violations, and based on that information an indication of common PSM program gaps can be assessed.

A ten-year span of OSHA PSM violation data was examined (2007-2017). The data shows the top four OSHA cited PSM sections (often referred to as elements) are (highest to lowest):

1) Process Safety Information
2) Mechanical Integrity
3) Operating Procedures
4) Process Hazard Analysis

Each of these elements have known solutions, and specific corrective actions can be easily implemented given the lessons learned, the current level of expertise that exists in the process safety industry and the cost-effective software tools that are now available.

Regulatory compliance is not the only goal of adequate PSM programs, nor is it the goal of this review. The purpose of reviewing the OSHA violation information is not meant to help facilities avoid OSHA citations, but to uncover gaps found by OSHA and bring awareness and attention to some common issues, hopefully leading to focused solutions for safer, more reliable facilities.

1.0 Introduction

There are approximately 24,000 sites that are subject to the OSHA PSM standard nationwide [2]. Facilities are subject to the standard if they process any of the 130+ chemicals listed in Appendix A (of the standard) in excess of the associated threshold quantities, also listed in Appendix A, on-site in one location. They are also subject to the standard if they process flammable liquids and/or gases in quantities equal to or greater than 10,000 pounds. OSHA defines "on site in one location" as areas under the control of an employer, or affiliated employers, that have control over any group of vessels that are interconnected, or in separate vessels that are located in close enough proximity that a HHC could be involved in a catastrophic release.

Exceptions to the applicability include [3]:

- ✓ Retail facilities
- ✓ Oil/gas well drilling or servicing operations
- ✓ Normally unoccupied remote facilities; the standard defines "remote" as a facility that is operated, maintained or serviced by employees who visit the facility periodically to check its operation, no permanent employees are stationed at the facility, and is geographically remote from all other buildings and persons.
- ✓ Hydrocarbon fuel used solely for workplace consumption as a fuel if such fuels are not part of a process containing another HHC covered by the standard
- ✓ Flammable liquids with a flashpoint below 100°F stored in atmospheric storage tanks or transferred, which are kept below their normal boiling point without benefit of chilling or refrigeration

The above information outlines specific threshold quantities that trigger, and exclude, the PSM regulatory requirement. It is strongly recommended that some form of PSM be practiced at plants that may not be subject to PSM, especially those that fall into the following categories:

- Plants that store quantities of HHCs close to the threshold quantities listed in Appendix A of the standard
- Plants that utilize hazardous chemicals not listed in Appendix A of the standard
- Processes exempt from the standard, for example, those listed as exceptions in the standard, plants that have runaway reaction potential, etc.

An overview of "PSM Lite", and related fire code requirements, can be found in these conference proceedings in the paper titled *Fire Code Chemical Process Safety Requirements*. Implementation of an abbreviated form of PSM should be evaluated on a case-by-case basis which will take into account various factors that would clarify the actual risk and the residual risk that would remain after programs are incorporated into the operations.

2.0 Violation and Corrective Action Evaluation

The PSM elements are listed in the 1910.119 standard in the following order/sections:

(c) Employee Participation
(d) Process Safety Information
(e) Process Hazard Analysis
(f) Operating Procedures
(g) Training
(h) Contractors
(i) Pre-startup Safety Review
(j) Mechanical Integrity
(k) Hot Work Permit
(l) Management of Change
(m) Incident Investigation
(n) Emergency Planning and Response
(o) Compliance Audits
(p) Trade Secrets

OSHA has been administering violations for the 1910.119 standard since 1992. For purposes of relevancy, the last 10 years of violation data was examined (2007-2017). These violations were issued when OSHA completed an inspection and/or accident investigation visit to a particular facility.

Each one of the elements listed above has subsections pertaining to the requirements for that element. For example, when OSHA administers a PSM violation during a visit it is listed as: 1910119 E03 I. The nomenclature correlates to a violation for section (e) Process Hazard Analysis (PHA), part 3(i), or (3) the process hazard analysis shall address (i) the hazards of the process. This particular example indicates that the PHA did not properly address any or all of the hazards of the process. The OSHA inspection information also includes other data such as penalty dollar amounts, dates and type of inspection. The information provided in this summary reviews a few segments of this data from years 2007-2017 [4].

3.0 Violation Data Interpretation

All OSHA PSM violations were tallied over the 10-year period (2007-2017), this included:

- ✓ All citation classifications types (serious, repeat, willful etc.)
- ✓ All types of inspections/visits (accident, planned etc.)
- ✓ All types of facilities

This review does not distinguish whether violations were given during the same visit or during separate visits.

The data shows that there are 4 elements, or 4 deficiencies, that clearly stand out as the most prevalent (see Figure 1):

Most Commonly Cited PSM Elements (highest to lowest):

1) Process Safety Information (PSI)
2) Mechanical Integrity (MI)
3) Operating Procedures (OP)
4) Process Hazard Analysis (PHA)

Total Monetary Amount (highest to lowest):

1) Mechanical Integrity
2) Process Safety Information
3) Operating Procedures
4) Process Hazard Analysis

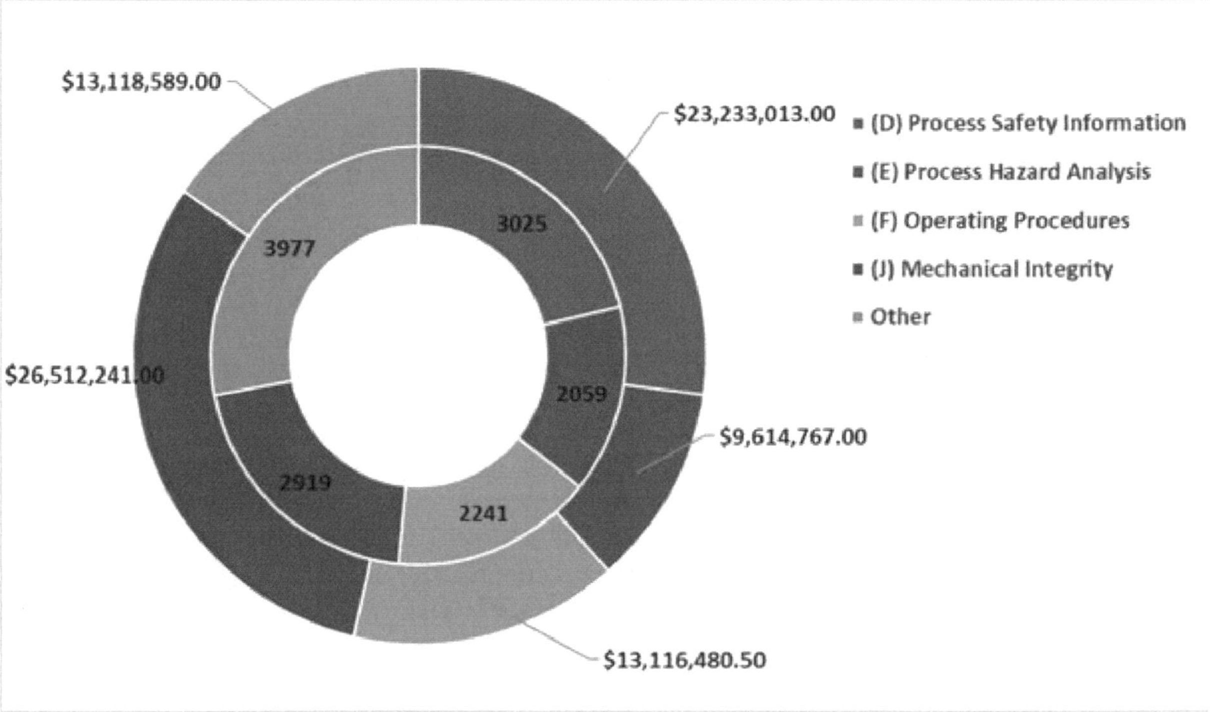

Figure 1: Total Number of OSHA PSM Violations and Total Citation Dollar Amounts by Element: 2007-2017

The top two elements (PSI and MI) have close to the same number of violations and the same goes for the third and fourth most common violations, giving the impression that the struggle to execute PSI and MI is about equal and OP and PHA both come in as close second.

It is alarming that PSI is the most cited element. Having correct and adequate PSI is the foundation for implementing, not only all PSM elements correctly, but any and all chemical safety precautions. PSI is crucial in completing an adequate PHA, the analysis that examines the hazards of the process and ensures that the proper safeguards are in place. The lack of PSI could lead to process hazards being missed, or incorrectly evaluated, during a PHA. Incomplete PSI can also lead to having inappropriate MI programs since some of these programs can be based on the hazards of the process, for example, more frequent inspections than the manufacturer's recommended practices based on the characteristics of the material being processed.

MI is the second most cited element but has the largest monetary amount, this could point to more instances of MI being noted within a single violation. OSHA will increase a fine for a single violation based on the number of "instances", or how many times they see a violation for the same element during the same visit. This review did not take into account the number of instances per violation.

MI can be difficult to implement, but new software solutions and training programs exist to help with proper execution. There are MI software products on the market that include capabilities to capture PSI, PHA and OPs, addressing all of the top-cited elements within one program.

It is helpful to utilize PSM experts for implementation and execution of PSM programs, as they are aware of the depth of information that is needed to meet the intent of the standard and for overall safety. They can often recommend products depending on your particular operation. Individuals dedicated to PSM will have access to resources that might not be available to others who are not embedded in the PSM industry. For instance, access to fee-based databases with previous incident data, one item that must be documented as part of a PHA. The expenditure for bringing in this type of expertise can provide a beneficial return on the investment when examining the average cost per violation and average number of violations per visit (Figures 2 and 3).

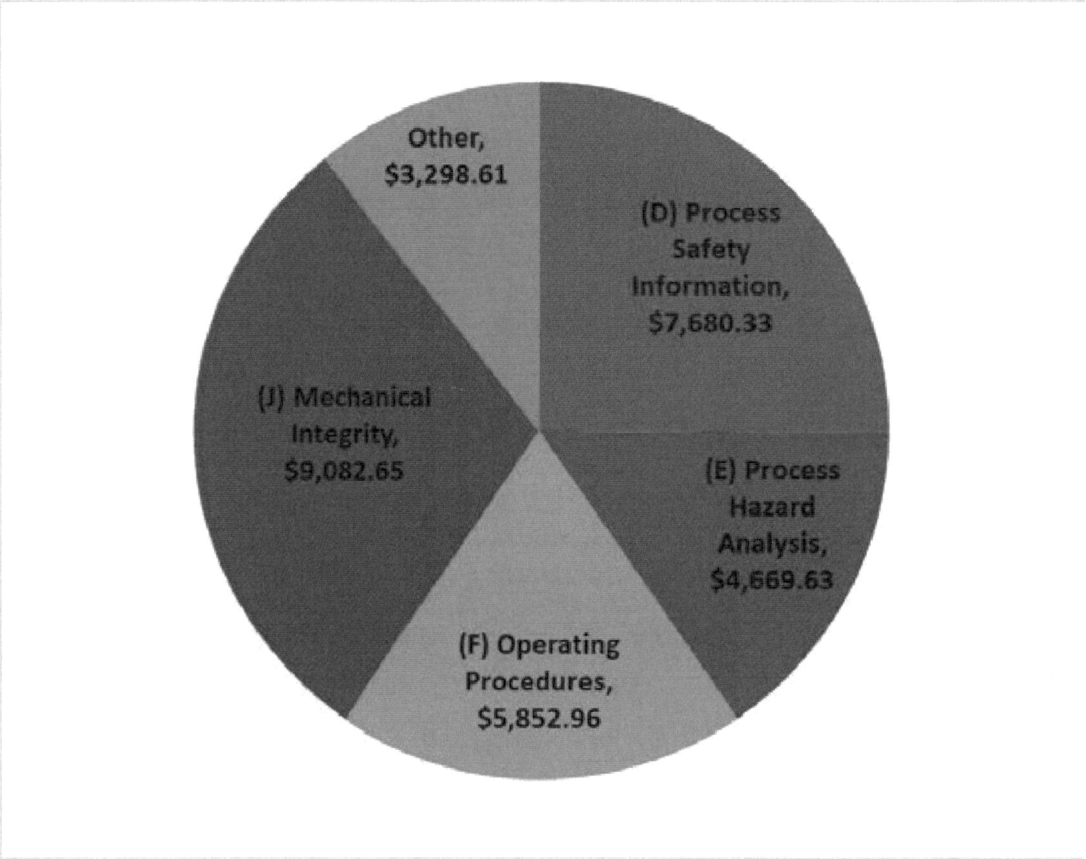

Figure 2: Average Cost per Violation by Element 2007-2017

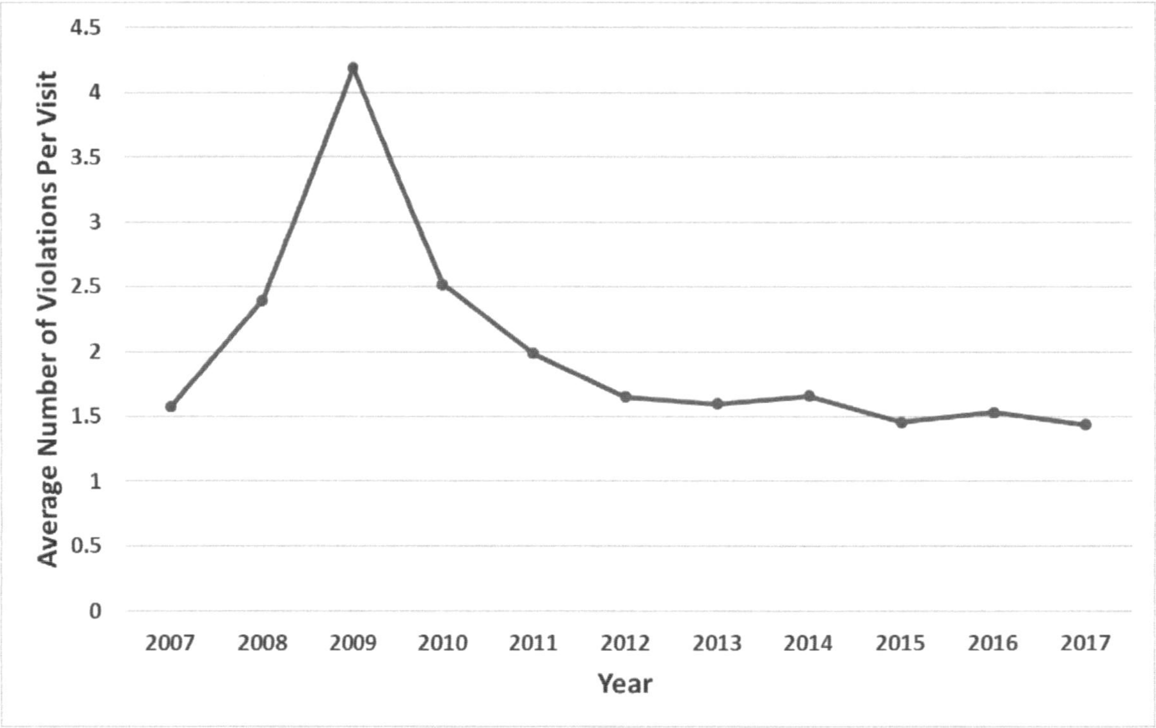

Figure 3: Average Number of Violations per Visit 2007-2017

(To note about Figure 3: the spike in the number of violations in 2009 could be due to the National Emphasis Program (NEP) in which OSHA extended to include a pilot on PSM-covered facilities from the 2007 NEP which only included refineries. The NEP went on from 2011 to only include chemical facilities and not petroleum refineries)

The citation costs are only one aspect to consider when reviewing the return on the investment. Other benefits that can be quickly realized once the programs have been established include but may not be limited to:

- ✓ Improvement in operating efficiencies
- ✓ Increased employee morale
- ✓ Lower turnover rates
- ✓ Safer working conditions
- ✓ Uptime gains
- ✓ Insurance benefits
- ✓ More consistent product quality
- ✓ Decreased likelihood of an incident

A deeper dive into the subsections of the most cited elements yielded the results in Figure 4.

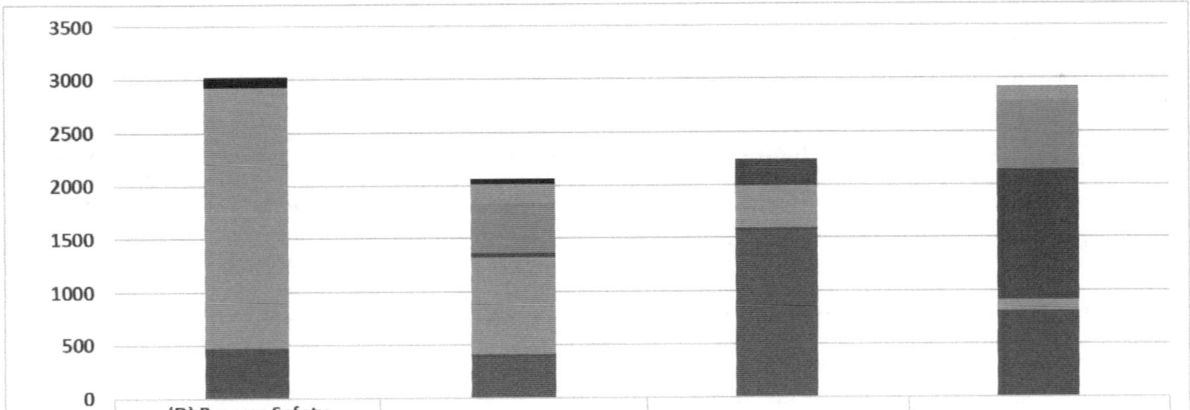

	(D) Process Safety Information	(E) Process Hazard Analysis	(F) Operating Procedures	(J) Mechanical Integrity
■ Blank	101			
■ 07		48		
■ 06		186		135
■ 05		463		640
■ 04		37	255	1235
■ 03	2452	918	391	105
■ 02	418	26	23	798
■ 01	54	381	1572	6

Figure 4: Sub-section Cited in the Top 4 OSHA Most Cited PSM Elements 2007-2017

Figure 4 shows that the biggest gaps over the last 10 years include the following sub-sections of these elements:

1. Process Safety Information (d)
    - Equipment Information (section 3)
2. Mechanical Integrity (j)
    - Written Maintenance Procedures (section 4)
    - Inspection and Testing (section 2)
    - Correcting Equipment Deficiencies (section 5)
3. Process Hazard Analysis (e)
    - Inadequate coverage of the hazards of the process (section 3)
4. Operating Procedures (f)
    - Develop written operating procedures (section 1)

There is a common theme among these four (4) subsection violations. While PSI was the most cited element, the most cited PSI sub-section is Equipment Information, which ties into the MI portion of PSM. Based on this, deficiencies in equipment related information and MI ranks very high on the list. MI related programs help to ensure that chemicals remain inside the vessels, pipes, etc. A review of the OSHA data shows that the industry may be struggling with proper execution of this critical piece of process safety.

As noted above, MI programs are complex and can be difficult to establish, but there are current solutions on the market to help correct these deficiencies, such as economical software solutions that improve compliance by organizing maintenance tasks. The right solution can alert staff to overdue maintenance, and store the maintenance history for each particular piece of equipment. Some tools can also track inventory of spare parts and link Management of Change documents related to the process and/or the equipment. Consultants versed in the regulations and MI techniques can help establish appropriate programs and find solutions useful to your specific operation.

Development of operating procedures for both process and maintenance activities has a high number of citations in the MI and OP elements. It is essential that these procedures are comprehensive, accessible, updated and easy to interpret. Mistakes in operating steps, or failure to recognize abnormal conditions, can quickly lead to catastrophic releases.

There are various writing and auditing techniques that can be used to ensure that these procedures include the correct information and address all stages of the process (start-up, normal conditions, abnormal conditions and shut down). Software tools currently on the market can also assist in organizing this critical information. This approach, while time consuming up-front, is time well spent in providing a safer and more efficient workplace.

Once authored, it is easier to audit the documents for accuracy and audit the operators using a behavior-based system to determine that procedures are being followed correctly.

Vast benefits can be gained using this approach, and many PSM consultants offer services to help author these documents and provide suggestions for document control and review processes.

4.0 Resources

Despite the intent for OSHA PSM to decrease the likelihood and severity of accidents, the OSHA data is showing that PSM gaps still exist in the most fundamental PSM elements. These gaps present an unacceptable risk. This is evidenced by the accidents that continue to occur. The various incident investigations performed by the U.S. Chemical Safety and Hazard Investigation Board often attribute inadequate PSM programs as the root cause.

There are many resources available for assistance with PSM implementation and improvement for plants of all types and sizes. There has been an increase in consulting organizations providing guidance and software solutions for PSM. The list of resources readily available to even the smallest of plants are much more prevalent than they were 10 years ago.

OSHA has published documents with guidelines for PSM for storage facilities and for small businesses. They also have a PSM resource page: https://www.osha.gov/SLTC/processsafetymanagement/. The Center for Chemical Process Safety has many publications, classes and tools to help with all aspects of PSM. Some of these resources are free, or provided at reasonable cost, and CCPS/AIChE (American Institute of Chemical Engineers) membership increases those benefits.

Small businesses often find it difficult to execute PSM. In 2016-2017, in almost the same number of inspections, OSHA administered 57 PSM related violations to companies with under 100 employees and only 18 violations to companies with over 100 employees. OSHA also points to a study that estimates "*employers with 1-25 employees are 47 times more likely to have a release and 17 times more likely to suffer an injury, per employee, than facilities with 1500 or more employees.*" [5]

There are a few PSM consulting firms that target small to medium companies, the companies that cannot dedicate resources to PSM, nor have the expertise in-house to meet the intent of the standard. Many of these consulting firms not only focus on OSHA compliance but becoming a partner to help embed PSM concepts into the plant's overall safety culture. Bringing in these third parties is worth the cost because they are embedded in the PSM realm and can bring along expertise from other industries, the latest information and years of experience.

There are cost effective software solutions, some aimed at small businesses from a price point and feature standpoint. These solutions can improve PSM while improving the bottom line through improved efficiency, production uptime gains, insurance benefits and violation and incident avoidance. The smaller chemical processing companies are yet to realize that these solutions exist at very reasonable prices, and the small upfront investments have very beneficial returns that easily justify the expenditure.

It is ideal to use the same consulting firm for various locations to standardize PSM solutions and audits across multiple sites that may have different risk. This type of PSM implementation helps to normalize the oversight and tracking to give a better picture of risk at the corporate level.

5.0 Conclusion

The research objective of this paper was to spot the potential for common gaps and issues in implementing PSM across the United States at all types of chemical processing industries using OSHA violation data, and to use that information to identify opportunities for solutions and preventative/corrective actions. The OSHA violation picture is only one piece of the gap analysis puzzle, but it is one place that collects enough data to increase awareness of PSM weaknesses nationwide regardless of the type of operation. This information can aid in creating education/awareness programs and campaigns aimed at improving PSM execution.

The top four cited violations are PSI, MI, PHA and OPs. Solutions to the deficiencies revealed by the OSHA data are well-known and PSM experts are aware of the many practical and feasible methods that exist to close these deficiency gaps. The return on investment on items outside of the safety aspect are usually realized soon after the solutions are put in place, including: more efficient and consistent work procedures (which can lead to improved product quality), increased up-time and insurance benefits to name only a few.

A review of the OSHA data can help quantify the costs that can be expected from OSHA PSM violations and citations. However, other impacts of violations and/or PSM incidents that are not so easily quantified include the business, legal, liability, company reputation and the impact that the injuries have on families. Sometimes plants are not able to rebuild in the same location or regain the customer share they may have lost due to an incident. Additionally, the OSHA violations are public record and can be accessed by lawyers, local authorities having jurisdiction, insurance companies, activists, etc. Companies may also find it difficult to obtain insurance due to OSHA violation or incident history. While these impacts are difficult to quantify, they must be considered in the overall cost to the business.

PSM implementation is a continuous improvement journey with new lessons learned, new technologies and new resources being created every year. Using the plethora of tools available can help close the deficiency gap.

References

[1] U.S Department of Labor OSHA 3132 Publication *Process Safety Management* 2000
[2] Washington State Department of Labor & Industries July 2010
[3] U.S Department of Labor OSHA 3133 *Publication Process Safety Management Guidelines for Compliance*, 1994
[4] OSHA Violation Data 2007-2017
[5] OSHA *Process Safety Management for Small Businesses* 2017

546725 Safety Performance Evaluation: A Benchmark of Practices across Various High-Risk Systems

The concept of performance has invaded all management spheres and imposed itself as a strong driver of decision making. Following the funding work of Kaplan and Norton (1996), performance appears as a multidimensional concept which evaluation and use require adapted tools and well calibrated decision processes.

Safety performance does not escape these requirements as it englobes various dimensions (workplace safety, process safety, environmental safety), can be approached through various evaluation paradigms (auto evaluation, external assessment through audits and KPIs) and may be used by several stakeholders (risk owner, regional and national authorities, public perceptions). Indeed, different combinations of these three dimensions allow for a large variety of performance governance paradigms carrying various understandings and practices with regard to the way performance should be assessed and used.

By comparing safety performance management in three different risk systems (process, railway and health) in Europe, this works aims at providing safety management actors with a description of the richness and variety of safety performance understandings and practices. Insights gained from this work should enhance safety managers' abilities to grasp the strengths and weaknesses of their approaches and explore the possibilities of a beneficial benchmarking.

19 AIChE Spring Meeting
15TH GLOBAL CONGRESS ON PROCESS SAFETY
NEW ORLEANS • 2019

Lessons learned from analyzing an explosion accident at Shanghai Secco chemical plant

Hanhai Dong
China University of Petroleum (East China)
No. 66, West Changjiang Road, Huangdao District, Qingdao, China
donghanhaiupc@126.com

Yi Liu
China University of Petroleum (East China)
No. 66, West Changjiang Road, Huangdao District, Qingdao, China
liuyi@upc.edu.cn

Dongfeng Zhao
China University of Petroleum (East China)
No. 66, West Changjiang Road, Huangdao District, Qingdao, China
zhaodf@vip.sina.com

Jian Chen
China University of Petroleum (East China)
No. 66, West Changjiang Road, Huangdao District, Qingdao, China
836771528@qq.com

Yanchun Wang
China University of Petroleum (East China)
No. 66, West Changjiang Road, Huangdao District, Qingdao, China
Wychun25@163.com

Wei Wang
China University of Petroleum (East China)
No. 66, West Changjiang Road, Huangdao District, Qingdao, China
15621097585@163.com

Prepared for Presentation at
American Institute of Chemical Engineers
2019 Spring Meeting and 15th Global Congress on Process Safety
New Orleans, LA
March 31 – April 3, 2019

Lessons learned from analyzing an explosion accident at Shanghai Secco chemical plant

Hanhai Dong
China University of Petroleum (East China)
No. 66, West Changjiang Road, Huangdao District, Qingdao, China
donghanhaiupc@126.com

Yi Liu
China University of Petroleum (East China)

Dongfeng Zhao
China University of Petroleum (East China)

Jian Chen
China University of Petroleum (East China)

Yanchun Wang
China University of Petroleum (East China)

Wei Wang
China University of Petroleum (East China)

Keywords: event sequence diagram; Multilayer Bow-tie Analysis; FLACS

Abstract

On May 12, 2018, six workers from Shanghai Aijinke Engineering Construction Services Co., Ltd. died on the spot of an explosion when they were maintaining a benzene tank in the tank farm at Shanghai Secco Petrochemical Co., Ltd. The accident was characterized as a disastrous accident caused by hot work in a confined space, which resulted in massive casualties and a direct economic loss of 11.66 million yuan. This paper aims to analyze the accident occurrence process in a chronological order firstly by using the event sequence diagram (ESD). Multilayer Bow-tie Analysis was adopted to explore and classify the causes of the explosion of the benzene tank, along with the corresponding consequences, and the preventive measures. The analyses show that the primary inducements were the unsafe behavior of on-site workers, hazardous conditions of in-tank substances, risky factors in the construction environment, and deficiencies in management. Casualties, fire and explosion, equipment damage, and property losses were all included as consequences. The FLACS software was used to simulate the variation of explosion pressure and explosion temperature varying with time when benzene vapor was ignited at the tank bottom during the accident, and a flow field diagram for the explosion was achieved correspondingly. The results obtained are of direct significance for investigation of similar accidents in the future and are instructive to equipment safety evaluation as well. Based on the qualitative and quantitative analyses, we emphasize three considerable reasons leading to the accident, which includes inapposite flammable gases detection points before the hot work, failing to test the flammable gases during the hot work, and deficiency of the Management of Change (MOC). Finally, the paper concludes with a summary of lessons learned from the accident as well as suggestions for accident prevention and safety improvement.

1 Introduction

The term confined space refers to closed or semi-closed facilities and places with limited entrance and exit, and poor ventilation, which may lack oxygen or contain flammable, explosive, toxic, or harmful substances. The main causes of accidents in a confined space include fire, explosion, poisoning, asphyxiation, and physical injury [1]. According to the statistics of the Ministry of Emergency Management of the People's Republic of China, from January 2012 to April 2018, a total of 64 accidents involving three or more deaths in a confined space occurred in China, resulting in 225 deaths and 82 injuries [2]. Among them, on April 8, 2014, three people were killed and two injured by an explosion during the repair and welding of the desulfurization tank at a factory in Wuhai, China. On March 18, 2015, the operators of a chemical company in Binzhou, China, suffered an explosion accident during an operation in a confined space, resulting in four deaths and two injuries. Hot work in a confined space is the common cause of the two accidents above.

Due to the poor ventilation, a large amount of flammable gas may be accumulated in a confined space. When construction workers perform hot work inside, they will face potential safety risks and problems, such as difficulty in escaping, delayed personnel rescue, and inconvenient communication. Hot work in a confined space can easily lead to fire, explosion, and personal injury accidents, which threaten not only the personal health of operators but also the property of enterprises.

Researchers have done a lot of risk analysis and exploration for the particular operating environment in a confined space. Selman et al. [3] analyzed the mortality rate of confined space operations in similarly industrialized countries from 0.05 to 0.08 per 100,000 workers and pointed out that the main risks are toxic atmospheric hazards and physical hazards. Selman et al. [4] divided the emergency rescue of confined space accidents into three levels: individual self-rescue, non-entry rescue, and progressive rescue. Botti et al. [5] applied the Internet of Things (IoT) technology and proposed a holistic approach to analyzing the procedure of confined space operations. Meanwhile, many studies adopted experimental and numerical simulation methods to observe the evolution of characteristic parameters during the diffusion and detonation of flammable gas in a confined space. The results obtained are helpful to the quantitative analysis of confined space accidents. Bang et al. [6] established a theoretical model to predict the pressure rise caused by the deflagration in a confined space. He et al. [7] presented an LES turbulence model to simulate the diffusion process of LPG in the underground confined space.

This paper mainly analyzes a disastrous accident caused by hot work in a confined space at Shanghai Secco Petrochemical Co., Ltd (Secco). On May 12, 2018, the workers from Shanghai Aijinke Engineering Construction Services Co., Ltd. (Aijinke) triggered the explosion of flammable gas when they were performing inspection and maintenance inside a benzene tank. The serious accident resulted in the immediate death of six on-site workers and direct economic loss of 11.66 million yuan. Through a series of qualitative and quantitative analysis methods, we reveal the root causes of the accident, simulate the development process of the explosion, and propose corresponding rectification suggestions. Here we want to emphasize that the reason why we write this paper is not to pursue responsibility and punishment but to reflect on how to avoid the recurrence of similar accidents.

2 The occurrence of the explosion accident

On the morning of May 12, 2018, Aijinke workers arrived at the utility tank farm of Secco to prepare for maintaining TK-75-0201 benzene tank (0201 benzene tank). Before the start of the operation, a Secco field operator used a hand-held gas detector to measure oxygen and flammable gas at the manhole outside the tank. The on-site supervisor of Aijinke, the on-site supervisor of Secco, and the on-duty monitor of Secco utility tank farm signed the work ticket successively without verifying the measurement data and checking the personal protective equipment (PPE) and explosion-proof tools of the on-site workers according to the requirements specified in the operating permit.

At 1:15 p.m., six workers of Aijinke continued to dismantle the floating pontoon in the tank and two workers worked outside for monitoring. At the scene, another Secco field operator was in charge of regular oxygen and flammable gas measurement. At 3:25 p.m., a sudden explosion occurred in the 0201 benzene tank.

After the explosion, Secco immediately launched their emergency response plan and reported the accident information to the response center of Shanghai Chemical Industry Zone Management Committee (SCIZMC), the fire brigade, the medical center, and the production operations monitoring center of Sinopec. At 3:50 p.m., the fire brigade put out the fire by spraying water. At 5:50 p.m., the on-site rescue work ended and a total of six Aijinke workers in the tank died on the spot. Figure 1a shows the scene of the 0201 benzene tank after the explosion.

Figure 1. The 0201 benzene tank and floating pontoons

2.1 Description of the timeline

The occurrence of accidents is a complex system problem, and it gives useful information on how accidents arise in practice [8]. In March 2018, Secco found that the breather valve of the 0201 benzene tank leaked slightly, which caused the VOC concentration to exceed the standard. After inspection, it was judged that the floating pontoon leaked. In order to find the root causes, we describe in detail the fundamental behaviors of the primary responsible party and the contractor prior to the explosion.

- On April 16, Secco established a maintenance order and a work list in the SAP system.
- On April 19, Secco began to empty the 0201 benzene tank, performing four days of steam dilution and eight days of nitrogen replacement.
- On April 24, Secco informed Aijinke to formulate maintenance plan.
- On April 26, Secco carried out the Job Hazard Analysis (JHA) around the maintenance work.
- From May 1 to 7, Secco opened the manhole of the tank for natural ventilation and inspection, and found that more than half of the floating pontoons had effusion.
- On May 8, the relevant departments of Secco held a special meeting and decided to change the original plan to replace all the floating pontoons.
- On May 9, Secco asked Aijinke workers to enter the tank and destructively dismantle all floating pontoons at the regular OSBL equipment maintenance meeting.
- From May 9 to 11, Aijinke workers entered the tank and dismantled the floating pontoons. A total of 38 floating pontoons were dismantled, as shown in Figure 1b.

2.2 *Analysis by event sequence diagram*

Event Sequence Diagram (ESD) is an accident scenario-oriented simulation method, which treats events as the basic research object, and takes into account the domino effect and risk transfer effect [9]. ESD is also an intuitive reflection of the accident scenario. It provides a more literal representation of the system state, indicating the behavior of crucial process variables, and even operator and hardware state changes [10].

In order to further illustrate and point out the critical links of the accident, this paper makes an intuitive description of events by using ESD. As shown in Figure 2, the basic events of accidents are divided into initial events, judge events, action state events, and termination events. At the same time, the path indicated by the dashed line represents the potential events that may occur in this accident, and helps to make a comprehensive and logical analysis of the whole process. The graphic symbols in the ESD are explained in the Appendix.

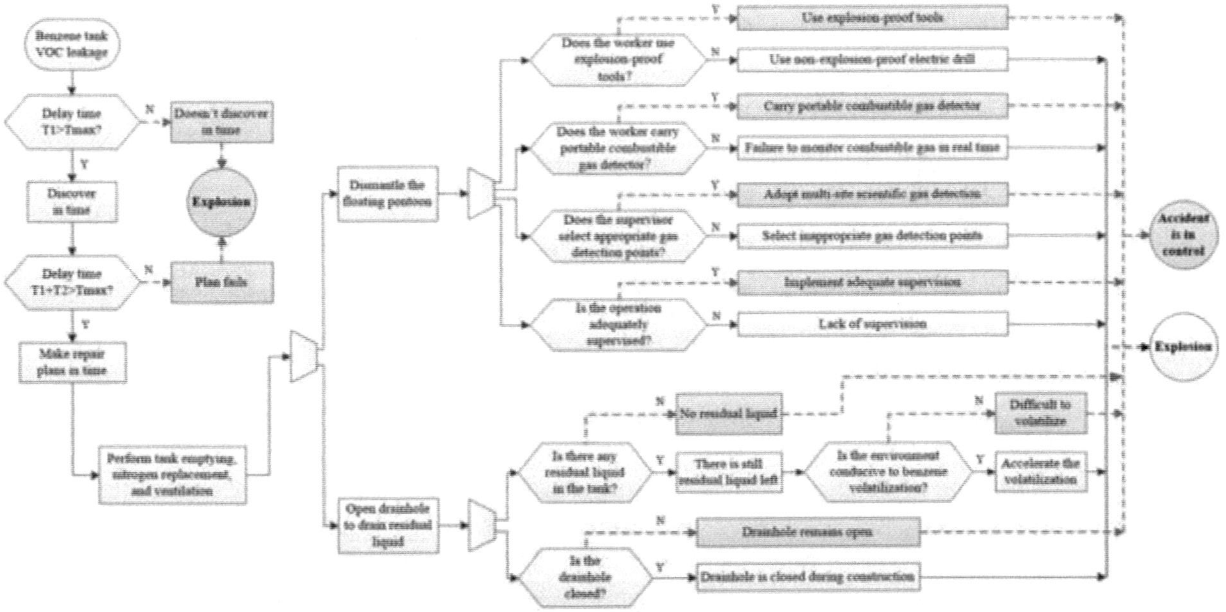

Figure 2. The ESD of the accident

As can be seen from the ESD, there are nine pivotal judgment nodes on the occurrence of the event chain, which can determine the direction of the incident. Among them, whether to discover leakage in time and whether to make a repair plan in time are the nodes at the time aspect. When the detection instrument and worker cannot find leakage and formulate the response plan in the maximum allowable delay time, they will directly lead to the explosion. Enterprises need to enhance equipment reliability and inspection frequency, so as to reduce emergency response time. Whether there are residues in the tank and whether environmental conditions are conducive to benzene volatilization are two nodes at the environmental aspect, which are also the leading causes of accidents. They require the workers to make a detailed hazard identification of the construction environment, and improve scientific detection. The critical nodes regarding human factors include whether the drainhole is closed, whether the on-site workers use explosion-proof tools, whether they carry portable combustible gas detector, whether they select appropriate detection points, and whether the supervisors take adequate monitoring [11]. Strengthening safety training before the operation, improving individual risk identification ability, perfecting risk control measures, and enhancing safety supervision level are essential means to avoid accidents caused by human factors.

All of these factors, organizational and operational, contributed to the accident, with a hierarchy of casualties among them. Through the comprehensive description, ESD can not only find out the nodes where the problem lies, but also guide the prevention of similar accidents. Therefore, enterprises should pay attention to the basic events expressed by nine judgment nodes and take appropriate interventions to block the development path of dangerous accidents.

3 Multilayer bow-tie analysis

3.1 The process of MBTA

Among many techniques available, bow-tie analysis is prevalent because it organically combines the two sophisticated risk analysis methods of Event Tree Analysis (ETA) and Fault Tree Analysis (FTA) [12], systematically summarizes the causes and consequences of accidents, and further integrates the prevention and mitigation of accidents or risks [13]. Bow-tie analysis has been proven to be a reliable and efficient tool, mainly used for process safety analysis, risk management, and accident risk analysis. It helps to understand which significant event combinations will lead to the top events, and which security barrier failures would escalate the top event to a certain consequence.

Traditional bow-tie analysis mainly includes top event, causes, consequences, preventive barriers, and mitigation barriers, which shows a lot of information. However, the single-level structure lacks logic and is not conducive to people's deep understanding. Thus, this paper establishes a Multilayer Bow-tie Analysis (MBTA), which resolves the top-event-related ETA and FTA individually into hierarchical structures, to deeply analyze the root causes, specify the severe consequences, and propose the corresponding safety barriers and scientific disposal methods. MBTA uses graphics to visually display the relationship between different levels and avoid repeated analysis effectively. Figure 3 shows the MBTA model, in which the code explanations are listed in Table 1.

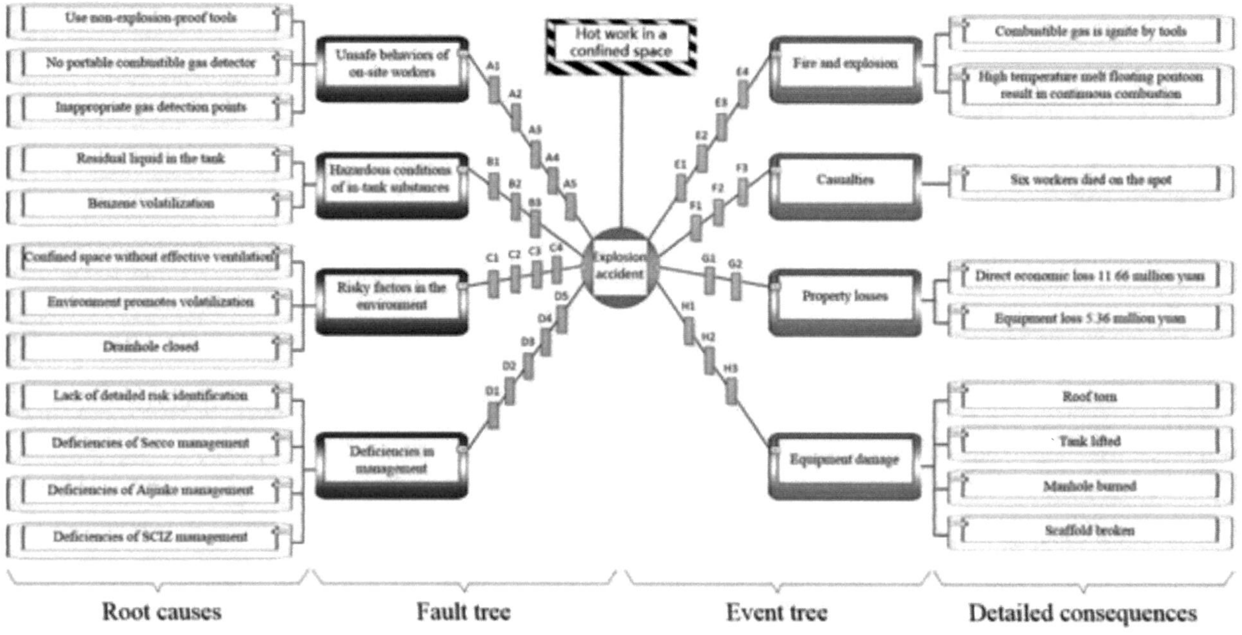

Figure 3. The MBTA model

Table 1. The explanations of MBTA

Code	Safety barrier	Code	Safety barrier
A1	Use explosion-proof tools	D4	Implement the main responsibility
A2	Carry portable combustible gas detector	D5	Improve contractor management
A3	Strengthen operation training	E1	Call the fire brigade in time

A4	Enhance supervision and assessment	E2	Spray water to extinguish fire
A5	Set up warning signs	E3	Fire isolation
B1	Remove the residual liquid	E4	Launch the emergency plan
B2	Hazardous material isolation	F1	Call for medical treatment
B3	Combustible gas detection in the tank	F2	Search and rescue the wounded
C1	Forced ventilation	F3	Launch the emergency plan
C2	Open drainhole	G1	Insurance claim
C3	Continuous detection during operation	G2	Launch the emergency plan
C4	Take environmental risk assessment	H1	Dangerous equipment isolation
D1	Compliance with rules and regulations	H2	Equipment repair
D2	Increase the intensity of punishment	H3	Equipment replacement
D3	Strengthen process safety management		

In the above MBTA model, we adopt the analytic hierarchy process (AHP) to extract the common characteristics of 12 causes of accidents and classify the causes into four categories: unsafe behaviors of on-site workers, hazardous conditions of in-tank substances, risk factors in the construction environment, and deficiencies in management. Moreover, we could conclude the direct and indirect reasons for the accident.

3.1.1 Direct reasons

Benzene is a colorless and transparent liquid with a strong aromatic odor. It has a boiling point of 80.1 ℃, a saturated vapor pressure of 10 kPa at 20 ℃, a flash point of -11 ℃, an explosion limit range of 1.2%-8.0%, and minimum ignition energy of only 0.20 mJ. Benzene is a highly flammable substance. The explosive mixture of benzene vapor and air can burn and explode when exposed to ignition source or high temperature.

At the construction site, the temperature is 24.7 ℃, and the pressure is 100.68 kPa. Under these conditions, benzene is easy to volatilize and form a gas explosion environment in a confined space. During the process of dismantling, residual benzene in the floating pontoons leaked onto the bottom of the tank and was not cleaned up in time. Because there was no effective ventilation in the tank, benzene vapor mixed with air to form an explosive environment and benzene concentration reached the explosion limit. The non-explosion-proof tools used by the operators produced ignition energy, which brought about the deflagration of the flammable gas. The high temperatures generated by the combustion melt the other aluminum alloy pontoons, resulting in sustained combustion for a short period.

3.1.2 Indirect reasons

Aijinke failed to abide by the relevant safety production rules and regulations. Managers did not fulfill their duties of safety management and supervision. Aijinke failed to give safety technical information to workers following the operating rules before the construction. Without implementing and supervising PPE, carrying portable gas detector, and using explosion-proof tools, operators were arranged to work in a confined space. When significant changes occurred in the job conditions, Aijinke did not promptly request workers to stop working and implement the Management of Change (MOC).

Secco had some omissions in safety risk management, professional management, and special operation management. In addition, Secco did not follow the safety production rules and regulations. On-site inspectors failed to carry out gas detection in accordance with specifications. After the change of operation, the manager did not ask the contractor to stop the process in time and revise the construction plan according to the regulations. Secco failed to report the maintenance plan to SCIZMC as required.

SCIZMC failed to implement the safety management system fully and did not discover that Secco performed maintenance operations without permission.

3.2 Lessons learned from MBTA

This paper puts forward a series of preventive measures, such as strengthening operation training, removing residual benzene in the tank, improving risk assessment of the environment, and enhancing process safety management. It is noticed that the environment in a confined space is harsh and there are many potential unsafe factors. Furthermore, the contractor's maintenance work cycle spans the statutory weekend. Based on the two reasons above, the enterprise should upgrade the level of hot work and adopt a broader system of operation approval and management. On-site workers need to stop working immediately when safety problems are found in the construction process and report to their superiors by relevant regulations.

Through refining the consequences of accidents, we summarized that detailed safety emergency plans and energy isolation methods are universally applicable in various measures to control and mitigate the impacts of accidents. They require Secco and Aijinke to formulate a comprehensive emergency response plan before, including fire extinguishing, personnel search and rescue, equipment maintenance, property damage remediation and so on. When an accident occurs, workers have to stay away from dangerous substances and places immediately.

As with many disasters, what ended up in a sequence of equipment damage and human errors started mostly as a management problem. Many unsafe practices were tolerated and even encouraged. The risk management of confined space operations should be divided into ten steps according to relevant regulations, namely training, pre-operation planning, job approval, hazardous energy isolation, ventilation, gas detection, personal protection and tool safety, warning signs, monitoring, and emergency response, so as to strengthen process safety management comprehensively. It is necessary to enhance the issuance of management of work tickets for hazardous operations such as hot work and confined space operations. Enterprises must strictly inspect illegal operations and supervise safety precaution measures to ensure that the operation process is safe and controllable. For changes occurring in the construction process, enterprises ought to strictly implement the MOC, analyze the potential risk and possible accidents, formulate corresponding safety measures, and carry out safety education for all operators.

MBTA provided a detailed analysis of the accident and established the corresponding safety barriers. However, there are still some certain subjective problems, and its content should also be flexible. Therefore, we encourage employees, accident investigators, case learners, and other relevant personnel to improve and develop the content of MBTA continually, thereby effectively expanding the application scope and effectiveness of MBTA.

4 Quantitative simulation of the explosion

The analysis of the accident should include not only qualitative analysis but also quantitative analysis [14]. We use FLACS software to simulate and analyze the occurrence of the benzene explosion. According to the relevant information, the 0201 benzene tank is an internal floating roof tank with a volume of 10,000 m^3. Its diameter is 30 m and height is 19 m. The design temperature range and pressure range are -10 – 65 ℃ and -0.49 –1.96 kPa, respectively. The tank adopts box-type aluminum alloy internal floating pontoons, and a manhole with a diameter of 0.6 m is set at the position of 0.75 m above the ground. Figure 4 shows the 0201 benzene tank model.

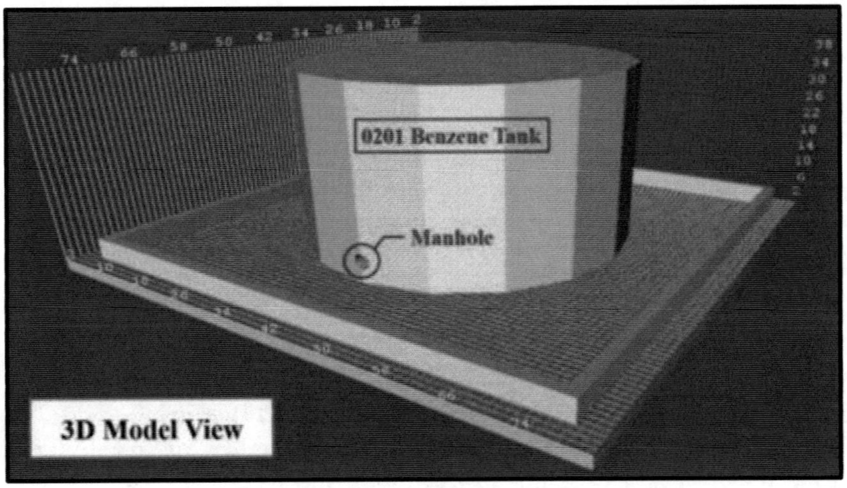

Figure 4. The 0201 benzene tank model

The simulated environmental conditions are fixed at the temperature of 24.7 ℃ and the pressure of 100.68 kPa, which are the same as the construction environment. We make reasonable assumptions: (a) The residual benzene in the tank evaporates continuously into the confined space and forms a uniform mixture of flammable gas. When the volatilization concentration of benzene reaches the lower explosion limit (LEL), the ignition source that persists during the construction operation detonates the flammable gas and results in an explosion. (b) Six on-site workers dismantled the float pontoons at the bottom of the tank, so the ignition source is set at the height of 0.75 meters from the geometric plane center, as shown in Figure 5 point I. There are several detection points of pressure and other characteristic parameters in the model of benzene tank, as shown in Figure 5, points M1-M5. (c) The benzene tank is windless, and the flammable gas accords with the real gas state equation. (d) The gas explosion process satisfies a series of fundamental equations, such as the continuity equation, momentum equation, and energy equation.

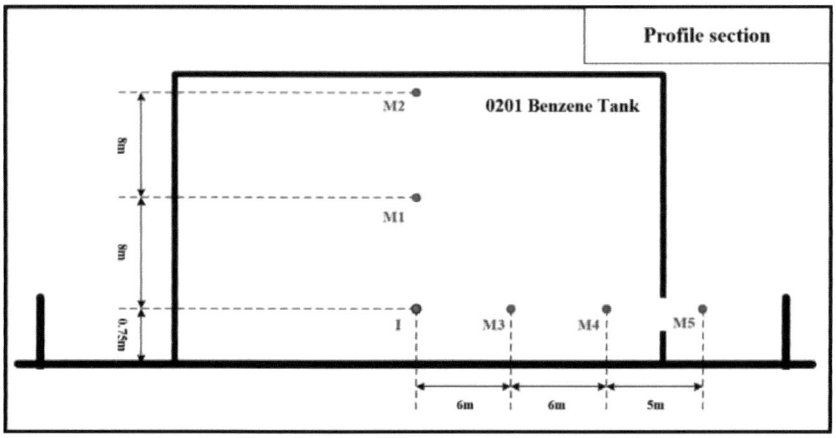

Figure 5. The profile section of the model

We selected the maximum profile section of the tank as the research object and intercepted the temperature field and pressure field at four different moments in the explosion process, as shown in Figure 6 and 7. The simulation results intuitively describe the development process of the explosion. When the flammable gas in the tank was ignited, the flame diffused around in the form of a spherical wave and reached the bottom of the tank at 0.35 s, as shown in Figure 6a and 7a. Since the radial distance of the tank is less than the vertical axial distance, the flame propagated horizontally to the wall, as shown in Figure 6b and 7b. At 1.52 s, the flame reached the manhole and sprayed out to the outside of the tank in the form of jet flow. During this period, the pressure and temperature increased continuously, which are the leading causes of the manhole being burned, as shown in Figure 8a. When the flame propagated to the top of the tank at 2.11 s, the uncombusted gas was recirculated by the viscous effect and blocking effect of the wall, which further caused the recirculation vortex. The recirculating vortex enlarged the contact area of the flame front, resulting in flame acceleration and pressure increase, as shown in Figure 6c and 7c. Figure 6d and 7d illustrate the flammable gas burned out at 3.24 s. Simultaneously, the explosion temperature and pressure reached the peak value contributing to the tearing of the roof and the lifting of the tank, as shown in Figure 8b and 7c.

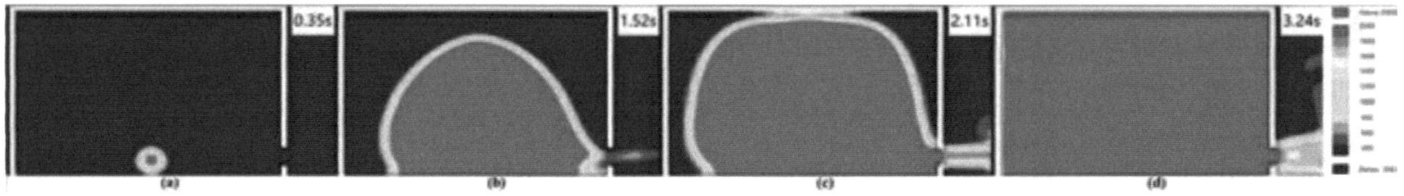

Figure 6. Temperature simulation of the explosion

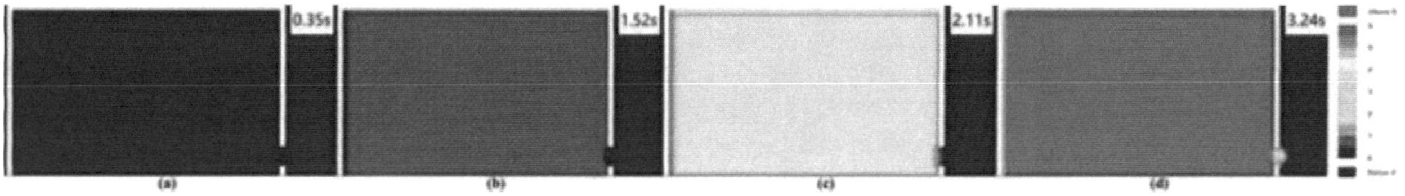

Figure 7. Pressure simulation of the explosion

Figure 8. The results of the accident

Figure 9 reveals the change of temperature and pressure recorded by the detection points M1-M5 during the explosion. Table 2 describes the data of the detection points, including maximum and minimum explosion pressures, maximum and minimum explosion temperatures, peak data arrival times and so on. Combining the data of detection points and the pressure field diagram in Figure 7, it is shown that the pressure in a confined space will be uniform immediately for a certain time and further proves the accuracy of the analysis of the flame propagation process. In particular, the external detection point M5 was affected by the propagation of explosion flame in the tank at 1.6 s, resulting in a peak explosion overpressure of 0.031 bar and a peak temperature of 1895 ℃. Subsequently, due to the pressure rise in the tank, the pressure at the detection point M5 dropped to negative pressure until the flammable gas burned out.

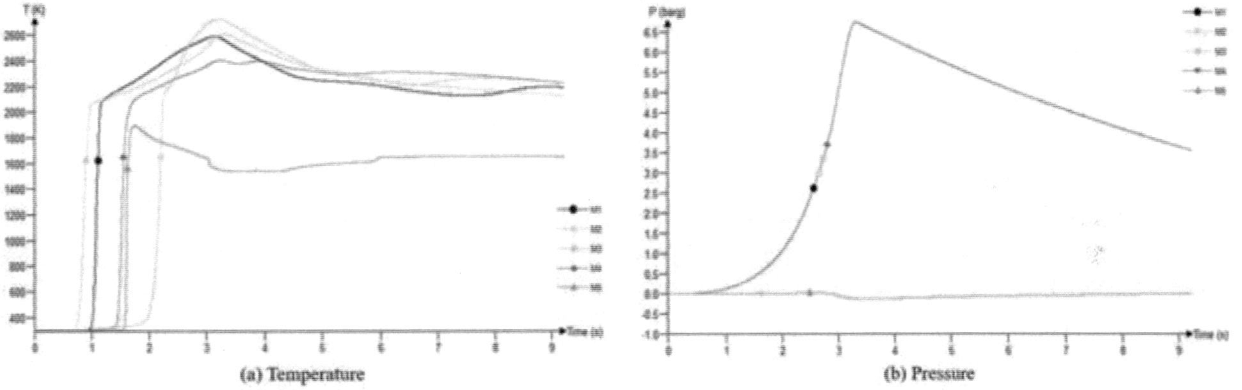

Figure 9. Parameters change of the detection points

Table 2. Data from the detection points

No.	Temperature				Pressure			
	Tmax/K	Time/s	Tmin/K	Time/s	Pmax/bar	Time/s	Pmin/bar	Time/s
M1	2589.4	3.11	298.1	0.04	6.75	3.24	0.00	0.04
M2	2732.2	3.24	298.1	0.04	6.76	3.24	0.00	0,04
M3	2609.3	3.28	298.1	0.00	6.76	3.24	0.00	0.00
M4	2306.2	3.25	298.1	0.00	6.75	3.24	0.00	0.00

M5	1895.5	1.68	298.0	0.69	0.05	1.65	-0.15	3.29

Although there is only a small amount of residual benzene, the benzene volatilization forms a flammable gas with the air inside the tank, which generates a substantial destructive power when ignited. The simulation results show that the flame propagation velocity of the flammable gas increases constantly, with the peak value reaching up to 41 m/s. Since the flame is propagated in the form of subsonic velocity, it can be judged that there are mainly the phenomena of deflagration occurred in the process of explosion [15]. Soloviova-Sokolova et al. [16] calculated the flame burning rate of the benzene vapor was 0.6 m/s at 298 K, 0.1 MPa and 0.6 eq., which clearly illustrated that the peak propagation velocity of the benzene vapor in a confined space would be several times higher. Meanwhile, the explosion pressure in a confined space is usually greater than that in an open area due to lack of heat loss and volume expansion.

When workers are engaged in hot work in a confined space, they should carry out comprehensive risk identification, select several representative flammable gas detection points, and detect the flammable gas in real time. Workers ought to stop work immediately and withdraw in an orderly manner once the flammable gas exceeds the standard. Through the analysis of the explosion process, the flame propagates to the wall in a very short time. Thermal radiation and explosive shock wave are the primary causes of casualties in the tank. Owing to the special working environment in a confined space, the phenomenon of mass casualties and injuries often occurs under the effect of high temperature and high pressure, which needs people's sufficient attention.

5 The subsequent event

The explosion accident of Secco has aroused widespread concern, and 22 responsible persons have also been punished. But unfortunately, lessons from the accident did not prevent a subsequent similar event. On November 26, 2018, three contractor workers of the Ningbo Engineering Company (NEC) and one worker of Secco inspected the discharge tank of cracking gas compressor in the olefin separation unit. Without ascertaining the concentration of oxygen in the tank, a contractor worker of NEC entered the tank for construction and fainted unexpectedly. A Secco worker immediately went into the tank to rescue him, but he also fell into the tank. The accident resulted in the death of two workers who had been rescued after asphyxia.

It is the second confined space operation accident in the same company within six months. The main causes of the accident are still the deficiency of contractor management and risk identification in a confined space. While expressing sadness, the public should profoundly reflect on why accidents continue to occur, what is the role of punishment and accountability, and how relevant regulations and reform measures can be truly implemented. Drawing lessons from the previous accidents is necessary. Enterprises ought to raise the safety awareness of their employees and use strict supervision to ensure safe production of chemical plants.

6 Conclusions

This paper provides an introduction of an accident caused by hot work in a confined space, together with the detailed ESD-based description of nine pivotal judgment nodes leading to this disastrous accident. It concludes that enterprises should take suitable intervention measures to block the pathway of the accident. The MBTA, covering the causes and consequences of the accident, offers proactive and reactive measures based on an intuitive analysis. Meanwhile, it emphasizes the importance in selection of appropriate detection points, real-time detection of in-tank operation, and implementation of MOC. Also, we apply the FLACS software into quantitative simulation of the explosion process in the tank. The results contribute to the guidance of accident investigation and safety measures selection. Finally, it suggests that enterprises adopt technical remedies and better safety management practices, since they are the essential basics for effective implementation of risk control measures.

Appendix

Symbol	Annotation
	Initial event: Beginning event of the ESD.
	Judge event: The event has two possible outcomes that determine whether subsequent events occur.
	Action state event: Describe the action state of the event.
	Termination event: Ending event of the ESD.
	Output AND gate: Input event lead to multiple output events occurring.

7 References

[1] D.Burlet-Vienney, Y.Chinniah, A.Bahloul and B.Roberge, Design and application of a 5 step risk assessment tool for confined space entries, Safety science 80 (2015), 144-155.

[2] The Office of the State Council Security Committee Notified 8 Cases of Blindly Rescuing in a Comfined Space Leading to Casualties Enlargement, Available at http://www.chinasafety.gov.cn/default.htm Accessed on June 8, 2018.

[3] J.Selman, J.Spickett, J.Jansz and B.Mullins, Safety science 109 (2018), 333-343.

[4] J.Selman, J.Spickett, J.Jansz and B.Mullins, Safety science 113 (2019), 78-90.

[5] L.Botti, V.Duraccio,M.G.Gnoni and C.Mora, A framework for preventing and managing risks in confined spaces through IOT technologies, ESREL (2015), 3209-3217.

[6] B.Bang, C.Ahn, J.Lee, Y.Kim, M.Lee, B.Horn, D.Malik, K.Thomas, S.James, A.Yarin and S.Yoon, Theoretical, numerical, and experimental investigation of pressure rise due to deflagration in confined spaces, International Journal of Thermal Sciences 120 (2017), 469-480.

[7] B.He, X.Jiang, G.Yang and J.Xu, A numerical simulation study on the formation and dispersion of flammable vapor cloud in underground confined space. Process Safety and Environmental Protection 107 (2017), 1-11.

[8] I.M.Shaluf, F.Ahmadun and A.M.Said, Fire incident at a refinery in West Malaysia: the causes and lessons learned, Journal of Loss prevention in the Process Industries 16.4 (2003), 297-303.

[9] S.Swaminathan and C.Smidts, The event sequence diagram framework for dynamic probabilistic risk assessment, Reliability Engineering & System Safety 63.1 (1999), 73-90.

[10] J.Zhou, G.Reniers and N.Khakzad, Application of event sequence diagram to evaluate emergency response actions during fire-induced domino effects, Reliability Engineering & System Safety 150 (2016), 202-209.

[11] E.Zarei, M.Yazdi, R.Abbassi and F.Khan, A hybrid model for human factor analysis in process accidents: FBN-HFACS, Journal of Loss Prevention in the Process Industries 57 (2019), 142-155.

[12] N.Khakzad, F.Khan, P.Amyotte, Dynamic risk analysis using bow-tie approach, Reliability Engineering & System Safety 104 (2012), 36-44.

[13] M.B.Mulcahy, C.Boylan, S.Sigmann and R.Stuart, Using bowtie methodology to support laboratory hazard identification, risk management, and incident analysis, Journal of Chemical Health and Safety 24.3 (2017), 14-20.

[14] C.Zhu, J.Zhu, L.Wang and M.S.Mannan, Lessons learned from analyzing a VCE accident at a chemical plant, Journal of Loss Prevention in the Process Industries 50 (2017), 397-402.

[15] I.C.Tolias and A.G.Venetsanos, An improved CFD model for vented deflagration simulations–Analysis of a medium-scale hydrogen experiment, International Journal of Hydrogen Energy 43.52 (2018), 23568-23584.

[16] J.V.Soloviov-Sokolova, V.A.Alekseev, S.S.Matveev, I.V.Chechet, S.G.Matveev and A.A.Konnov, Fuel 175 (2016), 302-309.

Strengthening Process Safety Culture throughout Learning from Each Other´s Vulnerabilities

Luciana Poli SIlva
PETROBRAS
Av. Republica do Chile, 65, 4th floor, aisle 402.
lucianapoli@petrobras.com.br

Prepared for Presentation at
American Institute of Chemical Engineers
2019 Spring Meeting and 15th Global Congress on Process Safety
New Orleans, LA
March 31 – April 3, 2019

Strengthening Process Safety Culture throughout Learning from Each Other´s Vulnerabilities

Luciana Poli Silva
PETROBRAS
Av. República do Chile, 65, 4th floor, aisle 402.
lucianapoli@petrobras.com.br

Keywords: andragogy, dialogue, leadership, safety leadership culture, trust, vulnerabilities.

Abstract

A fundamental approach to any program that targets the implementation of a process safety culture is to work within reality, going beyond procedures and compliance. To facilitate the emergence of the unspoken is essential to stimulate the dialogue among leaders that brings about their vulnerabilities and opens new perspectives to address them effectively. This paper presents a pilot program designed to increase the Process Safety Leadership Culture in a Brazilian Energy Company. Based on a benchmark made within the industry and in andragogic principles, the program created a trust ambiance among peers and inspired leaders to recognize and share their vulnerabilities, as well as to take actions towards the strengthen the role of being a safety culture leader.

1 Introduction

A lot has been said and written about Leadership Process Safety Culture, but there is a gap between what the compliance procedures dictate and what managers really do. When it comes to the Brazilian Energy Industry, the pressure leaders are under in order to achieve the results they are expected to can be considered significant. How to balance the production challenges with a safety culture that must be among the core values of the operation? Are these factors compatible?

The pilot program entitled '*Strengthening Process Safety Culture*' dealt with reality, using andragogic approaches to bring about the complex factors forty managers face in their routine, such as the existence of several levels separating top managers from the operational team, privileging safety over production goals, and being an effective representative of the Company´s policies. All these factors can jeopardize the safety culture from becoming stronger every day.

This paper describes the implementation of a methodology designed to create a safe environment, in order to build trust among participants and at the same time, make them reflect and dialogue about how they can evolve from compliance to commitment.

In the coming sections, we will present the central concepts used to develop the program – Culture, Safety Culture and Leadership - exploring its implication on an actual work environment.

In the sequence, we will talk about what it takes to make the unspoken, unwritten rules emerge. From there, how we stimulated the increase in the self-conscious level as well as in the collective one, building a matrix entitled 'Personnel Influence on Safety' to diagnose how one really sees himself concerning his leadership in safety culture.

We will then present the method used to address the issues managers pointed out, in order to transform the discussions into actions to strengthen the safety process leadership culture.

In the conclusion, we will present our findings and discuss possible future approaches to amplify the safety leadership culture further.

2 Safety Culture and Leadership: definitions and implications

There are many possible definitions for culture. Taylor (2015) says culture has to do with the behaviour patterns that are encouraged, discouraged or tolerated over time in an organization. Schein (2015) says that culture is "the total sum of everything an organization has learned in its history in dealing with the external problems – which would be goals, strategy, how we do things – and how it organizes itself internally."

Since we humans are gregarious social beings, we are naturally wired to fit in with our community. We use our perceptions to read what the tribe accepts and we adapt by behaving accordingly, because we want to fit in and be rewarded. Therefore, we reinforce the existing norms. This process is perpetuated over time; established behavior influences the behavior of new employees and so on.

Most cultural sources are non-verbal. Taylor (2015) says it is a myth to believe that culture reflects the values statements that are declared in the company's annual reports. Most commonly, the statement will be the organization´s intents, not the culture as it actually is.

If we want to change culture, it is necessary to have a clear direction and a focused, effort-driven leadership to lead people towards the goal culture. Culture has to do with what is encouraged and allowed. For Taylor (2015), three major sources send cultural messages: behaviours - of other people, especially important, recognized ones; symbols - artefacts, events and decisions that are meaningful to the community; and systems - mechanisms that manage people and tasks.

In this sense, culture has to do with the manifestation of what is valued in the organization.

According to Lafraia (2014), culture shapes reality. Cultural lenses structure the perception of the people involved in the organization.

When it comes to safety culture, its meaning may vary from one organization to another, or even in different departments of the same company. It is perfectly possible to have diverse lenses that translate the meaning and the importance of safety practices in diverse ways.

OGP, in its 452 Report (2013), says that a positive Safety Culture has to do with a culture in which safety is a core value for everyone who works in the organization, and it is pursued in an everyday basis. In order to build a Safety Culture, there are factors that must be considered, such as "leadership commitment to safety; employee involvement and motivation; responsibilities and accountability; production and bottom line pressures versus quality issues; and actions, or lack of action, to correct unsafe behaviours and unsafe conditions"; among others.

In order to build a consistent safety culture, leaders must educate their teams by demonstrating, by their practices, that some policies are fundamental. Safety culture should use this lens, rather than a command and control opaque one. Leaders must educate people to understand why safety is a core value that concerns the most precious gift: one´s life.

Safety culture is not about only following the rules, but making the team understand that the rules serve a greater principle. This approach can influence values and beliefs and change the individual and team´s behavior. The challenge leaders face is to conciliate the educational role with the policeman role, guaranteeing that compliance is required and everyone in the organization, starting by himself, values that as an unnegotiable principle.

In order to develop a strong safety culture, some elements must be considered (OGP, 2013):

Elements of a Strong Safety Culture

- an informed culture: the company has quality data concerning safety and uses it well;
- a reporting culture: people feel they can report safety concerns without worrying about being punished;
- a learning culture: the organisation actually learns from its unsafe conditions and changes them, banning any equipment, practice or whatever takes place that hazards safety;
- a flexible culture: the company changes its command chain if necessary to adapt;
- a just culture: people understand what is acceptable and unacceptable. What is not tolerated is dealt fairly.

For Peter Drucker (Cohen, 2018), leadership can be defined as the lifting of a man's vision to higher sights. Throughout leadership, people perform in higher standards, they are inspired and evolve in multiple levels.

When it comes to safety leadership, we can quote Ram Charan (2007) when he says that "a leader who does not produce other leaders is not a great leader." A safety leader should act in a way to enhance in everybody´s minds and actions a safety leadership commitment.

In order to make it possible, it is crucial that the leader himself is aware of his beliefs and constantly monitoring his behaviours. It takes time and a lot of effort to build a strong safety culture. Despite that, a safety culture is fragile and easily threatened by wrong messages. (OGP, 2013)

In the next section we explore the method used to address self-consciousness about safety.

3 Creating a trust environment that favors Safety Process Leadership Culture

According to Malcon Knowles, who develop the paradigm of andragogy, adults have a learning style characterized, among other factors, by changing one´s concepts according to new experiences one lives; the pursuit for autonomy; increasing one´s own motivational level to match the social role assumed; and acquiring knowledge to immediately solve practical issues (Muresan, 2013).

Considering this practical approach, we began the program to strengthen safety leadership culture by making each participant reflect on its daily practices concerning management, as well as the safety implications involved in routine decisions.

To start this reflection, we did a benchmark and decided to invite a senior top manager from another company in the Energy Industry to give a testimony. He made an honest statement about his team perception of his management practices. Concerning safety process, he mentioned that he realized that, in order to influence people, he must first change himself. He then shared with an amazed group of participants some practical examples of safety interventions he changed in his auditing procedures, for example, in order to be coherent with the message he wanted to pass forward.

The testimony had an impressive impact on the managers. A remarkable phrase that was repeated by managers was: "*Simplify. Be the Powepoint. Less Presentations, More You.*"

After the testimony, there was a presentation on the invited company's safety philosophy, about its journey into safety leadership. The practical steps mentioned were followed by videos with managers from different parts of the world talking about the way they see safety culture benefits.

With that background environment set, we forwarded to a self-perception diagnosis. The goal was to stimulate the increase in the individual conscious level as well as in the collective one. We used a matrix entitled 'Personnel Influence on Safety' that was obtained in the benchmark process with the above mentioned company.

It aimed to diagnose how one really sees himself concerning its leadership in safety culture. It involved two dimensions: the personal influence towards oneself; and the personal influence concerning other people´s safety.

The figure below illustrates the matrix:

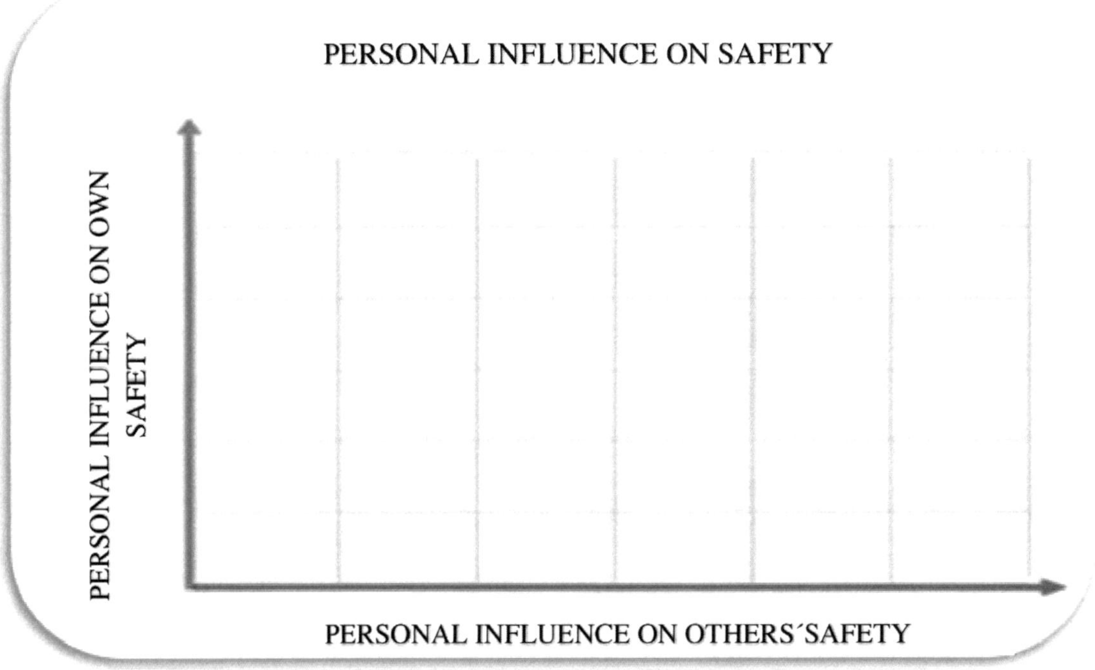

Figure 1. Personal Influence on Safety Matrix

We used the Self-perception as a tool to promote reflection and consciousness. We asked the participants to fill in the matrix in two different papers: one in his scrapbook (explained below) and another one on a plain sheet to hand over to the program facilitator. All the matrices were compiled and the result is shown in figure 2. As indicated, most participants initially evaluated themselves as influencing more themselves than others.

The coordination staff evaluated the matrix result as requiring attention from the participants. The staff observed the fact that some managers attending the program were obese, others worked extra hours, and most of them had reported having trouble balancing personal and professional life. Therefore, to promote deeper self-reflection, the consultant who was facilitating the program posed some questions to the participants regarding personal care. The remarks made were an invitation to honestly look at oneself and feel if the level of personal influence corresponds to what one would recommend to a loved person. Quoting Brené Brown (2015), the consultant said "Treat yourself like someone you love. Do you do it on a regular basis? Do you put yourself first?" Those remarks made some participants change their self-evaluation.

The intention was to call leaders ´attention to the fact that genuine influence on safety starts with their own safety, and goes beyond process or operational safety. To be legit and truly be a safety leader, one should take good care of oneself.

Figure 2. Compiled Personal Influence on Safety Matrix with the results from all the participants

We also developed a scrapbook that was given to each participant as a job-aid material to be used both during the program and after. The content of the scrapbook followed the program plan and was an invitation to self-analysis. The participants had special pages to do some journaling about what they had just listened to or dialogued about. There was also a special page dedicated to register insights. Figure 3 illustrates the scrapbook.

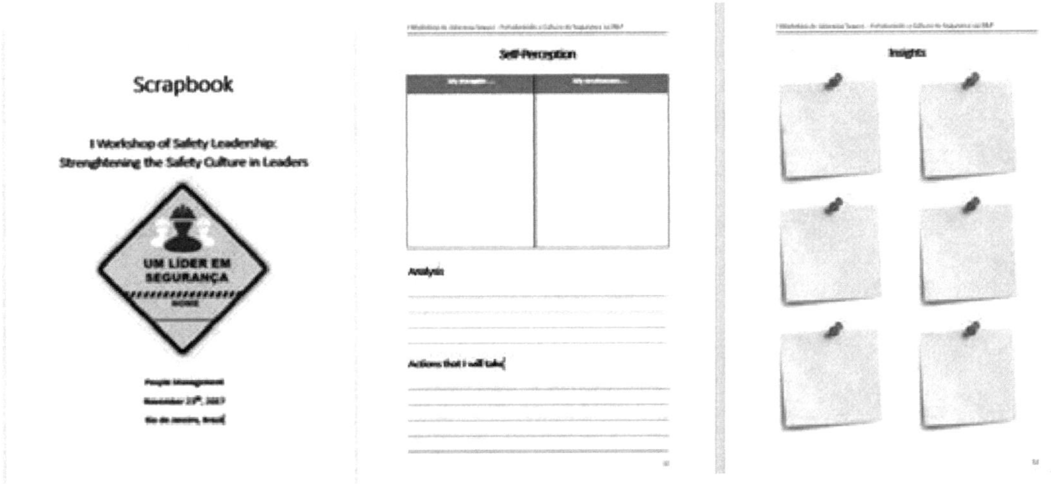

Figure 3. Pages from the Scrapbook used to stimulate self-consciousness and to plan actions

4 Exposing and Addressing Vulnerabilities

4.1 Control Room

We will now present the method used to address the issues managers pointed out, in order to transform the discussions into actions to strengthen the process safety leadership culture.

We prepared a special control room, where we designed a pathway decorated with images of accidents from the company and from the industry, and the number of fatalities. We also put an alarm sound playing during the route and videos showing accidents. These resources were intended to create an emotional ambiance to take the analysis to another level.

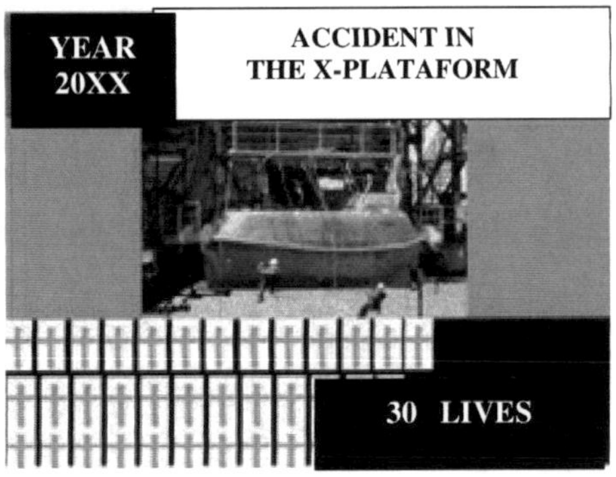

Figure 4. Picture of the Control Room and example of poster from accident used to decorate it.

Besides the planned journey passing several industry accidents, the Control Room was decorated with red material, to create an atmosphere of urgency. After going through the posters and watching the videos, the participants were invited to give their testimony of accidents they´ve been through.

The group tone got emotional as some of them shared unpleasant stories they had lived. The goal of inducing the reflection to an in-depth level was successfully achieved.

After that point, the participants went back to the regular room and dialogued about some questions that were in their scrapbook. The questions were related to the *Elements of a Strong Safety Culture* preconized by OGP (2013), mentioned in section 2 of this paper.

The methodology privileged social interaction among participants and the active listening of each other's considerations. Below is the list of the questions addressed to the participants.

Questions to Reflect and Dialogue:

1. Am I able to influence my operational team?
2. How many hierarchical levels are there between the operational team and me? To what extent does my speech actually makes sense to them?
3. How much of what is said by the operational team do I believe genuinely applies to me? Do I believe I receive good quality feedback?
4. Of the security actions implemented in my management, do I receive direct feedback on the effectiveness or otherwise of its implementation?
5. What degree of tolerance do I accept as a leader, regarding violations of safety standards?
6. Do I, as a leader, make it mandatory that the urgent inspection recommendations are executed immediately?
7. Do I, as a leader, take every opportunity to motivate, mentor and develop my team to enhance individual safety awareness and safety leadership?
8. Do I praise when I see a safe behavior?
9. Do I perform Behaviour Audits in a protocolled way, to meet the Scorecard, or do I do it because I believe in the effectiveness of what can arise from it?
10. Is Safety a value or is it a Priority for me as a leader?
11. Do I have the courage to step in and stop an activity when I see someone at risk? Do I always intervene?
12. Do my behaviours as a leader have safety as a foundation?
13. When I find that the environment is unsafe, do I privilege safety over other goals?
14. Safety comes first in my management?
15. How do I, as a leader, ensure that safety in my management is dynamic enough to ensure continuous improvement in operational procedures?
16. Thinking Safety as a chain, how can I, as a leader, be an effective link between operations and the company policy?

4.2 Results

In order to capture the precious discussions and share insights among the participants, in the final part of the workshop there was a collective presentation based on what each small group discussed about the questions above. Each group had to have at least one delegate, and others could participate if they wanted to.

Some of the findings reported were, for example: "*I have to go to the platform more often*"; "*Behaviour audits can be very useful, I should profit more from them*"; "*There are too many levels between my operational team and I. I have to get together with them on a regular basis.*"

After sharing insights, all managers filled in the Personal Influence on Safety Matrix again. The group reported that, this time, the self-perception was lower, since many managers realized they need to invest more efforts and more quality time in Safety, in both dimensions – with oneself and towards others.

After that, each participant revised and complemented his individual action plan (figure 3). They also wrote a personal commitment as a Safety Process Culture Leader.

At the very end of the program, we asked everyone to do a check-out by sharing with the entire group one action of the action plan. This intended to create a deeper commitment, as well as possible insights to others. Also, it was a practice that could be used with the managers team.

According to the group evaluation, presented in the word cloud below, the program succeeded in strengthening the safety culture in leaders, who mentioned as the top benefits from the program: (*Safety as a*) Value; Communication; Life; Commitment; Example; Integration; Consciousness; and Synergy.

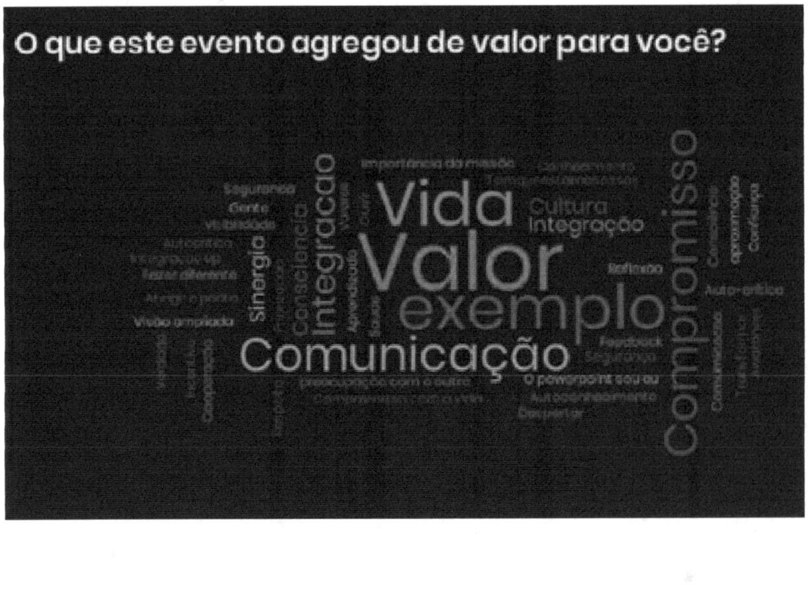

Figure 5. Evaluation: Word cloud containing the managers´ perception about the contribution brought by the program to participants.

5 Conclusions

The feedback given by the program´s participants was encouraging. It indicated that testimonies from partners, combined with relevant questions based on the daily problems managers face concerning safety, created a trust ambiance among managers.

The activities planned based on andragogic approaches favored dialogue and in-depth self-analysis. The use of the same instrument (Personal Influence on Safety Matrix) twice, as well as the individual scrapbook, facilitated the self-consciousness process and lead managers to review their initial assessment and to elaborate an action plan.

Leaders play a decisive role in strengthening the safety culture in the organizational world. Bringing about their vulnerabilities and having honest conversations with peers definitely helps managers to find ways to evolve from compliance to commitment.

Strengthening the safety leadership culture is a process that does not come to an end. It is an on-going process that must be continuously taken care of by leaders who value life and put safety first. Future projects that aim to implement effective actions in Safety Culture should take into consideration the company´s culture, as well as daily operational problems and safety as a core value to preserve life and business.

6 References

B. Brown, "*Daring Greatly: How the Courage to Be Vulnerable Transforms the Way We Live, Love, Parent, and Lead*", Penguin Random House, NY, 2015.

C. Migueles, J.R.B. Lafraia, G. C. Souza, "Criando o Hábito da Excelência: Compreendendo a Força da Cultura na Formação da Excelência em SMS", Quality Mark, Rio de Janeiro, 2014.

C. Taylor "*Walking the Talk: Building a Culture for Success"* Random House Business Books, London, 2015.

M. Muresan, "*Using Cybergogy and Andragogy Paradigms in Lifelong Learning*", 5th World Conference on Educational Sciences – WCES. Romania, 2013

OGP, *Shaping Safety Culture through Safety Leadership*. Report N° 452. International Association of Oil & gas Producers. October, 2013.

R. Charan, *Ram Charan on Leadership*,

Available at https://hbr.org/ideacast/2007/01/harvard-business-ideacast-27-r.html. Jan, 2007 (accessed Dec 4, 2018)

T. Kuppler, *Edgar Schein on Culture*,
Available at https://www.leadershipandchangemagazine.com/edgar-schein-on-culture/ Accessed on January 2, 2019.

W. Cohen, *Drucker's Top Tips to Successful Leadership*
Available at https://www.managementmattersnetwork.com/strategic-leadership/columns/druckers-top-tips-to-successful-leadership. Jun, 2018 (accessed Dec. 4, 2018)

7 Acknowledgments

I would like to thank Petrobras who fully supported this project, as well as Shell Brasil, for a very profitable partnership. A special thanks to my Human Resources, HSE and E&P work colleagues, who had a decisive role in making this project a reality.

548424 Development of Human Reliability Analysis Method for Oil and Gas Operations

The oil industry has grown in recent decades in terms of quantity of facilities and process complexity. However, human and material losses still occur due to major accidents at the facility. The analysis of these accidents reveals that many involve human failures that, if prevented, could avoid such accidents. These failures, in turn, can be identified, modeled and quantified through Human Reliability Analysis (HRA), which forms a basis for prioritization and development of safeguards for preventing or reducing the frequency of accidents. The most advanced and reliable HRA methods have been developed and applied in nuclear power plant operations, while the petroleum industry has usually applied Quantitative Risk Analysis (QRA) focusing on process safety in terms of technical aspects of the operation and equipment. A joint project from the B. John Garrick Institute for the Risk Sciences and a big company from the oil and gas industry is focusing on developing a HRA method focused on this industry operations, to bridge the existing gap between HRA and the Petroleum industry. The Phoenix HRA methodology was used as a basis, which has three layers i) a crew response tree (CRT), which models the interaction between the crew and the plant; ii) a human response model, modeled through fault trees, that identifies the possible crew failures modes (CFMs); and (iii) "contextual factors" known as performance influencing factors (PIFs), modeled through Bayesian networks. In addition to building on such a structure, the method development will rely on interviews with key personnel, visitations to installations, and analysis of past incidents data.

547327 Studying the Dependence of Operator Performance on Step Level Complexity

In the chemical, oil, and power industry, procedures are used by operators to perform their work correctly without compromising safety. Unfortunately, many procedures fail in adequately conveying information to their users. Heuristics tell us that technical jargon, vague language, and operator training all play a role in determining the effectiveness of procedures. To what extent these variables play a role is still not fully understood. Knowing this information will allow plants to more effectively write procedures. To counter this, researchers at Texas A&M's School of Public Health and Mary Kay O'Connor Process Safety Center are investigating the underlying factors that cause operators to deviate from procedures.

One possible factor may be that complex steps hinder an operator's ability to perform a task. This work aims to understand why certain steps are more complex than others. To do this, a set of procedures on how to operate a gas chromatography mass spectrometry (GCMS) machine is written. A study is performed in which participants use a set of procedures to analyze samples through gas chromatography. Using a scoring rubric, the complexity of each step is determined. The procedures have been designed to obtain an even distribution of complexity. Participants for this study are chemical engineering students with little to no prior experience in handling a GCMS machine. Thus, the experience, education, expertise, and training of the participants is controlled. The performance of each participant is reviewed and is grouped into one of four quadrants. For this study, a decrease in operator performance is expected as the complexity of each step increases.

548400 Prevent Them All or Stop Them Small: Tripod Best Practices in Reducing Chemical Accidents and Enhancing Response Preparedness

Prevent Them All or Stop Them Small: Tripod Best Practices in Reducing Chemical Accidents and Enhancing Response Preparedness

The connection between industry, government, and public safety, which we call the Tripod, is vital when implementing the 3-P's - Prevention, Protection, and Preparation. As identified by AICHE's Vision 2020 program, there are significant opportunities for organizations that address industrial process safety to work together. Responsible Collaboration between industry, regulatory and investigatory authorities, labor organizations, communities, and technical organizations promotes mutual understanding of risks, effects process safety and can reduce chemical accident risk in both small and large communities. Community based safety partnerships and leadership are fundamental to the success of such efforts.

For the past five years, a series of geographically focused Chemical and Ammonia Safety Days have cultivated and showcased chemical emergency prevention and planning efforts between Industry, NGOs such as the Ammonia Safety Training Institute (ASTI), and a myriad of public agencies including fire departments, health departments, LEPCs, SERCs, local regulators, and state and federal agencies such as EPA, CSB, DHS and OSHA. Within the context of Chemical Accidents and Releases, the focus of these efforts has been to Prevent Them All or Stop Them Small.

These Safety Day programs are designed for and with local community stakeholders from industry and first response agencies with a goal of seeding a permanent, engaged geographic collaboration for accident prevention and response readiness ownership and partnership. Agenda and curriculum elements include technical training focused on local risk, first responder field practicums and case studies, accident prevention program prioritization, practice and accountability, bowtie analyses of event propagation and management, and emergency response planning centered on the first 30 minutes and one-plan approach. The programs often include field demonstrations and exercises for fire departments and facility responders.

As a best practices demonstration, the presenters for this session will share the developed programs, recruitment methodologies and the demonstrated results of these efforts. In addition to Responsible Collaboration, other Vision 2020 principles embraced and promoted by the Safety Day Approach include Disciplined Adherence to Standards, Intentional Competency Development, Enhanced Application & Sharing of Lessons Learned, and Enhanced Stakeholder Knowledge. As a bonus, the societal theme of Harmonization of Standards is enhanced by the community of practice element of engaging federal, state and local regulators as well as industry stakeholders and process safety professionals.

548394 Risk Analysis Screening Tool – CCPS Project 280

A spreadsheet based Risk Analysis Screening Tool (RAST) is available through the Center for Chemical Process Safety (CCPS), courtesy of the Dow Chemical Company. The tool allows users to identify hazards and understand risks including causes, consequences and probabilities. RAST can be used for existing or proposed installations to facilitate MOC reviews, assess compliance with risk criteria, or evaluate proposed designs.

The users can input the type of equipment being used along with the equipment characteristics, chemical data, process conditions, and the plant layout. Using these inputs, the tool is used to help identify hazard scenarios and evaluate the consequences of representative loss of containment scenarios. RAST also allows the users to specify their own hazard scenarios based on their own knowledge of the process. Users can also input their own risk ranking matrix. With the hazard scenarios and the risk matrix, the user can select those scenarios of most concern and use the tool to create a Layer of Protection Analysis (LOPA) spreadsheet for the process.

As an example a comparison of RAST calculated scenario identification and consequence results to actual case studies from the CSB will be presented. Courses on how to use RAST, and how to adjust the pre-loaded default parameters are available.

550863 Impacts of Climate Change on Process Safety Management in Oil and Gas Operations

Impacts of Climate Change on Process Safety Management in Oil and Gas Operations

Chizaram Nwankwo

Coventry University

Project objectives and goals

The aim of this study is to analyse the impacts of climate change on process safety management in oil and gas operations. The objectives for this study include the following: -

- To evaluate past, present and future trends of climate change around various global regions
- To establish the relationship between climate change and process safety management in oil and gas operations
- To proffer adaptation and mitigation measures for climate change impacts on process safety management in oil and gas operations

In the last two decades, climate change has emerged as one of the most controversial topics that has been argued on several fronts (Hulme 2009). The idea of global warming and the negative impact of human-driven activity on the global weather has been fiercely debated (Cook *et al.* 2016). However, Field *et al.* (2012) argue that the facts and figures tend to reiterate the continuous occurrence and proliferation of climate change, as well as its increasing impacts in different industrial sectors. Indeed, the warming of the climate is unequivocal (Panteli and Mancarella 2015). According to IIGCC *et al.* (2017), 70% of global greenhouse gas (GHG) emissions are attributed to energy-related activities. The oil and gas industry contributes to 60% of these energy-related emissions due to production of fossil fuels as their main operational activity (Puig *et al.* 2015). Evidence from NASA (2018) suggests that there has been 1^0C increase in temperature which is close to the set mark in accordance to the Paris Agreement.

There are several case studies such as the flood in San Jacinto River, Houston Texas in 1984 and Coffeyville Refinery, Kansas in 2007. In the case of the San Jacinto flooding event in 1984, 8 pipelines were ruptured and 29 were undermined, 35,000 barrels of oil were spilled and ignited, while 547 people were injured and $23 million was incurred in damage costs (Matek 2017).

Hence, adaptive measures need to be applied to manage the severe consequences of climate change on oil and gas operations (Smith 2013). Such adaptive measures include more robust design specifications for oil and gas assets exposed to harsh climatic conditions (Cruz 2010). Ramasamy and Yusof (2015) suggest that this could be vital in offsetting the impacts of climate change on assets such as offshore rig platforms and equipment, transport pipelines, storage facilities etc. However, there is still little attention accorded to design and adaptation of oil and gas assets to offset the effects of climate change on them (Skjaerseth and Skodvin 2010). Therefore, this study seeks to analyse the impacts of climate change on deterioration of oil and gas assets.

Description of methods and results

Firstly, this study adopts a pragmatic paradigm since it explores the consequences of actions, is problem-centred, pluralistic, real-world practice oriented and allows for use of mixed models (Creswell and Creswell 2018). The area of investigation for this study is focused on impacts of climate change on oil and gas operations. The predominant approach taken for this study is a correlational research approach which tries to establish a relationship between climate change and oil and gas operations. The selected approach is due to the continuous occurrence of accidents in the oil and gas industry due to asset deterioration.

After thorough literature review of climate change trends and asset integrity issues in the oil and gas industry, this study identified a gap in knowledge that the oil and gas industry has not duly considered climate change during design of assets; hence posing a risk to oil and gas operations conducted under harsh climatic conditions. Therefore, the data collection for this study employed "concurrent triangulation" method of quantitative and qualitative data to ensure robust analysis of key variables such as climate change and oil and gas operations. Consequently, data collection tools for this study include questionnaires, interviews, archive data and documentary analysis. The data collection for this study was conducted over a period of 3 months.

Firstly, it involved piloting the questionnaire among academic and industry professionals to identify any errors, as well as testing and refining the data collection instrument. Next, the study was approved as a medium-risk project by Coventry University Ethics Committee, as primary and secondary data were collected from archive data, documents, questionnaires and interviews. The population sample for the study includes oil and gas industry professionals such as Mechanical Engineers, Facility Managers, Production and Well Engineers, Process Engineers, Health and Safety Officers, as well as Safety and Reliability Engineers. The questionnaires used for this study were designed using Bristol Online Survey (BOS) tool and covered questions around climate change and oil and gas operations. These data were predominantly collated and analysed using various software such as Zotero, Microsoft Excel and SPSS software. Statistical analysis conducted in this study include trend analysis, descriptive statistics, Pearson's correlation, Cronbach's alpha reliability analysis and Relative Importance Index (RII). Lastly, the findings of this study were benchmarked and justified with peer-reviewed academic literature.

Potential for application of results

This research could aid in curbing several accidents caused by asset deterioration due to increasing number of climate change events. It will help the oil and gas industry to avoid the enormous financial, legal and moral costs associated with accidents. Also, this study will provide a deeper insight on current climate change statistics and proffer adaptation and mitigation measures on how to deal with these problems. Similarly, it could broaden the scope of researchers to start incorporating impacts of climate change on various industrial sectors and subsectors asides the oil and gas industry.

References

Cook, J., Oreskes, N., Doran, P.T., Anderegg, W.R., Verheggen, B., Maibach, E.W., Carlton, J.S., Lewandowsky, S., Skuce, A.G., and Green, S.A. (2016) 'Consensus on Consensus: A Synthesis of Consensus Estimates on Human-Caused Global Warming'. *Environmental Research Letters* 11 (4), 048002

Creswell, J.W. and Creswell, D.J. (2018) *Research Design: Qualitative, Quantitative, and Mixed Methods Approaches*. 5th Edition (International Student Edition). Sage Publications

Cruz, A.M. (2010) *Vulnerability of Oil and Gas Infrastructure to Climate Change and Extreme Weather Events* [online] available from <http://indico.ictp.it/event/a09141/session/35/contribution/25/material/0/0.pdf> [5 March 2018]

Field, C.B., Barros, V., Stocker, T.F., and Dahe, Q. (2012) *Managing the Risks of Extreme Events and Disasters to Advance Climate Change Adaptation: Special Report of the Intergovernmental Panel on Climate Change*. Cambridge University Press

Hulme, M. (2009) *Why We Disagree about Climate Change: Understanding Controversy, Inaction and Opportunity*. Cambridge University Press

IIGCC, CERES, and IGCC (2017) *Global Climate Disclosure Framework for Oil & Gas Companies* [online] available from <http://www.iigcc.org/files/publication-files/Global_Climate_OG_DisclosureFramework.pdf> [14 November 2017]

Matek, B. (2017) *CHARACTERISTICS OF OIL AND GAS PIPELINE ACCIDENTS CAUSED BY CLIMATE CHANGE INTENSIFIED HURRICANES* [online] Baltimore, Maryland: Johns Hopkins University. available from <https://jscholarship.library.jhu.edu/bitstream/handle/1774.2/46088/Benjamin%20Matek.pdf?sequence=1&isAllowed=y> [26 June 2018]

NASA (2018) *Global Climate Change* [online] available from <https://climate.nasa.gov/> [9 May 2018]

Panteli, M. and Mancarella, P. (2015) 'Influence of Extreme Weather and Climate Change on the Resilience of Power Systems: Impacts and Possible Mitigation Strategies'. *Electric Power Systems Research* 127, 259–270

Puig, D., Haselip, J., and Naswa, P. (2015) *Adaptation to Climate Change in Colombia's Oil and Gas Industry: Recommendations to Promote Risk Management* [online] UDP-CRD-OLADE-01(EN). Denmark: UNEP DTU Partnership. available from <www.unepdtu.org> [19 September 2018]

Ramasamy, J. and Yusof, S.M. (2015) 'A Literature Review of Subsea Asset Integrity Framework for Project Execution Phase'. *Procedia Manufacturing* [online] 4, 79–88. available from <http://linkinghub.elsevier.com/retrieve/pii/S2351978915011336> [14 May 2017]

Skjaerseth, J. and Skodvin, T. (2010) *Climate Change and the Oil Industry: Common Problem, Different Strategies*. Manchester University Press

Smith, M.H. (2013) *Assessing Climate Change Risks and Opportunities for Investors: Oil and Gas Sector* [online] Analysis of Climate Change Risks and Opportunities. Technical Report. ANU and Investor Group on Climate Change. available here. [5 April 2018]

Further Information

Chizaram Nwankwo (E: nwankwo4@uni.coventry.ac.uk)

546246 Study of the Effects of Wind Speed and Surface Roughness Under Stable Atmospheric Condition Using Aloha Model and CFD Model

Study of the effects of wind speed and surface roughness under stable atmospheric condition using ALOHA model and CFD model

Many substances released in chemical processes are toxic, flammable, or even both. When these substances accidentally release to the atmosphere, human health and the environment could be threatened. As results, a reliable quantitative risk analysis tool is crucial to provide decision makers useful information on downwind concentration and hazards. ALOHA consequence model and Computational Fluid Dynamics (CFD) numerical method are widely used in chemical industries. Validation of these models under different scenarios is a necessary preliminary step before applying them to risk analysis. CFD models generally have acceptable concentration predictions on dense gas dispersion. ALOHA model is verified with DEGADIS model, but there is lack of study on effects of wind speed and surface roughness on dense gas dispersion. This paper compared field data from the Kit Fox experiments, which were CO_2 gas releases under stable atmospheric conditions with increased roughness, with CFD and ALOHA models' predictions. Also, statistical performance indicators were utilized to evaluate the accuracy of ALOHA model. Statistical performance indicators showed that ALOHA had systematic errors for dense gas dispersion with increased roughness under stable atmospheric conditions. Comparing effects of wind velocity profile and surface roughness values in CFD model and ALOHA model, this paper provided a better understanding of effects of key parameters in both ALOHA and CFD models. Furthermore, engineers would have better quantitative risk assessments under stable atmospheric conditions when they are using ALOHA consequence model and CFD model.

548222 Consequence Assessments and Collaboration to Streamline Project Siting

Collaboration is often sought after across disciplines, but all too often disregarded in favour of linear project development. For siting and consequence, as with other areas where design interacts with environment, health and safety constraints, this often leads to more costly design due to increased project iterations.

Stantec recently completed a consequence assessment for a compressor station to assess flammable extents during an emergency blowdown. At the same time, the client was experiencing siting issues due to sound emissions from the same process. The facility was grappling with competing factors: blowdown the facility in time to avoid catastrophic failure of their infrastructure, avoid excessive flammability extents from the blowdown stack location, and avoid extraneous noise exposure to nearby receptors. Using transient consequence modelling tools to model the blowdown sequence, Stantec was able to simultaneously and efficiently assess the flammable extents and sound emissions from the blowdown operations.

Consequence modelling for flammability was completed using a transient pipeline blowdown modelling tool to simulate the mass, energy, and momentum throughout the pipeline system. The model first simulated normal operating conditions, and then simulated the blowdown to estimate mass flow rate, liquid mass fraction, and release temperature from the blowdown stack. Several sizes of reducers upstream of the blowdown stack were introduced to estimate the pressure drops in the piping that would be required to appropriately reduce noise levels but also achieve full depressurization of the facility in time to meet safety criteria.

Consequence modelling of flammability was completed using SLAB, while the predicted change in sound pressure level from the blowdown stack was completed using Cadna acoustic modelling software. By combining these two analyses, the assessment costs streamlined for both required studies. In addition, the acoustics study benefited from additional insight provided by the consequence modelling tool.

548348 CFD Simulation of the Effect of Wind Speed and Release Elevation on Dispersion of Flammable Fluids at LNG Export Facilities

Facility siting studies for LNG export terminals are required by federal regulations to assess the potential hazard from accidental releases of flammable fluids. Federal regulations require under 49 CFR 193.2059 the evaluation of flammable vapor-gas dispersion distances from releases from flammable fluids in the facility for a range of representative wind conditions based on historical weather data expected to produce the maximum dispersion distances. A parametric CFD study will be conducted to evaluate not only the effect of wind speed but also the effect of the release elevation on the dispersion distance from flashing and jetting fluids. Hazard distances will be compared between the 2D software PHAST and the 3D software FLACS for a set of exemplar fluids found at LNG export facilities. The ability of the FLACS software to handle detailed 3D geometry provides for the evaluation of the effect of obstructions, congestion, confinement and mitigation measures on the dispersion distances. Release scenarios typical of LNG export facilities will be presented, with a focus on jetting and flashing releases of LNG, ethylene, propane, mixed refrigerants and heavy hydrocarbons.

546616 A Comparative Study of the Optimization Models for the Gas Detector Placement in Process Facilities

Gas detector network is an important layer of protection in process facilities for prevention gas release accidents of turning into fire and explosion disasters. But traditional standards just provide basic principles of the installation of gas detectors. Recently mathematical programming gradually became an important method to solve the problem. When using the mathematical programming method to optimize the detector layout, different optimization strategies will result in different optimization target models and detector layouts. In this study, three models have been developed and contrasted, namely minimal detection time P-Median model (MITP), minimal leakage concentration P-Median model (MICP), and minimal individual risk P-median model (MIRP), for the optimization of the gas detector placement based on stochastic programming. And as a basis, possible release scenarios were identified. In order to achieve the goal of using finite discrete scenarios to represent all scenarios effectively and quantitatively, clustering analysis was used to filter similar scenarios and select representative leak scenarios. The input data of these models was obtained from computational fluid dynamics. Furthermore, an adaptive particle swarm optimization (APSO) algorithm was applied to solve these models. The gas detector network optimization in a diesel hydrogenation unit was carried out as a case study. Results demonstrate that the MICP model is superior to the MITP model and MIRP model in describing the integrity of hazardous gas leakage and diffusion distribution. But the cumulative detection time of its optimized layout is slightly longer than that of MITP. Moreover, the control effect of MIRP model on personal risk caused by hazardous gas leakage was increased by 24.2%, although the minimum number of detectors was slightly higher than that of MICP model.

549411 Reducing Industrial Risk in a Converging It/OT World: 20 Experts Share Their Strategies

Rapid adoption of Industrial IoT requires tight integration between IT and OT. CISOs and corporate OT leaders are now scrambling to develop strategies, secure budget, and implement solutions to secure their mission-critical industrial control systems against burgeoning cyber threats.

Effective integration of IT and OT cybersecurity strategies is essential to address risks successfully. To secure industrial facilities and ensure safe, reliable production, IT and OT security teams – traditionally two separate disciplines with different priorities – must collaborate and share cybersecurity and risk management best practices.

In a series of interviews in late 2017 and early 2018, more than 20 different experts on the front lines of OT cybersecurity risk mitigation shared practical advice for making control systems more secure. These experts represented a diverse range of industries including oil and gas, chemicals and refining, and power generation.

In their interviews, they shared their insights into how to leverage IT/OT convergence and develop an effective OT cybersecurity program that meets challenges in real-world ICS environments. Technology plays an important role, but as these experts also point out, more technology isn't always the solution. Just as important is the need to establish effective communication and trust between cybersecurity professionals and OT engineers.

Security project teams must make a concerted effort to create alignment between business management, the cybersecurity community, and OT professionals. Without alignment, security initiatives will fail to attain the support required to ensure success.

In this presentation, attendees will learn:

- Practical, real-world recommendations from OT cybersecurity experts
- Best practices for breaking down barriers and building trust between cybersecurity professionals and OT engineers
- How to align business management and plant operations on cybersecurity strategies

549417 The Intersection of OT Security and Process Safety

The rise of cyberattacks on critical infrastructure creates troubling concerns for security, economic stability, and personnel and public safety. However, many companies are still relatively unprepared to successfully defend their facilities from cyberattacks launched by increasingly sophisticated threat actors. Safe production is at risk.

Safety focuses on incident prevention and measures risks to personnel. Cybersecurity focuses on protecting systems from threat actors and measures risk based on the threat landscape. The increased network connectivity that exists within industrial facilities today means that cybersecurity is now closely linked to the safe operation of industrial facilities and their processes. Successful defense of cyber-physical systems from nation-state attacks requires close coordination and collaboration between cybersecurity and process safety. Every safety initiative must now consider cyber risk. Every cybersecurity initiative must now consider safety.

Considerable overlap exists between risk analysis in safety and cybersecurity. Safety and cybersecurity disciplines can learn much from each other. However, safety and cybersecurity standards and corporate organizational structures remain siloed in far too many companies. This session focuses on the importance of executing a cohesive strategy for both safety and cybersecurity to ensure safe and reliable operations.

In this session, we will:

- Examine the existing cyber threat landscape
- Discuss how established process safety best practices can be applied to OT cybersecurity
- Provide practical guidance for improving cyber defenses and reducing safety and environmental cybersecurity risks

551385 Process Safety at Offshore Aerodromes (Helidecks)New Application for Process Safety

Offshore air transport is an importance activate of the oil & gas industry. In this way, the management of air operations is essential for maintaining the safety and reliability. The present work has the objectives of divulging the methods used by
Petrobras to audit maritime units with helideck and to demonstrate the contribution of such methods to the safety of process in the air operations. To achieve these objectives, the paper was based on the Center for Chemical Process Safety (CCPS) publication "Guidelines for the Safety of Risk-Based Processes", and aims to share the Bow tie methodology for the qualitative analyzes of risks inherent in Air Operations, focusing on the barriers typically faced by oil and gas industry, in order to contribute to the process by providing a graphical and structured approach to the required barriers, both to avoid catastrophic consequences and to prevent events recognized as precursors.

Thermodynamics and Computational Methods for Relief Valves

By Umesh Mathur, P.E.
umesh.mathur@aggreko.com , mathuru@gmail.com
Aggreko Process Services
3732 Magnolia Street, Pearland, Texas 77584, USA

15th Global Congress on Process Safety (AIChE, March, 2019, New Orleans)

1. INTRODUCTION

The computation of flow through a relief valve is one of the most important aspects of safety system design. This paper discusses the fundamental question of choosing the proper thermodynamic basis for such calculations. We also discuss the discrepancies we have observed in the literature on this crucial question and attempt to dispel some of the misconceptions we have found.

Our concern with this issue arose from a study of the relevant literature on this subject, which shows that the flow of a compressed gas (or gas-liquid mixture) through a rigidly held valve or hole in a pipeline must be treated as an irreversible, isenthalpic process. This reasoning flows simply from the First Law of Thermodynamics as applied to a flow process.

For example, as early as 1947, Hougen and Watson[1] discussed "Free Expansion" of gases (p. 551) and noted as follows:

"The unrestrained expansion of a gas is known as free expansion. Under conditions of no restraint no work is done, and under adiabatic conditions no heat is added. Free expansion under flow conditions is commonly known as throttling or as the Joule-Thompson effect referred to in Chapter XII.

The Joule-Thompson effect is measured experimentally by expanding the gas slowly and in steady flow through a well insulated porous plug; in this way, potential work is lost, and no heat is allowed to enter or leave the system through the walls. The fluid flows reversibly into and out of the process, but the expansion step is completely irreversible. Since no heat is added or lost, the process takes place under conditions of constant enthalpy, according to Equation (VII-11), page 207. However a change in internal energy results if any change in flow energy pV occurs."

Similarly, in a discussion of valves and throttles, Elliott and Lira[2] observe as follows:

*"A throttling device is used to reduce the pressure of a flowing fluid without extracting any shaft work and with negligible fluid acceleration. Throttling is also known as **Joule-Thompson expansion** in honor of the scientists who originally studied the thermodynamics. An example of a throttle is the kitchen faucet. Industrial valves are modeled as throttles....Changes in kinetic and potential energy are small relative to changes in enthalpy as we just discussed. When in doubt, the impacts of changes in velocity can be evaluated as described in Example 2.9. The amount of heat transfer is negligible in a throttle. The boundaries are not expanding, and there is also no mechanical device for transfer of work, so the work terms vanish. Therefore, a throttle is isenthalpic."*

Finally, in a discussion of nozzles, Klein and Nellis[3] observe that:

"...The potential energy terms of the inlet and outlet flows are not usually significant. Nozzles neither require nor produce power ($W_{in} = W_{out} = 0$) and are very nearly adiabatic. Note that the kinetic energy terms associated with the flows, particularly the outlet flow, cannot be neglected as the purpose of the nozzle is to increase the kinetic energy of the fluid. With these simplifications, an energy balance on the nozzle becomes:

$$H_{in} + \tfrac{1}{2}V_{in}^2 = H_{out} + \tfrac{1}{2}V_{out}^2 \qquad (4\text{-}54)"$$

In the light of these assertions by these noted authorities, and others[6][10], it is hard to understand how the flow through a relief valve, or a hole in a vessel that is open to the atmosphere, can be treated as isentropic (***reversible*** adiabatic) in nature, as opposed to the isenthalpic (***irreversible*** adiabatic) basis that is consistent with free expansion. This is because an isentropic expansion process would necessarily show an outlet enthalpy for the gas from the nozzle or valve (or an orifice in a punctured vessel) to have a lower enthalpy than the inlet, the difference in such enthalpies being equal to the work done, for example on a rocket or airplane. However, a relief valve or vessel remains stationary and absolutely no work is done during a free expansion by the gas flowing through.

Nevertheless, several standard references have used isentropic formulas for free expansion[4],[5] in stationary valves, or nozzles, from the inlet to the throat. We believe that this methodology is flawed as it violates the steady flow energy equation (i.e., the first law of thermodynamics written for an open system). This procedure would result in an erroneous prediction of outlet gas temperature when entropy is held constant. This, in turn, would predict an erroneous outlet gas density and thus, for a desired gas relief rate, the required cross-sectional area for the valve throat and also any conduit or pipe downstream of the valve.

The practical consequence of this anomaly in predicting outlet gas properties (especially the gas density at the throat) would be that the relief valve sizing could be incorrect. Therefore, it is important that the proper thermodynamic basis (isenthalpic expansion) be used in all such computations.

Conversely, for a given relief valve, errors in throat density based on an isentropic path would also affect the relief rate. Both types of errors could lead to potentially serious process safety concerns.

For these reasons, when calculating the release rate for gas at a high pressure through a relief valve, it is important to be able to use the proper basis for gas enthalpy (and therefore accurate computations of fluid thermodynamic properties). The same considerations apply to the gas release rate calculations resulting from a rupture in a pipeline or vessel containing a gaseous mixture under pressure.

To help clarify the proper methodology for a relief valve - or a hole in a pressurized vessel - the computations that should be performed are described below. We also show a few simple examples. Note that the possibility of sonic velocity at the throat, or choking, of the valve (or at the orifice in a ruptured vessel) is explicitly considered. In both such cases, the procedure described below is subject to the following assumptions:

- The opening can be represented by a convergent nozzle of exit area equal to the area of the throat of the relief valve, or hole in the case of pipeline rupture

- The compressed gas pipeline, or leaking container, is a large reservoir in which upstream pressure and temperature are maintained constant
- The compressed gas undergoes an irreversible, adiabatic process of expansion through the nozzle.

The flow through such a device from either (a) the inlet to the throat, or (b) from the throat to the outlet follows the steady flow energy balance equation (Equation 9.21, Himmelblau & Riggs[6]):

$$\Delta E = Q + W - \Delta(H + KE + PE) \qquad (1)$$

Where:

ΔE = energy accumulation (zero at steady state)

Q = net heat transfer into the system (very close to zero for a relief valve)

W = net work done by the system (zero for a relief valve)

ΔH = change in enthalpy of the system

ΔKE = change in kinetic energy of the system

ΔPE = change in potential energy of the system (zero for a relief valve)

Clearly, for steady flow through an orifice, valve (or relief valve), ΔE, Q, W, and ΔPE are all zero. Therefore, we are left with:

$\Delta(H + KE) = 0$, or

$\Delta H = -\Delta KE$, or

$$H_{out} = H_{in} - \Delta KE \qquad (2)$$

In other words, the outlet enthalpy is equal to the inlet enthalpy minus the change in kinetic energy in going through the throat or orifice. (It should be noted that the same logic would apply for discharge through an ordinary valve, a relief valve, or an orifice plate).

If there is no change in velocity, the outlet enthalpy will be the same as that of the inlet gas. This phenomenon is generally referred to as a Joule-Thompson (J-T) expansion. For an ideal gas, enthalpy is a function of temperature only; therefore, a J-T expansion would result in no change in temperature. However, for most real gases, a J-T expansion results in a decrease in temperature. There are a few notable exceptions, however, such as hydrogen or helium gas for which the J-T coefficient has the opposite sign[10]. On the other hand, if the gas speeds up, its kinetic energy will increase and, therefore, its enthalpy will decrease in accordance with equation (2) above. This kinetic energy increase would result in a decrease in outlet gas enthalpy, and therefore temperature, even for an ideal gas.

For ideal gases, the computation of enthalpy for ideal gases is straightforward as it is a function of temperature only. For real gases, however, enthalpy is also a function of pressure, temperature, and composition and use of an equation of state (EOS) becomes necessary[2] to calculate the enthalpy departure from the ideal, as discussed below.

COMPRESSIBILITY FACTOR AND ENTHALPY FOR REAL GASES

For real gases, enthalpy is found using a suitable equation of state (EOS), such as the Soave-Redlich-Kwong (SRK), Peng-Robinson (PR), or Lee-Kesler (LK). This procedure may be summarized as follows:

At the given upstream pressure (P), temperature (T), and composition, solve the EOS for the compressibility factor (z). This requires solving a cubic polynomial in z when using any cubic EOS, such as the SRK or PR, using either an explicit analytical or numerical trial and error method. For the LK, however, a trial and error procedure is always required as the LK EOS requires solving the 8-constant Benedict-Webb-Rubin[2] (BWR) equation, which is not a simple cubic in z, once for the simple fluid and again for the reference fluid.

Using the appropriate thermodynamic expressions for the EOS, use this value of z to calculate ΔH_d, the enthalpy departure from ideal.

Calculate the ideal gas enthalpy, H_0 (function of temperature T_0 only).

Find the real gas enthalpy using:

$$H_r = H_i - \Delta H_d \qquad (3)$$

Where:

H_r = real gas enthalpy
H_i = ideal gas enthalpy
ΔH_d = enthalpy departure from ideal

This procedure must be followed for calculating enthalpy at both upstream and downstream conditions.

The gas velocity at the throat may be sonic or sub-sonic. Also, as the gas speeds up in passing through the valve throat or hole (since the cross-sectional area is lower than that of the pipe), its kinetic energy increases and this results in a reduction of the enthalpy (Elliott & Lira[2], Eq. 2.53):

$$H_t = H_0 - \Delta KE \qquad (4)$$

Where:
H_0 = upstream gas enthalpy[J/kg]
H_t = exit gas enthalpy at the throat or hole[J/kg], and
ΔKE = change in kinetic energy[J/kg]

ΔKE is found as follows (Elliott & Lira's Eq. 2.53):

$$\Delta KE = (u_t^2 - u_0^2)/2 \qquad (5)$$

Where:

u_t = throat (or hole) velocity[m/s], and

u_0 = upstream velocity[m/s].

Clearly, the outlet and inlet enthalpies are equal only if the change in velocity is negligible.

RELEASE RATE CALCULATION

The computation of the release rate from a pipe rupture should be done subject to the following equations:

Bernoulli's Equation

This equation relates upstream and downstream conditions for an arbitrary flow geometry, as follows:

$$P_0 + \rho_0.u_0^2/2 + \rho_0.g.z_0 = P_e + \rho_e.u_t^2/2 + \rho_t.g.z_t + \Delta P_{friction} \quad (6)$$

Where the subscripts o and e refer to the inlet and exit planes, respectively, and

P	= pressure[Pa]
ρ	= density[kg/m^3]
u	= velocity[m/s]
g	= gravitational constant[m/s^2]
z	= elevation above datum[m]
$\Delta P_{friction}$	= friction losses[Pa]

The friction losses have been ignored for simplicity in this presentation.

Obviously, for a hole, or throttle (or relief) valve, the elevations z_0 and z_t are the same.

Sonic Velocity

The sonic velocity in any medium[2] may be expressed as:

$$u_s = (\gamma \,.\, z \,.\, R \,.\, T \,/\, M)^{0.5} \quad (7)$$

Where:

u_s	= sonic velocity[m/s]
γ	= gas C_p/C_v ratio[]
z	= gas compressibility factor[]
R	= gas constant = 8314.46[J/(kgmol.K)]
T	= temperature[K]
M	= gas molecular weight[kg/kgmol]

In Equation (7), properties at the throat must be used for choked flow calculations. For ideal gases, the compressibility factor z is equal to 1. For real gases, however, the compressibility factor z is generally not equal to 1 and can be found only using an appropriate EOS.

C_p, C_v, and $\gamma=C_p/C_v$ ratio

Similarly, for ideal gases, the value of γ is found from the value of C_p, which is a function of temperature only, and the value of C_v is found from the following:

$$C_v = C_p - R \quad (8)$$

For real gases, however, the value of γ again requires use of an equation of state to calculate the ΔC_{pd} and ΔC_{vd} departures from ideality. The real gas C_p and C_v values are found using the ideal values and these calculated departures from ideality, by subtraction, in a manner similar to that described above for enthalpy:

$$C_{p1} = C_{p0} - \Delta C_{pd} \quad (9)$$

Where
C_{p1} = real gas C_p[J/(kg.K)]
C_{p0} = ideal gas C_p[J/(kg.K)]
ΔC_{pd} = departure in C_p from ideal gas value[J/(kg.K)]

Similarly, $C_{v1} = C_{v0} - \Delta C_{vd}$ (10)

In Equation (10),
C_{v1} = real gas C_v[J/(kg.K)]
C_{v0} = ideal gas C_v[J/(kg.K)]
ΔC_{vd} = departure in C_v from ideal gas value[J/(kg.K)]

In a conduit, orifice, or valve, the actual gas velocity, u_e, can never exceed the sonic limit, u_s. Note that u_s, in general, is different at the upstream conditions compared to the value at throat conditions. The upstream and throat temperatures and pressures of the gas may be different owing to the kinetic energy effect, Equation (4).

At a sufficiently low upstream pipeline or vessel pressure, the hole velocity would be lower than the sonic velocity limit. Above some critical limit for the upstream pressure, however, the maximum discharge flow is limited by the sonic velocity criterion at the hole.

Density

Gas density is computed using:

$\rho = P . M / (z . R . T)$ (11)

Where:
ρ = density[kg/m^3]
P = pressure[Pa]
M = gas molecular weight[kg/kgmol]
z = gas compressibility factor[]
R = gas constant = 8314.46[J/(kgmol.K]
T = temperature[K]

The gas compressibility factor is 1 for ideal gases; it is calculated from the EOS for real gases.

1.2 Velocity

The velocity in a pipe or hole (or valve throat) is found using:

$u = W / (\rho . A)$ (12)

Where:
u = velocity[m/s]
W = mass flow rate[kg/s]
ρ = gas density[kg/m^3]
A = cross sectional area[m^2]

For a circular pipe, the cross-sectional area is defined as:

$A = \pi . D^2/4,$ (13)

Where:
D = inner diameter[m]

For non-circular cross sections, the equivalent diameter D_e is defined using the concept of the hydraulic radius which is defined as:

$D_e = 4$ (Cross-sectional-area/Wetted-perimeter) (14)

It can readily be seen that this definition holds for a circular pipe also: $4 . \pi . D^2 / 4 / (\pi . D) = D$

As noted earlier, for ideal gases, the compressibility factor z is equal to 1. For real gases, however, finding the gas density ρ requires first calculating z from the EOS. This is done using the procedure as described above.

It should be noted that conditions at the throat of a valve (or at the hole in a pipe) are not known. Therefore, the computation of the gas flow rate must be done using an iterative procedure. When using Microsoft Excel, this is best performed using the Solver Add-in. Solver is a non-linear optimization technique that can simultaneously adjust the unknown quantities and solve the problem by enforcing the constraints imposed by the Bernoulli Equation (6) for throat velocity, and the Steady-Flow Energy Equation (1) for hole enthalpy. Also, the calculated throat velocity from the Bernoulli equation can never exceed the sonic velocity limit, Equation (7), and this constraint must be added explicitly when the upstream pressure exceeds a threshold value.

When solving for an unknown gas flow rate, if the upstream pressure is sufficiently low, hole velocity could remain sub-sonic. However, if the upstream pressure is sufficiently high, the pressure at the hole and also the gas mass flow rate would vary in a manner that obeys the sonic velocity limit at the hole.

Example

CASE (A): Methane is flowing through a 300[mm] diameter, above-ground pipeline at an absolute pressure of 46[bar] and a temperature of 300[K]. The pipeline develops a hole with an equivalent diameter of 50[mm]. Calculate the mass rate of leakage to atmosphere and also the velocity of the gas jet emerging from the hole. The pressure inside the pipeline may be assumed to be constant owing to continuous supply from the source.

CASE (B): Rework Case (A) assuming that the pressure in the pipeline is 1.5[bar] instead of 46[bar].

Case (A) $P_o = 46$[bar]

PROBLEM: Solve for release rate from a hole in a pipeline - sonic velocity at throat.

WE SHOW THE FOLLOWING DETAILED SOLUTION FOR THE IDEAL GAS CASE ONLY. Note that the solution of the real gas case requires use of complex computer programs to solve the chosen equation of state (e.g., Lee-Kesler) iteratively for the required thermodynamic quantities. Since these routines are imbedded in an outer loop for the unknowns (T and P at the throat, and the mass flow rate), the computational burden is too great for hand calculations.

The POLYMATH software[7] was used for solving this problem. All variables are defined in Table 1. The problem statement for Case (A) for the ideal gas case is as follows:

 # Initial guesses

$T_t(0) = 300$ # Initial guess for T_t[K]
W(0) = 20 # Initial guess for W[Kg/s]
$P_t(0) = 46$ # Initial guess for P_t[bar]

 # Aly-Lee ideal gas enthalpy constants:
A = 33298 # Ideal gas enthalpy constant A
B = 79933 # Ideal gas enthalpy constant B
C = 2086.9 # Ideal gas enthalpy constant C
D = 41602 # Ideal gas enthalpy constant D
E = 991.96 # Ideal gas enthalpy constant E
F = -120280000 # Ideal gas enthalpy constant F

 # The following three functions must be made equal to zero by adjusting the three unknown variables W[kg/s], T_t[K], and P_t[bar]:
$f(T_t) = h_t - h_{t2}$
$f(W) = u_t - u_{t2}$
$f(P_t) = u_t - u_s$

 # Initial guess ranges for non-linear equations
W(min) = 1
W(max) = 50
$T_t(min) = 200$
$T_t(max) = 400$
$P_t(min) = 10$
$P_t(max) = 46$

This problem requires solving the three simultaneous, non-linear algebraic equations (defined above) by varying T_t (the hole temperature), W (the mass flow rate), and P_t (the hole pressure):
$f(T_t) = 0$
$f(W) = 0$
$f(P_t) = 0$

In this case, the upstream pressure is high enough to cause the velocity at the hole to reach the sonic limit. This problem requires solving three simultaneous non-linear algebraic equations. The POLYMATH solution shown below for Example A was obtained using a software package called POLYMATH[7]. Note that any other similar package can also be used. We verified this numerical solution using the MATHCAD[8] and EXCEL[9] software packages also.

NOTE: The numerical solutions for Cases (A) and (B) are shown in Table 1, following the problem statement for Case (B).

Case (B) $P_0 = 1.5$[bar]

PROBLEM: Solve for release rate from a hole in a pipeline. Assume that the throat velocity is sub-sonic. This means that the outlet pressure is fixed at 1[atm] = 1.01325[bar].

When the outlet pressure is fixed, we are left with two non-linear equations. Also, for consistency with the methodology in (A) above, we use a very large upstream diameter to make the inlet velocity very small.

NOTE: If the calculated outlet velocity exceeds the sonic limit, we would revert to the problem formulation under (A) above. (This would add an equation requiring outlet velocity to equal sonic velocity, and thus again require solving three simultaneous non-linear equations.)

As noted earlier, the solution of the real gas case requires use of complex computer programs to solve the chosen equation of state (Lee-Kesler, in these examples) for the required thermodynamic quantities.

In this case, the upstream pressure is low enough so that the velocity at the orifice does not reach the sonic limit. Accordingly, this problem requires solving only two simultaneous non-linear algebraic equations. The solution shown below was obtained using a software package called POLYMATH. Note that any other similar package can also be used. As with Case (A), we verified this numerical solution using the MATHCAD and EXCEL software packages also.

The POLYMATH problem statement for Case (B) is as follows:

$f(T_t) = h_t - h_{t2}$

$f(W) = u_t - u_{t2}$

 # Initial guesses

$T_t(0) = 300$ # Initial guess for T_t[K]

W(0) = 20 # Initial guess for W[Kg/s]

Using POLYMATH, the solution for Case (A) shows that the gas release rate, W, for an ideal gas is 17.16[kg/s]. As noted above, identical results were obtained using the Mathcad and Excel software packages.

TABLE 1: SONIC OR SUB-SONIC FLOW THROUGH A VALVE OR HOLE

VARIABLES AND EQUATIONS

Variable	Description & Units	Explicit equations
A	Ideal gas enthalpy constant A	A = 33298.
B	Ideal gas enthalpy constant B	B = 79933.
C	Ideal gas enthalpy constant C	C = 2086.9
D	Ideal gas enthalpy constant D	D = 41602.
E	Ideal gas enthalpy constant E	E = 991.96
F	Ideal gas enthalpy constant F	F = -120280000.
P_t	Throat pressure [bar]	P_t = 1.01325
kcal_j	Conversion: 1[kcal] = 4184[J]	kcal_j = 4184
M	Gas molecular weight[kg/kgmol]	M = 16.0
T_0	Inlet gas temperature[K]	T_0 = 300
P_0	Inlet gas pressure[bar]	P_0 = 1.5
z_0	Inlet gas compressibility factor[]	$z_0 = f(T_0,P_0,x)$
$(c_p/c_v)_0$	Inlet gas C_p/C_v ratio[]	$(c_p/c_v)_0 = 1.32$
R	Gas constant[J/(kgmol.K)]	R = 8314.46
D_0	Diameter of pipe[m]	D_0 = 0.3
D_t	Diameter of Throat[m]	D_t = 0.05
h_0	Inlet enthalpy[kcal/kg]	$h_0 = (A*T_0 + B*C /\tanh(C/T_0) - D*E / \tanh(E/T_0) + F)/M/kcal\_j$
rho_0	Inlet density[kg/m³]	$rho_0 = P_0*M*1.e5/R/T_0$
A_0	Pipeline cross-sectional area[m²]	$A_0 = 3.14159/4*D_0^2$
A_t	Throat flow area[m²]	$A_t = 3.14159/4*D_t^2$
h_t	Throat enthalpy[kcal/kg]	$h_t = (A*T_t + B*C/\tanh(C/T_t) - D*E/\tanh(E/T_t) + F)/M/kcal\_j$
u_0	Upstream velocity[m/s]	$u_0 = W/rho_0/A_0$
rho_t	Throat density[kg/m³]	$rho_t = P_t*M*1.e5/R/T_t$
u_t	Throat velocity[m/s]	$u_t = W/rho_t/A_t$
u_s	Sonic velocity at throat[m/s]	$u_s = (z_0*(c_p/c_v)_0*R*T_t/M)^{0.5}$
u_{t2}	Throat velocity from Bernoulli Eq.[m/s]	$u_{t2} = (2/rho_t*(P_0*1.e5 + rho_0*u_0^2/2 - P_t*1.e5))^{0.5}$
dh_{KE}	ΔKinetic Energy = $\Delta(u^2)/2$[kcal/kg]	$dh_{KE} = (u_{t2}^2 - u_0^2)/2/kcal\_j$
h_{t2}	Calc. enthalpy at throat = inlet enthalpy - $\Delta(u^2)/2$[kcal/kg]	$h_{t2} = h_0 - dh_{KE}$
T_t	Temperature at throat[K]	$f(T_t) = 0$
W	Mass flow rate[kg/s]	$f(W) = 0$
P_t	Pressure at throat[bar]	$f(P_t) = 0$

TABLE 1: SONIC OR SUB-SONIC FLOW THROUGH A VALVE OR HOLE (CONTINUED)

CASE (A) AND CASE (B) SOLUTIONS (USING POLYMATH)

Variable	CASE (A) Value (SONIC)	CASE (B) Value (SUB-SONIC)
A	33300.0	33300.0
B	79930.0	79930.0
C	2086.9	2086.9
D	41600.0	41600.0
E	991.96	991.96
F	-120300000.0	-120300000.0
P_t	27.71686	1.01325
kcal_j	4184.0	4184.0
M	16.0	16.0
T_0	300.0	300.0
P_0	46.00	1.50
z_0	1.00	1.00
$(c_p/c_v)_0$	1.32	1.32
R	8314.46	8314.46
D_0	0.30	0.30
D_t	0.05	0.05
h_0	226.2098	226.2098
rho_0	29.50683	0.96218
A_0	0.07069	0.07069
A_t	0.00196	0.00196
h_t	205.2641	210.3263
u_0	8.22527	7.71172
rho_t	20.86568	0.73254
u_t	418.7383	364.6543
u_s	418.7383	427.2981
u_{t2}	418.7383	364.6543
dh_{KE}	20.94576	15.88352
h_{t2}	205.2641	210.3263
T_t	255.6215	266.1833
W	17.15555	0.52433
P_t	27.71686	1.01325

CASE (A) RESULTS

This problem was also solved using the Lee-Kesler equation of state to calculate non-ideal gas properties, using a suitable computer program. This solution is provided below, with the ideal gas results quoted alongside for comparison:

CASE (A)

	IDEAL GAS	REAL GAS
Inlet pressure[bar]	46.00	46.00
Inlet temperature[K]	300.00	300.00
Calculated mass flow[kg/s]	17.28	18.25
Inlet velocity[m/s]	8.27	8.09
Throat velocity[m/s]	424.65	421.48
Sonic velocity at throat[m/s]	424.65	421.48
Inlet Enthalpy[kcal/kg]	225.61	214.87
Delta Kin. Energy[kcal/kg]	21.54	21.22
Outlet Enthalpy[kcal/kg]	204.07	193.64
Throat pressure[bar]	27.32	25.73
Throat temperature[K]	254.27	250.83

It should be noted that these results are quite similar for this case because the behavior of the gas (methane) is close to ideal at the conditions specified. However, these results would very likely differ significantly if, for example, the inlet temperature was considerably lower and the pressure much higher in which case the gas behavior would be markedly non-ideal.

CASE (B) RESUTS

Case (B) was also solved using the Lee-Kesler equation of state to calculate non-ideal gas properties. This solution is provided below, with the ideal gas results quoted alongside for comparison:

CASE (B)

	IDEAL GAS	REAL GAS
Inlet pressure[bar]	1.5	1.5
Inlet temperature[K]	300.0	300.0
Calculated mass flow[kg/s]	0.524	0.525
Inlet velocity[m/s]	7.71	7.70
Throat velocity[m/s]	364.7	364.2
Sonic velocity at throat[m/s]	435.06	435.3
Inlet enthalpy[kcal/kg]	226.21	226.21
Delta Kin. Energy[kcal/kg]	15.88	15.85
Outlet Enthalpy[kcal/kg]	210.33	210.36
Throat pressure[bar]	1.01325	1.01325
Throat temperature[K]	266.18	266.25

As might be expected, the gas mass flow rate is far lower for Case (B) compared to Case (A). Also, for Case (B), it was found that the conditions at the throat are virtually identical for the ideal versus the non-ideal gas.

2. ADDITIONAL EXAMPLES

To illustrate the importance of the issues discussed in this paper, we show a few additional examples for the differences in our numerical results compared to others.

2.1 Gas Release From A Hole In A Vessel

A 0.1[in] hole forms in a tank containing nitrogen at 200[psig] and 80[°F]. Determine the mass flow rate through this leak (Crowl[4], Example 4-4, p. 143).

The solution for this problem using the Lee-Kesler EOS was 4.26e-2[lb/s], compared to 3.86e-2[lb/s] in the reference (gas velocity at the throat was sonic). This difference is over 10% and not negligible.

2.2 Flow Through A Nozzle

Air at p_0=7.0e5[Pa] and T_0=293[K] discharges to the atmosphere. Compute the discharge mass flux, G, the pressure, Mach number, and velocity at the exit[5].

Using the Lee-Kesler EOS, the results are as follows:

Quantity	Lee-Kesler (this article)	Ideal gas (isentropic)[5]	Difference
Mass flux[kg/(m^2.s)]	1589	1650	-3.68%
Exit temperature[K]	238.5	244	-5.5 [K]
Exit pressure[Pa]	212041	370000	-42.7%
Mach number	1.0	1.0	-
Sonic velocity[m/s]	338	313	+8%
Velocity[m/s]	338	313	+8%

Note that the nozzle is not predicted to be choked when the downstream pressure is above 2.12[bar] (kinetic energy effects must be included), as opposed to 3.7[bar] in the original reference.

If a relief valve is to be designed for this scenario, the size would be under-predicted somewhat using the isentropic formulas used in the original reference. However, this discrepancy could be significantly larger in cases where an isentropic expansion incorrectly predicts partial condensation of the gas at the throat. In that situation, the density of the two-phase mixture at the throat could be far lower than that of the uncondensed gas (for the isenthalpic case) and the valve could be under-sized significantly.

3. CONCLUDING REMARKS

It has been shown that calculation of the release rate from a relief valve (or a hole in a vessel) should be done using an isenthalpic (i.e., irreversible adiabatic) path for the flow process. Using an isentropic thermodynamic path for this problem violates the First Law of Thermodynamics, as expressed for steady flow in a stationary relief valve or a hole in a pipeline. This could lead to an incorrect design both when sizing relief valves and estimating release rate from an existing valve or hole in a vessel.

In addition, it is important to use the proper equation of state to estimate real gas properties reliably when compared to the ideal gas assumption, especially when gas conditions are far from ideal. The Bernoulli equation (for throat pressure), and the sonic velocity limit (when flow is choked at the throat) must also be included for a complete definition of the problem. If the upstream pressure is low enough and the throat is un-choked, the isenthalpic methodology - as described in this paper - should again be followed to ensure a theoretically sound and consistent basis. Iterative computations for non-ideal cases (real gases) generally are too intensive for hand calculations and require use of computerized chemical engineering software.

REFERENCES

1. Hougen, O.A., K. M. Watson: "Chemical Process Principles" (John Wiley & Sons, New York, and Chapman and Hall, London, 1st Edition, Part II, p. 551, 1947, and 2nd Edition with R.A. Ragatz, Vol. II, p. 653, 1959)
2. Elliott, J.R., and C. T. Lira: "Introductory Chemical Engineering Thermodynamics" (Pearson Education, Inc. publishing as Prentice Hall, pp. 68-69, 2nd Edition, 2012)
3. Klein, S. and G. Nellis: "Thermodynamics" (Cambridge Univ. Press, 1st Edition, p. 168, 2011)
4. Crowl, D. A., and Louvar, J.F.: "Chemical Process Safety" (Pearson Education, 3rd Edition, pp. 130-131, 2011)
5. Green, D. W., and R.H. Perry: "Perry's Chemical Engineers' Handbook" (McGraw-Hill, 8th Edition, Section 6, pp. 23-24, 2008)
6. Himmelblau, D.M., and J.B. Riggs: "Basic Principles and Calculations in Chemical Engineering" (Pearson Education, 8th Edition, Fig. 9.18, Item 3, p. 525, 2012)
7. Cutlip, M.B., and M. Shacham: "Problem Solving in Chemical and Biochemical Engineering with POLYMATH, Excel, and MATLAB" (Prentice Hall, 2nd Edition, 2007)
8. Mathcad 2001 Professional: "Mathcad User's Guide With Reference manual", (Mathsoft Engineering and Education, Inc., 2001)
9. Freund, S.M., M. Jones, and J.L. Starks: "Microsoft Excel 2013: Complete (Shelly Cashman Series)", (Course Technology, 1st Edition, 2013)
10. Lewis, G.N., and M. Randall: "Thermodynamics", (McGraw-Hill, 2nd Edition, Revised by K.E. Pitzer and L. Brewer, p. 48, 1961)

548434 Engineered Solutions for Effective Flare System Analysis

Relief and flare system documentation is an important part of Occupational Safety and Health Administration (OSHA) Process Safety Management (PSM) 29 CFR 1910.119. Flare system analysis (sizing or rating) can be performed using commercially available software. Effective flare system analysis requires reliable Process Safety Information (PSI) such as process and instrumentation diagrams (P&IDs), relief system evaluation, Heat and Material Balance (H&MB), relief valve data sheets, isometrics, and disposal system loads. Good engineering methodologies can be implemented to effectively design header systems and rate existing header for any change. Some methodologies include use of dynamic simulation to predict realistic relief loads, reboiler pinch analysis for applicable systems, and selection of global relief scenarios. This paper will describes some methodologies and a combination of instrumented systems for effective flare system analysis. Effective flare analysis helps save capital expenditure spent on flare headers and possible flare load reduction yields reduced radiation footprint on a facility.

548364 Flares Are the End of the Line; How to Permit Correctly

INTRODUCTION

-

Flare is one of the key equipment in any refinery or LNG facility. Many relief valves relieve to flare, and if the flare is not permitted corrected – and functional - the plant is down. You cannot operate the plant without an active flare or flare field. The permitting of a flare is one of the prime activities in the permitting process. But the permit is dependent on the type of flare. This presentation covers various Federal Regulations involved in the permitting process, whether general or specific to each type of flare.

FLARE TYPES AND THEIR DESIGN FUCTURES

-

There are mainly three types of flares. The most common and most economical one is an elevated flare. There are elevated flares as high as 660 ft. The can be classified many ways; self-supported, guy wire supported, derrick supported, demountable type etc. One of the most important is the type of burner, whether a pipe tip or "furnace" burners. The second type is a ground flare with rows of furnace burners arranged in arrays at grade and surrounded with a wind fence. The last type is an enclosed flare with furnace burner enclosed with a tall stack. This type can be large as 70 ft in diameter and 150 ft tall. The flare type on a given project is selected based on the plot availability, location of the plant, sparing philosophy etc.

PERMITTING REGULATIONS

This presentation will discuss cover various Federal regulations that are applicable to all types of flares and the various parameter that the regulators are needing when apply for the permit. The presentation will cover the required destruction efficiency of the flaring gases, exit velocity for an elevated flare, alternate means of emission limitation for ground flares. In particular, the tip velocity, burner type, backpressure, emission testing and smokeless capacity are specific to each type of flare.

WHAT YOU WILL LEARN

-

- This paper will provide sufficient information to ensure the correct type of flare is selected and permitted
- This paper will cover various flares types employed in refinery and in LNG plants
- The limitations of these flare
- Importance of flare in any facility
- Application Federal Regulations
- Parameter required in the permitting applications
- Ground flare testing to establish minimum destruction efficiency
- Alternate means of emission limitation requirements in lieu of testing

548232 A Case Study on Application of HAZOP for a Brewery

Abstract

Beer is the most consumed alcoholic beverage in the world, and most popular beverage after water and tea. It's an alcoholic beverage produced by the saccharification of starch, often derived from the malted grains such as barley and wheat, and the fermentation of the resulting sugar. The brewing process can be broken into four main constituents before the bottling line, Mashing, Lautering, Wort boiling and Fermenting. Breweries commonly rely on ammonia refrigeration to keep their supplies cold. However, some brewers fail to realize that ammonia is one of the dangerous chemicals covered under 29 CFR 1910.119, OSHA's regulation for preventing or minimizing the release of toxic, reactive, flammable or explosive chemicals. "Safety matters at all levels", even for companies that may not be PSM level facilities.

A HAZOP study identifies hazards and operability problems. The study involves investigating how the plant might deviate from the design intent. It is based on the principle that several experts with different backgrounds can interact and identify more problems and failure scenarios working together and combining their results. This paper presents a review of consequences identified in brewery HAZOP. It also outlines some of the recommendations and action items identified from the case study of a brewery.

Keywords: Process Safety Management (PSM), PHA, HAZOP, Brewery, Ammonia, OSHA.

548410 A Basic Review of Fundamental of Thermodynamics and Its Application in the Analysis of Consequences and Accident Investigation

Once the stage of hazard identification has been concluded in a risk analysis, the next stage corresponds to select and specify the risk scenarios that will be modeled their effects or damages associated. This stage correspond to the consequences analysis and usually takes place in three basic steps: a) source term definition, b) dispersion model and d) effect or damage model.

Depending on the risk scenario to be modeled, several thermodynamic fundamentals or principles are applied in each of these steps supporting or sustaining the behavior and expected results of the consequences analysis model. In this work we propose to review the main thermodynamic fundamentals that support each of the models applied in each of the three steps that define a consequence analysis. This review of basic thermodynamic principles defines a set of different thermodynamic basic principles to apply to representative and different cases of consequence analysis. These basic thermodynamic principles are applied in the consequence modeling of a pair of cases of accident inventions to corroborate their validity.

548331 Evaluating Safe Location Criteria for Vents and Relief Devices Discharging to Atmosphere

In some process technologies, it may not be feasible to dispose the effluents from safety valves and vents to a closed system and discharging directly to the atmosphere can be the only alternative. Some examples include High Pressure Polyethylene Processes, Olefins and Storage units. The discharge can be directed to the atmosphere if:

- Local regulations permits it;
- Does not harm the environment;
- Does not harm the neighborhood;
- And is directed to a Safe Location.

The criteria for defining what a safe location is can vary, but it must include at least a consequence assessment of the flammable, toxic and thermal radiation effects. Some guidance can be found on industry standards and engineering references such as API 521 and the CCPS Guidelines for Pressure Relief and Effluent Handling Systems

One aspect to discharge hydrocarbons to a Safe Location is to evaluate the dispersion of the discharged hydrocarbons and ensure that it is diluted bellow the lower flammable before reaching the ground level or elevated structures. These areas are typically with people circulation and ignition sources. Another aspect that must also be evaluated is the thermal radiation generated due to an ignition of the discharge affecting personnel or damaging equipment or other assets. Lastly, if there is a possibility of the flammable mixture reaching confined spaces, the explosion possibility must also be assessed.

API 521 gives some guidance on chapter 5.8 for estimating the dispersion and concentration profile from a typical relief device discharge. It also describes that if the vapor relief velocity is high (> 30 m/s), the air entrance rate is also high and the vapors typically would be diluted below lower flammable limit in a very short distance.

This phenomenon explains the standard recommendation of locating the discharge of the relief device at least 3 m above any adjacent equipment, building or structure located within 15 m horizontally from the discharge point.

Experience has shown, however, that for some special cases these typical recommendations does not apply, such as:

- Heavy gas releases;
- Cryogenic releases;
- Control valve releases;
- Discharge piping with unconventional design (rain cap, horizontal, etc.);
- Discharge location within a building wake.

Adding to the complexity of the analysis, the relief devices can also be fitted with steam or other purge gas injection that could positively affect the dispersion.

After some case studies, which included the analysis of High Pressure Polyethylene plants Vent Stacks and Olefins unit relief and venting devices, Braskem developed an internal engineering standard with guidance and criteria for evaluating existing projects and new designs in order to ensure they comply with safe location. The guidance includes:

- Relief cases selection;
- Simplified and rigorous dispersion analysis;
- Consequence assessment endpoints;
- Thermal radiation effects to assets;
- Liquid, vapor and two-phase releases;
- Process Hazard Analysis;
- Diluting/snuffing steam;
- New systems and retrofitting criteria;

Applying the criteria in the developed standard allows the evaluation if the relief is in a safe location or not. If the discharge is not in a safe location, a risk ranking and recommendations to change the system must be made. One kind of a typical recommendation is to install the relief in a higher location. However, for some cases, achieving Safe Location is not possible and changes in the system are needed, such as relocating to a closed system or design changes to eliminate the relief scenarios associated with that relief device.

The development of an internal engineering guideline ensures a standard approach throughout the company for evaluating the Safe Location of hydrocarbon discharges and improves the quality of the risk evaluation during the PHA cycles.

548407 Experimental Study of Electrostatic Hazard inside Scrubber Column Using Response Surface Methodology

Experimental Study of Electrostatic Hazard inside Scrubber Column using Response Surface Methodology

Jingyao Wang[a,b], Yue Sun[a,b], Xiaodan Gao[a,b], Benjamin Wilhite[a,b], M. Sam Mannan [a,b]

[a]Mary Kay O'Connor Process Safety Center, Artie McFerrin Department of Chemical Engineering, Texas A&M University, College Station, Texas 77843-3122, USA

[b] Artie McFerrin Department of Chemical Engineering, Texas A&M University, 3122 TAMU, College Station, Texas 77843, United States

Abstract

Electrostatic ignition has triggered significant fire/explosion incidents in liquid processing industries including chemical, oil/gas, food, pharmaceutical and transportation. In this work, the impact of operating flowrate and temperature on the electrostatic energy accumulation in scrubber columns was studied. An *in-situ* charge measurement apparatus combined with a particle scattering laser function was installed inside a pilot-scale scrubber column, which can measure the aerosol particle size distribution, concentration and charge simultaneously. Four electrostatic ignition scenarios were considered including discharge from a single aerosol droplet, aerosol clouds, ungrounded conductor charged by the aerosol and ungrounded conductor charged by the liquid. Response surface methodology was applied to construct a quadratic empirical model correlating the electrostatic energy response with the flowrate and temperature. The obtained electrostatic response shows that corona from a charged single droplet was too small for safety concern. However, the ungrounded conductor polarized by charged aerosol or liquid can accumulate significant energy that is higher than the minimum ignition energy of some flammable chemicals. Temperature has two-sided effect where high temperature generates more concentrated charged aerosol but also facilitates the charge relaxation in the liquid phase. In all scenarios, increasing gas/liquid flowrate can cause significant electrostatic energy accumulation and can increase the hazards of many incidents. To reduce the electrostatic hazards, it is recommended to reduce the gas/liquid interaction flowrate and eliminate the presence of ungrounded conductors in liquid process operations.

Keywords: Scrubber column design, Electrostatic hazards, Response surface methodology, Process safety, Hazard identification, Aerosol hazards.

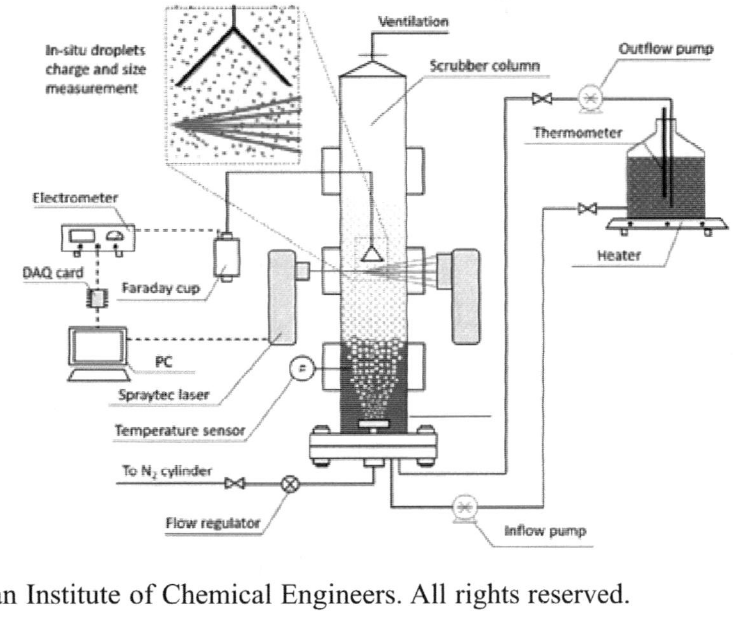

550852 Process Safety in Oil Refineries: Vacuum Distillation Column Hazop Study

Errors that can occur in the identification studies for hazards and risks which is one of the elements of a process safety management cause resulting major accident having undesirable long - term effects on human and environment. For preventing major accidents, identification and analysis of process hazards due to equipment failures, design or montage defects, improper maintenance, misuse, material defects, metal fatigue, corrosion, welding defects, human failure etc. are the main principles of process safety management. The aim of present study is to investigate possible accidental risks occurring in Vacuum Distillation Column of oil refining within contributing to the process safety frame.

In this cross - sectional study, semi quantitative HAZOP analysis was used for identifying accidental risks. By means of HAZOP studies 66 deviations, 238 deviation causes, 287 consequences, 651 barriers and 113 recommendation were identified. In the determination of accidental risk in the vacuum distillation column by applying semi quantitative HAZOP analysis method, effective results were obtained. The most common causes of deviations in the column have been found as control loop failure, leakage and mechanical failure. It is stated that the most common accidental risks are as a portion of 66.9 % operational deterioration, fire, explosion, release known as major accident risk. It is identified that the accidental risks mostly cause existence damage and the highest accidental risk occurs in the "vacuum column overhead system" and "bottom of the vacuum column system" 87.7 % of 113 recommendations are modification in the design, 5.3% is about modifications in the process safety management system.

551179 API Process Safety Site Assessment Program - Trends and Good Practices in Process Safety

The Process Safety Site Assessment Program is a collaborative industry effort managed by API and is focused on improving process safety in petroleum refining and petrochemical facilities. The program has conducted over 65 process safety assessments and 38 HF RP-751 assessments since its inception in 2012. The assessments are conducted by independent and credible third-party teams of industry-qualified process safety expert assessors using industry-developed protocols.

The presentation will review trends, both positive and low scoring, seen in the extensive benchmarking data collected. Good practices observed in the assessments will be identified and discussed. The emphasis will be to have the attendees leave with an understanding that typically some site(s) has developed a way to handle challenging areas of process safety.

545445 Emerging Trends in Process Safety Management Relating to Contractor Workforce Compliance

Following a Six Sigma Black Belt process improvement session conducted 2016 to 2018 for petrochemical complexes along the Gulf Coast the presentation identifies several emerging trends for Best Practices in industrial workforce compliance.

Safety Councils work for petrochemical plants and their contractors to facilitate innovative ways to reduce incident rates resulting in a 95% reduction in recordables. 2017 marked an evolution in Safety Council deliverables, slowly migrating away from site specific training to develop new workforce compliance and risk management processes. Working within the industry we developed several internationally recognized Best Practices for contractor compliance, risk management, labor cost management. ISTC collaborated with our partners bringing years of quantitative data to identify problems that are developing, have occurred in the workplace and formulating solutions, cost structures and most importantly compliance management to combat workforce issues. The presentation, best practices in contractor compliance highlights the journey from identifying the problems, and the associated issues with hiring, compliance, auditing, accident response and even Department of Homeland Security, CFATS inspections. Beyond the logistics of building solutions for a particular problem, the presentation reflects on how companies were able to share their issues among peers without violating confidentiality or company procedures to solve what was to be discovered as an industry trend. Using Six Sigma tools, the units were able to identify root problems, collaborate on a solution, model cost recovery and implement the solutions to the contractor workforce with a minimum of disruption to the business at hand.

Análisis Cuantitativo De Riesgo (QRA) Como Herramienta Para La Toma De Decisión Sobre La Infraestructura En Sistemas De Transporte De Fluidos Peligrosos

Daniel Mauricio Moncada Saavedra
CENIT -Transporte y Logística de Hidrocarburos
Cll 113 #7-80 Piso 13 Bogotá – Colombia
daniel.moncada@cenit-transporte.com

Maria Isabel Montoya Rodriguez
CENIT -Transporte y Logística de Hidrocarburos
Cll 113 #7-80 Piso 13 Bogotá – Colombia
maria.montoya@cenit-transporte.com

Prepared for Presentation at
American Institute of Chemical Engineers
2019 Spring Meeting and 15th Global Congress on Process Safety
New Orleans, LA
March 31 – April 3, 2019

Análisis Cuantitativo De Riesgo (QRA) Como Herramienta Para La Toma De Decisión Sobre La Infraestructura En Sistemas De Transporte De Fluidos Peligrosos"

Daniel Mauricio Moncada Saavedra
CENIT -Transporte y Logística de Hidrocarburos
Cll 113 #7-80 Piso 13 Bogotá – Colombia
daniel.moncada@cenit-transporte.com

Maria Isabel Montoya Rodriguez
CENIT -Transporte y Logística de Hidrocarburos
Cll 113 #7-80 Piso 13 Bogotá – Colombia
maria.montoya@cenit-transporte.com

Palabras clave:

Abstract

La pérdida de contención en activos utilizados para el transporte de hidrocarburos puede ocasionar impactos considerables a personas, medio ambiente, económicos y de reputación. Con el fin de mitigar la ocurrencia este tipo de eventos, se implementan estrategias sobre los activos, siendo la identificación y análisis de riesgo un elemento clave durante el proceso.

El presente análisis utilizó un enfoque cuantitativo, tanto para probabilidad como para consecuencia, con el fin de poder estimar el riesgo que genera la operación de una línea de transporte en un área de alta consecuencia clasificado como sitio identificado, planteando posibles alternativas de control y mitigación del riesgo.

Como resultado del estudio realizado se pudo establecer el nivel de riesgo actual asociado a la operación de la línea en el sitio de interés. Como actividades de mitigación para reducir el riesgo asociado se identificó: la reparación de los defectos, así como el realineamiento y reubicación de la tubería.

1 Introducción

Para el presente análisis se implementó un cálculo de la probabilidad de falla a partir del análisis de integridad estructural del sector de interés. Adicionalmente, se consideró el análisis de la susceptibilidad que se presenta a partir de las amenazas presentes, la cuales son identificadas a partir de las actividades de monitoreo e inspección desarrolladas sobre la infraestructura. El cálculo de la interacción entre las amenazas a la integridad; así como el resultado final de probabilidad de falla, se puede realizar mediante diferentes enfoques, el presente estudio utilizo cálculo de probabilidad de excedencia y aplicación de redes bayesianas para la estimación.

Para el cálculo de la consecuencia se buscó verificar la posible afectación a personas, producto de la ignición del producto transportado. Dentro de los sucesos finales se consideró: incendio de piscina, incendio por chorro de fuego, llamarada y la sobrepresión debida a la explosión. El cálculo de estos efectos se realizó en el sitio donde se presenta la perdida de contención, así como también para el caso en el que el hidrocarburo es transportado por un cuerpo de agua o se desplaza en tierra.

Una vez determinada tanto la probabilidad de que se desarrollen los sucesos finales identificados como la consecuencia que cada uno de ellos puede generar, se procedió a realizar el cálculo del riesgo expresando el resultado en términos de riesgo social e individual, comparando el resultado contra criterios de tolerabilidad referentes en la industria. Para la condición actual establecida se identificaron alternativas las cuales buscaban disminuir el nivel de riesgo, implementando actividades de mitigación del nivel de riesgo.

2 Análisis de Riesgo Cuantitativo

3 Frecuencia Final de Ocurrencia de los Sucesos Finales

En el sector bajo interés se tienen que la línea es susceptible a las siguientes amenazas:

3.1 Análisis de Condición.

En este aparte se relacionan los resultados de análisis de probabilidad de falla, obtenidos en el sector de interés, PK 43+000 al PK 45+000 del poliducto, para las amenazas de corrosión externa, corrosión interna, geotecnia y daños por terceros involuntarios:

3.1.1 Corrosión Externa (CE).

A continuación, se describe el estado de salud de las barreras de CE del poliducto en el sector solicitado (ver Ilustración 1), así como el análisis de probabilidad de excedencia de las anomalías ILI identificadas con la corrida del 2014. En la Ilustración 2 se observa la disponibilidad de la URPC y se reporta que la tubería cuenta con recubrimiento en buen estado. Bajo las condiciones y premisas consideradas para el cálculo y proyección de la POE (Probabilidad de excedencia; entre ellas una rata de corrosión de 2-3 mpy aproximadamente; no hay riesgo de excedencia de los criterios de profundidad y ERF (Factor Estimado de Reparación) de las anomalías ILI hasta el año 2021 en el sector 43+590 y 42+746 (Anomalías ILI con DR(Distancia de registro): 2646.76, 3490.08, 3492.38, 3492.71) (Ilustración 2).

Ilustración 1. Perfil de riesgo y salud de barreras CE.

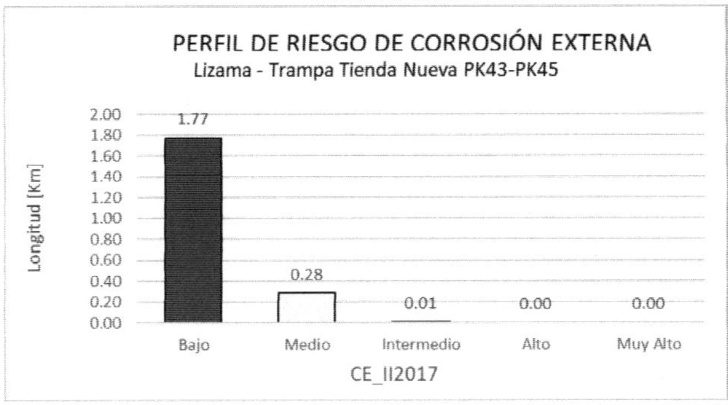

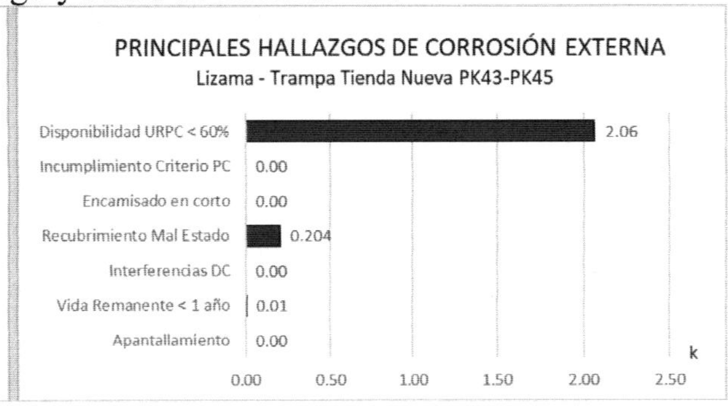

Ilustración 2. Análisis de POE de anomalías de CE en el tiempo.

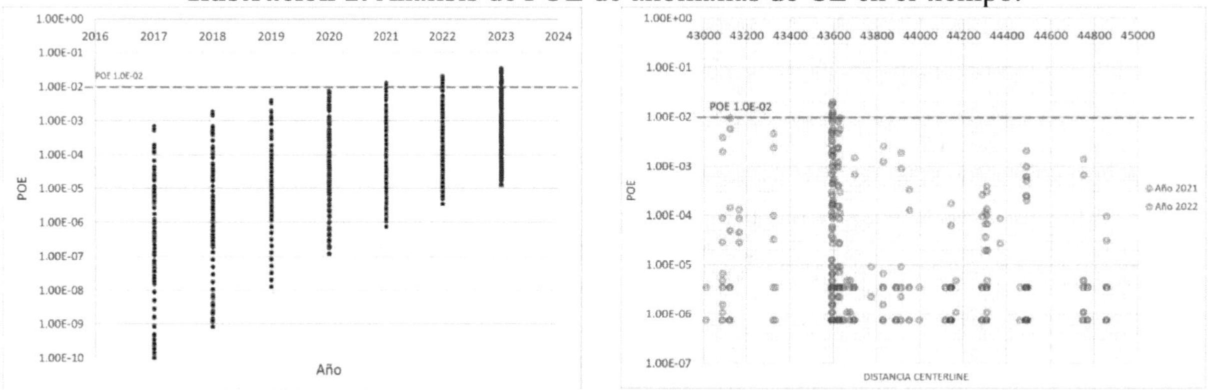

3.1.2 *Corrosión interna (CI).*

A continuación, se describe el estado de salud de las barreras de CI de la tubería en el sector solicitado (ver Ilustración 3), así como el análisis de probabilidad de excedencia de las anomalías ILI identificadas en la corrida del 2014. De acuerdo con los resultados de valoración de riesgo, se presenta un correcto estado de salud de las barreras de corrosión interna. Adicionalmente, de acuerdo con las condiciones y premisas consideradas para el cálculo y proyección de la POE, no hay riesgo de excedencia de los criterios de profundidad y ERF de las anomalías ILI en el mediano plazo (Ver Ilustración 4).

Ilustración 3. Perfil de riesgo y salud de barreras CI.

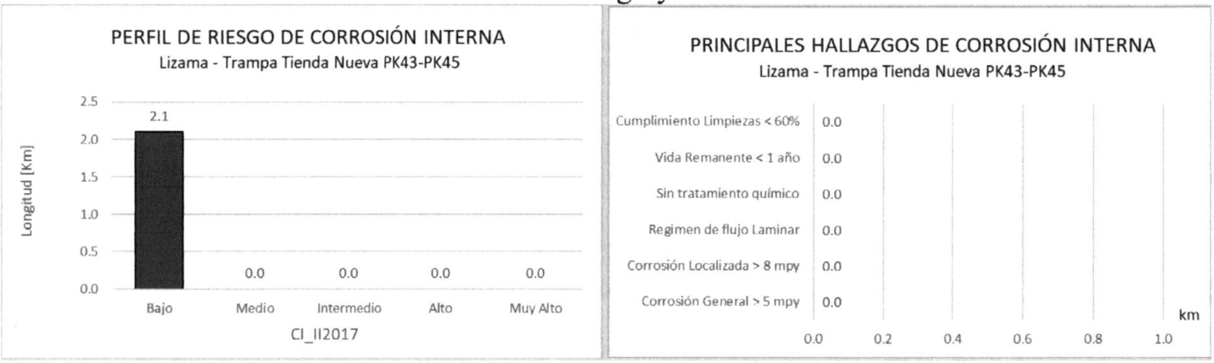

Ilustración 4. Análisis de POE de anomalías de CI en el tiempo.

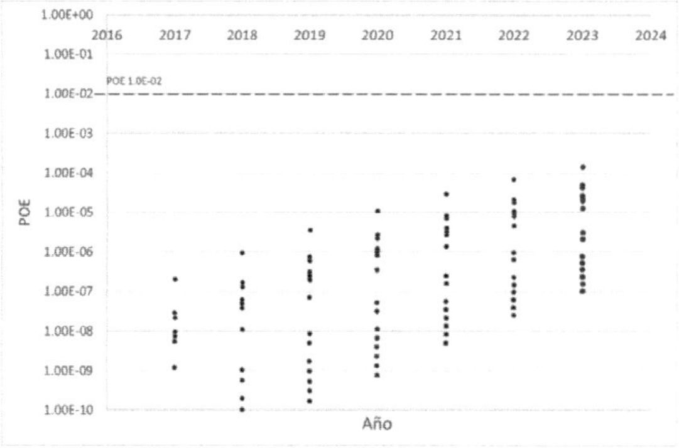

3.1.3 Geotecnia.

En el sector de interés se reporta hallazgo de geotecnia en la distancia centerline 44643 – 44703 (PK 44+400), donde se presenta proceso erosivo incipiente. Actualmente, no se identifica avance de la condición, razón por la cual, se continuará seguimiento a través de la inspección visual del derecho de vía (IVDDV).

Ilustración 5. Perfil de riesgo y salud de barreras de Geotecnia.

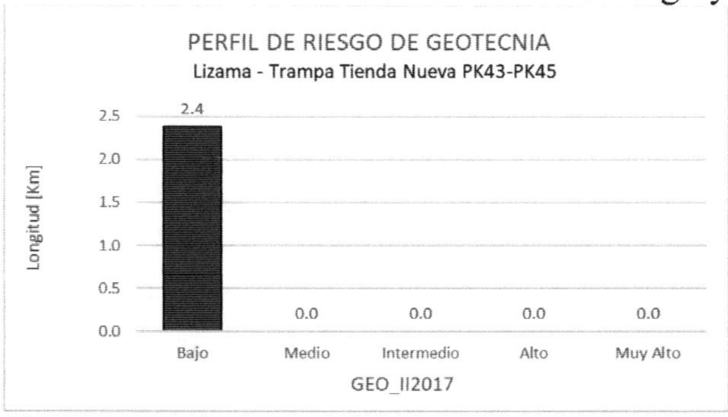

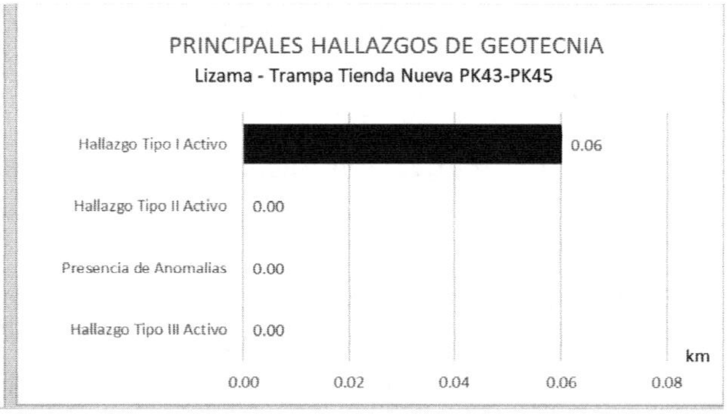

3.1.4 *Daños por Terceros Involuntarios (DPT).*

Debido a la condición de la tubería y a la gestión realizada por el operador para el control de la amenaza, ésta presenta una susceptibilidad baja de ser golpeada por un tercero sobre el derecho de vía. A continuación, se relaciona el perfil de riesgo de la amenaza de DPT en el sector analizado:

Ilustración 6. Perfil de riesgo y salud de barreras de Daños por Terceros Involuntarios.

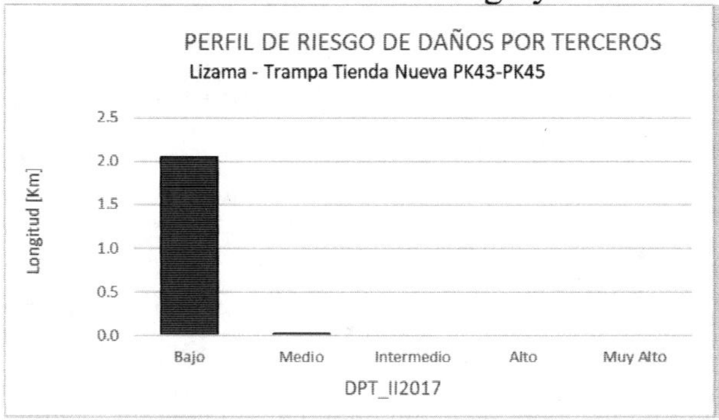

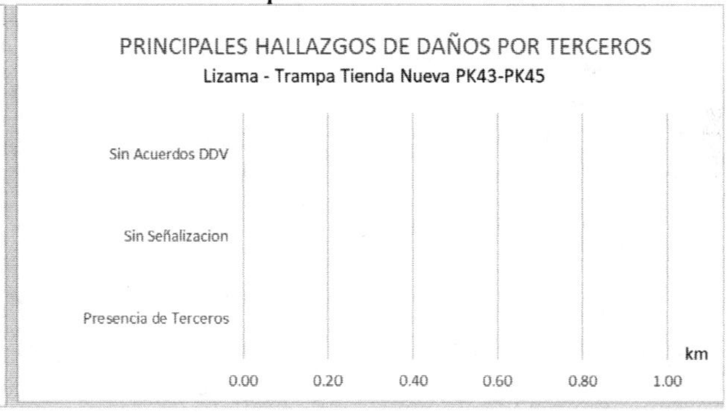

De acuerdo con el análisis ejecutado, el evento más probable obedece a una pérdida de contención generada por la amenaza de corrosión externa; donde el análisis estructural identifica que el modo de falla se daría por fuga. Sin embargo, las anomalías encontradas no presentan riesgo de excedencia de los criterios de profundidad y ERF hasta el año 2021.

Con el fin de poder estimar la probabilidad de falla de la línea en el sector de interés, se realizó el análisis estructural considerando principalmente defectos por perdida de espesor, bajo estas premisas se obtuvo un valor de probabilidad de excedencia de los criterios de profundidad y ERF de 7.09E-4. Del análisis de condición se concluye que el iniciador más probable en este sector es la falla de la tubería por fuga asociado a corrosión externa, para el cual a continuación se presenta el análisis de riesgo.

4 Análisis de consecuencia

A continuación, se presenta el análisis de consecuencia para los sitios en los cuales se ha identificado que se puede llegar a presentar una pérdida de contención.

4.1 Identificación de escenarios

4.1.1 Selección de sustancias.

En el poliducto se transportan las siguientes sustancias: Gasolina, GLP, Jet A1 y B2

4.1.2 Identificación de iniciadores.

Los iniciadores representan las situaciones iniciales de pérdidas de contención o exposición que se pueden presentar en los das diferentes equipos. Los iniciadores pueden ser de dos tipos, genéricos o específicos. Los iniciadores genéricos, involucran una pérdida de contención debida a causas no específicas del proceso u operación, como pueden ser corrosión, errores de construcción, fallas de la soldadura, etc. Los iniciadores específicos se caracterizan por tener causas concretas debido a las condiciones de proceso, el diseño, los materiales utilizados, etc.

A continuación, se describen los iniciadores tenidos en cuenta para cada uno de los sitios en los cuales se identificó la inestabilidad de taludes.

- **Sector PK 43+590.** Este sitio tiene la menor distancia desde la infraestructura a la escuela bajo análisis, 30 metros aproximadamente, sumado a esto en las cercanías de este se ha identificado la presencia de anomalías las cuales, bajo las condiciones actuales de operación y deterioro hacia el año 2021, pueden exceder los límites de profundidad critica a la falla.

- **Sector PK 43+805 – PK 44+110.** En este sitio se tiene que ante la posible pérdida de contención del producto se puede llegar a presentar la formación de una ruta de derrame en tierra, por la cual se desplazaría el producto hacia el interior de la Escuela.

4.1.3 Identificación de sucesos finales.

La identificación de los sucesos finales que pueden tener lugar después de que se da el iniciador, se debe realizar utilizando los árboles de eventos o sucesos. Dichos árboles permiten ver el comportamiento de la sustancia tras la descarga y estimar la probabilidad de ocurrencia de cada suceso, teniendo en cuenta los sucesos intermedios que condicionan la evolución del suceso final.

Esta identificación de escenarios se lista en Tabla 11 para los escenarios que involucran sustancias inflamables tales como Gasolina o Nafta. Los posibles sucesos finales identificados en el poliducto son: derrame continental, incendio de piscina, incendio de chorro y llamarada.

Tabla 1. Identificación de escenarios.

Escenarios	Descripción	Iniciadores	Sustancia	Tipo de pérdida
Afectación directa				
E1	Poliducto	L3	Gasolina	Incendio de piscina
				Incendio de chorro
				Llamarada
E2	Poliducto	L3	GLP	Incendio de piscina
				Incendio de chorro
				Llamarada

Escenarios	Descripción	Iniciadores	Sustancia	Tipo de pérdida
Afectación directa				
Afectación indirecta				
E3	Poliducto	L3	Gasolina	Incendio de piscina
				Llamarada

4.2 Estimación del área de afectación por los efectos que involucran producto

4.2.1 Condiciones para la simulación.

Para determinar las consecuencias, es necesario definir las condiciones propias del entorno de la infraestructura que servirán como base para la simulación de los efectos. Estos parámetros son presentados en la Tabla 2. Las estabilidades atmosféricas fueron determinadas a partir de la rosa de vientos basada en datos estadísticos y modelos meteorológicos, la cual es presentada en la **¡Error! No se encuentra el origen de la referencia.**.

Tabla 2. Condiciones atmosféricas

Parámetro	Valor	Estación de Referencia
Temperatura promedio (ºC)	31	Colosal
Humedad relativa promedio (%)	83	
Rugosidad el terreno (m)	0,08	
Estabilidad atmosférica más probable y velocidad del viento (m/s)	2F	
Estabilidad atmosférica F y velocidad del viento (m/s)	2F, 5D	

4.2.2 Distancias de afectación directa

Las distancias de afectación representan las longitudes dentro de las cuales se pueden manifestar los efectos correspondientes a los niveles de afectación, para cada suceso final (incendios de piscina o llamaradas). Se denominan distancias de afectación directas, ya que los sucesos finales de que dependen se originan y desarrollan su efecto directamente en el sitio en el que se produce la pérdida de contención del producto

4.2.3 Estimación de Descargas.

Para la simulación de los posibles efectos producto de una pérdida de contención, es necesario conocer cuánto producto podría derramarse en cada punto de estudio sobre la infraestructura. El volumen derramado corresponde al estimado teniendo en cuenta el llenado de la tubería, las condiciones de operación y el tiempo de respuesta operativo, es decir, el tiempo de detección más el tiempo de actuación (generalmente el seccionamiento del sistema o parada de bombeo).

Tabla 3. Tiempos de actuación

Equipo	Tiempo de Detección (min)	Tiempo de Actuación (min)
	Fuga	**Fuga**
Poliducto	30	180

Tabla 4. Volumen y Tasa de descarga

Volumen (US bls)	Tasa de descarga (bls/hr)
59	22

4.2.4 *Simulación de Efectos de los Sucesos Finales.*

En este numeral se presentan los resultados de las distancias de afectación directa, obtenidas para cada uno de los niveles de afectación asociados a los sucesos finales. En la **¡Error! No se encuentra el origen de la referencia.** se presentan las distancias de afectación obtenidas para el evento incendio de piscina, es importante resaltar que este depende en gran medida del área de empozamiento que se forma una vez se produce la descarga del producto.

En la **¡Error! No se encuentra el origen de la referencia.** se presentan las distancias de afectación para incendio de chorro obtenidas durante las simulaciones, es importante resaltar que la distancia obtenida depende de las propiedades químicas de la sustancia y de las condiciones en las cuales se produce la descarga.

Tabla 5. Distancia de afectación incendio de piscina.

Equipo / Abscisado	Iniciador	Sustancia	Volumen Derrame	Niveles de Afectación por Incendio						
				Incendio de Piscina						
				1.6 Kw/m^2	5.0 Kw/m^2	7.3 Kw/m^2	9.5 Kw/m^2	14.5 Kw/m^2	20.9 Kw/m^2	37.5 Kw/m^2
			(Bbl)	Distancia (m)						
K43+590	Rotura mínima	GLP	59	46,94	28,63	24,58	21,98	18,01	14,55	8,38
K43+590	Rotura mínima	Gasolina	59	94,63	49,69	35,17	27,20	20,98	20,98	-

Tabla 6. Distancia de afectación incendio JET FIRE.

Equipo / Abscisado	Iniciador	Sustancia	Volumen Derrame	Niveles de Afectación por Incendio						
				Chorro de fuego						
				1.6 Kw/m^2	5.0 Kw/m^2	7.3 Kw/m^2	9.5 Kw/m^2	14.5 Kw/m^2	20.9 Kw/m^2	37.5 Kw/m^2
			(Bbl)	Distancia (m)						
K43+590	Rotura mínima	GLP	59	45,9	33,61	30,93	29,33	27,13	25,23	23,42
K43+590	Rotura mínima	Gasolina	59	44,43	31,20	28,46	26,82	24,58	22,96	20,84

En la **¡Error! No se encuentra el origen de la referencia.** se presentan las distancias de afectación por llamarada; las cuales están relacionadas con la volatilidad del producto junto con las condiciones climáticas de las zonas del evento.

Tabla 7. Distancia de afectación Llamarada.

Equipo / Abscisado	Iniciador	Sustancia	Volumen de Derrame	Niveles de Afectación por Llamarada más probable		Niveles de Afectación por Llamarada más estable	
				LII	LII/2	LII	LII/2
			(Bbl)	Distancia (m)		Distancia (m)	
K43+590	Rotura mínima	GLP	59	24,12	58,53	26,48	66,7
K43+590	Rotura mínima	Gasolina	59	-	-	-	-

4.2.5 *Distancias de Afectación Indirecta*

Las distancias de afectación indirecta representan las longitudes de potencial afectación por sucesos finales que se originan en una trayectoria o ruta de derrame (en tierra o en agua) o, a partir de estancamientos o acumulaciones de producto. En este numeral se presentan los resultados de las distancias de afectación obtenidas para cada uno de los niveles de afectación asociados a los sucesos finales que se pueden presentar en la ruta de derrame. Estos resultados corresponden a la radiación generada por piscina de fuego, evento que en los cuerpos de agua depende de la sección del cuerpo de agua, así como por el frente de llama generado por el volumen derramado.

En la Tabla **8**8 **y ¡Error! No se encuentra el origen de la referencia.**9 se referencian los corredores de afectación generados por incendio de piscina sobre las rutas de derrame en tierra identificadas en el poliducto, en el sector comprendido entre los PK 43+000 - PK 45+000.

Tabla 8. Distancias de afectación indirectas POOL FIRE.

Abscisado	Medio de Transporte	Sustancia	Incendio						
			Incendio en piscina						
			1.6 Kw/m²	5.0 Kw/m²	7.3 Kw/m²	9 Kw/m²	14.5 Kw/m²	20.9 Kw/m²	37.5 Kw/m²
			Distancia (m)						
Km 43 - 45	Ruta en tierra	Gasolina	82,02	54,68	27,34	24,61	19,14	13,67	8,20

Tabla 9. Distancias de afectación indirecta FLASH FIRE

Abscisado	Medio de Transporte	Sustancia	Niveles de afectación por llamarada más probable		Niveles de afectación por llamarada más estable	
			LII	LII/2	LII	LII/2
			Distancia (m)		Distancia (m)	
Km 43- 45	Ruta en tierra	Gasolina	2,14	4,74	8,64	32,24

4.2.6 Áreas de Afectación

En sistemas de ductos, el área de afectación directa se define por un corredor medido desde el eje central del ducto. Este corredor está dado por la mayor distancia probable de afectación obtenida en la simulación de efectos para todas las sustancias analizadas. Por otro lado, el área de afectación indirecta está dada por varios corredores a lo largo de las posibles rutas de derrame tanto en suelo como en agua. En la Ilustración **7**7 se presenta el área de afectación identificada.

Ilustración 7. Área de afectación identificada.

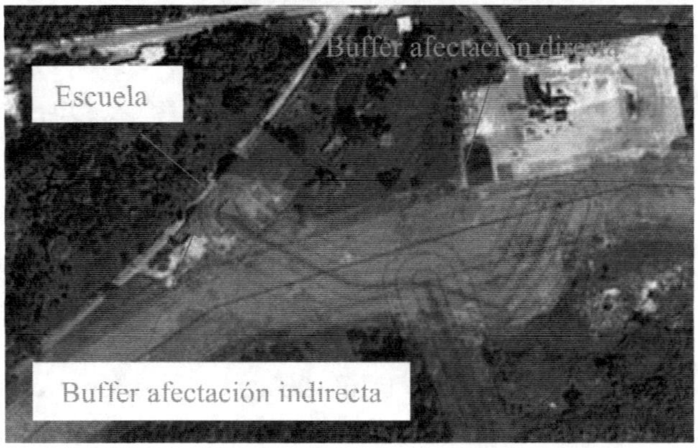

4.2.7 Identificación de Elementos Vulnerables

Para calcular el riesgo de la población respecto de la posible afectación producida por la pérdida de contención descrita, se requiere el número y localización de las personas, en la Tabla 100 se presenta la población identificada al interior de la escuela.

Tabla 10. Presencia de población Escuela

Presencia de Población							
Escuela - Dia				**Escuela - Noche**			
Niños	63	Profesores	6	Niños	0	Profesores	2

5 Evaluación del Riesgo.

A continuación, se presentan los resultados del análisis de riesgo el cual tuvo en cuenta la información presentada en los numerales anteriores asociada a las posibles causas de falla, los modos en que puede darse cada falla identificada, la frecuencia de ocurrencia de la falla, la probabilidad de ocurrencia de los sucesos finales, y finalmente, la estimación de la magnitud de las consecuencias de tales sucesos finales en el entorno en términos cuantitativos.

5.1 Riesgo Individual.

Los resultados de riesgo individual expresados en número de fatalidades por año (o año-1), sí permiten establecer zonas de aceptabilidad del riesgo según los diferentes criterios que se utilizan en la industria. En la Ilustración 8 se presentan los resultados de riesgo individual en el sector comprendido entre los PK 43+000 y 45+000.

Ilustración 8. Riesgo Individual.

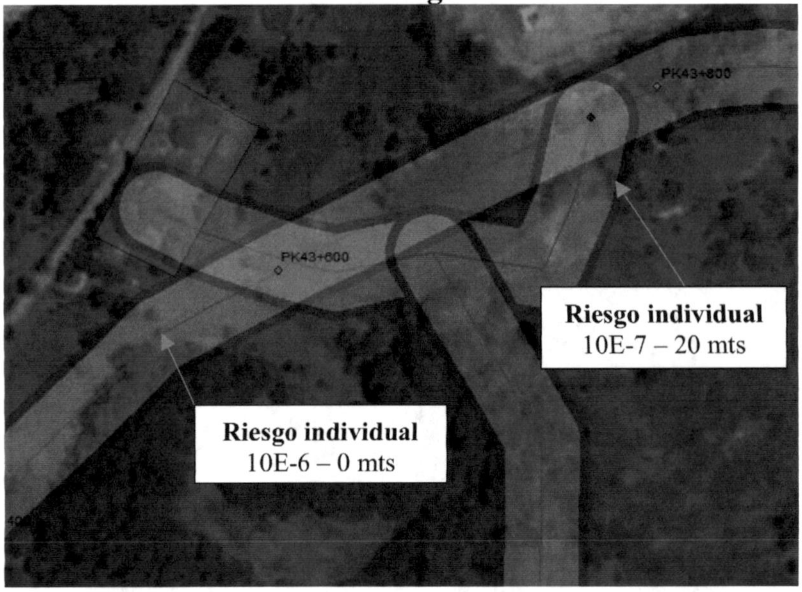

5.2 Riesgo Social

El riesgo social representa el número de fatalidades que pueden llegar a ocurrir en los sitios evaluados en el poliducto en el PK 43+000 – PK 45+000, al revisar la información contenida en la Ilustración 9 se puede apreciar como el riesgo se mantiene en la región tolerable, muy cercana al límite inferior ALARP. El evento dimensionante corresponde a una llamarada en la ruta de derrame, con el ducto transportando gasolina, la cual puede llegar a afectar a ocho (8) personas.

Ilustración 9. Riesgo social Escuela PK 43+000 – PK 45+000.

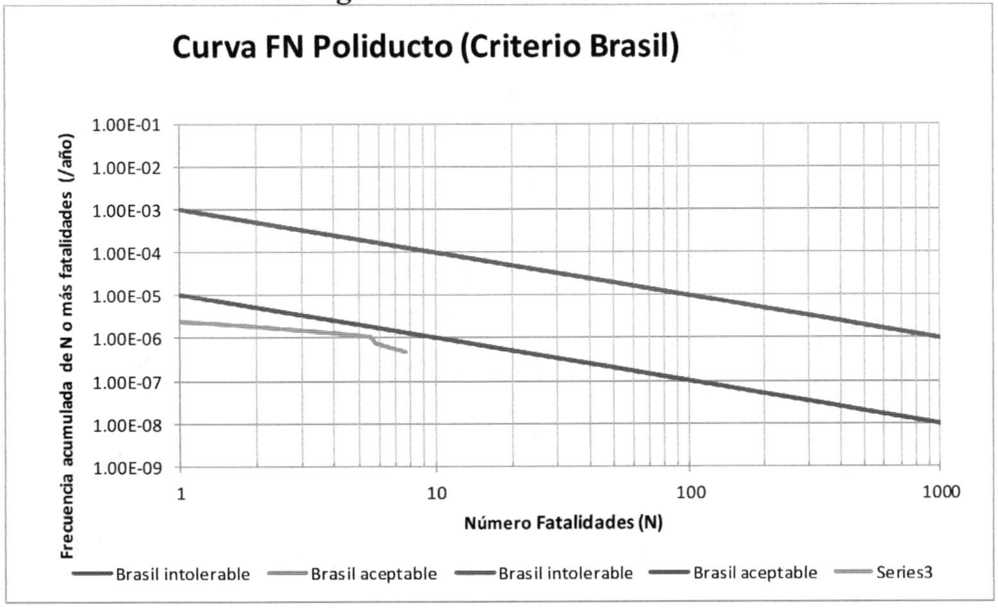

6 Alternativas de mitigación de Riesgo

Considerando los resultados del análisis de riesgo cuantitativo sobre el Poliducto para el sector PK 43+000 - PK 45+000, se tiene que ante una rotura total el evento dimensionaste corresponde a una llamarada con una distancia de afectación de 135 metros, para el caso de una fuga por corrosión el evento dimensionaste corresponde a una llamarada con una distancia de afectación de 32,24 metros, finalmente como resultado de una perdida de contención se puede llegar presentar la formación de una ruta de derrame en tierra la cual ingresaría a la escuela, con una distancia de afectación de 8.6 metros generada por una llamarada, tal como se presenta en la Ilustración 10.

Ilustración 70. Distancias de afectación

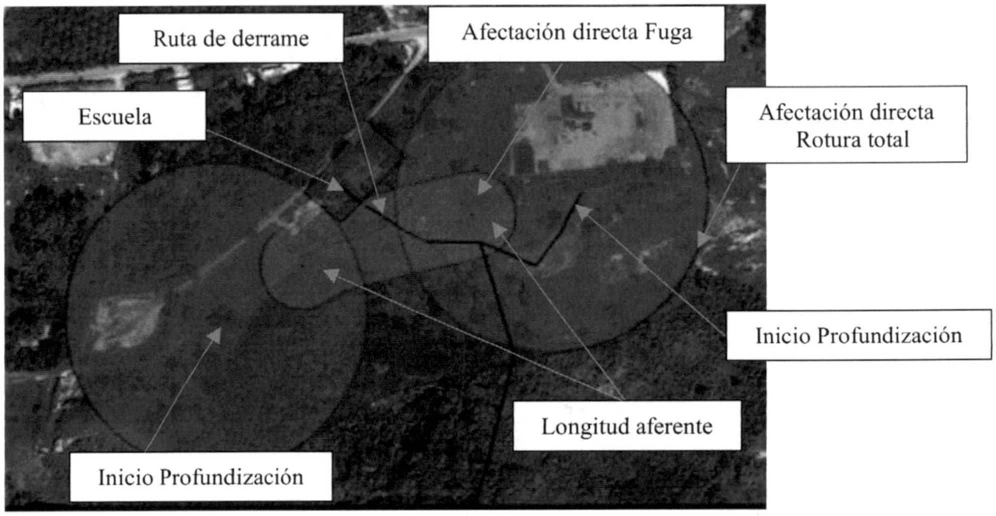

Teniendo en cuenta que el evento corresponde al presentado en la ruta de derrame, se hace necesario considerar la evaluación de alternativas de mitigación de riesgo. Con base en lo anterior, se consideró que la actividad de realineación y profundización cubra la longitud aferente de esta ruta, la cual es de 155 metros e inicia en la coordenada 74°33'35.466"O - 6°9'8.213"N y finaliza la coordenada 74°38'15.025"O - 6°9'8.737"N. Lo anterior, tiene como objeto de eliminar la ruta de derrame actual que conduce a la escuela en caso de una falla y perdida de contención, dado que el vertimiento se genere en sentido contrario al actual.

Sumado a lo anterior es necesario considerar una distancia de protección adicional para el dado caso en el que se llegue a presentar la rotura total de la línea, esta distancia se establece a partir del radio de impacto del evento llamarada, localizando los puntos sobre la línea que no presentan afectación a la escuela. A partir del análisis se estableció que esta distancia es de 316 metros e inicia en la coordenada 74°31'33,57146"O - 6°9'8,22442"N y finaliza en la coordenada 74°32'28,98241"O - 6°9'9,70756"N, esta distancia cubre la longitud aferente descrita anteriormente.

Dado que la alternativa de mitigación tiene como fin eliminar la afectación indirecta y el impacto directo sobre la escuela a continuación se presentan dos alternativas de trazados propuestos, que aseguran que las rutas de derrame en caso de una perdida de contención tendrán dirección opuesta a la ubicación de la escuela.

Ilustración 80. Opción 1 realineamiento entre PK 43 + 695 a PK 44 + 205.

Ilustración 91. Opción 2 realineamiento entre PK 43 + 625 a PK 44 + 460 de 752 metros.

En la ilustración 12 se presenta las dos alternativas propuestas y la distancia de separación del trazado actual de tubería, así como de la escuela.

Ilustración 102. Alternativas propuestas para mitigación de riesgo

7 Conclusiones

Del análisis realizado se tienen las siguientes conclusiones.

- El máximo número de fatalidades a generar es de ocho (8) y es generado por una llamarada en la ruta de derrame en tierra la cual, en el dado caso de que se presente el evento, ingresaría a las instalaciones de la escuela.
- La mayor susceptibilidad de la línea es por la amenaza de corrosión externa, especialmente en algunos sectores en los cuales el recubrimiento se encuentra en mal estado, esta condición junto con las anomalías encontradas en la línea, impactaron la frecuencia de falla establecida para el sector.
- En términos generales el nivel de riesgo se encuentra en la zona tolerable, muy cercana al límite inferior ALARP.

8 Recomendaciones.

- Es recomendable realizar actividades tendientes a disminuir el volumen disponible a desplazar en la ruta de derrame, para ello se puede realizar la profundización de la línea lo cual además de disminuir el efecto de los eventos por JET FIRE y FLASH FIRE disminuye la cantidad de producto disponible a desplazar al favorecer la evaporación de este.

- A partir del análisis realizado se tiene que la distancia a realinear y profundizar es:

  Opción 1 realineamiento entre PK 43 + 695 a PK 44 + 205 con una longitud de 516 metros
  Opción 2 realineamiento entre PK 43 + 625 a PK 44 + 460 con una longitud de 752 metros.

- Con ambas alternativas se elimina la afectación indirecta y la afectación directa, sobre la escuela, estas distancias cubren todos los posibles eventos que se pueden presentar y tienen afectación sobre la escuela en mención.

- Se pueden construir cortacorrientes los cuales eviten que la ruta de derrame ingrese o haga su tránsito en las cercanías de la escuela, junto con la construcción de estructuras de retención las cuales detengan el avance del producto mientras se realizan actividades de recuperación de este.

- Es recomendable rehabilitar el recubrimiento en aquellos sectores en los cuales se encuentra en mal estado, junto con la atención de aquellas anomalías que sean críticas.

9 Referencias

[1] Fingas, M. (2011). Oil Spill Science and Technology. Kidlington, UK: Gulf Professional Publishing Elsevier.

[2] D.O.T. Cornell Law School. Obtenido de 49 CFR 192.905 - How does an operator identify a high consequence area?: https://www.law.cornell.edu/cfr/text/49/192.905 Accesado 15 Enero, 2019

[3] D.O.T. Cornell Law School. Obtenido de 49 CFR 192.5 - Class locations: https://www.law.cornell.edu/cfr/text/49/192.5 Accesado 15 Enero, 2019

[4] Yellow Book. TNO. Methods for the calculation of Physical Effects Due to releases of hazardous materials (liquids and gases) – Third edition Second revised print 2005

[5] Purple Book. TNO. Guidelines for quantitative risk assessment – First edition 1999/2005

Referencias no citadas:

[6] ASME.ASME B31.8S. Managing System Integrity of Gas.New York, USA.2005.

[7] ASME.PIPELINE TRANSPORTATION SYSTEMS FOR LIQUID HYDROCARBONS AND OTHER LIQUIDS ASME B31.4-2002. New York, USA.2002.

[8] API.API Std 1160. Managing System Integrity for Hazardous Liquid Pipelines. Washington, USA.2008.132 Pags.

548036 Best Practice in MOC and Pssr Automation for Opex

Management of Change (MOC) and Pre Start-up Safety Review (PSSR) are still the most challenging elements of OSHA's Process Safety Management (PSM) standard and the EPA's Risk Management Program (RMP) rule. Now MOC and PSSR are requirements of the US Department of Interior's Bureau of Safety and Environmental Enforcement's Safety and Environmental Management Systems (SEMS). Even though the PSM standard has been around since 1992 and the industry has been managing change for several decades, we still can get it wrong, sometimes with disastrous results. The mindful or diligent efforts of many working in concert are necessary to ensure that change is identified, analyzed and executed in a quality way. But just when we get the process on paper and get the workforce up-to-speed, we have employee turnover, complacency sets in and sometimes there is regulatory change.

Many high-profile companies have addressed this challenge by creating and automating standard processes across their manufacturing or production facilities. MOC and PSSR are two of these major processes that have been successfully automated or **digitized**. Technology has been used to **automate** each step of the MOC workflow, enabling the collection of change ideas, formalizing the analysis of the ideas, and documenting the approval and execution of the change. Reports and dashboards can provide visibility of the health or current state of your MOC and PSSR processes. The use of **software** can enable the standardization of best practices.

On a well-meaning quest to standardize on a MOC/PSSR software solution and process, many companies are stifling creativity and innovation. Operational Excellence cannot be achieved in such ways. This paper discusses how the chemical process industry has defined the MOC and PSSR best management practices and how they should be automated for operational excellence.

Chemical Process Safety Requirements in the Fire Code

Ronald Melucci
Engineering Planning and Management, Inc.
959 Concord Street
Framingham, MA 01701
rbm@epm-inc.com

Prepared for Presentation at
American Institute of Chemical Engineers
2019 Spring Meeting and 15th Global Congress on Process Safety
New Orleans, LA
March 31-April 3, 2019

Ronald Melucci
Engineering Planning and Management, Inc.
959 Concord Street
Framingham, MA 01701
rbm@epm-inc.com
Keywords: NFPA Fire Code, International Fire Code, Process Safety

Chemical Process Safety Requirements of the Fire Code

Abstract

When most Chemical Process Safety professionals think about regulation and enforcement, they think about the applicable Occupational Safety and Health Administration (OSHA) and Environmental Protection Agency (EPA) regulations. However, the majority of US states adopt NFPA 1 or the International Fire Code (IFC) as their state fire code which contain requirements for chemical process safety. These requirements are applicable for processes of all sizes and levels of complexity below PSM and RMP thresholds and they address both fire and non-fire related chemical hazards. These requirements are enforced by municipal or state fire officials/marshals, who are becoming more aware of these requirements and chemical process safety in general. This paper will review some of the code requirements with a focus on requirements that may be commonly overlooked in the industry.

Introduction

The OSHA Process Safety Management (PSM) and EPA Risk Management Program (RMP) regulations, enforced by OSHA and EPA, respectively, are familiar to most safety professionals who work in industries that include processing of large quantities of chemicals. However, the applicable fire code requirements related to chemical process safety are often more restrictive than the federal regulations and are generally lesser known to such safety professionals.

The two model fire codes adopted by states in the US, are: *NFPA 1, Fire Code,* published by The National Fire Protection Association (NFPA), and the *International Fire Code* (IFC) published by the International Code Council (ICC). The states that do not adopt the IFC will typically use NFPA 1 and each state may also institute their own NFPA 1 amendments. Local amendments are almost always more restrictive than the model code verbiage, and many states have legislation that prohibits fire code amendments that reduce the level of safety/standard of care intended by the model fire code.

Both of these model fire codes contain many requirements that are similar to PSM and RMP regulations and contain references to them for additional guidance. Generally speaking, the model fire codes do not regulate all hazardous materials, but focus on hazards that are acutely dangerous to people or property. The PSM and RMP regulations address these hazards as well but additionally include materials that present a risk of chronic health or environmental hazards.

However, many states and municipalities have local fire code amendments that often include lower chemical quantity thresholds, such as those of the Commonwealth of Massachusetts as described in the following sections. The enforcement of these requirements is therefore not limited to OSHA and the EPA, as is commonly believed. The requirements are also enforceable by the local (municipal) fire authorities through the fire code references to these regulations.

Fire Code Requirements

The Commonwealth of Massachusetts currently adopts NFPA 1 – 2015 edition, with amendments, as 527 CMR 1.00: *Massachusetts Comprehensive Fire Safety Code*. Since 2006, 527 CMR has included chemical process safety requirements that are more restrictive than the federal regulations. The impetus for this was several accidents that occurred involving chemical process quantities below the OSHA and EPA thresholds. These Massachusetts (MA) amendments have been added to Chapter 60 of NFPA 1 – 2015 edition and include specific reference to OSHA PSM and EPA RMP regulations. Chemical processes with quantities above the PSM/RMP thresholds are defined in MA as "Category 5" processes. However, the 527 CMR amendments have also defined processes with lower quantities, i.e. Category 1, 2, 3 and 4, each with specific and unique requirements as described herein.

The NFPA 1 committee is currently reviewing proposals to adopt requirements with a "tiered" quantity threshold approach similar to MA. This is in recognition of the fact that severe hazards often exist, and severe accidents have occurred, in chemical processes with quantities below the PSM/RMP thresholds. It is likely that such a "tiered" approach will be incorporated into NFPA 1 and subsequently adopted by many US states. A preview of the applicable NFPA 1 and 527 CMR requirements are summarized herein.

Massachusetts Amendments to NFPA 1, Section 1.12.8.34.2 introduces the requirements as: *Process or Processing of Any Hazardous Material at any Facility* and it states the following:

(1) "A permit shall be required for the process or processing of any hazardous material at any facility identified in this *Code* as Category 2 through Category 5, as described in Chapter 60.
(2) The Authority Having Jurisdiction (AHJ) may require technical assistance in accordance with *1.15* to evaluate the adequacy of a Category 3 or Category 4 facility process safety conditions, programs, procedures, and practices undertaken at the facility, but only after a notice of denial has been properly served upon the person making application."

As part of the MA amendments to NFPA 1 – 2015 edition, Section 60.8, ***Hazardous Material Process or Processing,*** was added to 527 CMR. Section 60.8 applies to both new and existing facilities that process hazardous materials, with a list of exceptions such as:

- Motor vehicle service stations
- Refrigeration systems using a refrigerant other than ammonia or LPG,
- Certain wastewater treatment operations
- Storage of hazardous materials in atmospheric vessels.

Definitions:

NFPA 1 defines "**Hazardous Materials**" as "A chemical or substance that is classified as a physical hazard material or a health hazard material, whether the chemical or substance is in usable or waste condition."

OSHA and EPA define "**Process**" as "Any use, storage, manufacturing, handling, or the on-site movement of hazardous chemicals, or combination of these activities." The MA state fire code defines "Process" as "A sequence of operations in which the sequence can be inclusive of physical operations such as heating, cooling, mixing, distilling, compressing, and pressurizing, and chemical operations, such as polymerization, oxidation, reduction, and other chemical reaction processes. The sequence can involve but is not limited to: preparation, separation, combination, purification, or any actions that cause a change in state, energy content, or chemical composition."

MA Amendments to NFPA 1 Section 60.8 categorizes hazardous material processes into five (5) categories, as follows:

Table 1: Hazardous Material Threshold Quantities in the MA Amendments to NFPA 1

Category	**Quantity (gallons)**
Category 1	< 2.5
Category 2	2.5 – 60
Category 3	60 – 300 (or Group H)
Category 4	300 and not Category 5
Category 5	≥ the threshold quantities stated in 29 CFR 1910.119

The MA Amendments also add Section 60.8.5.1 to NFPA 1, which requires that facilities operating hazardous material processes must maintain the required documents and procedures for periodic inspection and review by the head of the Fire Department to remain in compliance with the state fire code. Requirements for each process category are as follows:

Table 2: Outline of Regulatory Requirements in the MA Amendments to NFPA 1

Category	Requirements
Category 1 (Vessel Capacity ≤ 2.5 gal)	• OSHA 1910.1450 (Occupational exposure to hazardous chemicals) • OSHA 1910.1200 (Hazard communication to employees) • Policy per NFPA 1, Ch. 66 (Flammable liquids) and Ch. 67 (Flammable solids) • Requirements of NFPA 1, Section 60.1 – 60.5
Category 2 (Vessel Capacity ≤ 60 gal)	• All requirements of Category 1 • Permit required per NFPA 1, Sections 1.12.8.34.2 and 60.8.1.1
Category 3 (Vessel Capacity ≤ 300 gal OR Group H Occupancy)	• All requirements of Category 1 & 2 • Category 3 Hazard Evaluation • Requirements of NFPA 1, Section 60.6
Category 4 (Vessel Capacity ≥ 300 gals AND not Category 5)	• All requirements of Category 1-3 • Limited Safety Program (PSM "lite")
Category 5 (Vessel Capacity ≥ the threshold quantities stated in 29 CFR 1910.119)	• All requirements of Category 1-4 • Process Safety Management (PSM) Program 29 CFR 1910.119 (OSHA) • Risk Management Program (RMP) 40 CFR Part 68 (EPA)

Category 1 – Requirement Details

Category 1 processes, as stated in Table 2 above, must possess documentation of compliance with OSHA 1910.1450, *Occupational Exposure to Hazardous Chemicals in Laboratories*, OSHA 1910.1200, *Hazard Communication,* maintain and implement a policy in compliance with NFPA 1, Ch. 66, *Flammable and Combustible Liquids*, and Ch. 67, *Flammable Solids*, demonstrate compliance with NFPA 1, Section 60.1.5.1, *Emergency Action Plan*, and Sections 60.5.1.4.3.2(5) through 60.5.1.4.3.2(9) and comply with permitting requirements in Sections 1.12 and 60.8.4 of the MA Amendments to NFPA 1.

OSHA 1910.1450 and 1910.1200

Compliance with OSHA 1910.1450 requires facilities to set and adhere to exposure limits, monitoring plans, hygiene plans, precautions, training, medical plans, and other plans, as found to be acceptable by the AHJ, to react to and minimize potential exposure to hazardous chemicals. OSHA 1910.1200 requires hazard communication methods for employees such as inventory documentation, hazard programs, labels, safety data sheets, and training.

NFPA Chapter 66 and 67

In addition to demonstrating compliance with OSHA standards, facilities meeting Category 1 requirements must implement and maintain a written policy in compliance NFPA 1, Ch. 66 and Ch. 67.Chapter 66 is divided into 29 sections, most of which contain multiple requirements. Section 66.6 contains the prescriptive requirements for fire and explosion prevention and risk control, and Section 66.6.4 goes on to list requirements related to a facility's hazards analysis, which is required to be reviewed to ensure that fire and explosion hazards are addressed by fire prevention, fire control, and emergency action plans. The AHJ should be consulted with for the development of the hazards analysis to ensure that the extent of the fire prevention and control is deemed acceptable or an engineering evaluation of the operation and application of fire protection and process engineering principles is soundly applied. The hazards analysis must include, but not be limited to, analysis of the following:

(1) Fire and explosion hazards
(2) Emergency relief from process vessels
(3) Applicable facility design requirements in Chapters 17, 18, 19, 28, and 29 of NFPA 30
(4) Applicable requirements for liquid handlings, transfer, and use, as covered in Chapters 17, 18, 19, 28, and 29 of NFPA 30
(5) Local conditions/exposures
(6) Emergency response capabilities of the local emergency services

Additional guidance for conducting a fire risk assessment can be found in NFPA 551, *Guide for the Evaluation of Fire Risk Assessments.* Facility management must repeat the hazards analysis when any of the following occur:

(1) Changes occur in the materials in process
(2) Changes occur in process equipment
(3) Changes occur in process control
(4) Changes occur in operating procedures or assignments

This is consistent with the requirements of OSHA 1910.119 Sections (e) and (l). Not all sections of Chapter 66 may be related to processes; however, Section 66.17 is for processing facilities. Section 66.17 includes prescriptive requirements for processing facilities such as location of process vessels and equipment (Section 66.17.4), containment, drainage, and spill control (Section 66.17.10), and management of operations hazards (Section 66.17.15). Later sections of Chapter 66 include additional requirements for dispensing, handling, transfer, and use of liquids (processes) (Section 66.18) and specific operations such as recirculating heat transfer system, vapor recovery and vapor processing system, solvent distillation units, and cooking oil storage tank systems in commercial kitchens (Section 66.19).

NFPA 1 Chapter 60

In addition to Chapters 66 and 67 of NFPA 1, Category 1 processes must also adhere to the requirements of 60.1 – 60.5 of NFPA 1 and the MA Amendments to NFPA 1. Depending on whether the quantities of materials at the facility exceed the MAQ or not, determines which requirements of Section 60.1 are applicable (Section 60.1.3).

Section 60.1.5 of NFPA 1 requires that an emergency action plan, consistent with the available equipment and personnel, shall be established to respond to fire and other emergencies in accordance with requirements set forth in NFPA 1. Section 10.8 of NFPA 1 includes additional requirements and guidance on generating emergency action plans, such as what must be included in the plans, purpose, responsibilities, etc. Section 60.4 of NFPA 1 contains requirements related to permissible storage and use locations and sets prescriptive requirements on the maximum allowable quantities for each individual control areas based on the area's occupancy type.

Section 60.5 contains the fundamental requirements for hazardous materials, including requirements for personnel training, ignition sources, systems, equipment, and processes (Section 60.5.1.4, 60.5.1.5, and 60.5.1.6 respectively). In Massachusetts, Section 60.5 includes multiple requirements that come from both NFPA 1 and the MA Amendments to NFPA 1. Despite Category 1 processes having a vessel capacity much less than Category 2-5 processes (Category 1 processes have a vessel capacity ≤ 2.5 gallons), there are numerous NFPA and state fire code requirements that must be followed. Therefore, the importance of fire protection engineers and their understanding of the NFPA and state fire codes cannot be understated for processing facilities. Each subsequent category process contains additional fire code fire protection/chemical process safety requirements of increasing complexity.

Category 2: Requirement Details

Category 2 processes must comply with all Category 1 requirements and additionally submit for a permit to the AHJ, as required per 1.12.8.34.2 and 60.8.1.1 of MA Amendments to NFPA 1. Permits submitted to the AHJ must comply with the requirements of Section 1.12 of NFPA 1 – 2015 edition. The nature of the operations and materials at the facility determine the permit requirements. Depending on the nature of the process, the AHJ may require technical assistance to evaluate the adequacy of facility process safety conditions, programs, procedures, and practices undertaken at the facility, but only after a notice of denial has been properly served upon the person making application.

Category 3: Requirement Details

While meeting all the requirements of Category 1 and 2 processes, Category 3 processes must also provide the following documentation and evaluations:

- Compliance with the requirements of Category 1 and 2 processes;
- Category 3 hazard evaluation; - per Section 60.8.2
- Implementation of process safety controls to mitigate hazards of normal and abnormal conditions identified in the hazard evaluation;
- Maintain documentation for review for a minimum of two (2) years;
- Third-party review (if required by the fire department/AHJ);
- Post-Incident Analysis per 60.8.6 (to be initiated within 48 hours of incident);

MA Amendment to NFPA 1, Section 60.8.2.2 provides requirements and details for the hazard evaluation that is required for Category 3 processes. **Hazard evaluations** are meant to identify hazards, including adjacent vessels that contain hazardous materials, and determine the required preventative, protective, and safety control measures in conformance with **r**ecognized **a**nd **g**enerally **a**ccepted **g**ood **e**ngineering (and safe work) **p**ractices (RAGAGEP) associated with a particular process or condition. Section 60.8.2.5 requires that a person with *specialized knowledge* beyond that of an average person, about risk assessment, process hazard analysis, and/or process safety management principles, for the process or processes being evaluated *conduct the necessary evaluations*. Section 60.8.2.5 further emphasizes the importance of both chemical process safety specialists and fire protection engineers for analysis of hazardous processing facilities.

In addition to generating a hazard evaluation, Category 3 processing facilities (vessel capacity ≤ 300 gal OR Group H Occupancy) must also meet the requirements of NFPA 1, Section 60.6, *Emergency Action Planning, Fire Risk Control and Chemical Hazard Requirements for Industrial Processes.* Where quantities of materials in use require compliance with Protection Level 1, Protection Level 2, Protection Level 3, or Protection Level 4 based on materials exceeding the Maximum Allowable Quantity (MAQ) in the following categories shall comply with the requirements of Chapter 7 of NFPA 400, *Hazardous Materials Code*:

(1) Unpackaged organic peroxide formulations that are capable of explosive decomposition in their unpackaged state
(2) Oxidizer Class 3 and Class 4: solids and liquids
(3) Pyrophoric solids, liquids, and gases
(4) Unstable reactive Class 3 and Class 4: solids, liquids, and gases
(5) Highly toxic solids, liquids, and gases
(6) Water-reactive liquids, Class 3

NFPA 400, Chapter 7, *Emergency Planning, Fire Risk Control, and Chemical Hazard Requirements for Industrial Processes,* contains requirements and guidance on topics such as:

- Process review and plan preparation
- Operating and maintenance procedures
- Safety reviews
- Incident investigation plans

Section 7.2.3 contains specific requirements for Emergency Action Planning that facilities must comply with. Some requirements of Section 7.2.3 include:

- Procedures for notifying and coordinating with the fire department and other applicable agencies
- Procedures and schedules for conducting drills
- Alternative measures for occupant safety
- Maintenance of fire protection equipment

Lastly, in lieu of complying with Chapter 60 in its entirety, occupancies containing high hazard Level 1 to high hazard Level 5 contents shall be permitted to comply with Chapter 10 of NFPA 400, as stated in NFPA 1, Section 60.7. MA Amendments to NFPA 1 further states that complying with NFPA 1, Section 60.7, is subject to an independent review in accordance with Section 1.15 and a copy, including its recommendations, shall be submitted to the Building Official.

Category 4: Requirement Details

Category 4 processing facilities (vessel capacity ≥ 300 gal) must meet:

All the requirements of Category 1, 2, and 3 processes

And also create and maintain a **Limited Safety Program (PSM "lite")**. The Limited Safety Program documents evaluation, policy and/or required procedures to ensure compliance with the following requirements of the MA Amendment to NFPA 1:

(1) Process information including, but not limited to, MSDS for the chemicals and products being processed, process chemistry, piping and instrumentation diagram, safety relief design, process control safety alarms and interlocks;

(2) Facility suitability including, but not limited to, the *Building Code* compliance, electrical hazard classification, ventilation design, fire alarm and fire protection, spill containment and control;

(3) A process hazard safety analysis including, but not limited to, effects in the event of failure, suitable administrative and engineering controls to minimize failure and to control unanticipated releases, and emergency responses to safeguard life and property;

(4) Written procedures, including routine operating and maintenance, as well as precautionary, shut-down and emergency response measures;

(5) A written training program for operating and maintenance personnel and outside contractors whose work or activity may affect process safety;

(6) A written records management protocol, which tracks any changes, including but not limited to changes to chemicals, equipment, operating procedures training program. Such records shall include the date of such change and the name of the manager responsible for such change; and an internal review at a maximum every three years.

These requirements are similar to those found in OSHA 29 CFR 1910.119, *Process Safety Management of Highly Hazardous Chemicals*.

Category 5: Requirement Details

Category 5 processing facilities must provide evaluation and documentation per 29 CFR 1910.119 OSHA, *Process Safety Management of Highly Hazardous Chemicals*, and 40 CFR Part 68 EPA, *Risk Management Program Rule*, all while meeting the requirements for Categories 1-4. OSHA's *Process Safety Management* (PSM) *of Highly Hazardous Chemicals* was issued in February 1992 and contains 14 elements that focus mainly on worker safety and catastrophic loss of containment. The Process Hazard Analysis (PHA) is the "heart" of the standard and is the most time consuming. The PHA must be completed every five (5) years and *is required to be completed by someone knowledgeable with the PHA methodology*. Additionally, compliance audits for the standard must be completed every three (3) years and may be completed in-house or by a third party (preferred). EPA's *Risk Management Program* (RMP) *Rule* consists of 12 elements and is focused on public safety and the environment. The program contains Chemical Accident Prevention Provisions (CAPP) and Accidental Release Prevention/Risk Management Plan Rule. The RMP must be submitted to the EPA every fire (5) years or if there are any changes.

Conclusion

Some fire code requirements for chemical process safety apply to processes with quantities much lower than the PSM/RPM thresholds. Failure to comply with these requirements can result in fire department citations, fines, remedial work and in extreme cases, shutdown of operations. Between 2013 and 2018, 214 OSHA PSM violations were recorded in Massachusetts alone, which resulted in over $485,000 in penalties. There are far more non-PSM facilities than PSM facilities in MA, so the magnitude of the problem can be seen by simple extrapolation. Failing to comply with the fire code and the federal requirements can also result in significant property loss, injuries and fatalities with much higher costs than the violation penalties. Compliance with the local fire code and federal regulations should be determined by experienced chemical process safety specialists and fire protection engineers to minimize the risk of significant loss.

Fire protection engineering and science is its own discipline and chemical process safety is its own discipline usually practiced by the chemical engineering field. Finding ways to create awareness and bridge the gap between the two disciplines, and sharing chemical process safety management concepts with those in the fire protection arena, will likely improve the oversight, understanding and enforcement of process safety resulting in better overall safety. There may be a way of joining these disciplines into one overarching safety category and therefore giving a better sense of how interrelated they really are.

546769 Poly-Butadiene Formation in Olefins Debutanizer Condenser

Abstract Title : Poly-butadiene formation in Olefins Debutanizer condenser

Discipline : Process Safety

Topic : Poly-butadiene formation in Debutanizer condenser

Author Name : Ali Bazroun

Participating team: Fahad AlShammari, and Eldridge Monette, Bertrand, Da, Cruz, Fahad AlSaad, Abdullah Al-Ismail

ABSTRACT

In one of Sabic olefins plant Plant, the condenser of the debutanizer tower was fouling. This was evident through the pressure increase in the tower and the cooling water control valve reaching the maximum opening (69%) limited by the mechanical stopper. The condenser process service is butadiene, which has the tendency of polymerizing and formation of popcorn, which could plug relief piping or accumulate in the condenser until the vessel shell is over stressed causing loss of containment and leading to fire or explosion.

The risk assessment team evaluated that the risk as significant, meaning that it could result into fatality and multiple lost time injury with a likelihood of 10E-3. With the propagating shell deformation, the likelihood could increase to L3 (10E-2) over a longer period. The risk could become C2-L3 and remain Significant. To ensure effective risk mitigation, a major plant modification was executed by constructing a new process line that bypasses the debutanizer tower and sends the material to the flare. The project execution was remarkable as it was completed in about a week. 15 tons of C4 is being flared per. This, however, allowed the olefins unit to mitigate the risk effectively without any interruption of the production. The change was not done before in the local industry, and it reflected how effective measures can be taken to mitigate risk through strong leadership and excellent engineering. Interim mitigations, moreover, were in place until the modification was completed: for example, inhibitor injection in the exchanger to inhibit the popcorn formation as soon as possible, and an emergency procedure in case loss of containment. Without the effective mitigation, Sabic operation team would have had to shutdown the olefins unit for over 2 months. This olefins unit feeds two poly ethylene that would shutdown consequently, causing significant losses over the mentioned period.

After isolating the exchanger, borescope test proved the formation of polymer. A root cause analysis was initiated, along with cleaning the tower and fabricating a new exchanger. The details of the RCA will be shared in the full paper.

19 AIChE Spring Meeting
15TH GLOBAL CONGRESS ON PROCESS SAFETY
NEW ORLEANS • 2019

Industrial process plant safety by collaboration of a Safety Instrumented System and a Dynamic Simulator

Tadaaki Ando
Yokogawa Corporation of America
12530 West Airport Blvd, Sugar Land, TX 77478, USA
Tadaaki.Ando@us.yokogawa.com
(Tadaaki.Andou@jp.yokogawa.com)

Yasuhiko Yamashiro
Yokogawa Electric Corporation
2-9-32 Nakacho, Musashino-shi, Tokyo, 180-8750 Japan
Yasuhiko.Yamashiro@jp.yokogawa.com

Prepared for Presentation at
American Institute of Chemical Engineers
2019 Spring Meeting and 15th Global Congress on Process Safety
New Orleans, LA
March 31 – April 3, 2019

Industrial process plant safety by collaboration of a Safety Instrumented System and a Dynamic Simulator

Tadaaki Ando
Yokogawa Corporation of America
12530 West Airport Blvd, Sugar Land, TX 77478, USA
Tadaaki.Ando@us.yokogawa.com
(Tadaaki.Andou@jp.yokogawa.com)

Yasuhiko Yamashiro
Yokogawa Electric Corporation

Keywords: Industrial process plant, SIS (Safety Instrumented System), Dynamic Simulator

Abstract

Safer HMI design for operators' quick and appropriate response is very important. Allowed maximum delay for operator's action from the alarm is not same all the cases because it is depends on many factors, i.e. ambient temperature, density of material, O2, and etc. So design of SIS includes HMI should be based "worst case". However, the 'worst case' is difficult to be found.

A "Dynamic Plant Simulator" can precisely simulate behaviors of the real process by visualizing how a hazard is diffused at various stages of a chemical reaction. And combination of DCS/SIS/HMI simulator with "Dynamic Plant Simulator", safety logic design is optimized about trip point and alarm threshold, and HMI design is modified safer manner.

The effective result of application of "cyber model by Dynamic Process Simulator" to a vinyl acetate monomer (VAM) plant and its value for operator training will be discussed.

1 Introduction

Industrial process plants in Japan have been seeking for the highest productivity and optimization of their processes, same with any process plant all over the world. These plants were renovated and expanded as time passed for the purpose, the plants themselves became complicated and troublesome as a result. Few years ago, severe process plant incidents occurred in Japan. The most probable causes of these severe incidents are out of "incorrect judgement and operation" and "shortage of prediction for technical aspects" during exceptional manual work that has been done by incidents. Understanding the hazardous situation by risk analysis and designing safety measures utilizing a SIS (Safety Instrumented System) are important and effective for resolving "incorrect judgement and operation" and "shortage of prediction of technical aspect". Safety measure includes human intervention via SIS man-machine interface which showing alarm, event, and process data.

The safety measure and its function are developed on the desk by brainstorming or so-called "hazard and operability study (HAZOP)." The verification of the safety function hardware could be established by reliability based analysis, SIL calculation in other word. However, the verification of safety logic is the issue because it has only been verified "on the desk". Especially the determination of the criteria for the emergency shutdown is the most important, but it is not possible to be confirmed with "try and error" on a running plant.

2 Safe HMI Design

Safer HMI design with considering different conditions and situations in the real process for operators' quick and appropriate response is very important. To design it, knowing the behavior of the process itself and related factors which affect the process is necessary.

For example, a reaction speed varies on the ambient temperature, density of material, O_2, catalyst, and etc., so allowed maximum delay for operator's action from the alarm is not same all the cases. So design of SIS includes HMI should consider "worst case". However, finding the worst-case scenario with how much is the allowed operator reaction is difficult because of many factors.

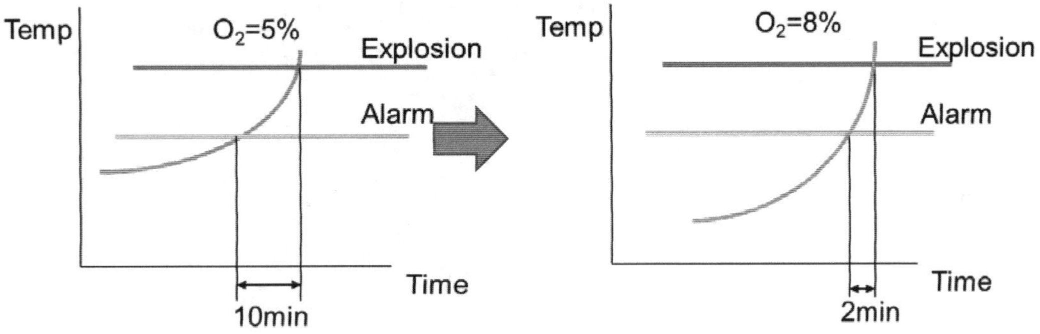

Fig.1 Example: Difference of O_2 density vs. allowed reaction time

Another example is the 'thermal runaway'. Even if feeding of material, heat, and catalyst were stopped by ESD because of a high temperature of a reactor, reaction continues due to a feed-forward loop of heating, means reaction generates heat to continue the reaction itself.

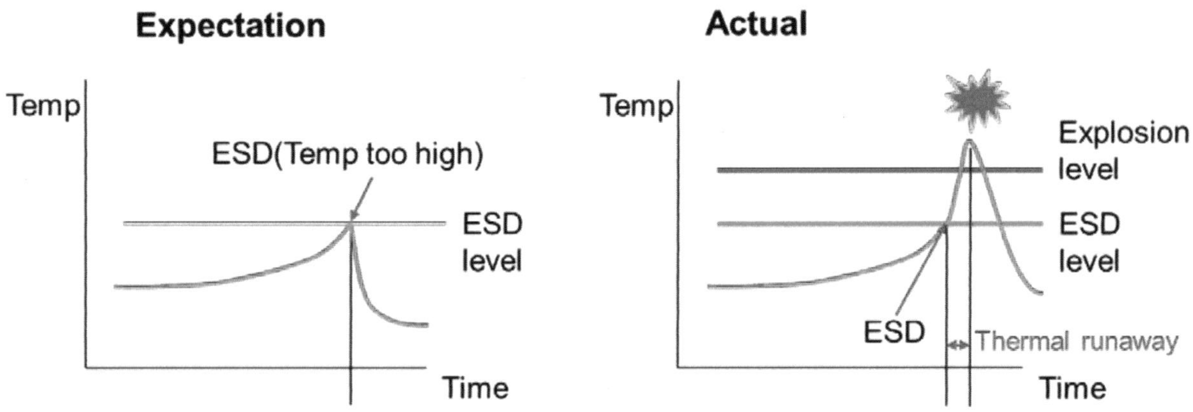

Fig.2 Even after ESD, thermal runaway leads the process to explosion

Thus, considered 'worst case' could not be a 'worst', or could be 'too much pessimistic'.

A "Dynamic Plant Simulator" can precisely simulate behaviors of the real process by visualizing how a hazard is diffused at various stages of a chemical reaction. And combination of DCS/SIS/HMI simulator with "Dynamic Plant Simulator", safety logic design is optimized about trip point and alarm threshold, and HMI design is modified safer manner.

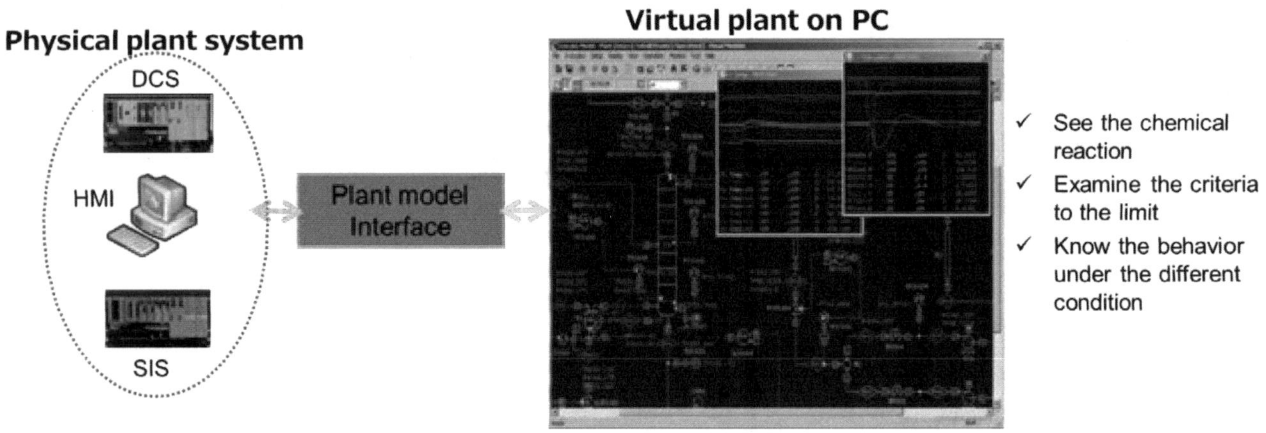

Fig.3 Dynamic Plant Simulator

3 "Dynamic Plant Simulator" at VAM (Vinyl Acetate Monomer) process

"Dynamic Plant Simulator" with "DCS/SIS/HMI simulator" was applied for several shutdown conditions of VAM plant. "Dynamic Plant Simulator" can simulate a variation of factors like ambient temperature and thermal runaway, because is based on a theory of chemical reaction. And "DCS/SIS/HMI simulator" is copying a real function and operation of DCS/SIS/HMI.

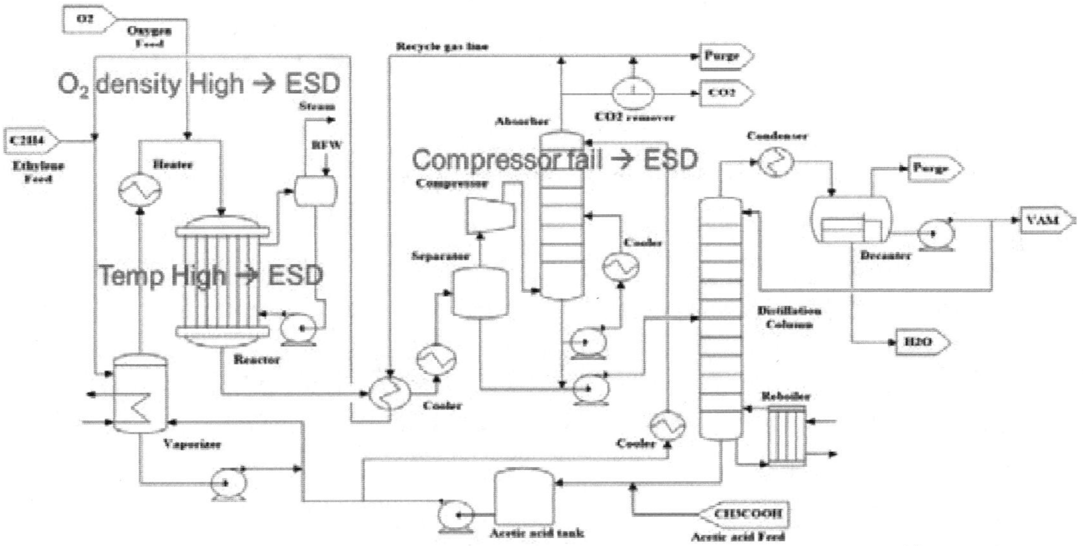

(Main Reaction) $C_2H_4 + CH_3COOH + 1/2O_2 \rightarrow CH_2{=}CHOCOCH_3 + H_2O$

(Side Reaction) $C_2H_4 + 3O_2 \rightarrow 2CO_2 + 2H_2O$

Fig.4 VAM process model

3.1 Shutdown by high temperature to avoid a damage to reactor

Reactor temperature must be monitored and controlled lower than 200 Degrees C (392 Degrees Fahrenheit) to avoid a damage of the reactor. If the temperature reached to a set point for emergency shutdown, emergency shutdown action stops feeding of Ethylene and O_2.

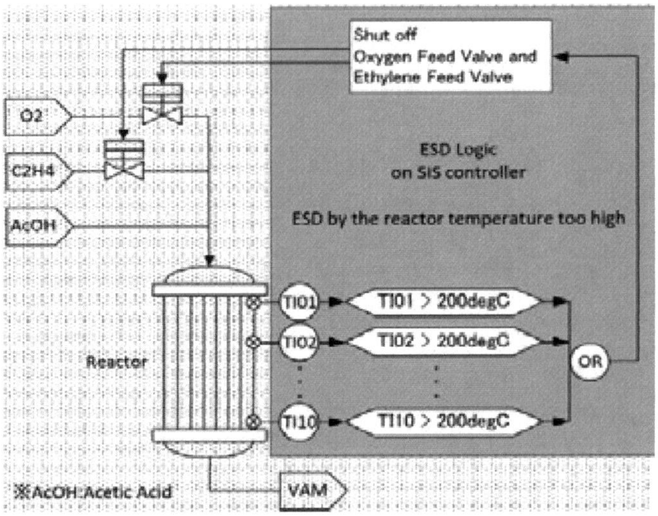

Fig.5 ESD logic for high temperature shutdown

When 190 Degrees C (374 Degrees Fahrenheit) is selected as the set point to shutdown, the result of simulation is, the reactor temperature must become over 200 Degrees C due to a runaway reaction in the reactor. It means reactor must be damaged. When 180 Degrees C (356 Degrees Fahrenheit) is selected, the simulation shows the reactor temperature can be kept lower than 200 Degrees C, so no damage to the reactor.

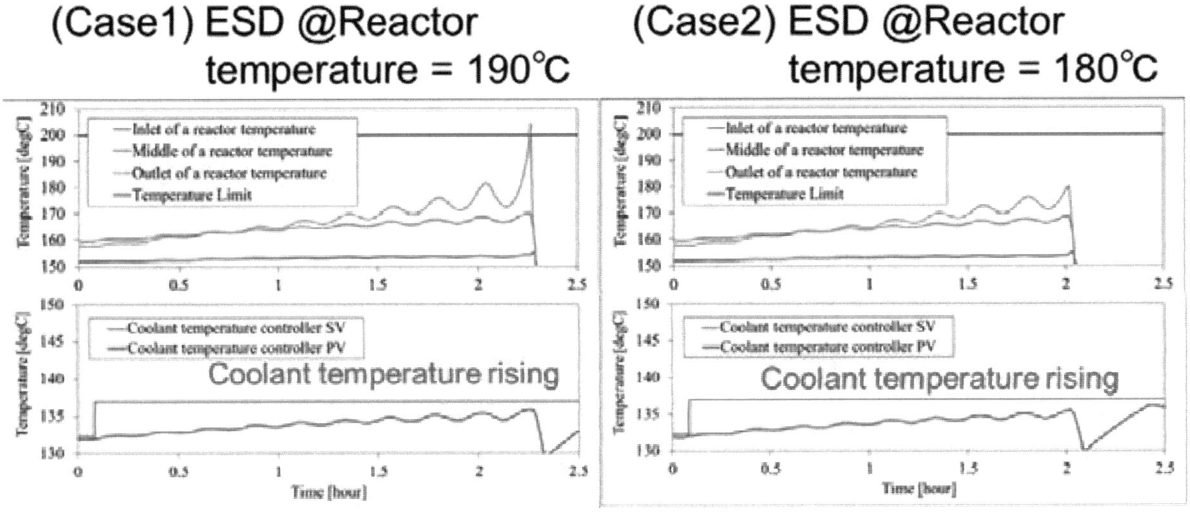

Fig.6 Simulation results

And simulation suggested the reactor temperature must be monitored in multipoint, middle of the reactor and top of the reactor. When the reaction temperature is lower than 160 Degrees C (320 Degrees Fahrenheit), temperature of a middle of the reactor is higher than other points, but when the reaction temperature is higher than 160 Degrees C, the temperature of top of the reactor is higher than other points.

3.2 Shutdown by high O_2 density to avoid explosion

If the density of O_2 in the reactor exceeds 8%, there is a risk of explosion. However, the O_2 density varies depends on other factors, i.e. feed pressure of ethylene to the reactor. Simulation has been done about O_2 density when the feed pressure of ethylene is suddenly changed. Simulation of some cases of dumping the feed pressure to lower side which increase the density of O_2 in the reactor were done.

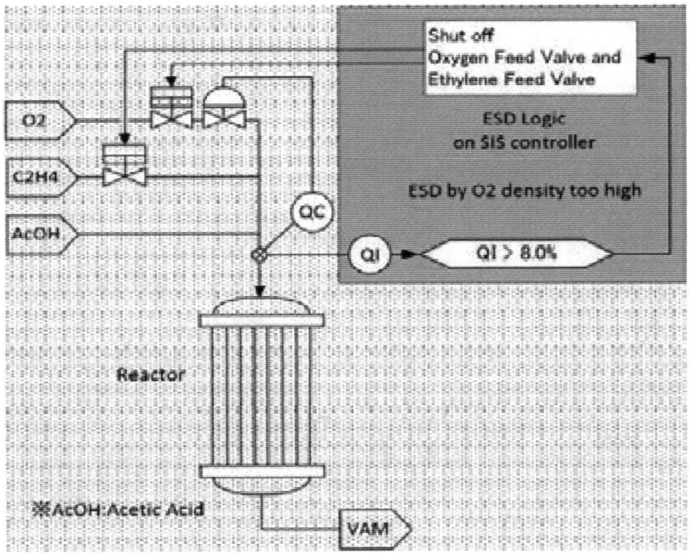

Fig.7 ESD logic for high O_2 density shutdown

When the feed pressure of ethylene dumped 10%, density of O_2 in the reactor doesn't reach to 8%. So there is no risk of explosion. However, the dumping of ethylene feed pressure is 20%, O_2 density will be higher than 8%, then an emergency shutdown must be activated. Shutdown does affect to business.

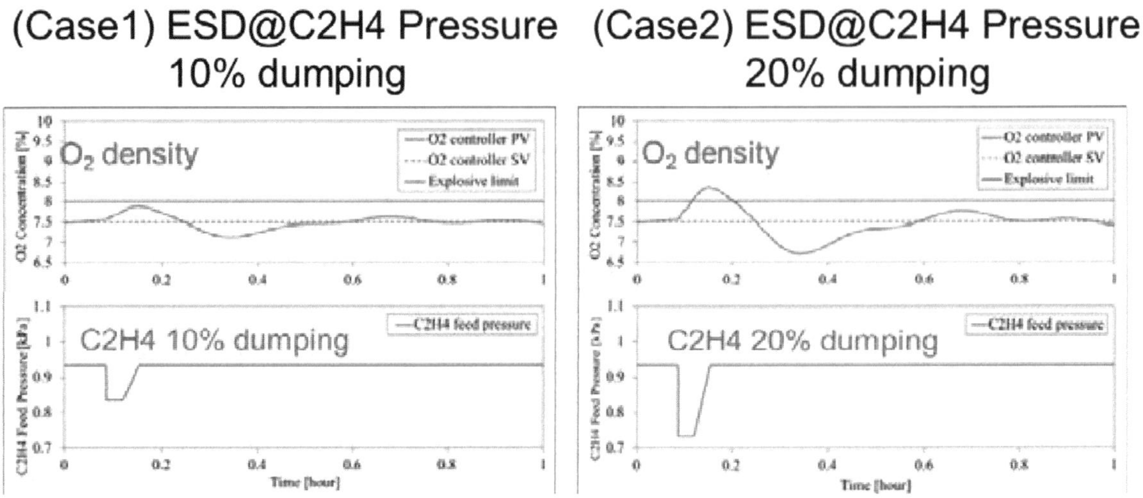

Fig.8 Simulation results

If there is a possibility of dumping ethylene feeding pressure, feeding O_2 in normal operation should be lower to avoid the risk of reactor explosion. On the other hand, lower O_2 density leads lower reaction speed, less production. Finding a best balance with safety and productivity is the key.

Using dynamic process simulator, many factors of process can be manipulated not only within normal operation range, but also close to abnormal, emergency situation or more un-realistic range, and results can be confirmed. This type of "try-and-error" approach is not possible on the real process, but possible in the "cyber" world to find a best balance of setting for maximize production with keeping or improving safety.

4 "Dynamic Plant Simulator" for Operator training

The training of operators is important in many reasons. Especially training for response to critical / emergency situation is indispensable, because operator rarely or never face such a situation and consequent secondary action to DCS/SIS or response to alarm flooding in a real plant. Therefore operator are impossible to familiarize with their required response in right procedure to critical / emergency situation. If operators are not capable or not knowing how to response such a situation, risk must propagate real hazard or accident. Furthermore, operators need to re-start the process after confirming the cause of emergency doesn't exist anymore and checking all the conditions of materials are steady, status of equipment are ready to re-start. This procedures are also not normal for operators.

"Dynamic Plant Simulator" realizes operator training with "equal to real" environment. Operators will be trained for correct and timely manner response until proficient. And operators will have more understanding for process itself by playing on the simulator.

5 Conclusion

The effectiveness of "Dynamic Process Simulator" for safer design of DCS/SIS/HMI is confirmed on VAM plant. The best balance of safety and productivity can be found and optimization of safety action is created by simulation. The created new design of DCS/SIS/HMI is possible to use an actual system directly. And operator training with "Dynamic Process Simulator" will increase a total safety of the plant.

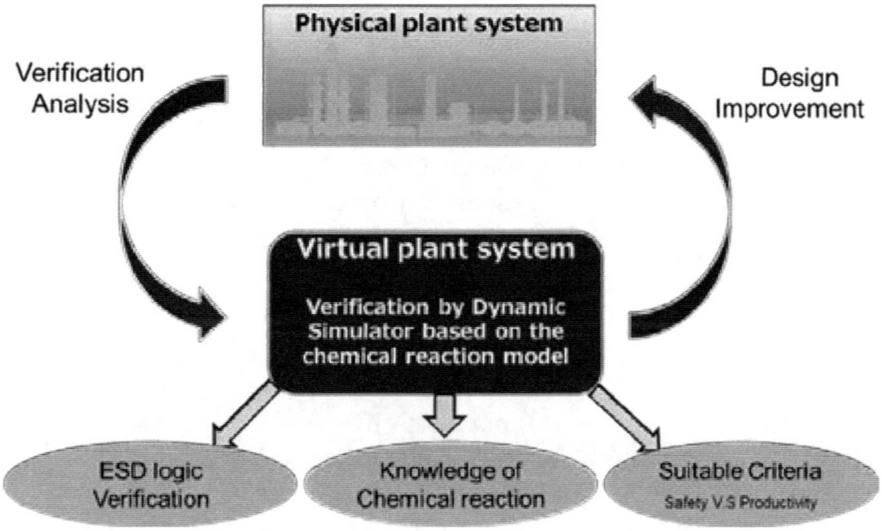

Fig.9:Virtual plant system realizes well balanced plant operation

Analysis of Chemical Effluent Discharge through the Emergency Relief System using Air-Water Properties

Basar Ozar
Jensen Hughes
600 W Fulton St. Suite 500
Chicago, IL 60661
bozar@jensenhughes.com

Jens Conzen
Jensen Hughes
600 W Fulton St. Suite 500
Chicago, IL 60661
jconzen@jensenhughes.com

Damian D. Stefanczyk
Jensen Hughes
3610 Commerce Dr. Suite 817
Baltimore, MD 21227
dstefanczyk@jensenhughes.com

Prepared for Presentation at
American Institute of Chemical Engineers
2019 Spring Meeting and 15th Global Congress on Process Safety
New Orleans, LA
March 31 – April 3, 2019

Analysis of Chemical Effluent Discharge through the Emergency Relief System using Air-Water Properties

Basar Ozar
Jensen Hughes
600 W Fulton St. Suite 500
Chicago, IL 60661
bozar@jensenhughes.com

Jens Conzen
Jensen Hughes
600 W Fulton St. Suite 500
Chicago, IL 60661
jconzen@jensenhughes.com

Damian D. Stefanczyk
Jensen Hughes
3610 Commerce Dr. Suite 817
Baltimore, MD 21227
dstefanczyk@jensenhughes.com

Keywords: two-phase flow, hydrodynamic load, emergency relief system, chemical effluent

Abstract

A sudden opening of a pressure relief device (typically a pressure relief valve or a rupture disk) during an overpressure event will cause the chemical effluent to accelerate through the emergency relief system (ERS) piping. The rapid change of momentum of the effluent creates transient hydrodynamic loads on the piping. ASME B31 in the United States and European Directive 2014/68/EU requires these transient loads to be considered along with other applicable loads (i.e., dead weight, thermal) in the system design. Additionally, the pressure distribution along the piping may be of importance, particularly when other vessels are connected to the same ERS. The back pressures generated on these vessels should be evaluated and compared with the vessels' MAWP (maximum allowable working pressure) as well as the set pressure of any isolation devices. As such, a detailed time history of the effluent pressure and hydrodynamic loads are required to perform a proper evaluation. The effluent may be composed of proprietary chemical mixtures containing various materials. Therefore, the exact material properties of the effluent during the transient might not always be known. In this study, a previously published methodology is used for calculating the dynamics of the effluent and the resultant hydrodynamic loads for a typical ERS by using the RELAP5 (Reactor Excursion and Leak Analysis Program) software. Air-water properties are assumed for the effluent. Key output parameters, such as pressure drop and hydrodynamic loads, are then compared with established hand calculation methods using the actual effluent properties. Given the mechanical nature of the hydrodynamic load event, it was found that the use of air-water properties provides a reasonable representation of the important characteristics of the actual effluent discharge in mechanical design space.

1 Introduction

Pressure build up in a system may occur as a result of a process upset, instrument failure or fire. These upset scenarios can result in either, or both, vaporization due to exothermic runaway reactions or generation of gases due to the decomposition process, which are of particular interest in terms of system and public safety. The extent of the potential hazards caused by such incidents are generally determined by the rate and amount of the energy being generated, the chemical substance makeup, and the quantity of the effluent.

The vessels or systems, which contain such chemicals (e.g. chemical reactors or storage tanks), are typically protected by an ERS that prevents from over-pressurization of the equipment as a consequence of the vaporization of chemicals and/or generation of gases. The process for designing of an ERS was previously discussed by Conzen and Wood (Ref. [1]) and Ozar (Ref. [2]). The design process consists of a process hazard analysis (PHA), defining credible and limiting upset scenarios, adiabatic calorimetry testing, vent sizing calculations, fluid transient analysis, pipe stress analysis, and a dispersion analysis. The central task in the design process is the fluid transient analysis (Refs. [1] and [2]), which is the main focus of this study.

The ERS is typically equipped with either a rupture disk or a pressure relief valve (PRV). Some systems may have both pieces of equipment. The sudden opening of the relief device during an overpressure event will typically result in a two-phase discharge and cause the chemical effluent to accelerate through the ERS piping. The rapid change of momentum of the effluent creates transient hydrodynamic loads on the piping. ASME B31 (Ref. [3]) in the United States and European Directive 2014/68/EU (Ref. [4]) requires these transient loads to be considered along with other applicable loads (i.e., dead weight, thermal) in the system design. Additionally, the pressure distribution along the piping may be of importance, particularly when other vessels are connected to the same ERS. The back pressures generated on these vessels should be evaluated and compared with the vessels' MAWP as well as the set pressure of any isolation devices. As such, a detailed time history of the effluent pressure and hydrodynamic loads are required to perform a proper evaluation.

The two-phase behavior of the chemical effluent has been studied extensively. Following earlier developments by Epstein et al. (Ref. [5]) and Grolmes and Leung (Ref. [6]), a simplified correlation was suggested by Leung (Ref. [7]) to predict flashing choked flow in a frictionless duct or nozzle. The correlating parameter, ω, was based on an isenthalpic (constant enthalpy) assumption.

Leung and Grolmes (Ref. [8]) integrated the momentum equation for homogenous two-phase flow in a horizontal duct using "ω" as a closure relationship for two-phase mixture specific density. Leung and Epstein (Ref. [9]) extended the derivation to an inclined duct configuration. Leung (Ref. [10]) provided the derivation of the reaction forces during a two-phase discharge from ERS piping.

Computational methods have also been used by various industries to investigate two-phase flow behavior in pipes and the hydrodynamic loads due to a sudden acceleration or deceleration of the fluid and the propagation of pressure waves. RELAP5 is a tool for analyzing small-break loss of coolant accidents and system transients in nuclear power plants. The program has the capability to model one dimensional single and two-phase fluid flow and heat transfer problems. The software has undergone extensive review and validation by the US Nuclear Regulatory Commission (Ref. [11]). It has also been benchmarked (Refs. [12], [13] and [14]) for predicting pressure wave propagation and hydrodynamic loads in piping systems. Refs. [15] and [16] provide a discussion about some of the nuclear power plant applications. However, RELAP5 is limited to non-condensable gas-water and steam-water two-phase flow applications.

This study provides a discussion about using the analytical methods developed by Leung and Grolmes (Ref. [8]) and Leung (Ref. [10]) to demonstrate the similarity between the two-phase mixture of common industrial chemical effluents and air/steam-water two-phase flow. Computational and analytical calculations for three different discharge scenarios are also provided. The comparisons between the computational and analytical results are used to demonstrate the applicability of using air/steam-water properties to simulate chemical effluent discharge to reasonably predict the hydrodynamic loads and pressure distribution in the ERS following the actuation of a pressure relief device.

2 Theoretical Background

The governing steady-state conservation equations for one-dimensional homogenous two-phase flow in a horizontal duct with a constant flow area are (Ref. [8]):

Mass

$$G = \text{constant} \tag{1}$$

Momentum

$$vdP + G^2\left(vdv + \frac{4fv^2dL}{2D}\right) = 0 \tag{2}$$

Energy

$$h_0 = h + \frac{G^2v^2}{2} \tag{3}$$

G, v, P, f, L, D and h are the mass flux, specific volume, pressure, two-phase friction factor, length of the duct, hydraulic diameter of the duct, and specific enthalpy, respectively. The specific enthalpy and specific volume are expressed in terms of quality (x) and phasic properties, i.e., $h_0 = h_{f0} + x_o h_{fg0}$, $h = h_f + xh_{fg}$, and $v = v_f + xv_{fg}$. Subscript 0 denotes stagnation conditions and fg the difference between vapor and liquid properties.

Epstein et al. (Ref. [5]) and Grolmes and Leung (Ref. [6]) suggested a simple relation between the two-phase specific volume and pressure as given by:

$$\frac{v}{v_0} - 1 = \omega\left(\frac{P_0}{P} - 1\right), \tag{4}$$

where ω is a non-dimensional parameter. Leung (Ref. [7]) correlated ω for flashing choked flow in a frictionless duct or nozzle. The correlating parameter was shown to be in the following simplified form:

$$\omega = \frac{x_0 v_{fg0}}{v_0} + \frac{C_{f0} T_0 P_0}{v_0} \left(\frac{v_{fg0}}{h_{fg0}} \right)^2 \tag{5}$$

C_{f0} is the heat capacity of the liquid. The last term on the right-hand side is zero for non-flashing gas-liquid two-phase flow and the non-dimensional parameter, ω, is provided by:

$$\omega = \frac{x_0 v_{fg0}}{v_0} \tag{6}$$

Leung and Grolmes (Ref. [8]) considered a two-phase horizontal discharge from a large reservoir (Figure 1). They integrated Equation 2 over the length, L, by substituting Equations 4 and 5 and using the isenthalpic (adiabatic and reversible) flow approximation. The momentum equations after integration yields to:

$$N \equiv 4f\frac{L}{D} = \frac{2}{G^{*2}}\left[\frac{\eta_1 - \eta_2}{1-\omega} + \frac{\omega}{(1-\omega)^2} ln\left[\frac{(1-\omega)\eta_2 + \omega}{(1-\omega)\eta_1 + \omega}\right]\right] - 2ln\left[\frac{(1-\omega)\eta_2 + \omega}{(1-\omega)\eta_1 + \omega}\left(\frac{\eta_1}{\eta_2}\right)\right] \tag{7}$$

The non-dimensional parameters are defined by:

$$\eta_1 \equiv \frac{P_1}{P_0}, \eta_2 \equiv \frac{P_2}{P_0}, \eta_3 \equiv \frac{P_3}{P_0}, \; G^* \equiv G/\sqrt{\frac{P_0}{v_0}}, N \equiv 4f\frac{L}{D} \tag{8}$$

η, $G^*$and f are the dimensionless pressure, dimensionless mass flux and two-phase friction factor, respectively. Leung and Grolmes (Ref. [8]) recommended a range between 0.003 and 0.005 for the friction factor.

G^* and η_1 are related via Bernoulli's equation for compressible fluid (Ref. [8]):

$$G^* = \frac{\sqrt{-2[\omega ln\eta_1 + (1-\eta_1)\omega - (1-\eta_1)]}}{\omega\left(\frac{1-\eta_1}{\eta_1}\right) + 1} \tag{9}$$

The exit condition can be either subcritical, in which case $\eta_2 = \eta_3$, or critical where the local choking statement would require (Leung and Grolmes, Ref. [8]):

$$G^* = \frac{\eta_2}{\sqrt{\omega}} \tag{10}$$

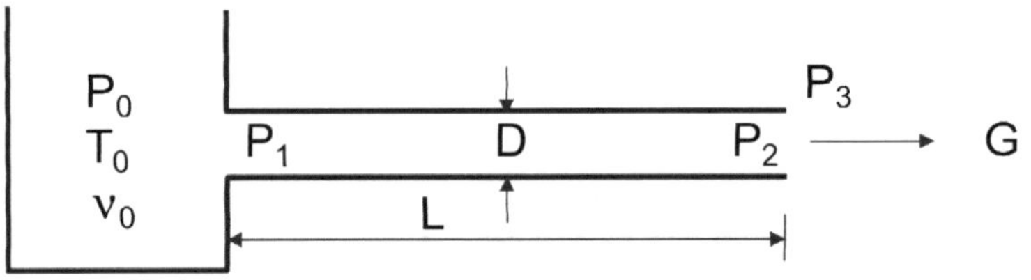

Figure 1. Two-phase horizontal discharge from a large reservoir.

With $\eta_1 = \eta_2 = \eta$, the above two equations satisfy the following transcendental equation (Leung and Grolmes, Ref. [8]):

$$\eta^2 + (\omega^2 - 2\omega)(1-\eta)^2 + 2\omega^2 ln\eta + 2\omega^2(1-\eta) = 0 \quad (11)$$

Similarly, the generalized thrust expression (T_R) is provided by Leung (Ref. [10]):

$$\frac{T_R}{P_0 A_e} = \frac{G^2}{P_0 \rho_0}\left(\frac{v_2}{v_0}\right) + \left(\frac{P_2}{P_0} - \frac{P_3}{P_0}\right) \quad (12)$$

The first term represents the fluid momentum expulsion rate and the second term represents the exit plane pressure (imbalance) force. For the case of unchoked exit, the exit pressure imbalance term vanishes, and the trust is only dependent on the exit momentum rate (Ref. [10]):

$$\frac{T_R}{P_0 A_e} = \frac{2[-\omega ln\eta_2 - (\omega - 1)(1-\eta_2)]}{\omega\left(\frac{1}{\eta_2} - 1\right) + 1} \quad (13)$$

This approximate analysis shows that for two-phase flow in ducts (e.g. piping systems) ω is the sole independent variable and hence is the scaling parameter.

3 Transient Fluid Analysis using RELAP5

3.1 Description of the benchmark scenarios

A chemical reactor with a volume of 2.7 m^3 is considered for the benchmark scenarios. The reactor vessel is rated for 3 bars-g. The reactor is equipped with an emergency relief nozzle and a rupture disk of 41.1 mm inner diameter (DN40, 3.6 mm thickness). The nozzle is located near the top of the vessel. The piping downstream of the rupture disk has an inner diameter of 80.9 mm (4 mm thickness) and is 20 m long. It is a straight and horizontal section of piping.

Three different relief scenarios are used for the benchmarks. A summary of the postulated scenarios is provided below:

- Scenario 1: Polyurethane (PUR) and toluene diisocyanate mixture at 60 °C.
- Scenario 2: N-methyl pyrrolidone, formic acid, trimellitic anhydride and methylene diphenyl diisocyanate (MDI) at 90 °C.
- Scenario 3: Acetone under-fire scenario with vessel initially filled to 20% of its volume.

The chemical mixtures for Scenarios 1 and 2 are gas (CO_2, carbon dioxide) generating reactions. Alternatively, the scenarios are also called gassy reactions. A foamy mixture is expected inside the reactor vessel for these scenarios. Therefore, the chemical effluent discharging through the ERS is a two-phase gas-liquid mixture.

Scenario 3 is the boiling of a solvent due to a fire load applied on the vessel. The level of the boiling mixture is low, and liquid acetone is not entrained into the ERS. Only acetone vapor discharges through the ERS for this scenario.

3.2 Description of the computer model

The calculation was performed using the RELAP5/MOD3.3 (Patch 05) computer code, which is a light water reactor transient analysis code developed for the United States Nuclear Regulatory Commission (US NRC). It is suitable for the simulation of a wide variety of hydraulic and thermal transients in both nuclear and nonnuclear systems involving mixtures of steam, water, non-condensable gases and solutes under single phase and two-phase conditions (Ref. [17]).

A RELAP5 model was developed for the simulation of the venting scenarios from the reactor vessel into the ambient. The model was used to determine the hydrodynamic loads resulting from the sudden opening of the rupture disk. The mass flow rate of the two-phase effluent and the pressure distribution along the ERS piping were also calculated. Figure 2 provides an overview of the nodalization diagram of the model.

Time dependent volumes (TDVs) were used for the pressure boundary conditions and the thermo-physical condition of the fluid at those boundaries. Pipe components (P) were used to model the piping. Single junctions (SJs) connected the components such as pipes with pipes or pipes with TDVs. The valve junctions (V) was similar to the SJs but with the additional ability of the junction to open and close.

The length-to-diameter ratios (L/D) were set close to 1.0 for each pipe node and the time step was selected to be below the acoustic Courant limit. This is recommended for simulations involving propagation of acoustic waves and more generally for models where rapid fluid momentum changes are anticipated. The rupture disc was assumed to open linearly over a period of 10 msec, which is consistent with typical data (Ref. [18]). A steady-state calculation was performed for the first 5 seconds of each simulation to initialize the model. The transient portion of the simulation was initiated at 5 seconds, when each rupture disc started to open and was fully open at 5.01 sec. No heat transfer through the pipe walls was considered for the hydrodynamic fluid analysis. This is often a reasonable assumption for industrial processes due to the short duration of the relief event.

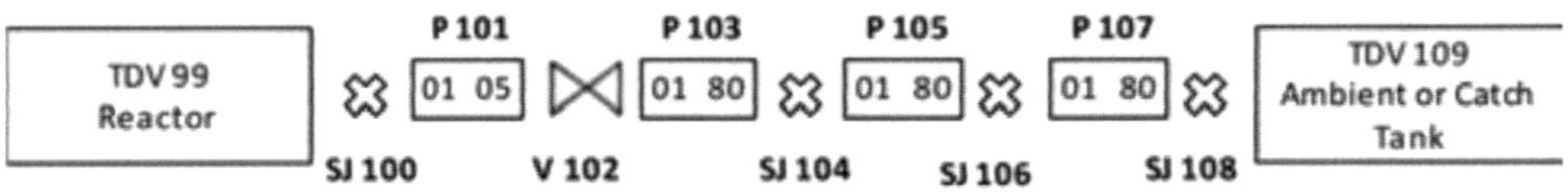

Figure 2. Nodalization diagram for the RELAP5 model.

The hydrodynamic loads (also commonly referred to as fluid force) on the modeled piping were computed through command files by using the RELAP5 output data. The generalized force equation in one-dimensional form can be resolved by a pipe segment bounded by two elbows as shown in Equation (14) (Ref. [19]). Note that the sign convention for the force is considered positive in the nominal direction of the flow as defined within the physical system.

$$F_W = -\frac{d}{dt}\int \rho A V dx \tag{14}$$

F_W is the hydrodynamic load, ρ is the fluid density, A if the inner flow area of the pipe, V is the fluid velocity, and x is the distance along the pipe. For pipe segments with an open end an additional force component needs to be considered, which is regarded as the thrust load. Examples of open ended pipe segments are pipes that connect to pressure vessels, large manifolds, or a catch tank. The total thrust load is calculated as shown in Equation 15, which is made up of the momentum flux (due to fluid momentum expulsion) and pressure terms:

$$F_T = -(P_E - P_A)A - \dot{m}_E V_E, \tag{15}$$

where P_E, P_A, $\dot{m}_E$ and V_E are the pressure at the pipe exit, pressure of the reservoir the open piping is connected to, and the mass flow rate and fluid velocity through the pipe exit, respectively.

3.3 Boundary conditions for the RELAP5 model

A summary of the upstream boundary (reactor vessel) conditions for the RELAP5 models of the three scenarios are provided in Table 1. 10% over-pressure (10% greater than the set pressure) was used for Scenarios 1 and 2, which is typical for relief system sizing of reactive chemicals. 21% over-pressure was used for Scenario 3 per API 520's (Ref. [20]) recommendation for under-fire scenarios. Two-phase gas liquid discharge was assumed for Scenario 1 and 2, where foamy mixtures inside the reactor vessel were established. Single-phase gas discharge was assumed for Scenario 3. The gas-liquid mixture of chemical effluent was simulated by using the air-water properties. The quality values of the air-water mixtures were adjusted so that the ω value of the gas-liquid mixture of the chemical and that of the air-water mixture would match. The quality and the ω value of chemical mixtures were determined based on the initial fill ratio and the homogenous two-phase mixture assumptions. A single-phase acetone vapor release scenario (Scenario 3) was simulated by using air. The determination of the choking condition at the rupture disk was made based on the analytical model, which was described in Section 2. The downstream boundary conditions were set to be at typical atmospheric conditions (air at 20 °C and 1 bar-a).

Table 1. Upstream Boundary Conditions for the RELAP5 Model

Upstream Boundary Conditions	**Scenario 1**	**Scenario 2**	**Scenario 3**
Pressure	4.4 bar-a	4.4 bar-a	4.8 bar-a
Flow regime	Two-phase gas-liquid	Two-phase gas-liquid	Single phase vapor
Simulant	Air-water	Air-water	Air
ω	0.085	0.336	Not calculated
Choking at the burst disk	No	Yes	Yes

4 Results and discussion

The time history of the mass flow rate at the rupture disk calculated using RELAP5 for Scenario 1 is provided in Figure 3. The flow quickly accelerates following the opening of the rupture disk. The effluent experiences the frictional and form losses only for the sections of the piping, which it has already occupied at a given time. Therefore, the transient mass flow rate is slightly greater than the steady-state mass flow rate while the effluent is filling up the ERS piping. Once the entire ERS piping is filled, the mass flow rate achieves a steady-state value.

Figure 4 shows the pressure-time histories downstream of the rupture disk and at the end of the ERS piping for Scenario 1. The pressure downstream of the rupture disk gradually increases to its steady-state value as the effluent fills the piping. The pressure at the exit of the piping remains near atmospheric pressure (downstream boundary condition).

Figure 5 shows the dynamic load calculated on the ERS piping. The total load and its constituents are provided. There is a spike for the wave load as the rupture disk opens. This spike is caused by the depressurization wave traveling from the rupture disk to the reactor vessel. Typically, this section of the piping is filled with gas or vapor for which the speed of sound is well known. Therefore, the initial spike is captured accurately and the speed of sound of the two-phase mixture (which is typically unknown) does not have a significant role on the results. The wave load continues to be the dominant mechanism of the load as the effluent continues to fill up the ERS piping. Once the piping is filled up, the flow momentum term replaces the wave load. The flow momentum is due to the discharge of the effluent from the ERS piping into the ambient or a catch tank and it is the first term on the right-hand side of Equation 15. The pressure force is zero since the exit of the ERS piping is at the same pressure as the ambient; there is not choking at the exit of the piping.

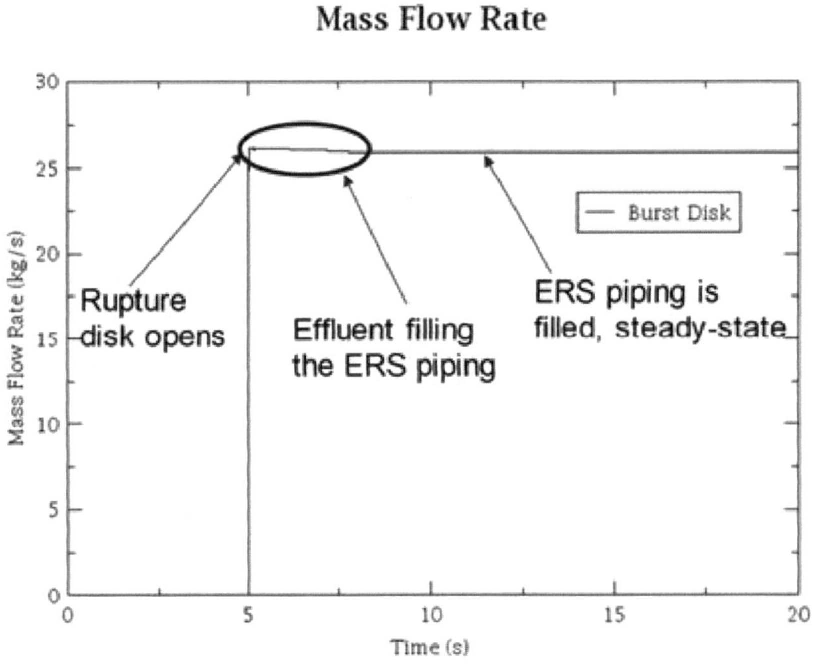

Figure 3. Mass flow rate for Scenario 1.

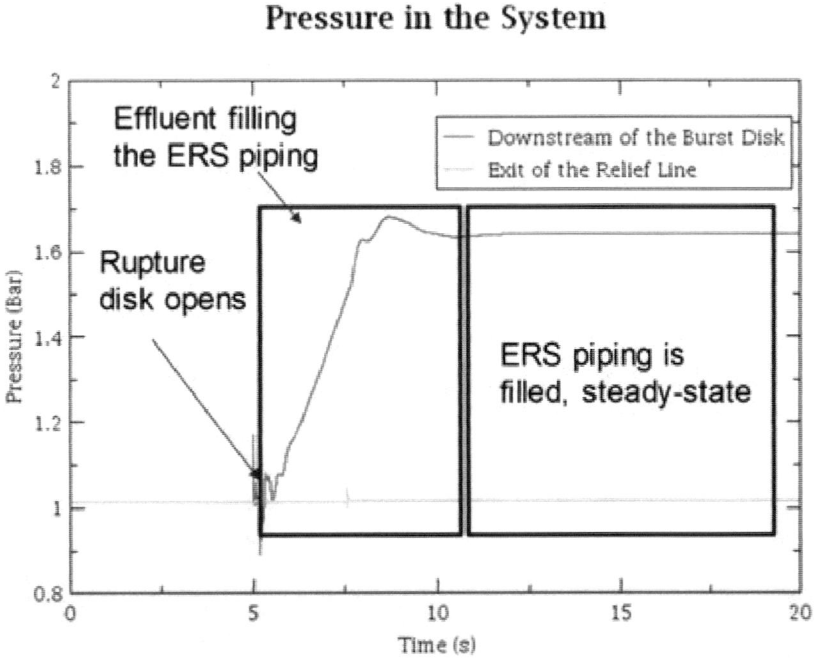

Figure 4. Pressure in the ERS piping for Scenario 1.

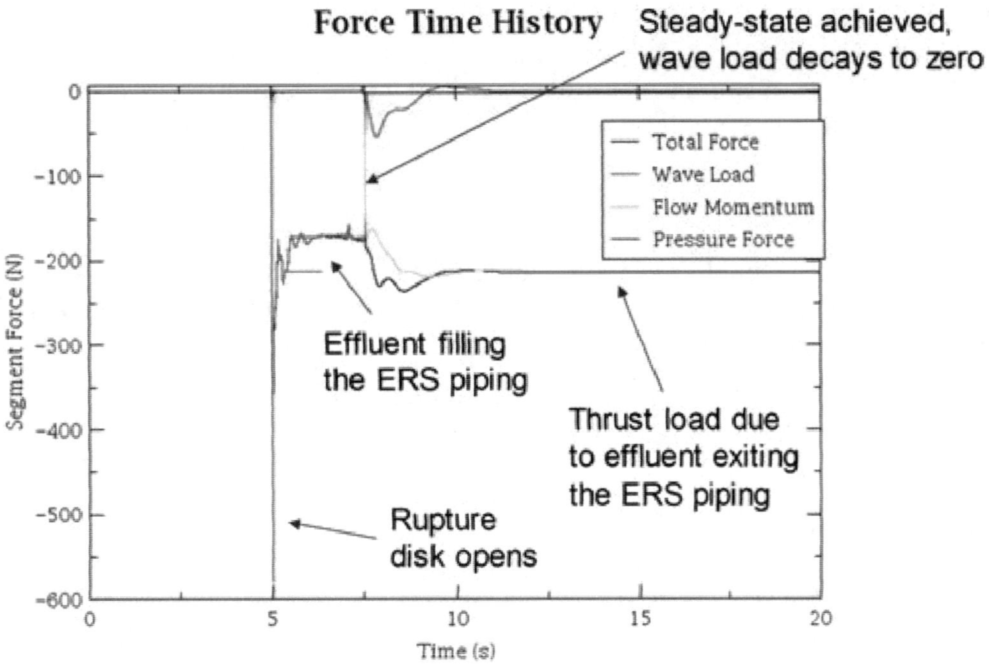

Figure 5. Dynamic loads for Scenario 1.

It is interesting to note that the wave load (while the ERS piping is filling up) and the load due to flow momentum (at steady-state) are almost similar. This is explained using the mathematical relations for the wave and thrust loads. The wave load for an effluent while filling up an empty piping (using Equation 14) is mathematically same as the flow momentum term in Equation 15.

The steady-state pressure, mass flow and thrust load values were compared between RELAP5 calculations and the analytical model, which was provided in Section 2. Actual chemical properties were used for the theoretical model. The comparisons between the two methods are provided in Table 2. The results demonstrate that the two-phase chemical effluent can be simulated with air/steam-water properties as long as the ω values are matched for these mixtures.

Table 2. Comparison between calculations using air-water (RELAP5) and actual two-phase chemical properties

Parameter	Scenario 1	Scenario 2	Scenario 3
Pressure downstream of the burst disk (bar-a)	1.6\1.7	1.6\1.7	1.9\1.6
Mass flow rate (kg/s)	25.9\26.1	16.3\16.4	1.2\1.2
Steady-state thrust load (N)	212.5\191.4	176.3\199.9	219.3\235.0

Black: RELAP5 results Red: Theoretical model

In the chemical manufacturing industries, the effluent can be composed of proprietary chemical mixtures containing various materials. Therefore, it can be challenging to know the exact material properties at any given time during an emergency relief event. The method presented in this study requires only the stagnation properties of the chemicals at the venting conditions for predicting key parameters, such as the pressure distribution and wave loads during the transient. These properties are typically established by adiabatic calorimetric testing or similar characterization of the chemicals. The method presented in this study may also be used with other commercially available software, which can predict transient fluid dynamics and wave loads.

5 Conclusions

In this study, a previously published methodology is used for calculating the dynamics of the effluent and the resultant hydrodynamic loads for a typical ERS by using the RELAP5 software and assuming air-water properties for the effluent. Key output parameters, such as pressure drop and hydrodynamic loads, are then compared with established analytical methods using the actual effluent properties. The following conclusions were observed:

- The fluid transient analysis is a key aspect of the design of the emergency relief systems. It provides the wave loads required for the dynamic pipe stress analysis. It also provides key parameters such as the pressure distribution, mass and volumetric flow rates through the system, and wave loads. These key parameters are often used to determine back pressure effects on other vessels, for chemical dispersion analysis and dynamic pipe stress analysis.

- The analytical approach shows that there is a similarity between the air/steam-water and the gas/vapor-liquid chemical mixtures if the "ω" values of the two mixtures are matched. This allows for an emergency relief event to be simulated with air/steam-water properties.

- Similar pressure distribution in the piping system and loads can be achieved through the similitude of "ω". Therefore, "ω" of the air/steam-water mixture should be matched with the "ω" of the chemical effluent at the stagnation conditions. The quality of the air/steam-water mixture can be adjusted until a similar "ω" is achieved.

- The sonic velocity of the effluent is typically not known. Also, it is practically difficult to match the sonic velocity of the air/steam-water mixture with the sonic velocity of the chemical effluent, which is a major requirement for tracking pressure waves and wave loads. This study demonstrates that the emergency relief scenarios are pipe filling events, which only require the momentum flux to be calculated for a reasonable prediction of the initial wave load. Therefore, the sonic velocity of the effluent is not required to capture the initial wave load.

- The effluent can be composed of proprietary chemical mixtures containing various materials. Therefore, it can be challenging to know the exact material properties at any given time during the transient. The method presented in this study may be used with most commercially available software and air/steam-water properties for a reasonable representation of the important characteristics of the actual effluent discharge in mechanical design space.

- The study needs to be extended further to identify applicability and limiting conditions for this method.

6 References

[1] Conzen, J., Wood. R.G., (2015), "Modeling and Computation of Reaction Forces on Relief Piping During Depressurization", AIChE 11th Global Congress of Process Safety.

[2] Ozar, B., (2018), "Analysis of Chemical Effluent Discharge through the Emergency Relief System using Stream-Water Properties," DIERS Fall Meeting, 15-17 October, Houston, USA.

[3] ASME. *ASME Code for Pressure Piping, B31.* The American Society of Mechanical Engineers, New York, NY, 2016.

[4] "Directive 2014/68/EU of the European Parliament and of the Council," Official Journal of the European Communities, May 2014.

[5] Epstein, M., Henry, R.E., Midvidy, W., Pauls, R., (1983), "One-dimensional Modeling of Two-phase Jet Expansion and Impingement," Thermal-hydraulics of Nuclear Reactors, 2nd International Topical Meeting Nuclear Reactor Thermal-hydraulics, Santa Barbara, CA.

[6] Grolmes, M.A., Leung, J.C., (1984), "Scaling Considerations for Two-phase Critical Flow," Multi-phase Flow Heat Transfer III, Part A: Fundamentals, T.N. Veziroglu and A.E. Bergles, eds., 549, Elsevier, Amsterdam.

[7] Leung, J.C., (1986), "A Generalized Correlation for One-Component Homogenous Equilibrium Flashing Choked Flow," American Institute of Chemical Engineers, Vol. 23, pp. 1743-1746, October.

[8] Leung, J.C., Grolmes, M.A., (1987), "The Discharge of Two-phase Flashing Flow in a Horizontal Duct," American Institute of Chemical Engineers Journal, Vol. (33), p. 524.

[9] Leung J.C., Epstein, M., (1990) "The Discharge of Two-phase Flashing Flow from an Inclined Duct," Transactions of the ASME, Vol. 112, pp. 524-527, May.

[10] Leung, J.C., (1992), "Reaction Forces During Two-phase Discharges," American Society of Mechanical Engineers Journal of Fluids Engineering, Vol. (114), p. 689-692.

[11] Information Systems Laboratories, (2016), "RELAP5/MOD3.3 (Patch03) code manual volume I to VIII, NUREG/CR-5535/Rev P5-Vol I to VIII", prepared for USNRC.

[12] International Agreement Report, (2007), "Simulation of the Propagation of Pressure Waves in Piping Systems with RELAP5/MOD 3.2.2: Comparison of Computed and Measured Results", TÜV SÜD - NUREG/IA-0206, June.

[13] Sokolowski, L., Koszela, Z., (2012), "RELAP5 Capability to Predict Pressure Wave Propagation Phenomena in Single- and Two-Phase Flow Conditions", Journal of Power Technologies 92 (3), page 150 – 165.

[14] Miller, J.S., Ramsden, K.B., (2005), "Validation of a Thermal Hydraulic Computer Code to Perform Two-Phase Multi-Component Force Calculations for Structural Evaluations", Proceedings of 13th International Conference on Nuclear Engineering (ICONE13), 16-20 May 2005, Beijing, China.

[15] Ozar, B., Henry, C.E., Harvill, R., Norton, D.A., (2012), "Analysis for Low Pressure Cooling Injection System Suction Hydrodynamics for a Boiling Water Reactor," Proceedings of 20th International Conference on Nuclear Engineering (ICONE20), July 30-August 3, 2012, Anaheim, CA.

[16] Henry, C.E., Lim, J., Ozar, B., (2012), "Investigation of Hydrodynamic Loads Associated with Pyrotechnic Valve Actuation," Proceedings of 20th International Conference on Nuclear Engineering (ICONE20), July 30-August 3, 2012, Anaheim, CA.

[17] Riemke, R.A., Davis, C.B., Schultz, R.R., (2009), "Recent Hydrodynamics Improvements to the RELAP5-3D Code", Proceedings of the 17th International Conference on Nuclear Engineering, ICONE17-75380, July 12-16.

[18] Health and Safety Executive, "Offshore Technology Report 2002/023: Testing and Analysis of Relief Device Opening Times," Pipeline Simulation and Integrity Ltd., Yorkshire/UK.

[19] Moody, F.J., (1990), "Introduction to Unsteady Thermofluid Mechanics", John Wiley and Sons, New York, Section 7.4, p. 365-366.

[20] API Standard 520, (2008), "Sizing, Selection, and Installation of Pressure-relieving Devices in Refineries," 8th edition.

547437 Evaluation of the Thermal Stability of Styrene in Contact with Impurities

Safe handling and storage of reactive monomers have long been an issue for polymer manufactures. The unsaturated double bond may self-react to initiate unintended polymerization reactions under many scenarios, for example, completely depletion of inhibitors, exposure to fire or external heat, and in contact with impurities. Usually, a trace amount of incompatible species is able to trigger unintended reactions and lead to catastrophic consequences. Thus, understanding how the impurities promote self-reaction is expected to provide valuable guidance to safe monomer-handling practices. Our goal in this study is to quantify the contamination effects of some most common contamination chemicals on styrene self-polymerization. Thermal analysis tools including DSC and ARSST are applied to examine the exothermic characteristics of styrene mixed with contaminating substances at different concentration levels and mixing ratios. From the preliminary results, we observed the onset of self-polymerization shifted to a lower temperature when in contacted with water and also alkaline. A trace amount of concentrated strong acid initiated unwanted exothermic activity even at room temperature. Self-heating rate was also elevated when diluted acid was present. Additional research is needed to understand what is the impact in of other types and concentrations of contaminants such as rust, metal ions. Our study demonstrates significant implications in the prevention of runaway incidents during transportation and storage of styrene.

549038 A Mixed-Integer Programming Framework for Placement of Fire Detectors in Chemical Processing Facilities

In chemical processing plants, gas and fire detectors provide an important layer of protection for personnel. However, given the large variety of chemical and physical sensors and number of factors influencing their capabilities to successfully detect hazards, the decision of where to place sensors is challenging. Gas detectors (e.g. ultrasonic, catalytic bead) must be in contact with gas leaks to be effective, while fire detectors (e.g. point infrared, open path infrared, optical) require a visual path to the fire, free of obstructions, for successful detection. Numerical optimization techniques provide an opportunity to overcome these challenges and determine effective sensor placement.

The optimal gas detector placement problem was formulated by [1] as a stochastic program based on the p-median facility location problem and relied on scenarios generated from CFD simulations within a petrochemical facility. Their results showed that coupling coverage with expected detection time using a scenario-based method outperformed the existing coverage-only approaches and has inspired formulations based on MINLP approaches for non-uniform failure probabilities [2].

But while gas detectors must be placed in the path of gas leaks for effective detection, fire detectors must be placed within unobstructed visual distance of flames. The optimal placement of visual fire detectors in processing plants can be particularly difficult due to the large number of physical obstructions (e.g. piping, valves, tanks).

This work focuses on an optimization formulation for placement of fire (visual) detectors. The scenario-based, p-median approach demonstrated by [1] for gas detectors is translated to a maximum coverage model for fire detectors in a 3D space with consideration to failure probabilities. We show that this approach allows for the placement of fire detectors for a fixed sensor budget while under uncertainty. We demonstrate the effectiveness of this formulation on several case studies. We further discuss how the framework is readily scalable for larger problem sizes and amenable to various objectives.

We would like to thank the Purdue Process Safety and Assurance Center for their support in this work.

References:

[1] Legg, S. W., Benavides-Serrano, A. J., Siirola, J. D., Watson, J. P., Davis, S. G., Bratteteig, A., and Laird, C. D. (2012). A stochastic programming approach for gas detector placement using CFD-based dispersion simulations. Computers and Chemical Engineering, 47:194–201.

[2] Liu, J., and Laird, C.D., A Global Stochastic Programming Approach for the Optimal Placement of Gas Detectors with Nonuniform Unavailabilities, to appear in Journal of Loss Prevention in the Process Industries, 2017.

546855 Advance Alarm Management Applied to DCS Improvement in a Petrochemical Plant: A Case of Study

Alarms use in process industry is fundamental for avoiding process accidents in which people, the environment or process equipment could be damage. Currently, Carbon Black production plants are equipped with distributed control systems (DCS) for setting alarms values and to provide indication of abnormal events to the control rooms operators. However, the basic control system could not meet the needs against the different production conditions and the presence of improper alarms could send a bad signal to control rooms and it could be the start of a process safety event.

In the present study, an alarm system assessment was performed in a petrochemical plant, located at Mamonal industrial park, in Cartagena – Colombia; based on data analysis, it was observed that when process conditions changes occured, the DCS was not be able to adjust alarms settings automatically, according to the load fluctuation and/or process conditions. In order to solve this problem, in first place the DCS performance was studied and it was reported an alarms list, which provided information regarding to the alarm flooding phenomena, alarm chattering and others alarm system issues.

On the other hand, the information earned and alarm rationalization methodology were used for establishing an action plan in order to improve DCS performance. Results showed the utility of these tools and it was identified the need to perform betters in the alarm system logic to avoid flooding alarm phenomena and to make operations safer.

548361 Overcoming Motivational Biases in Hazops

Hazard and Operability Studies (HAZOPs) are one method of assessing the risk in a facility. Based on the risk ranking of a scenario, recommendations may be required to reduce the risk. Often, these recommendations require additional capital expenditure and/or operational resources. In aging plants where resources are not readily available or in greenfield projects where there is a tight timeline to start up, there is sometimes a bias to risk rank scenarios such that there is no need to add additional safeguards. Motivational biases can outweigh common sense such that scenarios are not ranked according to their real risk. We have developed a method to overcome some of these motivational biases with the use of a polling system to engage all participants in the HAZOP. Participants are asked to vote anonymously on the severity and likelihood for a given scenario, then the results are discussed. During this discussion time, arguments for all the different risk rankings are discussed, including those voted on by the minority in the room. This facilitates an open discourse about the real risks for a given scenario rather than shifting the attention to the most vocal person in the room, who may be strongly influenced by a desire to reduce spending or minimize the effort required for maintenance. Following this discussion a second poll is performed to reach a consensus. We have found this to be an effective method for reducing motivational biases and ensuring that the right decisions are made based on the most logical argument instead of based on budget or time constraints. Tracking polling results increases the accountability of participants and makes them less likely to skew the risk ranking. The requirement for consensus ensures that all opinions are heard and considered, even if they are not popular. Together these make the polling system an effective method for overcoming motivational biases in HAZOPs and ensuring that safe decisions are made.

546515 Integrating Culture into Process Safety Assurance

Purpose: PETRONAS Chemicals Group Berhad (PCG) has embarked on a group-wide culture transformation, introducing the concept of the Results Pyramid® and a Culture of Accountability® i.e. a workplace culture where people take accountability to think and act in the manner necessary to achieve results.

Methodology: This paper will discuss two such activities anchoring on the Results Pyramid® that have been deployed across several plants and units within PCG.

The first activity is an enhancement of existing assurance programmes, where culture subject matter experts are assigned to observe plant leadership teams and front-liners during interviews and site verifications conducted by the technical assessors. The observations are synthesised into belief statements, which are presented together with the technical findings.

The second activity is Power Up, a session designed to incorporate powerful and specific feedback exchange between various teams, culminating in a specific list of actions to improve delivery of each team's activities in order to achieve the shared key results. This workshop can be done either as a follow-up to an assurance activity, in preparation for a project/turnaround activity or as a part of a periodic performance review.

Results: Examples of observations, feedback exchanges and specific actions from these activities will be shared, together with the impact on process safety leading and lagging indicators. Recommended practices and lessons learnt for continuous improvement will also be included.

Conclusion: Applying the Results Pyramid® concept to the design and implementation of Process Safety-related programmes has helped to identify the underlying leadership, mindset and behaviour issues impacting their success. By doing so, more specific interventions can be put in place to ensure full compliance to safety requirements and maximise effectiveness of these programmes.

548442 Emergency Plans Integration By Diferent Companies

In Brazil, different companies with pollution potential must have an emergency plan. However, each company analyzes its risks and contingency measures separately, based on the Conama 398 law. With the establishment of the Area Plan, companies from the same territory with pollution potential caused by oil in waters began to analyze how to combat a major accident quickly and effective together. This work shows how the integration of the individual contingency plans of each company and the cooperative result of the area plan in territory of significant scope and considerable environmental sensitivity.

No Brasil, diferentes empresas com potencial de poluição devem ter um plano de emergência. No entanto cada empresa analisa seus riscos e medidas de contingência separadamente, com base na lei Conama 398. Com a instituição do Plano de Área, empresas do mesmo território com potencial de poluição causadas por óleo em águas passaram a analisar como atender conjuntamente de forma rápida e eficaz a um grande acidente em seu território de abrangência. Este trabalho mostra como foi estruturada e organizada a integração dos planos de emergência individuais de cada empresa e o resultado cooperativo do plano de área em território de abrangência significativa e considerável sensibilidade ambiental.

548111 Moving Back to Flammable Refrigerants - Facility Siting and Personnel Safety Considerations

"Everything old is new again – Paul Allen"

Not just a catchy tune from the 1970's, this jingle summarizes a current technological shift in the world of refrigeration. Historically, refrigeration cycles used toxic and/or flammable materials as the working fluid. As the technology matured and made its way into transportation and consumer applications, the working fluid also evolved. Halogenated hydrocarbons, such as chlorofluorocarbons (CFCs), were developed as replacements to conventional refrigerants such as ammonia and propane. CFCs, which may have toxic effects, are desired for their tunable working range and for being non-flammable. However, the environmental drawbacks to the use of CFCs has been well researched and documented. As we embark into the future, where climate change and environmental quality are at the forefront of concern to consumers and industry alike, we are seeing a move back to flammable refrigerants. Everything old is, indeed, new again.

New construction and expansion projects can certainly consider and plan for the use of flammable refrigerants in commercial and industry facility design and siting, but what about retrofit applications? What are the risks and hazards associated with converting a CFC refrigeration system, designed and sited for the use of non-flammable refrigerants, to a system now utilizing flammable refrigerants? Beyond the need for additional fire protection systems, what risks are posed to the facility and personnel? In this paper, we seek to answer these questions by comparing the hazards and risks of mid-scale refrigeration systems using either non-flammable or flammable refrigerants. We explore how facility siting constraints may change with respect to the use of flammable refrigerants, and perform a quantitative risk assessment (QRA) for an example retrofit project in order to assess the potential for increased risk that should be considered in such projects.

548210 Making Sense of Risk Results

Risk based studies are very useful, but at the same time, the results are extremely complicated. While we have a basic set of results that can be used to illustrate the risks to people, these are relatively simple for risk practitioners and sometimes too complicated for site managers.

FN curves show the risk to all people from all sources, and individual risk contours illustrate the risk from all scenarios to people across specific locations, but how do we translate these results into actions. Generally, the action is to relocate people to lower risk areas, or upgrade buildings to mitigate the impacts identified. However, this is not always feasible. What if the site footprint is limited; there are public receptors who cannot be relocated; or building upgrades are cost-prohibitive or infeasible?

We must better quantify and evaluate the **sources** of risk rather than focusing on mitigating the impacts from the events. We need to evaluate the type of risks, the potential initiating events, and then we can assess the potential methods to effectively mitigate the risks. Existing risk result presentations are often very poor at illustrating the sources of risk. To effectively identify the sources of risk requires that we carefully dissect the risk results, while maintaining their validity and intent.

This paper will focus on how we find these risk drivers and make enough sense of them to create actionable recommendations that protect people.

551205 Ammonia Dispersion Modeling for Ammonia Storage with Multi-Scenario Failure System

PT Pupuk Kalimantan Timur is one of the largest fertilizer companies in Southeast Asia where located in Bontang City, East Kalimantan Province, Indonesia. The total production capacity of PT Pupuk Kalimantan Timur are 2.7 million Tons / year Ammonia, 3.43 million Tons / year Urea and 350 thousand tons / year NPK.

To accommodate ammonia production, PT Pupuk Kalimantan Timur currently operates 3 ammonia tanks with a capacity of 2 x 25,000 tons and 1 x 50,000 tons. In order to manage the hazard risk contained in the ammonia tank, it is necessary to do a modeling to map the potential and ammonia leakage potential areas and the impact of their exposure to the workforce and the community around the factory area.

The results of this modeling are expected to be used by Pupuk Kalimantan Timur's management in carrying out corrective actions for occupational health and safety performance, especially in controlling the potential hazard of ammonia, especially in ammonia tanks.

In this paper modeling is done using the AERMOD® Modeling System software. AERMOD Modeling System was chosen because it can be used on:

1. U.S. EPA regulatory Gaussian plume air dispersion model.
2. For near-field impacts (50 km max).
3. Prefered model for complex terrain.
4. Improved treatment of urban effects.

For modeling, we choose the AS-T-101 ammonia tank which has the largest capacity, which is 50,000 tons. Ammonia distribution models will be simulated based on weather conditions and ammonia leak source scenarios that can occur in the AS-T-101 ammonia storage tank system. Ammonia leakage sources will be displayed in 5 scenarios, i.e:

1. 1 Vent opens
2. 1 PSV poping
3. 2 PSV poping
4. 3 PSV poping
5. 3 PSV and 1 vent poping

The AERMOD Modeling System program is used to make ammonia distribution models for each event scenario assuming an exposure radius of up to 10 km. Based on the results of the calculation and modeling of the potential distribution of ammonia from AS-T-101 using the AERMOD Modeling System, the highest exposure of ammonia is in the 5th scenario, which is up to 36611 μg / M3.

546560 De-Mystifying the Concepts of Hazardous Area Classification

Purpose

The purpose of this topic is to address and clarify major items of confusion associated with hazardous area classification assessments and compliance auditing in classified locations. The ultimate objective is developing documentation that provides a framework for the control of electrical energy ignition sources in environments made hazardous by the potential presence of flammable vapors.

Background

In February of 1992 Occupational Safety and Health Administration (OSHA) 29 CFR 1910.119 Process Safety Management of Highly Hazardous Chemicals became law in the United States and confusion reigned supreme. Over the next several years industry professionals began the process of sorting through the responsibilities, regulations and impending deadlines that had been placed squarely in their laps. Acronyms such as PSM (Process Safety Management), PSI (Process Safety Information), RAGAGEP (Recognized and Generally Accepted Good Engineering Practices), MOC (Management of Change), and PHA (Process Hazards Analysis) became commonplace in the vocabulary of process safety professionals nationwide.

Employers now were required to meet stringent deadlines regarding the creation and management of a plethora of process information associated with equipment in the process covered by this law, i.e. any covered process. This information would be used to support the exhaustive efforts needed to create a baseline PHA for each covered process by May of 1997. In addition OSHA required all PHA's be placed on a five (5) year life cycle requiring their review and revalidation. Along with the revalidation of PHA's came the task of also revalidating all process safety information on the same five (5) year cycle.

As a result, every willing and able bodied employee was thrust into the quagmire of supporting, participating in, facilitating, or conducting PHA's. The required information used to support PHA's included the materials of construction, Piping and Instrument Diagrams (P&ID's), Relief System design and design basis, ventilation system design, material and energy balances, safety shutdown systems, and last but not least **electrical classification**. Yes, "electrical classification" was and still is legally considered an integral part of required process safety information.

This abstract addresses the following and provides an interactive discussion regarding the three (3) major myths surrounding electrical classification:

- assigning a classification to a location
- proper use of the classification information
- keeping the documentation evergreen

  The following topics will be addressed and clarified:

- What Is Hazardous / Electrical Classification?
- The Process of a Hazardous Area Classification Assessment
- The Concept of Electrical vs Hazardous Area Classification
- The Compliance Assessment (where does it fit in the process of controlling electrical energy ignition sources)
- Common Compliance Assessment Issues

546108 Maintaining Safety Excellence By the Implementation of Process Safety Inspection in Badak LNG

Badak LNG as one of the most experienced LNG Producers in the world, has undergone safety culture change throughout its operation. The goal of this change is to achieve excellent safety performances, indicated by the lower rate of Lost Time Injury Frequency (LTIF), lower number of Loss of Primary Containment (LOPC) Tier-1, High Potential Incident and multiple Train trip. This change was triggered by the occurrence of a fatal incident in 2005 as well as lost time injury incident in 2006. Since then, Badak LNG has integrated Process Safety Management (PSM) into its Company Management System, strengthened the roles and responsibilities of Managements, and pursued for interdependent safety behaviour of each personnel. One of PSM implementations is through process safety inspection.

Process safety inspection is an activity to assess the current equipment integrity based on visual observations compared with equipment's standard condition. The standard condition checklist for process safety inspection in Badak LNG is based on Physical Barrier Assessment tools, owned by Det Norske Veritas (DNV). The assessed items are divided into 5 categories: Inherent Safety, Hazard Prevention Measure, Hazard Control, Mitigations, and Support System. The overall score is determined by total number of standard items in all categories compared to the total number of verified items in all categories. The minimum acceptable score is at 80%.

During the year of 2012-2017, the average score of process safety inspections in Badak LNG is 90%. As a result of these inspections, Badak LNG had also identified the need of improvement in some items such as blinds, pipework and flanges, passive fire protection, fire & gas detection system and lock-open/lock-close block valves. In overall, by the implementation of PSM, Badak LNG has aligned on the right path to maintain its safety excellence achievement in LNG business, such as zero LTIF since 2007, zero multiple Train trip since 2012, zero LOPC Tier-1 since 2016, and zero High Potential Incident in 2017.

548411 Using HAZOP/LOPA to Create an Effective Mechanical Integrity Program

Many people view the conduct of a HAZOP/LOPA to address regulatory requirements as a chore, and stop there. However, the implementation of a quality HAZOP/LOPA has the potential to provide a framework for addressing numerous safety and operational optimization issues at plants, including the formulation/refinement of the Mechanical Integrity Program. The purpose of this paper is to focus on the mechanical integrity program, illustrate how a quality HAZOP/LOPA can support the effective implementation of some of the new requirements for California Refineries, and optimize key elements of an effective Mechanical Integrity Program, e.g.:

- Inspection/testing methods

- Testing intervals

- Maintenance outage periods

- Repair prioritization and allowable outage

- Identification of low priority equipment

544678 *"Identifying Critical Process Parameter to Ensure Asset Integrity" or "Pulse of Mechanical Integrity Management System"*

The management of equipment begins with an awareness that ageing is not about how old the equipment is, but is about what is known about its condition and the factors that influence the onset, evolution and mitigation of its degradation. The presentation provides an overview of Integrity Operating Window (IOW) parameters of any process that are key to understand the asset degradation and prevent Process Safety Incidents.

IOW are the situations in the process industries where a process parameters (temp., pressure, dosing rate, conc. etc...) when gets out of operating windows could result in significant loss of primary containment leading to potential severe consequences.

Arrazi - SABIC developed an IOW system that covers IOW identification, associated risk, mitigation (interim – long term), stewardship/management of IOWs and elimination.

The presentation shall cover Insight concept of the Topic, major Process Safety accidents related to breeching IOWs, actual example, IOW Management Process, its influence on PM plans etc.

three key bullet points that we can include in the agenda:

- Knowing operating limits with associated mechanical integrity or process safety risk to ensuring proper actions to be in operating envelope
- Developing the engineering leadership to systematically guide the operators, superintendents and engineers to be in safe zones
- Dynamic inspection frequency based on plant conditions and sharing some industrial major incidents for this critical item

19 AIChE Spring Meeting
15TH GLOBAL CONGRESS ON PROCESS SAFETY
NEW ORLEANS • 2019

The Practice of Process Safety Analysis and Management System Based on Safety Reliability
(THE NEW TITLE: The PHAMS system in SINOPEC)

Jiang Xue
SINOPEC Research Institute of Safety Engineering,
QingDao,China
Jiangxue.qday@sinopec.com

The PHAMS system in SINOPEC

Jiang Xue
SINOPEC Research Institute of Safety Engineering,
QingDao,China
Jiangxue.qday@sinopec.com

Keywords: Safety Barriers Reliability, PHAMS, HALOPA, SIL, Safety Management Online.

Abstract

The big data technique has been in trend for the process safety assessment. In China, it's the very need to transfer the qualitative safety management to quantitative safety management, which should rely on the professional risk analysis technique, as well as the standard safety reliability database. Based on the above, it's very important to integrate several professional risk analysis technique, as HAZOP, LOPA, SIL etc. to realize the integral risk identification and evaluation, grading and management plan making, supporting by the proprietary SINOPEC risk metrics and grading standards, as well as safety equipment reliability database,which is developed as the Process Safety Analysis and Management System(PHAMS) for SINOPEC, to support all SINOPEC branches to make dynamic risk analysis and management online

1 Introduction

With the smart factory's development, the static, periodic risk assessment and management can not satisfy the identification and real-time safety control of plant's risk events. The PHAMS which establishes the plant-wide dynamic risk analysis, can effectively solve the risk rating and monitoring of the unit and realize the plant-wide or corporate-wide management optimization.

The key data for HALOPA, SIL and Management module in PHAMS is the dependent protection barriers, which we rely on RBI correction, Personnel exposure probability and ignition probability to optimize the six mainly safety barriers' failure probability, simultaneously, do use of the Bayesian Correction Modelling to verify the instant failure probability on site, as well as the SIL identification, which can validly save the grading cost on time and personnel, and all of these are towards to the deeper research on dynamic risk analysis for PHAMS.

2 HALOPA Analysis System

2.1 SINOPEC RISK MARTIX

According to international common matrix criteria, as well as statistics from the SINOPEC accident data and the maintenance data, the frequency of risk occurrence and severity classes are coupled to the 8×7 matrix by a linear exponential model, and be assigned risk index to each risk level.

SINOPEC RISK MARTIX		Occurrence Frequency							
		1	2	3	4	5	6	7	8
		$\le 10^{-6}$/y	10^{-6}-10^{-5}/y	10^{-5}-10^{-4}/y	10^{-4}-10^{-3}/y	10^{-3}-10^{-2}/y	10^{-2}-10^{-1}/y	10^{-1}-1/y	>1/y
Severity Classes	A	1	1	2	3	5	7	10	15
	B	2	2	3	5	7	10	15	23
	C	2	3	5	7	11	16	23	35
	D	5	8	12	17	25	37	55	81
	E	7	10	15	22	32	46	68	100
	F	10	15	20	30	43	64	94	138
	G	15	20	29	43	63	93	136	200

Figure 1 SINOPEC Risk Matrixp[1]

Based on this matrix, SINOPEC has established four-level risk list for establishing targeted management of branch network. For example, the executive board of the group is in charge of the red-level risk list, the branch directors is in charge of the orange-level risk list, etc.

2.2 HALOPA Analysis Modelling

2.2.1 System risk analysis based on equipment operation cycle

The HALOPA analysis model can be used to identify the hazard of the device or process unit, clarifying the initial event, process deviation, independent protective layer and consequence events, combined with the event tree analysis method to analysis the reduction influence of the occurrence frequency of initial events. The protection layer performs a failure frequency evaluation role to determine the final frequency of the on-site device risk events during each operating cycle under the coupling reduction of the existing protective layers at each level [2].

An important assumption of the model is that each protection layer is independent, that is, the failure of a single protection layer will not affect the risk reduction impact of other existing protection layers [3]. Each time a protective layer takes effect, the frequency of accidents is reduced by the risk reduction of the protective layer.

When the frequency of the initial event does not exceed 1 time/year or is less than twice the test frequency of the first protective layer, the evaluation of the consequence frequency can be referred to the following formula [4]:

$$f_I^C = f_I \times P_{enable} \times \prod_{j=1}^{J} PFD_{Ij} \quad (1)$$

where

f_I=Occurrence frequency of initial event I annually;

Penable=Moderation factors by enabling conditions and other conditional probabilities.

PFDIj=Failure probability of dependent protection layer j of initial event I.

When the frequency of the initial event exceeds 1 time/year or exceeds the test frequency of the first protective layer twice, the first protection layer failure shall be used as the initial event, and the calculation formula of the consequence frequency is as follows:

$$f_I^C = f_{\text{IPL\_first}} \times P_{enable} \times \prod_{j=2}^{J} PFD_{Ij} \qquad (2)$$

where

$\mathbf{f_{IPL\_first}}$= The failure probability of 1st dependent protection layer

The HALOPA analysis model in PHAMS designed a modified calculation model for equipment failure frequency and personnel exposure probability, and realized BPCS, SIS, physical protection, leakage storage and collection, fire and explosion suppression for equipment operation cycle and device in service. The system's six major types of protection layer failure frequency have the correction factors of risk reduction, and realize system risk assessment based on the device production cycle. For example, according to the system risk assessment of a diesel hydrogenation unit based on the PHAMS system, on-site data collection of various protective layers and key correction conditions under one cycle of the driving operation to date, online preliminary analysis of the risk overall value and grading of the feeding unit The results are as follows:

Figure 2 HALOPA Analytical Function Online

2.2.2 Dynamic risk assessment based on real-time data of safety control systems
According to the protection layer frequency analysis model, the failure probability of the protection layer lowering the frequency of a certain initial event can be regarded as θ , and the a priori risk probability distribution function is:

$$f_0(\theta) \propto \theta^a(1-\theta)^b \quad (3)$$

Its likelihood function $L(E|\theta)$ is

$$L(E|\theta) \propto \theta^m(1-\theta)^{n-m} \quad (4)$$

where

a, b=the prior distribution parameter of the failure probability of the protective layer;
m=The number of the posterior cumulative event occurred when the protective layer is valid;
n=The number of the posterior cumulative event occurred when the protective layer is failed.

With the posterior probability distribution function corrected by dynamic risk analysis as$f_1(\theta|E)$, the posterior probability distribution can be obtained by:

$$f_1(\theta|E) = \frac{f_0(\theta)L(E|\theta)}{\int f_0(\theta)L(E|\theta)d\theta} \propto \theta^a(1-\theta)^b\theta^m(1-\theta)^{n-m} \propto \theta^{a+m}(1-\theta)^{b+n-m} \quad (5)$$

Since m and n are parameters of the age change and the posterior probability distribution function seems to be continuous, the dynamic failure frequency of the protective layer at a certain moment can be obtained by the following formula[4-6].

$$P_i(\theta) = \int_{i-1}^{i} f(\theta)d\theta = (1-F_i(\theta|E)) - (1-F_{i-1}(\theta|E)) = F_{i-1}(\theta|E) - F_i(\theta|E) \quad (6)$$

In where $F_i(\theta|E)$ is $f_i(\theta|E)$ the cumulative failure frequency function as

$$F_i(\theta|E) = \int_0^i f(\theta|E)d\theta \quad (7)$$

And then bring in the following formula to get the dynamic risk frequency of the device or process unit:

$$f_I^C = f_I \times P_{enable} \times \prod_{j=1}^{J} P(\theta)_{Ij} \quad (8)$$

and

$$f_I^C = f_{IPL\_first} \times P_{enable} \times \prod_{j=2}^{J} P(\theta)_{Ij} \quad (9)$$

Where

$P(\theta)_{Ij}$=Posterior failure probability value of the jth independent protection layer

Based on this, the PHAMS system realizes the real-time update of the protection layer risk reduction frequency, and intelligently analysis the overall risk value and grading of the unit to plant.

3 SIL Intelligence Identification Online System

3.1 SIL Identification Modelling

The analysis data of the SIS protection layer of the HALOPA evaluation system in PHAM is updated and interconnected by means of the safety reliability database. After the HALOPA dangerous event using the same safety instrument loop is automatically brought into the SIL identification, the PHAMS system takes the same kind of consequences. C, to calculate without the cumulative frequency F^C before the SIF

$$F^C = \sum_{i=1}^{n} f_i^C \qquad (10)$$

where

$$f_I^C = f_I \times P_{enable} \times \beta \times \prod_{j=1}^{J} PFD_{IJ} \times \prod P_{condition} \qquad (11)$$

$$\beta = \frac{2^N - 1}{N} \qquad (12)$$

where

f_I^C= The occurrence frequency of the dangerous event I with consequence C

β= the common factors caused by several SIFs

N=the number of SIFs for the dangerous event I

Based on the modelling, the PHAMS system can intelligently group common risk events and assess the risk reduction factor RRF of SIF in each group [7] :

$$\text{RRF} \geq \frac{F^C}{TMEL} \qquad (13)$$

and

$$1 \times R_{\text{Factor}} < RRF_{SIL1} \leq 10 \times R_{\text{Factor}} \qquad (14)$$

$$10 \times R_{\text{Factor}} < RRF_{SIL2} \leq 100 \times R_{\text{Factor}} \qquad (15)$$

$$100 \times R_{\text{Factor}} < RRF_{SIL3} \leq 1000 \times R_{\text{Factor}} \qquad (16)$$

where

R_{Factor}= risk reduction factor between1 to 10.

Based on the SIL identification results of each group, the optimal solution is intelligently selected as the RRF of this safety instrument loop.

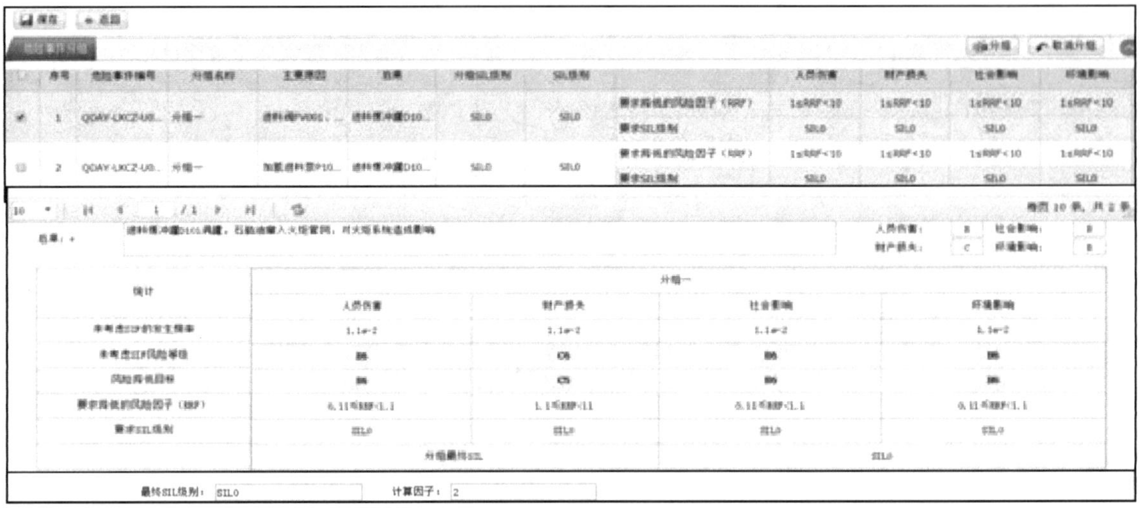

Figure 3 SIL Intelligence Identification Online

3.2 The SIL Identification Based on SINOPEC Risk Matrix in PHAMS

With the aid of data warehouse technology, PHAMS system can compare HALOPA assessment data which involved with SIS protection layers of the initial situation, the consequences classes and non SIS protective layer of data, and automatically take the residual risk of the events into the risk matrix. Then based on SINOPEC risk matrix, the SIL identification module can intelligently couple and assess the residual risk and RRF.

according to Sinopec's safety management focusing on high severity impact, PHAMS sets an independent TMEL requirements (as shown in the following table), and can automatically selects the most serious impact event as the final SIS identification data of this event. .

Table 1 TMEL of SIL Intelligence Identification[8-10]

Severity Classes	*Health & Safety (r/y)*	*Property Damage (r/y)*	*Social Impact (r/y)*
A	$\leq 1\times10^{-1}$	≤ 1	≤ 1
B	$\leq 1\times10^{-2}$	$\leq 1\times10^{-1}$	$\leq 1\times10^{-1}$
C	$\leq 1\times10^{-3}$	$\leq 1\times10^{-2}$	$\leq 1\times10^{-2}$
D	$\leq 1\times10^{-5}$	$\leq 1\times10^{-4}$	$\leq 1\times10^{-4}$
E	$\leq 1\times10^{-6}$	$\leq 1\times10^{-5}$	$\leq 1\times10^{-5}$
F	$\leq 1\times10^{-7}$	$\leq 1\times10^{-6}$	$\leq 1\times10^{-6}$
G	$\leq 1\times10^{-7}$	$\leq 1\times10^{-6}$	$\leq 1\times10^{-6}$

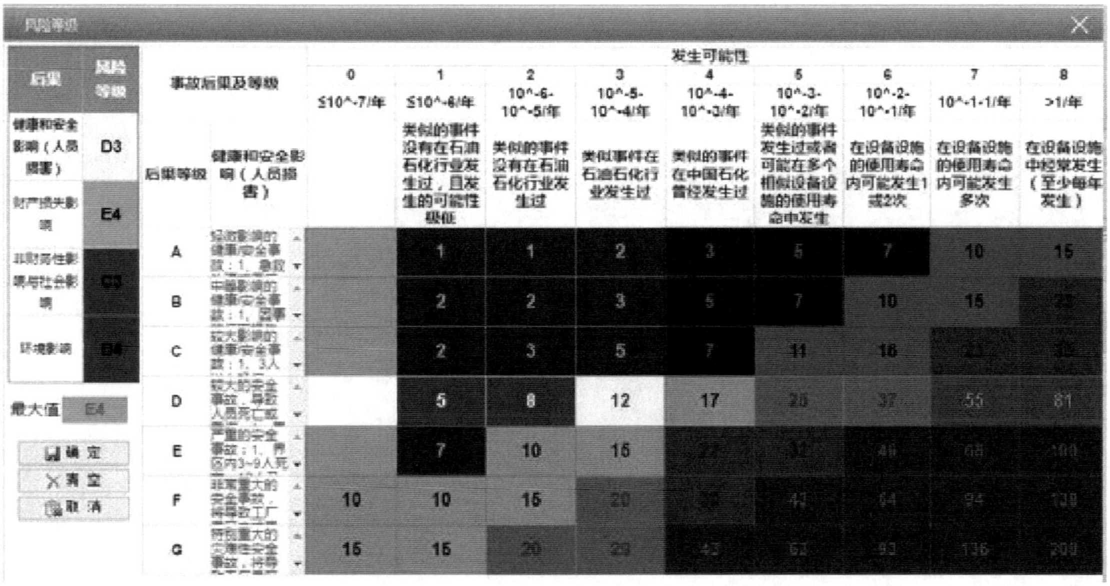

Figure 4 SIL Intelligence Identification on SINOPEC Risk Matrix

On the basis of SIL identification, the PHAMS system can also implement SIL verification based on multi-factor model, and evaluate the structural constraints, PFDavg and MTTF respectively. The verification result data is consistent with the accuracy requirements of IEC61508 and IEC61511.

4 Risk Management and Decision Support System

With the powerful safety and reliability database and expert experience database of PHAMS system, as well as the data mining of the result data of HALOPA and SIL system, the Risk Management and Decision Support System can implement the cumulative risk frequency calculation of independent devices and plants in construction or in service, to assess the overall risk based on different impact severities through the linear comparison [11-12].

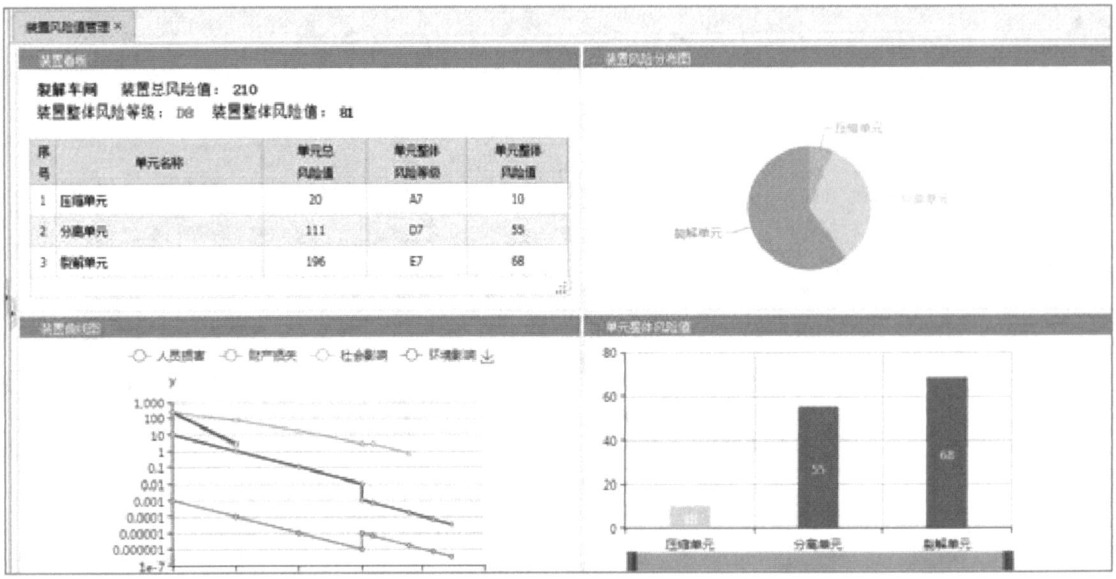

Figure 5 Overall Risk Rating and Risk Event Distribution Analysis of Ethylene Cracking Facility

5 Conclusion

1. PHAMS system sets up 9 categories of safety reliability database, expert experience data and Sinopec risk matrix, etc, which can support unified SINOPEC risk assessment and management standards, risk analysis template, process data storage and data mining for PHAMS system , as well as the interconnection and real-time update of data of each system.
2. HALOPA system implements plant overall risk analysis based on protection layer data, and performs data coupling evaluation from static analysis based on equipment cycle and dynamic risk correction to realize real-time evaluation of risk assessment technology and device operating conditions. The dynamic risk analysis research will be further strengthened to realize the system risk prediction function.
3. The SIL identificaiton and verification system can realize the evaluation of safety instrument function of in service and in construction. In the long run, it helps to improve the safety management level of the plant's overall safety instrument function
4. The risk management and decision support system can realize the overall risk assessment and risk distribution based on various evaluation results, through in-depth mining and statistical analysis, the PHAMS can realize the big data analysis management of Sinopec branches' risk management.

6 References

[1] The Bureau of Safe Production Supervision and Administration of SINOPEC. Guide for Safety Risk Identification and Assessment of SINOPEC[S].2018

[2] AQ/T 3054-2015. Guide for Application of Protective Layer Analysis Methods. Chapter 8： Scenery Frequency Calculation [S].

[3] Khan,Seyed Javad Hashemi,Nicola Paltrinieri,Paul Amyotte,Valerio Cozzani,Genserik Reniers.

Dynamic risk management: a contemporary approach to process safety management[J]. Faisal Current Opinion in Chemical Engineering.

[4] CCPS. Risk-based Process Safety [M]. 2013.

[5] Maryam Kalantarnia,Faisal Khan,Kelly Hawboldt.Dynamic risk assessment using failure assessment and Bayesian theory[J].Journal of Loss Prevention in the Process Industries . 2009 (5)

[6] Rakesh Ranjan,Sonam Singh,Satyanshu K. Upadhyay. A Bayes analysis of a competing risk model based on gamma and exponential failures[J] . Reliability Engineering and System Safety . 2015

[7]Zugravu-Soilita N. How does Foreign Direct Investment Affect Pollution? Toward a Better Understanding of the Direct and Conditional Effects[J]. Environmental & Resource Economics, 2017:1-46.

[8] Khakzad N, Yu H, Paltrinieri N, et al. Chapter 5 – Reactive Approaches of Probability Update Based on Bayesian Methods[J]. Dynamic Risk Analysis in the Chemical & Petroleum Industry, 2016:51-61.

[9] Villa V, Paltrinieri N, Khan F, et al. Towards dynamic risk analysis: A review of the risk assessment approach and its limitations in the chemical process industry[J]. Safety Science, 2016, 89:77-93.

[10] Mark Adrian van Staalduinen,Faisal Khan,Veeresh Gadag. SVAPP methodology: A predictive security vulnerability assessment modeling method[J]. Journal of Loss Prevention in the Process Industries.

[11] Podofillini L, Dang V N. Conventional and dynamic safety analysis: Comparison on a chemical batch reactor[J]. Reliability Engineering & System Safety, 2012, 106(4):146-159.

[12] R A F. Problems with risk analysis in the chemical industry. A detailed examination of the theoretical and practical problems faced by the risk analyst in the study of a chemical plant[J]. Plant/operations Progress, 1983, 2(3):185–190.

549550 Process Hazard Analysis

Process Safety Management System starts with managing the process related risks and from the first step of establishing a new process unit that is needed to evaluate the risks and define the necessary measurements. Process Hazard Analysis methodologies are the systematic tools which help to evaluate the risks in a process. For different needs there are several methodologies like DOW FETI, HAZOP, What if, HAZID, Fault Tree and Event Tree Analysis, Scenario Modelling, Facility Siting etc. Some of these methodologies are semi quantitative some are quantitative methodologies. In order to define and prevent major industrial accidents in our refinery we use DOW FETI, HAZOP, FTA and ETA and Scenario modelling. Moreover for some kind of projects we may also apply What if and HAZID analysis instead of HAZOP depending on the complexity. Facility Siting is another tool for risk related studies which is used to evaluate the risks for buildings in order to protect people inside the buildings. The workflow for a full unit PHA study starts with DOW FETI calculation which is used for risk ranking of al static equipment inside the unit. DOW FETI calculation is done in accordance with the equipment temperature, pressure, hold-up amounts, components and also reaction types if there is any. Then risk assessment process continues with HAZOP studies which is for scenario identification. In HAZOP, first step is chopping the unit into small parts to be evaluated which is called nodes. After that all nodes are evaluated using P&IDs and PFDs by keywords like temperature, pressure, flow, level etc. HAZOP helps to understand what kind of scenarios are possible inside the unit and these critical scenarios are taken through FTA and ETA studies. HAZOP is a semi quantitative risk assessment method where FTA and ETA are quantitative methods. FTA is applied for critical scenarios in order to find out the probability of top event then ETA is then applied in order to find out the probability of scenarios like fire, explosion, toxic dispersion etc. caused by this top event. FTA and ETA analysis all use the statistical PFD and failure rate values of all initiating events and barriers defined in HAZOP for the related scenario. Scenario Modelling is the following step which is used for understanding the effect of the probable scenario. The most probable scenarios coming from ETA is modelled to see the effect zones. These effect zones are used to plan Emergency Response Studies for all these potential major accidents.

548517 Site-Wide Asset Integrity Analytics & Benchmarking

All too often, reliability engineers focus on the equipment that is deemed as most critical or covered by some regulatory reference, while sometimes forgetting about those areas where something as simple as water can cause major issues. Non-destructive testing technicians and API inspectors alike can help facilities gather and compile data to perform predictive analytics and even benchmark sites with similar assets. This data can be used to identify assets where others may be seeing issues. This also allows facilities to compare reliability engineering effectiveness. This presentation will share some sample results of non-destructive testing and inspections of both regulatory required and non-required assets in the fuel ethanol production sector as well as an example of how that data can be assembled and shared across any sector to educate and learn from each other. Afterall, there are no secrets in safety!

548330 Reliability, Availability, and Maintainability (RAM). What Does It Provide?What Should You be Aware of before You Conduct a RAM Study?

Reliability, Availability and Maintainability (RAM) studies can significantly add value to your Process Safety Management (PSM) activities, especially with regards to Mechanical Integrity (MI). RAM studies are an excellent way to determine which equipment limits you from achieving optimum availability and throughput. Using industry failure rate data or even site-specific data can help you identify areas for improvement and determine how best to maintain your equipment. But RAM studies can go beyond purely theoretical studies by using the experience and knowledge gained from operating and maintaining similar facilities to develop a pragmatic approach to solving mechanical integrity-related issues. This paper will cover the pros and cons of site versus industry data and will present a case example of utilizing experience to better define the proposed problems and solutions. If done correctly, RAM studies can be more than statistics; they can be roadmaps to process reliability improvements.

546707 Driving Asset Integrity and Process Safety Consistency across a Multinational Organization

Mexichem is in the process of implementing world-class process safety management (PSM) and Asset Integrity Management (AIM) programs globally for their facilities. This is a major undertaking as Mexichem, to a large extent, has grown through acquisitions. Therefore, each plant has a safety culture and management system unique to that asset. As we all know, it's difficult to drive global consistency when dealing with different types of operations in various regions throughout the world under differing regulatory regimes. This is especially difficult when there is no common understanding of process safety. One of the first steps Mexichem has taken is to develop a "fit-for-purpose" management system with an emphasis on developing a common understanding of process safety and ensuring the culture is robust enough to support the system when it is implemented.

To aid Mexichem in this overwhelming task, Risktec Solutions were engaged to develop and deliver a bespoke set of interactive training workshops on a variety of PSM / AIM topics. The first of which was related to the identification and management of safety critical equipment (SCE). This presentation will outline the plan and methodology Mexichem and Risktec have used to transfer such knowledge and discuss some of the successes and challenges along with way.

548216 Alarm Floods and Their Connection to Plant Incidents and Near Misses

Investigations of many catastrophic plant events have shown alarm flooding to have been significant distractions for the operators just prior to the incident. Alarm floods have also been determined as a cause or contributing factor to smaller incidents and near misses.

In fact, many of the more recent findings by U.S. Chemical Safety and Hazard Investigative Board (CSB) include alarm management as one of the key issues in the investigative reports.

Even after several years of trying, many plants still struggle with controlling alarm floods. Many also have problems generating consistent and accurate alarm metric reports for the entire plant. Now that the ISA 18.2 Alarm Performance Metrics are considered "generally accepted good engineering practices," we should make haste to demonstrate where each of our plants fall (particularly the PSM plants) versus these metrics before the regulating authority moves in this area. If we are not able to prove we meet the metrics (particularly for alarm flooding) under all operating conditions, we should remediate.

Static rationalization can reduce your average number of alarms, but without controlling the alarm floods, there is no help for the operator when he needs it the most. For most, we must begin the process by recognizing that alarm floods occur most of all during a change of process state.

The means of controlling alarm floods has been available to us for many years. Unfortunately, only a select few have implemented the solution and are achieving the desired results today. These plants have achieved equal to or better than ISA 18.2 since before it was published.

This session will cover the justification for true alarm management from the safety and economic perspective with several practical steps that everyone should take if they are not currently meeting the ISA 18.2 specifications during all operating conditions.

548318 Do Incident Data & Lessons Learned Really Facilitate Process Improvement?

Regulations are a set of guidelines for the industry for protecting the people and environment, when it comes to process safety systems. While some companies tend to find that complying with regulatory requirements are always a challenge, on the other hand some tend to go overboard in their bid to stay compliant. In spite of the route taken, there are incidents – major, minor and near misses.

Other than the accumulation of this data, the investigations also result in lessons learned. Over the past four decades, lot of incident data and lessons learned has accumulated, been assimilated, and also analyzed. Complementing that are the audits, system reviews, hazard identification exercises, changes to regulations, and knowledge sharing forums.

It is a very efficient and effective mechanism to leverage these vast sources of current information developed, and implement its learnings, in the operations of existing processes. While the core concepts of safety are universal across the various industrial sectors, its application may differ. But once again, the lessons learned databases are effective sources of making this leveraging process more efficient.

This paper will discuss the some case studies of how the concept of lessons learned and the knowledge of the incident data has been leveraged by the authors in various manners ensuring success from one industry to another – example (1) Learnings from an incident due to gaps in the Emergency Response Plan, bringing about success at another chemical manufacturing site; (2) Incident data from a steel plant, when analyzed and applied to a chemical plant resulted in improved of effectiveness.

546673 Structured & Scrutinized Approach Towards HSE Excellence; A Success Story

Author: Usman Ali, Process Safety Engineer

Co-Author: M Ahsan Sarfraz, Unit Manager HSE

This paper studies the practices & programs that support reporting of Near misses, minor unsafe conditions, incidents and quantitative risk assessment strategies in our organization for analyzing and learning from past events of operational failures, injuries, safety and environmental incidents.

A near miss is a leading indicator to an incident that, if scrutinized appropriately, can prevent injuries and accidents. Some persons may be tempted to write off near misses as "no harm, no damage" situations, but organizations who track near misses, determine how and why they occurred, and take corrective action can prevent similar or more serious incidents from occurring in the future. Fatima Fertilizers Limited believes that, "We don't look at what happened; we look at what could have happened".

Keeping in mind the importance of near miss reporting, FFL initiates drive and implement systems for

1. Capturing and increasing near miss reporting at whole site.
2. Logging data for analysis and using this valuable data for identifying gaps in existing process safety system.
3. Implementing sustainable improvement plans and safety campaigns (Safe hands, Safe Routines, Job Cycle Checks, Safe Drive etc) in our process safety systems for making our site a safer workplace.

These actions has improved the site performance which is reflected as Site has now over 45 Safe Million Man Hours, Total Record-able Injury Rate (TRIR) is below 0.2 since last 03 years and No Loss time injury since last 06 years.

548162 Events Report Database Analysis for SHE/PSM Programs Improvements

Incident databases contain records of events that have occurred in a petrochemical facility during recent years. The most of chemical industries report databases are filled with accident valuable data (frequency and severity). Harnessing data mining to explore incident databases, it was used data mining to analyze the incidents database of the studied plant. Inferent approaches are ineffective for low frequency, high consequence events because of their rarity. Given this information limitation, this work uses database analyses to forecast incident frequencies, their relevant causes, equipment involved, and their possible consequences, in specific petrochemical plants.

Databases information analysis, also help to avoid future accidents, thereby reducing the risk. More specifically, this study presents dynamic analyses of incidents in a petrochemical plant ubicated in Mamonal industrial park, in Cartagena- Colombia. Differences in the chemical-specific predictions with the overall predictions are demonstrated. Furthermore, value theory is used for consequence of rare events by formulating distributions for events over a threshold value. Finally, the accidents number is used to estimate the capital at risk within an industry utilizing the frequency and loss-severity distributions.

In conclusion, it was recommended to update de maintenance program (predictive, preventive and corrective) in cases of abnormal changes in production rates, comparing the current production levels, with the analyzed plant production forecast. Also, it was suggested to improve the report culture, due to gaps founded regarding to events classification.

549483 The Process Safety Analysis and Management System in Sinopec

The big data technique has been in trend for the process safety assessment. In china, it's the very need to transfer the qualitative safety management to quantitative safety management, which should rely on the professional risk analysis technique, as well as the standard safety reliability database. Based on the above, it's very important to integrate several professional risk analysis technique, as HAZOP, LOPA, SIL,BOWTIE etc. to realize the integral risk identification and evaluation, grading and management plan making, supporting by the proprietary SINOPEC risk metrics and grading standards, as well as safety equipment reliability database,which is developed as the Process Safety Analysis and Management System for SINOPEC, to support all SINOPEC branches to make dynamic risk analysis and management online

The key data for HAZOP, SIL and Management module in PHAMS is the dependent protection barriers, which we do focusing on the failure probability, at the same time, we do use of the Bayesian Correction Modelling to verify the failure probability of safety equipent and protection barriers on site, which can validly save the grading cost on time and personnel, all of this is towards to the deeper research on dynamic risk analysis for PHAMS.

Thanks to the PHAMS, we has set up and implemented several standards on HAZOP, LOPA and SIL for SINOPEC.

552170 Analysis and Improvement of Safety Integrity Level Considering Human Factors

Safety Instrumented System (SIS) is an irreplaceable active protection layer that prevents the propagation of process accidents and plays an important role in reducing the risk of accidents. In addition to the sensor response triggering the SIS, a human emergency button is often provided in practice to ensure that the SIS is triggered by personnel in the event of the sensor part failure. Due to human factors added, the emergency button will improve the reliability of the SIS. However, there is no research to study the human reliability of the SIS upon sensor part failure in a continuous operation chemical plant, and how human factors can improve and affect the overall reliability of the SIS still remain undiscovered. In this paper, a verification methodology was proposed which integrated with Cognitive Reliability and Error Analysis Method (CREAM) and Reliability Block Diagram (RBD) to emphasize human reliability analysis of the SIS verification. The operator behavior is subdivided into several steps and the Human Error Possibility (HEP) is calculated by a modified CREAM. Then the RBD method is used to verify the SIS reliability considering human factors. An SIS system of one reacting furnace was taken as a case study using this method to show how human factors can improve the reliability of the SIS. The proposed methodology allows SIS verification considering human factors to improve safety integrity level in real practice thus help stakeholders saving costs and making more decisions on SIS implementation.

548383 Getting Incident Investigations Right

When safety investigations are conducted like criminal investigations, they tend to be backward-looking and focused on learning who to blame, rather than forward-looking with a focus on what to blame, so that similar incidents can be avoided.

Unfortunately, many internal incident investigations are led by people who learned what they know about incident investigations from television police dramas, so conduct them like criminal investigations. To avoid this, some organizations arrange for third-party investigations. One advantage of a third-party investigation is that nothing is off the table. Also, there is no presumption that the cause is already known. While the viewpoint of every investigator is informed by their experience and conceptions of causation, so never "objective", third-party investigations can bring a valuable perspective to the primary objective of incident investigations: learning how to avoid similar events in the future.

After decades of conducting incident investigations and training incident investigators, we've learned that there are certain pitfalls that often snare investigators, especially internal or infrequent investigators. We discuss some of those traps—from deciding which incidents to investigate and the unrecognized influence of conceptions of causation, to the nature and value of recommendations that are often made.

548448 Using Feedback to Improve the Quality of Your Incident Investigations

Trevor Kletz once explained that it was relatively easy for him to create his monthly Lessons Learned Bulletins, because ICI had a requirement that investigation reports be complete and well-written. As a result, he could easily understand what had happened and could explain it in simple language. Several years ago, Honeywell's Performance Materials & Technologies division embarked on a specific effort to improve the quality of our incident investigations, and of the learnings flowing from them. The purpose was – and remains – to reduce the number of process safety events by learning from smaller events and applying those lessons broadly. One important aspect of this broader program was to provide a "critique" of several important aspects of the investigations of our process safety incidents. The aspects we chose to focus on first included things like "was a proper team formed to examine the aspects of the incident" and "did the team collect the right information, including creating a timeline of the events leading up to the incident", "did they identify the cause(s)" plus several others. The degree to which the investigation complies with each criterion is evaluated by a team of seasoned professionals, and an Investigation Maturity Index is fed back to the investigators as well as reported upwards to leadership.

In this paper we will explain the criteria we use, the work process of the evaluation team as well as some of the insights and improvements we have made as a result.

19 AIChE Spring Meeting
15TH GLOBAL CONGRESS ON PROCESS SAFETY
NEW ORLEANS • 2019

Cultural Cause Analysis™

Mapping process safety incident and performance data to cultural causal factors

Laura O. Jackson
ABS Group
1525 Wilson Blvd., Suite 625 | Arlington, Virginia 22209 USA
ljackson@abs-group.com

Steve Arendt
ABS Group
1701 City Plaza Drive | Spring, Texas 77389 USA
sarendt@abs-group.com

Prepared for Presentation at
American Institute of Chemical Engineers
2019 Spring Meeting and 15th Global Congress on Process Safety
New Orleans, Louisiana
March 31 – April 3, 2019

Disclaimer

*ABS Group refers to various subsidiaries of ABS Group of Companies, Inc., including ABSG Consulting Inc. These concepts, methods, and tools were developed by ABSG Consulting Inc. (ABS Group). This methodology and the corresponding tools are meant to assist users in identifying (1) cultural causes of incidents that have occurred and (2) cultural issues that exist and may lead to accidents. The cultural cause analysis tool is not intended to eliminate all risks and does not guarantee against the occurrence of any incident. Any user of this tool agrees to defend, indemnify, and hold harmless ABS Group, its affiliates, directors, agents, employees, and subcontractors, against any claims of loss, injury, death, or damage of any nature whatsoever, including, but not limited to, claims for contribution and indemnification with respect to third-party claims unless directly caused by the sole negligence of ABS Group. This tool may be used by other parties, but it may not contain sufficient information for purposes of other parties and uses. This tool has been developed using a degree of care and skill ordinarily exercised, under similar circumstances, by reputable consultants practicing in this field at this time. No other warranty, express or implied, is made with respect to this method or tool. All ownership rights and interests in this tool remain with ABS Group.

Cultural Cause Analysis™

Mapping process safety incident and performance data to cultural causal factors

Laura O. Jackson
ABS Group
1525 Wilson Blvd., Suite 625, Arlington, Virginia 22209 USA
ljackson@abs-group.com

Steve Arendt
ABS Group
1701 City Plaza Drive | Spring, Texas 77389 USA
sarendt@abs-group.com

Keywords: Cultural Cause Analysis™ (CCA); Cultural Cause Map™; Root Cause Analysis (RCA); Root Cause Map™; Incident Investigation; Process Safety Management (PSM); Health, Safety, and Environment (HSE, ESH, SHE); Center for Chemical Process Safety (CCPS); Risk Based Process Safety (RBPS); safety culture; culture assessment; safety; quality; reliability; and ABS Group.

Abstract

Cultural Cause Analysis™ (CCA) is a new approach for evaluating cultural causes for more effective and efficient incident investigations and root cause analyses (RCAs). This approach was developed by the same authors as the Center for Chemical Process Safety's book, *Guidelines for Risk Based Process Safety*, which is recognized globally by high-hazard industries as the benchmark for process safety management (PSM) practices. CCA is a reactive analysis method that supports determining the underlying cultural issues that led to an incident (a near miss, accident, or series of chronic events). Specifically, CCA supports organizations in assessing the underlying behaviors, actions, and cultural issues that allowed management system weaknesses (root causes) to exist and led to equipment or human performance gaps that caused an incident(s). Corrective actions addressing these cultural issues, if effective, are valuable because they positively impact seemingly unrelated areas of performance, including safety, quality, reliability, etc. CCA is an extension of root cause analysis (RCA) and is completed at the same time as RCA or afterwards. The CCA method works universally with available RCA models. The Cultural Cause Map™ is a tool which supports CCA and enhances organizations' ability to systematically trend data related to cultural issues. CCA is a reactive analysis methodology, occurring like RCA after an incident. While many organizations proactively assess their culture (e.g., safety culture assessments), this next-generation RCA method and tool support organizations in reactively assessing and understanding cultural issues post incident to avoid repetition and improve organizational performance.

1 Cultural Cause Analysis™

Organizations want to understand what happened after an accident has occurred because it is vital to ensure it will not recur. In the 1960s, the introduction of the "5-Why's" method from the automotive industry spurred deeper thinking. In the 1970s, as government agencies sought higher reliability and better occupational safety standards in energy and space travel, they formalized the root cause analysis (RCA) method to understand management system weaknesses in standards, policies, and administrative controls leading to accidents. RCA now encompasses (1) most industries, (2) issues, including safety, reliability, and quality, and (3) analysis of near misses, chronic issues, and accidents. This paper introduces the next evolution in the RCA discipline, Cultural Cause Analysis™ (CCA). CCA dives further into understanding the organizational tendencies, behaviors, actions, and individual cultural causes that created the environment that allowed the root causes to occur and exist, which inevitably led to the incident(s). This greater depth of understanding strengthens an organization's ability to influence the underlying cultural issues to avoid repetitive negative results in the organization's performance.

1.1 Industry's Need to Understand Cultural Causes of Incidents

> *"We never want to have an accident like this occur again. So, how can we really learn from this?"*
>
> *"Our corrective actions and controls, developed from a thorough RCA, do not seem to correct the underlying drivers of behavior. So, now what?"*
>
> *"How can we achieve a sustainable 'drive to zero' process safety and Health, Safety, and Environmental (HSE) performance?"*
>
> *"We want to have a mature culture. What does a mature culture look like, where are we, and how do we get to the desired maturity?"*

These are the questions ABS Group response teams have documented from over 40 years of conducting on-site investigations following major process safety incidents such as an explosion, oil spill, or chemical release. Response teams support organizations through the incident investigation and response to executive and regulatory inquiry. The teams then complete an RCA, which includes data collection and interviews, and report these findings to the organization. Our teams have observed an inability in RCA to dive deeper into the drivers of performance. Wanting to support clients' need to (1) achieve sustained safety, quality, and reliability and (2) mature, ABS Group has continued to work beyond root causes to understand the cultural causes of incidents. After more than a decade of development, we have matured this approach and created the **Cultural Cause Analysis™ (CCA)** tool.

The desire for CCA is driven by many factors, both positive and negative. The *negative* drivers of CCA are that (1) accidents continue to occur and (2) the recommendations from RCAs that focus on correcting company standards, policies, and administrative controls do not create the desired long-term change. As a result, similar issues may occur again. The *positive* drivers of CCA are (1) the desire to create lasting changes post-accident, (2) the demand from stakeholders for sustained outstanding HSE performance, and (3) a drive to leverage HSE data in a more meaningful way.

A growing need exists to (1) understand cultural issues behind undesired behaviors, (2) develop corrective actions to influence the organizational culture in positive ways, (3) correct management system weaknesses, and (4) reduce the potential for chronic performance gaps and associated incidents. CCA is the answer to that need.

1.2 The Vision

1.2.1 <u>Fulfillment of the safety culture vision</u>

The need for deeper analysis of culture is not a surprise; it was the process safety management (PSM) vision. The following is from the closing remarks in CCPS's *Guidelines for Risk Based Process Safety*:

> *Widespread use of risk based process safety metrics will enable companies to correlate off-normal events to management system failures and dysfunctional behaviors, ultimately focusing on accident prevention by improving the culture at the base of the accident/incident pyramid. Companies will understand that "learning*

lower on the pyramid" leads directly to safer, more productive, and more profitable operations.

The PSM accident pyramid is in Figure 1.

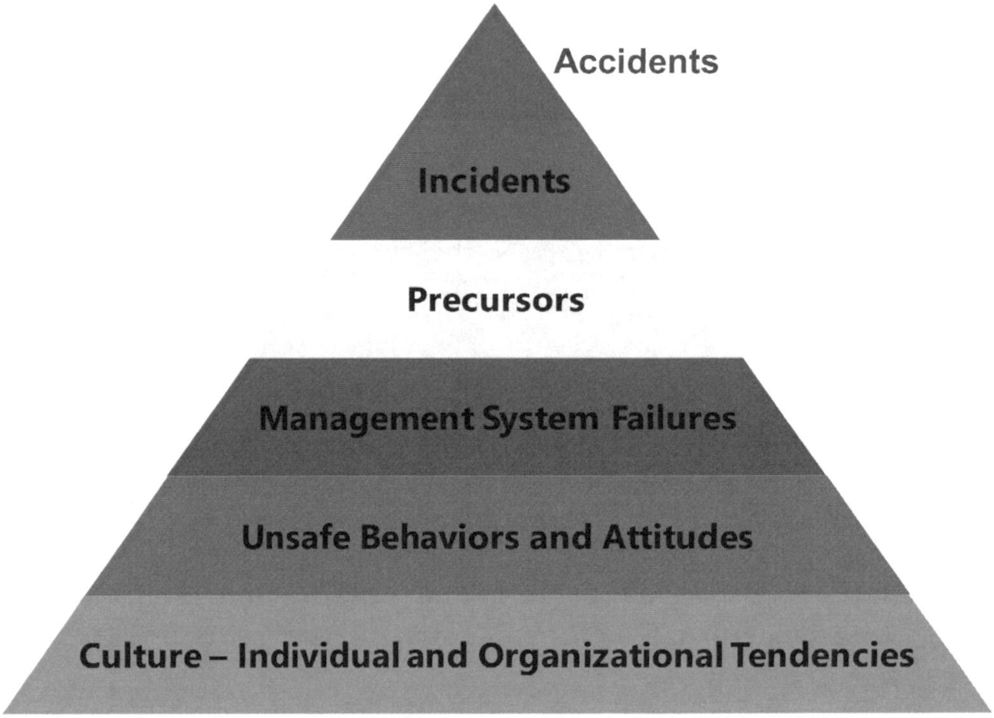

Figure 1. The PSM Accident Pyramid

At the base of this pyramid, poor individual and organizational tendencies [green] manifest into unsafe actions and behaviors [pink]. These cultural issues create the environment in which policies and standards are either not created, not sufficient, or not enforced [blue]. And those insufficient management systems allow for poor performance in people and equipment, leading to accidents [red]. While this pyramid specifically focuses on safety, the same principles apply to quality and reliability, or any other performance aspect that can be defined through its culture.

1.2.2 Fulfillment of the RCA vision

The need for deeper analysis of culture is not a surprise; it was the RCA vision. In 2005, the *Root Cause Analysis Handbook: A Guide to Effective Incident Investigation* (second edition) included Figure 2, which shows the "different organizational levels and the corresponding level of learning that can be accomplished based on an incident."

The lowest level of the triangle shows the organizational culture issues [black], leading to the management system issues (e.g., standards and policies that are nonexistent, deficient, or not enforced) [grey], which inevitably lead to performance gaps in both people and equipment at the top of the triangle [white].

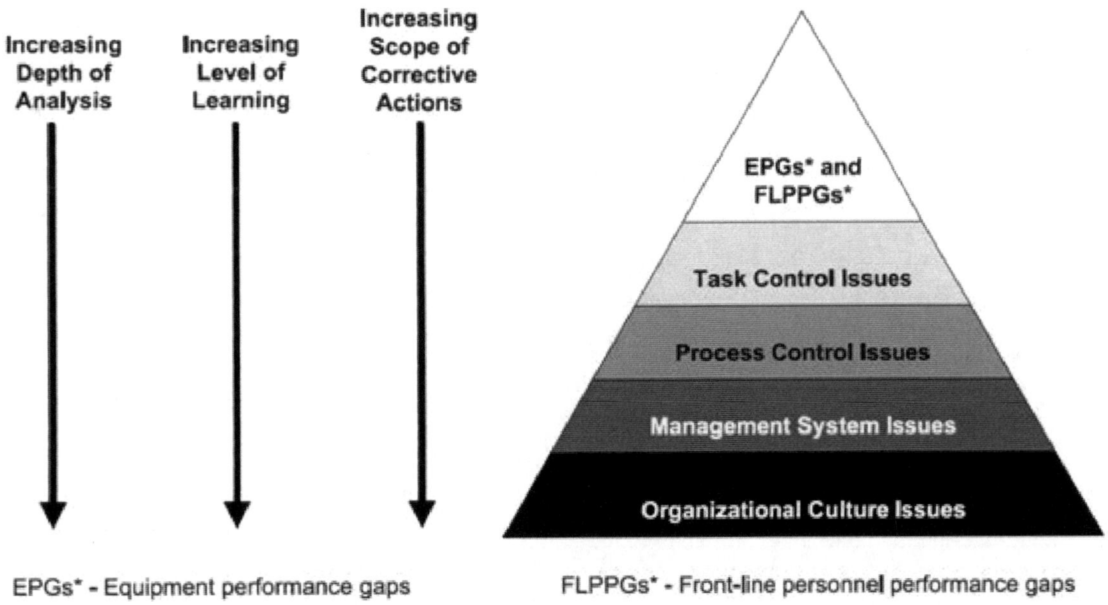

Figure 2. The RCA Triangle

The *Root Cause Analysis Handbook: A Guide to Efficient and Effective Incident Investigation* (third edition) discusses that existing root cause methods are effective at tackling the management system issues, but advancements are needed to understand and influence change at the next level, the cultural cause level. The arrows on the left side of the image depict the increasing depth of analysis, learning, and corrective actions that occur when evaluating incidents lower in the RCA triangle.

1.3 The Appeal of Proactive and Reactive Culture Analysis

Industry is making progress on evaluating culture *proactively*, as in "what cultural weaknesses can we identify that might lead to an issue?" However, industry generally does not attempt to identify and affect cultural issues *reactively* (i.e., after an incident).

PSM focuses on the prevention of, preparedness for, mitigation of, response to, and restoration from catastrophic releases of chemicals or energy from a process associated with a facility. This speaks to both proactive and reactive management, and yet industry focuses its evaluation of culture predominantly on *proactive* culture assessments rather than *reactively* performing analyses of cultural causes of incidents that do occur. States such as California are leading in this area of proactive safety culture assessment by regulating it. For example, the State has promulgated California General Industry Safety Order 5981.1 - Process Safety Management for Petroleum Refineries requirements to improve process safety performance. In the future, performing an analysis of cultural drivers to major accidents (at a minimum) should be expected as well.

All of us – and our organizations – want to do the right thing in the right way at the right time, all the time – even when no one is looking. And yet, culture is hard to measure and more difficult to change. Culture is the result of all the actions and inactions in institutional and workforce memory. Organizational culture is a complex combination of individual cultures and environments. The time for *culture* analysis both before *and* after incidents is now.

1.4 The Approach

1.4.1 Goals of CCA

The goals of CCA are to identify and resolve the true underlying undesirable behaviors, attitudes, and cultural causes of losses that occur for:

- individual major events (as they occur)
- chronic minor events (as indicated in risk profiling)
- near misses with high potential severity
- any performance gaps that reveal a cultural weakness

1.4.2 Definition of cultural causes

Cultural causes are the core causes of an event that can be reasonably identified, and management can correct or positively influence.

Typically, cultural causes are the absence, neglect, or deficiencies of an individual's or organization's demonstrated value system, which create an environment where management system (e.g., policies, procedures, and administrative control) weaknesses occur, leading to issues with human actions and equipment performance.

1.4.3 CCA connection to existing methods

Organizations desire normal operations with people and equipment functioning in harmony, behaving as they should, and yielding the desired results. However, even in high reliability operations, near misses and accidents occur, people do not perform as they should, and equipment fails. Sometimes the issues coalesce to a point of catastrophe. Other times, the consequences are minimal, but the chronic repeat nature of the issues can cause greater pain to the organization. No matter how the incident occurs, if the high-potential (risk profile) of the near miss or accident is high enough, then analysis is needed to help assure that it does not occur again. Figure 3 depicts the progression of the level of details associated with using increasingly more detailed incident investigation approaches.

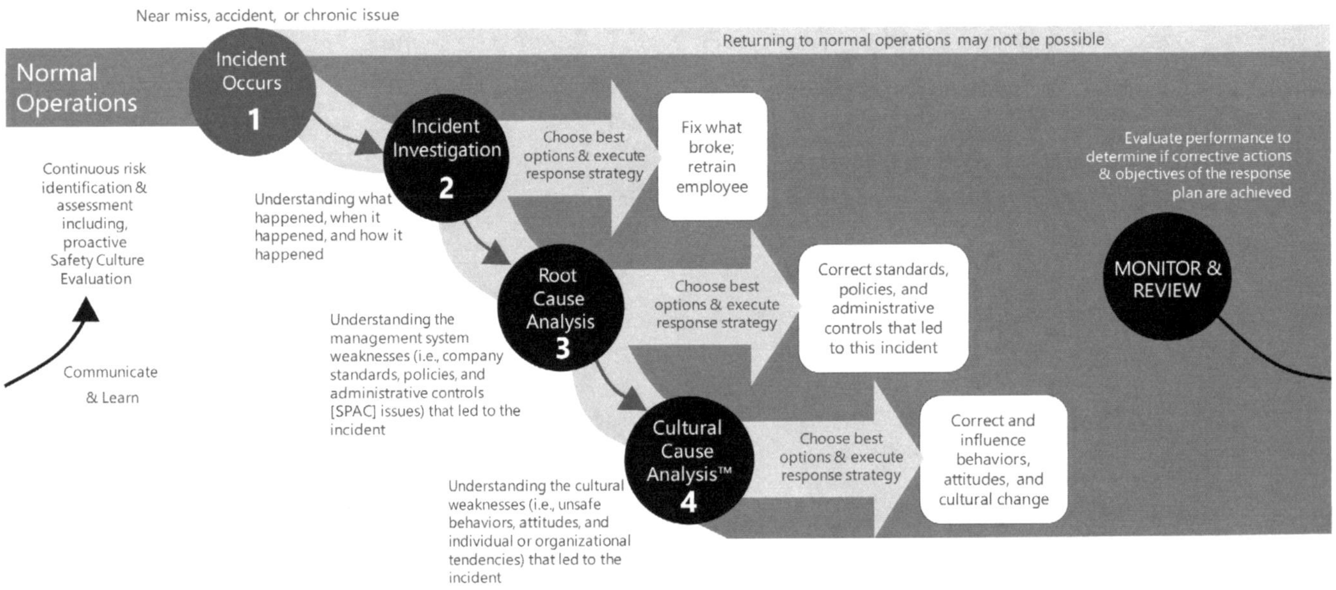

Figure 3. Increasing Depth of Understanding Post Incident

1.4.4 Benefits of CCA

There are several benefits from performing CCA. Using CCA and implementing the resulting recommendations can:

1. Keep the same and similar events from recurring
2. Keep other seemingly unrelated future events that would have shared the same underlying cultural cause from occurring
3. Save time and money by avoiding implementing recommendations that do not address the underlying attitudes, behavior, and cultural issues
4. Achieve quick results from the cultural cause initiative
5. Demonstrate excellence by meeting and exceeding regulatory requirements for incident investigations and RCA
6. Promote cultural change within the organization

1.4.5 Optimal approach for implementation

Like any program or process, an effective rollout is vital to the success of the new approach. The following three-step approach is recommended:

(1) Focus the initial CCA teams on tackling (a) high-priority chronic items identified in the high-level risk profile or (b) more significant sporadic problems

(2) Work with program management to structure a tailored CCA program (possibly integrated with the site's incident investigation and RCA program)

(3) Provide training sessions on the customized CCA program and associated Cultural Cause Analysis™ tools

1.5 The Analysis Technique

This analysis technique has two major benefits. First, CCA is built from the foundation of CCPS's RBPS Safety Culture Essential Features and can be aligned for use with other culture frameworks. Secondly, CCA is compatible with any conventional RCA methodology, with no tailoring needed.

1.5.1 *Comparison to CCPS's new book on safety culture*

CCA and its corresponding tool, the Cultural Cause Map™, are based on the Center for Chemical Process Safety's 2007 book, *Guidelines for Risk Based Process Safety*, which is recognized globally by high-hazard industries as the benchmark for process safety management (PSM) practices. CCPS recently published (2018) new culture core principles. The essential features of CCPS's RBPS (red book) are very similar to the new core principles. Much of the essential features of the two models are substantively identical, as shown in Table 1.

Table 1. Benchmarking with CCPS Culture Core Principles

CCPS's *Guidelines for Risk Based Process Safety* Culture Features (2007)	Comparison	CCPS's *Essential Practices for Creating Strengthening, and Sustaining Process Safety Culture* Core Principles (2018)
1. Process safety must be a core value	Word differences only	1. Establish the imperative for process safety
2. Provide strong leadership everywhere	Identical	2. Provide strong leadership
3. Enforce standards of performance and accountability	Word differences only	9. Combat the normalization of deviance
4. Formalize the culture approach	Combined into one	10. Learn to assess and advance the culture
5. Maintain a sense of vulnerability	Identical	5. Maintain a sense of vulnerability
6. Empower individuals	Identical	7. Empower individuals
7. Defer to expertise	Identical	8. Defer to expertise
8. Ensure open and effective communications	Word differences only	4. Ensure open and frank communications
9. Establish a safe questioning and learning environment	Possible gap →	6. Understand and act upon hazards/risks
10. Foster mutual trust	Identical	3. Foster mutual trust
11. Responsiveness to process safety issues	Possible gap →	
12. Provide continuous monitoring of performance	Combined into one	10. Learn to assess and advance the culture

1.6 The Tool

The tool ABS Group developed for CCA is intended to be used after an RCA has been completed or in tandem with an RCA. The style and function of the Cultural Cause Map™ are similar to the Root Cause Map™, progressing from the top to the bottom with an increasing depth of understanding. The Root Cause Map™, Figure 4, is a proven tool after being leveraged for over 20 years in thousands of investigations in a variety of industries. As issues lower on both the Root Cause Map™ and Cultural Cause Map™, Figure 5, are understood and corrected, the corrective actions can be more impactful for the organization in preventing future issues that were seemingly unrelated.

1.6.1 The Root Cause Map™

The Root Cause Map™ flows from top to bottom and is oriented with more equipment-based issues on the left and more personnel-based issues on the right. The top portion of the map begins with a causal factor (equipment or front-line personnel performance gap). For example, an equipment issue might be "the bolts failed on the coupling" and a personnel issue might be "the employee did not chock the wheels of the delivery truck." The map continues downward as the question "why" is asked until an intermediate cause level is reached; for example, "the bolt failed due to a design issue" or "the procedure did not require the employee to chock the wheels." Then the path transitions to the bottom left of the map where the root causes (deficiencies in standards, policies, and administrative controls) are identified.

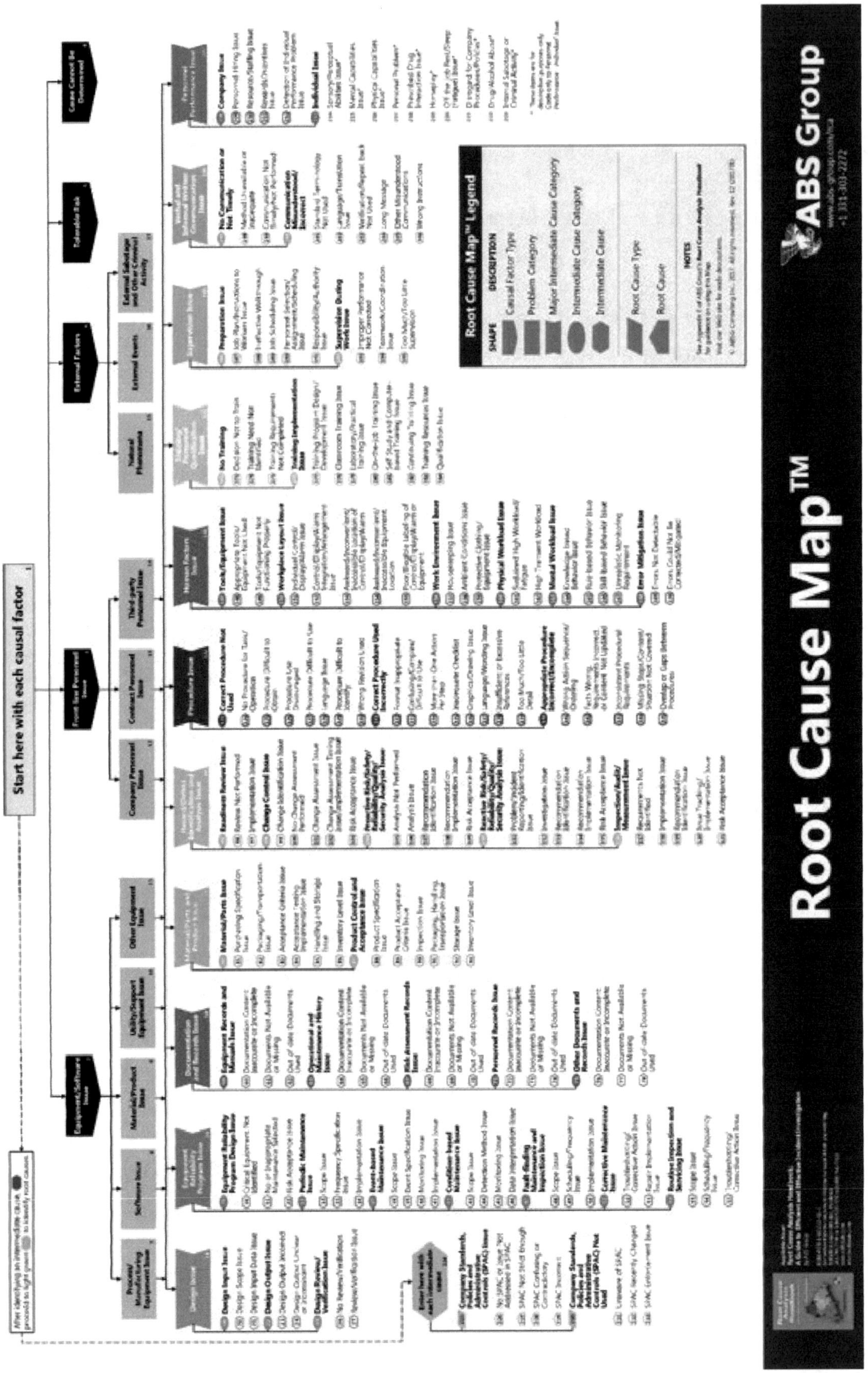

Figure 4. Root Cause Map

1.6.2 *The Cultural Cause Map™*

The Cultural Cause Map™ is divided into four layers of cultural causes. Like the RCA Map, the flow of the Cultural Cause Map™ is also from top (Layer 1) to bottom (Layer 4). Layer 1 is predominantly focused on performance action issues. Layer 2 targets issues related to situational awareness. Layer 3 addresses factors related to adoption of leadership values by organization personnel. Layer 4 is related to the values expressed by leadership that drive the organization. While it is possible to have a strong organizational culture without any documented policies and procedures or a poor culture with comprehensive documentation, our model is built on the principle that both the culture and its documentation are important.

The case studies depicted in the section following the Cultural Cause Map™ are simplified to only show one path through the map. Actual analysis is more complex with a one-to-many relationship. A CCA involves analyzing and documenting the entirety of that complex relationship. The aggregate of the causal information is valuable. For example, after the analysis is completed, a user may observe that "mutual trust issues" were the predominant cultural issue that led to deficiencies in management systems, which ultimately caused an incident to occur.

A key advantage of routinely performing CCAs on near misses, accidents, and chronic events using this map is providing more comprehensive and accurate data trends. The data collected from this trending are a predominant source of input into proactive culture evaluation and performance management.

1.7 Case Studies

Below are two case studies that demonstrate the application, flexibility, and applicability of the CCA methodology and tool. These studies are based on real experiences that occurred in ABS Group's support to organizations. For simplicity, while the case studies had multiple causal factors, this discussion focuses on only one causal factor and limiting intermediate causes and root causes for each case study.

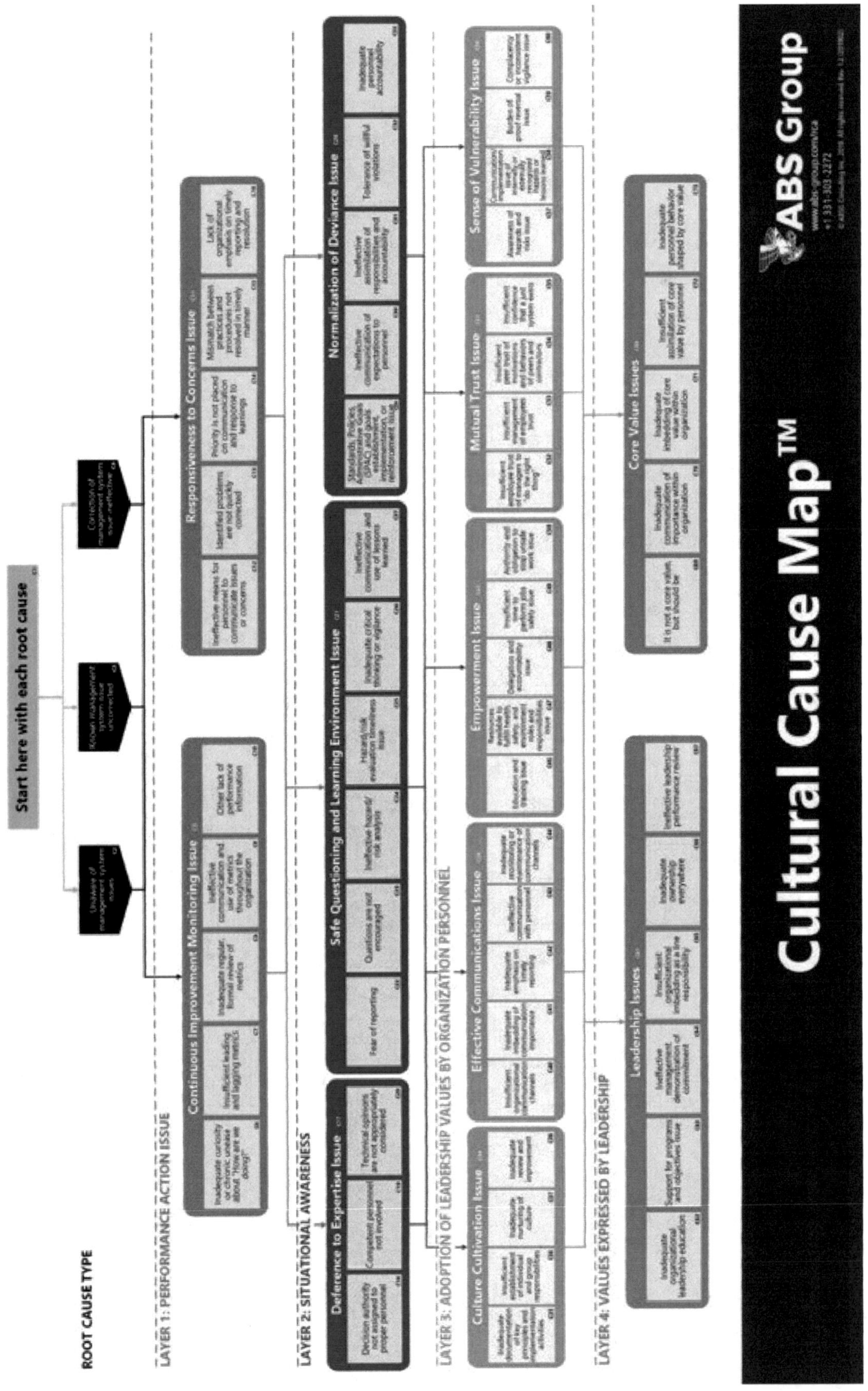

Figure 5. Cultural Cause Map™

1.7.1 *Case Study 1: The "Lucky" Explosion (significant accident, but not catastrophic, that triggered a quick look at cultural causes)*

At a process facility, an unrecognized change to a gas compressor inlet/outlet led to a "lucky" explosion that caused no injuries because everyone was at lunch. Someone repositioned the chemical additive injection point from the suction side of the compressor to the discharge side (a physical distance of about 1 meter). No one considered this to be a change in the process, so no one thought to look at the properties of the chemical additive, which is very corrosive above 50°C. The process temperature was 40°C at the suction side and 135°C at the discharge side. After about 2 hours in operation, the discharge piping failed due to accelerated corrosion caused by the additive.

The **accident** was the explosion of a gas compressor system at the process facility that led to shutdown of the unit for an extended period and financial losses.

The company's incident investigation identified the following **causal factor** (i.e., performance gap[s] either by a person or piece of equipment that caused the incident, allowed the incident to occur, or allowed the consequences of the incident to be worse than they might have been):

- Worker repositioned the chemical additive injection to a new location (Root Cause Map™ Path: 3, 12)

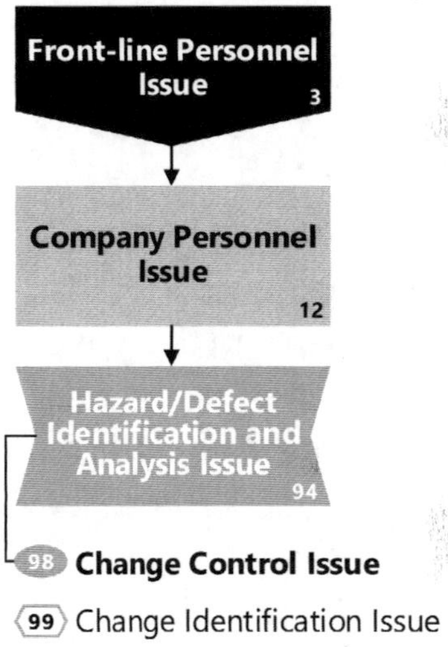

Figure 6. Path to Intermediate Cause

The **intermediate causes** (i.e., underlying reason[s] why the causal factor occurred, but not deep enough to be a root cause) were determined to be:

- Failure to recognize and evaluate the modification as a change (Root Cause Map™ Path: 3, 12, 94, 98, 99 [as shown in Figure 6]) and

- The organization's investigation procedure did not require incidents related to management of change (MOC) issues to be routed through the MOC coordinator. Had the organization routed incidents through the MOC coordinator, the organization believes the MOC coordinator would have seen the increase in related incidents and taken action to support front-line personnel to identify actions that needed MOC. (Root Cause Map™ Path: 3, 12, 122, 140, 144 [as shown in Figure 7])

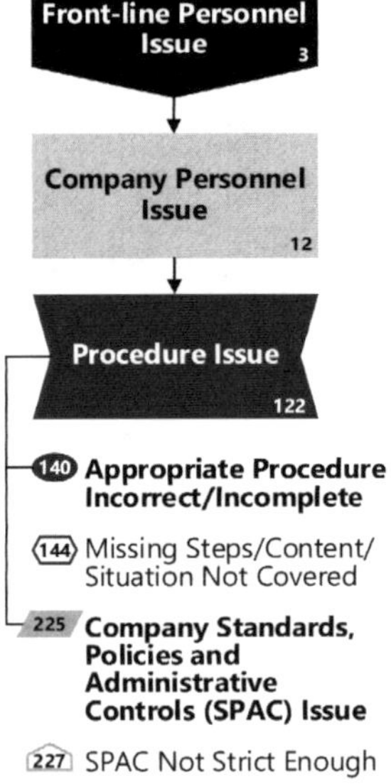

Figure 7. Root Cause Path

The company RCA identified the following **root causes** (i.e., management system weaknesses) of the incident:

- Failure to account for change type during the development of the MOC procedure (Root Cause Map™ Path: 3, 12, 94, 98, 99, 225, 227) [Not shown in a figure]
- Because investigation reports and findings related to MOC issues were not routed to the MOC coordinator, he/she was not aware of the need to improve the MOC management system based on the prior incidents. (Root Cause Map™ Path: 3, 12, 122, 140, 144, 225, 227 [as shown in Figure 7])

In addition to the immediate repairs required, the company also took the following **corrective actions** to address the root causes by:

- Revising the MOC procedure to address the change type involved in the incident
- Revising the investigation process to have all incidents that involve MOC issues routed to the MOC coordinator for learning assessment
- Providing refresher training for all personnel on incident lessons and the MOC process - in particular, recognizing changes

These corrective actions, if effective, addressed the direct technical intermediate and root causes. At the time, the investigators did not evaluate any underlying company cultural issues.

Sometime later, by request, ABS Group completed a quality assurance (QA) check of several incident investigations for the company. In our review, we came upon this incident report and another similar incident report. Given the significance of the incidents that occurred and that there were two similar incidents, we identified the need to analyze the cultural causes. The resources used for this quick-look CCA were one consultant, two employees, and 4 hours. The CCA identified the following **cultural causes**:

Root Cause Type –
Unaware of the MOC management system issue (Cultural Cause Map™ Path: C2)

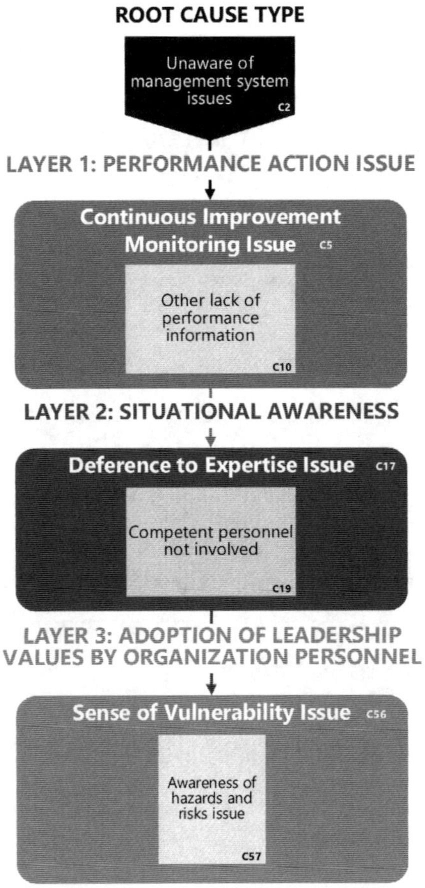

Figure 8. Cultural Cause Path

Cultural Cause Layer 1 –
Continuous Improvement Monitoring Issue – Other lack of performance information (Cultural Cause Map™ Path: C2, C5, C10)

Cultural Cause Layer 2 –
Deference to expertise issue – competent personnel not involved. The MOC coordinator was not involved in assessing MOC-related incident findings (Cultural Cause Map™ Path: C2, C5, C10, C17, C19)

Cultural Cause Layer 3 –
Sense of vulnerability – awareness of hazards and risks issue. There are two issues here: (1) the incident investigation findings were not proactively shared with the MOC coordinator and (2) the MOC coordinator did not actively seek out the incident findings *(Cultural Cause Map™ Path: C2, C5, C10, C17, C19, C56, C57) as depicted in Figure 8.*

Below are the other cultural causes reached at Layer 3 (from other Cultural Cause Map™ paths)

- *Lack of a sense of vulnerability* – Personnel were not disposed to consider the hazards of "standard" maintenance and other chemicals. They had become complacent about all but the most significant direct hazards they faced.

- *Failure to empower individuals* – Operations leadership reduced refresher training during the past 3 years due to cost reductions.

- *Failure of leadership* – There was leadership failure, resulting from inadequate support for programs and objectives.

In summary, had the initial RCA team dug deeper into the underlying causes using the CCA method and Culture Cause Map™, the team would have uncovered several cultural causes that needed to be addressed to help ensure prevention of similar events.

1.7.2 <u>Case Study 2: BP Texas City</u> (catastrophic accident that triggered an in-depth review of cultural causes)

The BP Texas City accident had numerous technical causes and associated root causes. The organizational findings "imply" probable cultural causes. Using the Cultural Cause Map™, ABS Group traced the cultural cause path, starting with the technical finding of "Failure to Follow Start-Up Procedure" to:

The **accident** involved the ignition of a hydrocarbon vapor cloud that exploded at the octane-boosting isomerization unit, killing 15 people, injuring hundreds, and crippling the refinery.

A **causal factor** of the incident was determined to be that the blowdown drum was overfilled with highly flammable liquid hydrocarbons (Root Cause Map™ Path: 3, 12).

An **intermediate cause** identified per the Chemical Safety Board (CSB) Technical Finding was an employee failed to follow the startup procedure because the procedure use was not encouraged (Root Cause Map™ Path: 3, 12, 122, 123, 126).

A **root cause** identified by CSB was that the enforcement of policies was not strict enough because cost-cutting and production pressures impaired process safety performance (Root Cause Map™ Path: 3, 12, 122, 123, 126, 230, 233). This path is depicted in Figure 9.

The cultural cause path of this root cause is outlined below, using the Cultural Cause Map™. One path through the map is depicted in Figure 10. The **cultural causes** that allowed this root cause to occur are:

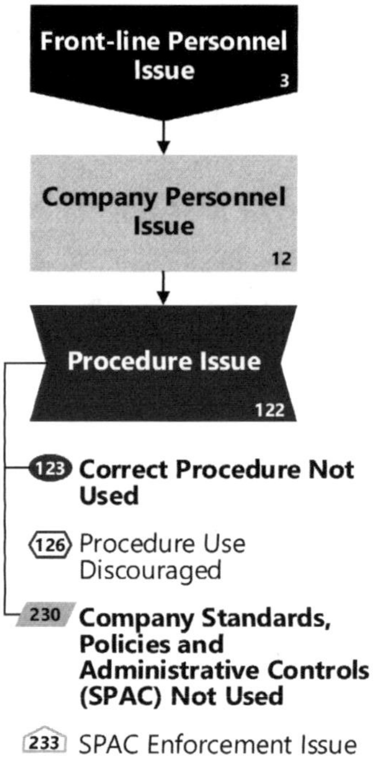

Figure 9. Root Cause Path

Root Cause Type –
The eroding management system enforcement of following procedures was an uncorrected known issue (Cultural Cause Map™ Path: C3)

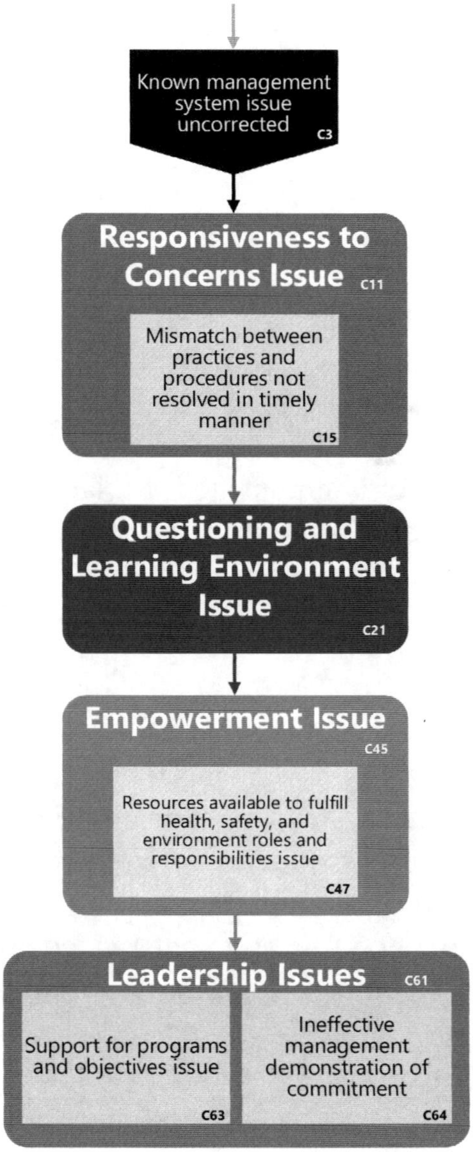

Figure 10. Cultural Cause Path

Cultural Cause Layer 1 –
Two cultural issues from Layer 1 led to the root cause:

Continuous improvement monitoring issue – Inadequate curiosity or chronic unease about "How are we doing?" The company was measuring process safety success/performance by occupational safety metrics (Cultural Cause Map™ Path: C3, C5, C6)

Responsiveness to concerns issue – Mismatch between practices and procedures not resolved in a timely manner (Cultural Cause Map™ Path: C3, C11, C15)

Cultural Cause Layer 2 –
Drilling down into understanding why "C15 – responsiveness issues" happened were due to two issues:

(1) Normalization of deviance – Tolerance of willful violations and inadequate personnel accountability (Cultural Cause Map™ Path: C3, C11, C15, C28, C32 & C33 [not pictured])

(2) Questioning and learning environment – There was a general lack of questioning, including all of the identified notes on this topic (C22 – fear of reporting, C23 – questions not encouraged, C24 – ineffective hazard/risk analysis, C25 – hazard/risk evaluation timeliness issues, C26 – inadequate critical thinking or vigilance, and C27 – ineffective communication and use of lessons learned). These are trended as: (Cultural Cause Map™ Path: C3, C11, C15, C21, C22 through C27)

Cultural Cause Layer 3 –
Drilling down into understanding why "C21 – questioning environment issue" happened was because of:

Empowerment issue – Resources available to fulfill health, safety, and environment roles and responsibilities issue (Cultural Cause Map™ Path: C3, C11, C15, C21, C45, C47)

Cultural Cause Layer 4 –
Finally, the core cultural cause is:

A leadership issue – Support for programs and objectives and ineffective management demonstration of commitment (Cultural Cause Map™ Path: C3, C11, C15, C21, C45, C47, C61, C63 & 64). This path is depicted in 10.

In summary, leadership's lack of support and commitment led to lack of empowerment, which resulted in a poor questioning and learning environment and normalization of deviance that led to a chronic mismatch between practices and procedures (i.e., the enforcement issue as a root cause, management system weakness).

1.8 Application of Data Analytics

Today, organizations spend extensive resources to collect and manage enterprise data to support a wide variety of operational and business functions. While these data are created to support other decisions, we recognize an untapped potential exists to use these data to support CCA. Performance management related to safety is not new, but applications related to more complex algorithms, business intelligence, timely (real-time) monitoring, and leveraging advanced techniques like machine-learning and big data are the next phase. Organizations are taking first steps to leverage artificial intelligence and data tools to find cultural issues, patterns, etc. For example, an organization might leverage big data applications to assess safety predominance in the culture and to trend it over time through monitoring safety-related content companywide through email and new or revised content on the topic of safety all of which can be paired with tracked information such as the number of documented near misses.

Modern data science tools are capable of extracting, integrating, and analyzing previously inaccessible and siloed data. ABS Group is already supporting organizations to use these data to monitor, analyze, and manage their culture as it evolves. Monitoring culture and ultimately predicting safety performance are no longer impossible tasks. Data science tools can support the proactive evaluation of the organization's culture (driving forces) and forensic analyses into "what went wrong?" and "why it went wrong" when an accident does occur. Use of these technologies are the vision for the future of CCA, and for some companies, it is already a reality.

1.9 Conclusion

Implementing effective corrective actions that address not only root causes but cultural causes is the best approach for achieving sustainable "drive to zero" process safety and HSE performance improvements, superior quality, and sustained reliability. If your organization is already using both proactive and reactive analyses of culture and achieving sustained performance, then bravo! If not, then the gauntlet is thrown down. Take RCA to the next level and dig deeper into understanding cultural issues by applying these CCA methods and the Cultural Cause Map™.

2 References

ABS Consulting, *Root Cause Analysis Handbook: A Guide to Effective Incident Investigation, 2005.*

ABS Consulting, *Root Cause Analysis Handbook, A Guide to Efficient and Effective Incident Investigation*, 2008.

Baker, James et al., 2007. "The Report of the BP U.S. Refineries Independent Safety Review Panel" http://www.safetyreviewpanel.com/, January 30, 2007.

Center for Chemical Process Safety (written by ABS Group under contract to CCPS), Guidelines for Risk Based Process Safety, 2007.

U.S. Chemical Safety and Hazard Investigation Board, BP America Refinery Explosion, REPORT NO. 2005-04-I-TX, March 2007.

3 About the Authors

ABS Group is a trusted technical services advisor with safety, risk, and integrity management at the core of what we do. ABS Group began providing technical services in 1971, and over the decades, we have expanded to offer comprehensive risk management solutions to an even broader spectrum of industries that power, fuel, and regulate our world. ABS Group is an experienced leader in incident investigations, RCA, and CCA and can support you as much or as little as you need: augmenting your staff with onsite support; being your on-call incident investigators and RCA and CCA response teams; developing and implementing management systems; performing specific analyses; providing software and other resource materials; designing and implementing big data applications in support of informed decision making; and training and coaching your staff.

Laura Jackson is Director, Advanced Training Solutions, and an expert risk management advisory professional. She specializes in developing innovative applications of risk, safety, quality, and reliability management to help major corporations, government agencies, and nonprofits make effective use of their limited resources to achieve organizational performance goals. Ms. Jackson is a recognized pioneer and subject matter expert of CCA, ERM, and RCA, and a thought leader in organizational excellence and business intelligence (BI). Ms. Jackson leads management projects that involve industry experts, risk modeling, data science, and technology-based solutions.

Steve Arendt is Vice President, Global Oil, Gas & Chemicals, with 42 years' experience in process safety and risk management in upstream, midstream, and downstream oil and gas, petrochemical, and related processing industries. He is a recognized thought leader with global expertise in PSM/HSE, organizational culture, and risk communication, with over 80 articles and 12 books to his credit. Mr. Arendt is a CCPS Fellow, a Mary Kay O'Conner Process Safety Center Merit Awardee, and is the 2019 AIChE Walton/Miller award winner recognizing his lifetime achievements in loss prevention and process safety.

Incorporating Process Safety Throughout the Lifecycle of Major Capital Projects (MCP)

Bebe Kueh
Chevron North America Exploration and Production Company
1500, Louisiana Street, Houston, Texas
Bebe.kueh@chevron.com

Prepared for Presentation at
American Institute of Chemical Engineers
2019 Spring Meeting and 15th Global Congress on Process Safety
New Orleans, LA
March 31 – April 3, 2019

Incorporating Process Safety throughout the Lifecycle of Major Capital Projects (MCP)

Bebe Kueh
Chevron North America Exploration and Production Company
1500, Louisiana Street, Houston, Texas
Bebe.kueh@chevron.com

Keywords: process safety, major capital project, concept selection, detailed design, process safety challenges, process safety opportunities, lessons learned

Abstract

Over the past decades, process safety has emerged as one of the focus areas in Major Capital Projects (MCP) to ensure the project is not only producing and profitable but above all, is safe to operate within its design life. This translates to the bottom line of as no loss of life, no loss of production due to a major incident and a more reliable plant. When we know better, we design better. However, it can be challenging to incorporate process safety in MCP when an incident averted is not measured in terms of cost savings or production efficiency. This paper aims to address the key challenges and opportunities in process safety throughout the lifecycle of Major Capital Projects (MCP) in the oil and gas industry.

1 Major Capital Projects (MCP)

The lifecycle of a Major Capital Project (MCP) in the oil and gas industry consists of different phases, from opportunity identification, concept selection, detailed design, and execution to operations. An example of a MCP simplified roadmap in Chevron is provided in Figure 1.

Each phase provides governance and validation that is required to ensure the Project objectives are met prior to the next phase. The level of design maturity and investment increases with each phase and it is vital to ensure that all the critical elements are considered as the MCP progresses.

Historically, the amount of effort invested in each MCP is measured by cost, schedule, quality and technology. Over the past decades, the yardstick of successful MCP has extended to include process safety to ensure the project is not only producing and profitable but above all, safe to operate within its design life. This translates to the bottom line of as no loss of life, no loss of production due to a major incident and a more reliable plant. When we know better, we design better.

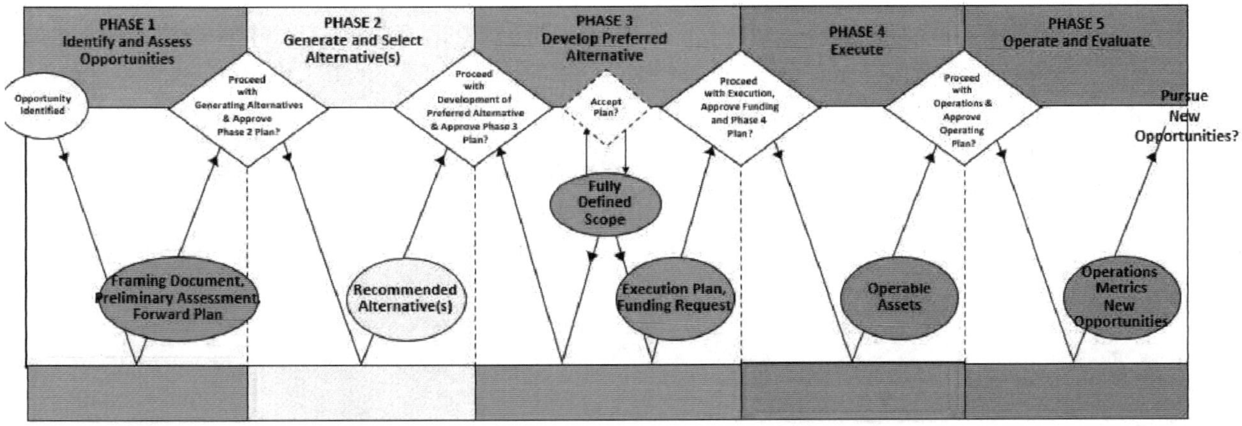

Figure 1: Chevron Project Management and Execution Process (CPDEP) simplified roadmap

2 The importance of Process Safety in MCP

Process Safety is a product of good engineering philosophies, design and operational discipline to reduce major incidents that can lead to a fire, explosion or toxic release. The aim of process safety is to mitigate the impact to personnel, assets and the environment, due to an incident. Process safety incidents can have catastrophic effects as demonstrated in Ekofisk Bravo (1971, Norway), Piper Alpha (1988, UK), Esso Longford (1998, Australia), BP Texas Refinery (2005, US), Deepwater Horizon (2010, US) and many others. With every process incident, the industry has more to learn and implement for future design improvements.

Once an MCP is successfully executed, the facility is expected to operate throughout its design life of 20 to 30 years. It is important that process safety is incorporated in every phase of an MCP to ensure that the facility is designed to safely operate throughout its design life.

It should be acknowledged that while MCPs have implemented design changes to avoid incidents in the past, there are also increasingly new technologies and more challenging terrains that could significantly change the hazard and risk profiles of a MCP. Figure 2 provides an illustration of decision making considerations based on different project types and risk profiles.

For Decision Type A, the risk is well understood. The design is generic and most concerns can be addressed with codes and standards, good practices and some engineering judgment. This is representative of a conventional offshore platform, onshore refinery and terminals.

For Decision Type B, the project may have some lifecycle implications, uncertainties or deviation. The decision-making process will include additional engineering considerations and risk-based analysis. This decision type can be related to a brownfield project with a new technology on an ageing facility or conventional facilities to be constructed in an unfamiliar region.

For Decision Type C, the project is challenging or novel with significant tradeoffs, stakeholder perceptions and large uncertainties. This will require various means of calibration and a precautionary approach. The precautionary approach is an assessment technique that is inherently conservative due to lack of information or uncertainty. Some examples of this include floating LNG, subsea compression, high pressure high temperature (HPHT) offshore facilities, ultra deepwater and challenging terrain.

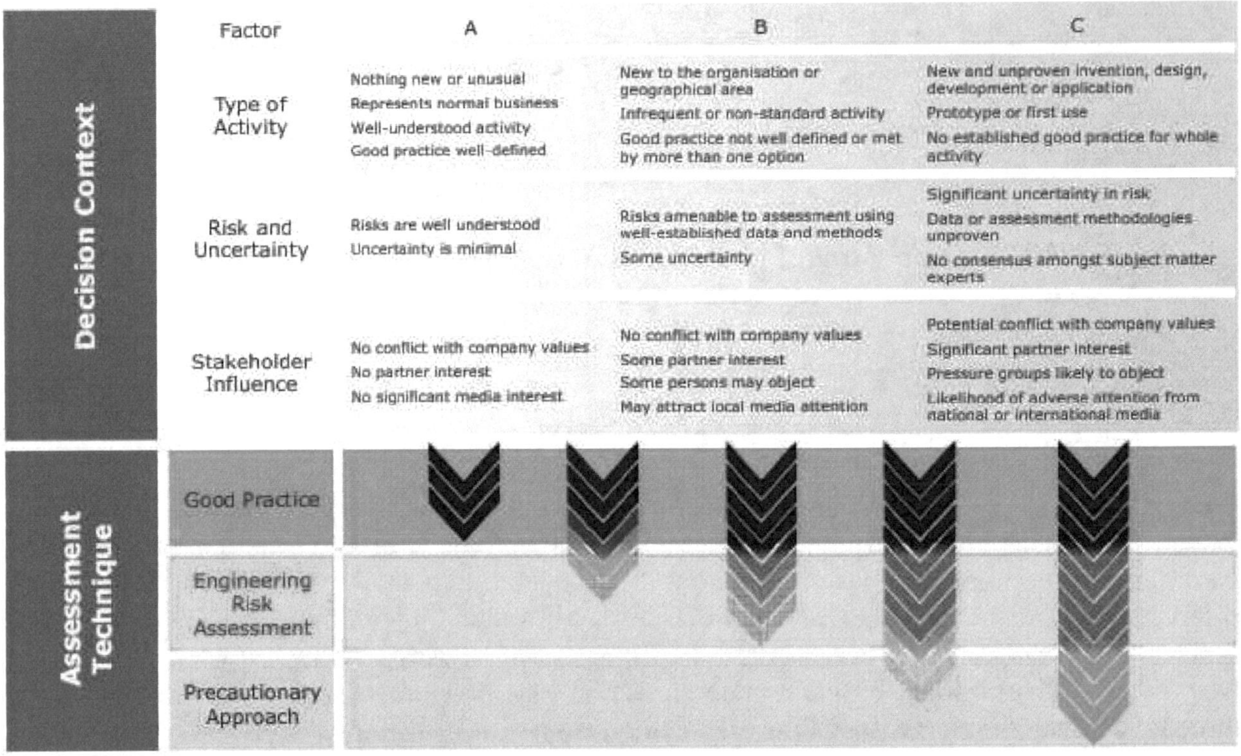

Figure 2: Risk related decision support framework [1]

Each MCP is unique in terms of facility type, process conditions, production capacity, equipment, layout and environmental conditions. A strategy to incorporate process safety in each MCP phase shall be appropriately scaled to the complexity and progress of the project to ensure fit-for-purpose, effective and compliant risk reduction measures.

2.1 Key Process Safety Aspects

The classification of process safety aspects can vary depending on organization approach. For the purpose of this paper, 5 main process safety aspects that are significant in defining safer design for MCP are selected for discussion. The 5 aspects are: qualitative risk, quantitative risk, human factor engineering, fire protection and process safety interface with other disciplines.

Figure 3 provides the summary of the 5 main process safety aspects and its impact on MCP and Operations.

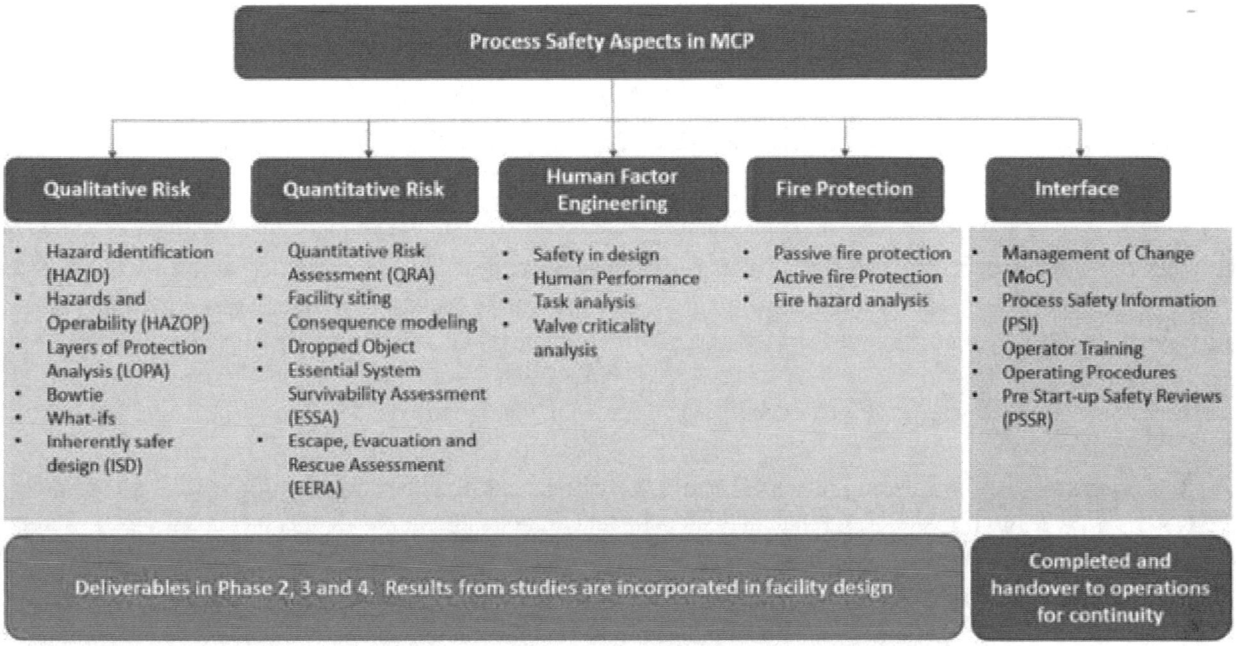

Figure 3: Process safety aspects in MCP

2.1.1 Qualitative risk

Most qualitative risk studies are carried out with a multi-discipline team to identify potential hazards and gaps in the design based on collective judgement, experience and lesson learned. Qualitative risk is one of the first safety studies in an MCP and the level of granularity of the study increases as the MCP progresses. Qualitative risk methodologies throughout MCP include:

- HAZID provides a high-level understanding of risk profile across different activities
- Bowtie is an effective tool to communicate and increase awareness on process safety to personnel across organization
- HAZOP and LOPA are used to identify safeguards and potential gaps in protection against consequence
- ISD allows team to brainstorm for safer design through the concept of elimination, minimization, substitution, moderation and simplification.

2.1.2 Quantitative risk

Quantitative risk analysis is defined as safety studies that utilize the combination of engineering calculations, mathematical modelling and existing databases to predict the safety consequence and risk profiles of a facility. Quantitative studies in an MCP allow:

- Concept selection based on sensitivity analysis to compare incremental risk across options

- Determination of the consequence of a fire, explosion and toxic dispersion scenario to aid design e.g. separation distance, inventory, equipment sizing and design
- Location of critical equipment in a safer area during early design phase
- Identification of protection for areas with high consequence e.g. fire wall, blast wall, relocate equipment

2.1.3 Human Factor Engineering (HFE)

Human factor engineering is the combination of engineering and human behavior studies to ensure that the design of a product, system and/or process is carried out in a way that optimizes human and overall system performance, safety and comfort. Early integration of HFE ensures that:

- The facility is designed to consider the flow of work and reduce the likelihood for human error
- Operation and maintenance is efficient, resulting in cheaper and simpler solution compared to a later change

2.1.4 Fire Protection

The selection of the type of fire protection shall be based on fire and explosion assessment to determine effective mitigation measures to prevent the escalation of fires (in credible scenarios). Early consideration of fire protection in an MCP allows:

- Effective and optimized design of firewater equipment and distribution
- Simplification of fire protection design and maintenance throughout the design life
- Effective placement of fire and gas detection that increase reliability and improve safety

2.1.5 Interface

The Interface aspect element includes all other process safety deliverables including but not limited to: process safety information (PSI), safety critical operating procedures, pre start-up safety reviews (PSSR), operator training and management of change (MoC). These interface activities are developed in different phases of an MCP and are the main deliverable for an MCP to handover to Operations.

3 Incorporating process safety in each MCP phase

Planning and timing is key in incorporating process safety throughout the lifecycle of an MCP. There can be a process safety analysis that is beneficial in a certain MCP phase and there are analyses that bring minimal benefit when carried out at the wrong time. Figure 4 below is the Chevron Project Delivery and Execution Process (CPDEP) value chart that shows the value identification and value realization across different MCP Phases.

Figure 4 shows the value identification and realization of each MCP phase. The maximum value is derived in the earlier phases, from Phase 1 to Phase 3, where there is major influence on the engineering design. The process safety opportunities in Phase 4 and Phase 5 will be primarily driven by the newly identified hazards, management of change or deviation during fabrication.

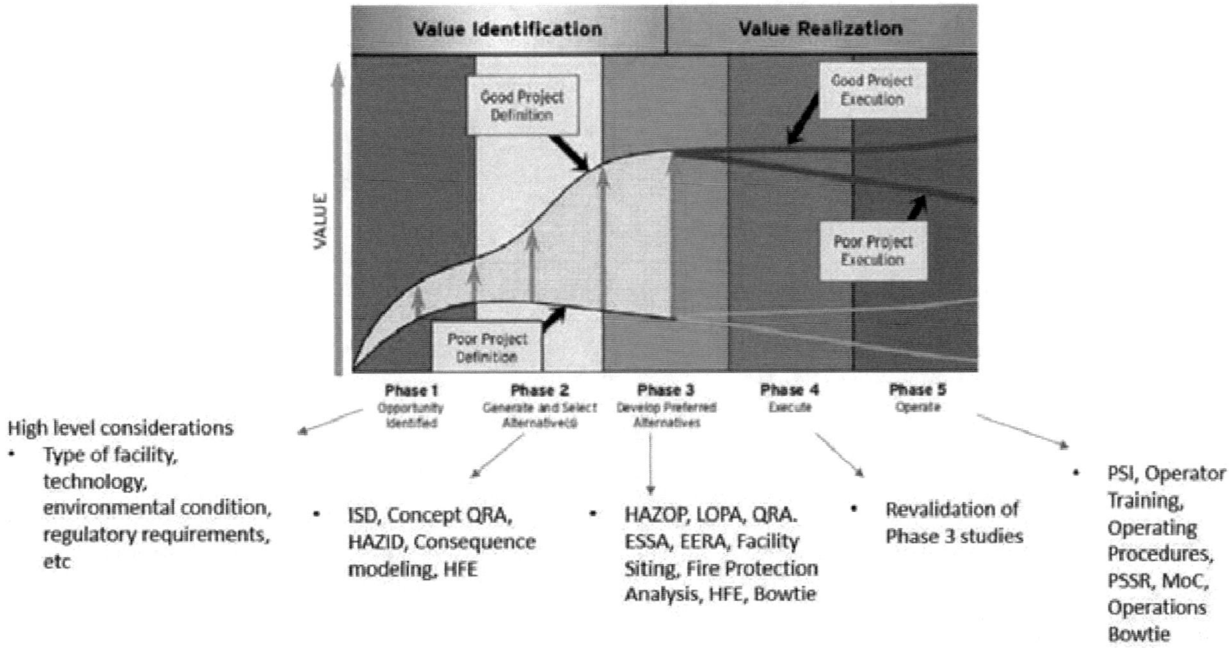

Figure 4: CPDEP value chart and process safety studies in each phase

4 Key challenges and opportunities for process safety implementation in MCP

It can be challenging to incorporate process safety in an MCP when an incident averted is not measured in terms of cost savings or production efficiency. The reality is process safety incidents can only be avoided with deliberate effort during the design, execution and operation phases. The following table provides a summary of challenges from experience on various MCPs and the opportunities identified to address the challenges.

	Challenges/Misconceptions	Opportunities	Examples
1	Process safety considerations will lead to design complexity and cost increase *"Gold plated facility"*	Process safety may not necessarily add equipment and there are instances where less equipment is safer, and less inventory is sufficient. What you don't have, can't leak (Trevor Kletz, 1978)	Application of inherently safer design (ISD) strategies in Phase 2 to Phase 3 is vital to ensure hazards are minimized or eliminated in the design. The hierarchy of ISD are: • Elimination or minimization of hazards • Substitution of a hazardous materials with a less hazardous materials • Moderation of process conditions • Simplification of process, operations, design and/or maintenance to reduce risk
2	Safety Culture of MCP *"It has not happened for the past 30 years, it won't happen"*	Safety culture in an MCP is a combination of safety culture of the Operator (solely operated or joint venture) and contractors. The safety culture is only as strong as the weakest link, and it takes time to build. The common goal of safe design is achievable if every organization within the MCP shares the same values and safety culture.	Consistent communication and engagement sessions to ensure the contractors and fabricators are on board with the safety culture and expectation of the Operator/MCP. For example: inclusion of process safety expectations and verification of process safety competency gap from contractors. In Chevron, the safety culture is part of the 10 Tenets of Operation that are based on 2 principles: • Do it safely or not at all • There is always time to do it right

<table>
<tr>
<td>3</td>
<td>Lack of process safety understanding</td>
<td>Building Process Safety competency to increase the understanding of process risk within the project scope. The competency in process safety will increase the understanding of process safety, leading to a better use of risk assessment tool to aid design and decision making.</td>
<td>
Introduction of specific process safety training in early MCP phases can increase process safety awareness and understanding of risk to improve design and limit changes as the MCP progresses.

One of the most effective ways to communicate risk and concurrently build process safety awareness is through simple illustration tools, such as Bowties. A Bowtie identifies the hazards, causes and safeguards that are required to be in place to prevent the consequence and can be applied from Phase 2 to Phase 5, depending on the hazard and level of details. Bowties also have the versatility to include critical position, equipment and activities, allowing deep dives and verification as the project progresses.

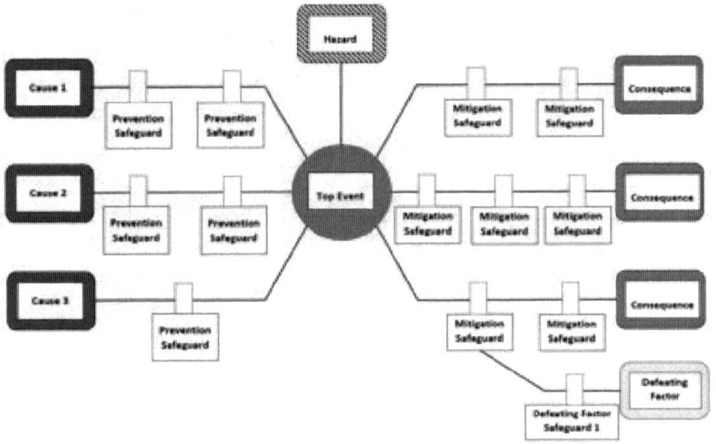

It is important to define roles and responsibilities for each safety critical role in an MCP to ensure risk discussion goes through to the appropriate party and proper approval. The roles and responsibilities can be mapped to a process safety competency matrix to identify mandatory training and assessment by their role and level of technical authority.
</td>
</tr>
</table>

4	Process safety requirements and demonstration of safe operation can vary with different geographical locations, e.g. safety case regime vs regulatory regime *"It must be safe as I am following the code"*	In the US, onshore facilities are governed under Occupational Safety and Health Administration (OSHA) and offshore facilities are governed under Bureau of Safety and Environment Enforcement (BSEE) and US Coast Guard, depending on type of facilities and area of concern. In Commonwealth countries, the requirement for process safety and demonstration of safe operation is governed under the safety case regime. Regulations and codes are a result of lessons learned and by following the applicable codes it is possible to mitigate specific risks. The opportunity is: there is value in learning other regulatory regimes and applying those aspects that are useful to other MCPs.	One of the examples is the use of Quantitative Risk Assessment (QRA). In a safety case regime, a QRA is used to identify risk contributors and demonstrate that the risk profile is managed to as low as reasonable practicable (ALARP). In the US regulatory regime, a QRA is not mandatory. However, some companies have adopted QRA as a tool in Phase 2 and Phase 3 to identify major risk contributors and utilize the information to better design the facilities. The figure below shows an example of a QRA result for an explosion from a process module and its impact to an occupied building. 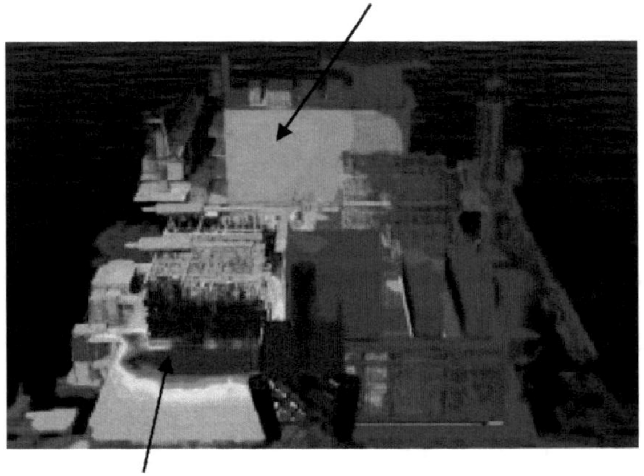 By understanding the extent of the overpressure, the design team can consider and evaluate mitigation measures in the design, e.g. increase separation distance, change equipment layout, strengthen the building or reduce process inventory.

5	Perception of risk *"I think" vs "I know"*	Decision making can be improved with better understanding of potential hazards, consequence and risk of any given scenario. The opportunity to improve perception of risk is to move from "I think" to "I know" by concentrating on the high consequence or high likelihood event. This will allow an MCP to reduce uncertainties, increase understanding and increase value in deliverables. 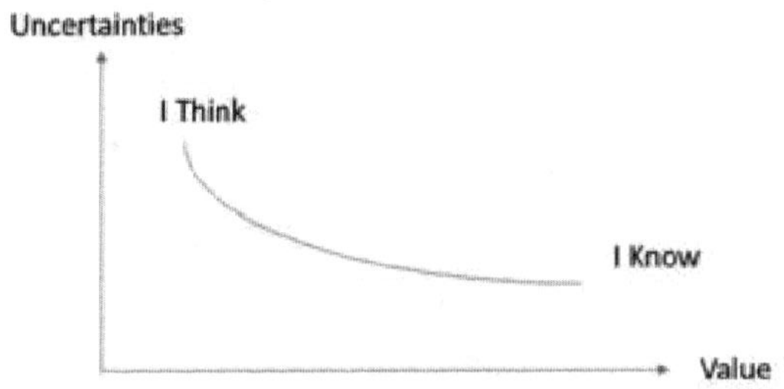	One of the ways to clarify the perception of risk is to ask leading questions to better understand the basis of the perception e.g.: - What is the concern? - What are the incidents that relate to this condition? - What is the root cause of the event? Once the perception is clarified, the hazards can be managed
6	Stakeholders management	The expectations of stakeholders can vary based on their operational experience and personal preferences. Potential conflicts can result in them advocating for different approaches for process safety implementation in design. Continuous engagement and work as one team to ensure consistent message, expectations and alignment with Project Leadership, the Operator, Co-owners, Contractors and others.	One example in Chevron MCPs is the use of Vision and Breakthrough session as the platform to increase communication, engage, understand and align expectations of stakeholders. Vision and Breakthrough session is carried out most effectively in early Phase 2 with the involvement from internal and external stakeholders. The objective of the session is to align the vision of success within the project team and identify three to five focus areas for the MCP to be successful. For each focus area, the project team will identify the potential challenges and formulate breakthrough strategies to effectively achieve the vision of the MCP. The use of Vision and Breakthrough session allows a more effective management and alignment of stakeholders from the early phase of MCP.

<table>
<tr>
<td>7</td>
<td>Failure to learn from incidents

"What happened there will not happen here, we operate differently"</td>
<td>Sharing of lessons learned is important at multiple levels: within an organization (from one business unit to another), within industry (from one organization to another) and across industries.

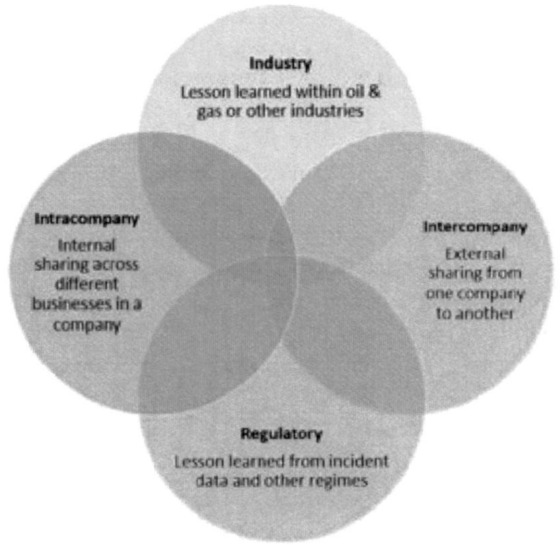
</td>
<td>Safety learnings or alerts can be included in design meetings to identify potential hazards and mitigation measures. An MCP lessons learned champion is important to ensure the relevant lessons learned are included in a lesson learned register to be tracked for implementation.

Most major engineering design companies provide services to different Operators and the same process design may be replicated in different facilities. As such, it is in the best interest of the industry to share the lessons learned.

A few good examples of industry sharing are:<ul><li>Bureau of Safety and Environmental Enforcement (BSEE) – Safety Alerts, Safety Bulletins and Incident Statistics</li><li>Center for Chemical Process Safety (CCPS) – Process Safety Beacon</li><li>Health and Safety Executive (HSE) – Safety Alerts and Notices</li><li>Occupational Safety and Health Administration (OSHA) – Hazard Alerts</li></ul>Internally within a company, there can be sharing across business units via:<ul><li>Safety Bulletins</li><li>Internal forums</li></ul>In addition, proper recording of incident investigations and root cause analysis in a facility is important for future projects to improve on their execution.</td>
</tr>
</table>

5 Conclusion

Process safety consideration is important in MCP and it can be challenging to incorporate process safety in MCP when the benefits of process safety to MCP is not well understood. An effective approach to integrate process safety in the design of an oil and gas facility is to incorporate process safety throughout the lifecycle of MCP. Some of the key takeaways of successful incorporation of process safety in MCPs are:

- Process safety awareness and addressing misconception on process safety
  Process safety awareness provides the premise for incorporating process safety in MCP and allows project team to better understand the concept and application of process safety. Process safety awareness can be introduced through project onboarding, safety moments and specific process safety training. Better understanding of process safety is the first important step to address misconception on process safety considerations and prevent process safety studies from being a tick the box exercise.

- Timing and scale of process safety studies in MCP
  Not all MCP have the same design and risk profile. A roadmap for process safety deliverables required for each phase of MCP can provide guidance to identify the appropriate timing, level of details and scale of study based on the complexity of the facility. For example, in Chevron, a preliminary HAZOP is carried out in Phase 3 to identify potential design improvements and the HAZOP is revalidated in Phase 4 to ensure previous recommendations are implemented and to identify any changes that may occur since Phase 3. The methodology for the HAZOP is the same but the timing, level of details and scale of the study changes as the MCP progresses.

- Tools and resources
  The aim to incorporate process safety in every phase of an MCP is to ensure that the facility is designed to safely operate throughout its design life. This can be achieved through identification of key process safety aspects in MCP and use of fit for purpose tools, applied by a competent practitioner. Different organization may apply different tools, models or methodologies in their process safety studies. The key is to ensure the quality of the process safety studies through consistent application of the tool by a competent practitioner and involving the right people (multi-disciplines) in the assessment.

- Communication and stakeholder management
  Application of process safety consideration in each MCP phase should not be viewed in silo, rather each process safety study is inter-related in the overall process safety management of MCP. Continuous process safety conversation and engagement is important to communicate the benefits of process safety considerations in design improvements and ensure all stakeholders maintain the momentum on process safety journey of the MCP.

- Execution
  Assessment of process safety implementation at the end of each MCP phase is useful to provide a feedback on the effectiveness of process safety incorporation in MCP

deliverables from end of Phase 1 to Phase 5. There may be value-added process safety studies that are beneficial to be carried out earlier in MCP and other studies that are not beneficial to the MCP. This MCP feedback loop should be an ongoing process to help identify improvements opportunities for better integration of process safety in future MCPs.

The reality is process safety incidents can only be avoided with deliberate effort during the design, execution and operation phases

"If you think process safety is expensive, try an accident." Process safety implementation does not have to be expensive if incorporated early in an MCP. When we know better, we design better and potentially have a simpler, safer and optimized facility throughout its design life.

6 References

[1] Oil & Gas UK. *Guidance on Risk Related Decision Making*, Issue 2, 2014

[2] CCPS, *Guidelines for Hazard Evaluation Procedures*, 3rd Edition, Center for Chemical Process Safety, American Institute of Chemical Engineers, New York, NY, 2008.

Embracing Complexity and Managing Risk:

PSM Challenges in Specialty Chemical Innovation

Lauren Moyer
Principal Chemical Engineer
Plant Protection
Tennessee Operations
Eastman
P. O. Box 1972
Kingsport, TN 37662
lmoyer@eastman.com

Prepared for Presentation at
American Institute of Chemical Engineers
2019 Spring Meeting and 15th Global Congress on Process Safety
New Orleans, LA
March 31 – April 3, 2019

Embracing Complexity and Managing Risk:

PSM Challenges in Specialty Chemical Innovation

Lauren Moyer
Principal Chemical Engineer
Plant Protection
Tennessee Operations
Eastman
P. O. Box 1972
Kingsport, TN 37662
lmoyer@eastman.com

Keywords: Hazard Identification and Risk Analysis, Process Design & Development, and Specialty Chemicals

Abstract

Success in the specialty chemical business is dependent upon bringing highly specialized products to market quickly, often to meet very specific customer needs. This paper will explore Eastman's recent shift to focus on innovation-driven growth as a specialty materials company and some lessons learned in managing risk in the development and scale-up process. In an ideal project plan, a stage-gate approach can ensure that safety and environmental risk are managed at the right level before a project is advanced to the next phase of development. When developing specialty products, the project phases can begin to overlap as product, process, and business case development may occur simultaneously in the effort to quickly deliver a chemical to fulfill a specific customer application. To navigate this complexity, a scale-up program must begin by managing safety risk at the earliest point of development, bench scale lab work. This involves looking beyond proper PPE and glassware safety and having an awareness of both intended and unintended chemistry and determining key chemical hazard properties to be used in future scale-up work. As the project is moved from bench scale to piloting and commercialization, wasted time and resources can be avoided by establishing "ground rules" for hazard identification and risk management. This could include site or facility specific restrictions such as area electrical classification or facility siting limitations, or more general design requirements for use of inherently safer design or hazard identification techniques in process development. Effective risk management can be tailored to allow for rapid process development through "right-sizing" of hazard identification and risk assessment techniques. Processes can be screened based on chemical properties, process conditions, or type of chemistry to determine an appropriate risk assessment methodology and the resources that should be involved. Working to improve in each of these areas allows development projects to function more smoothly and to safely and quickly bring new products to market.

1 Introduction

The specialty materials industry is built on bringing highly customized products into the market often to meet a very specific customer need. Success in this industry is accomplished through strong and well-organized innovation and scale-up capabilities, from the research lab through pilot, demonstration, and commercial scale facilities.

Frequently new product development is accomplished through some form of phased gate or stage-gate process. Each phase in the product development increases the business and/or technical knowledge until the project is completed and ready for full-scale commercialization [1]. Similarly, chemical process scale-up phases, such as discovery, lab-scale development, and pilot plant demonstration, serve to provide increasing amounts of technical information to facilitate a commercial scale plant design. Ideally overall project stage-gate and chemical process scale-up activities are linked together such that scale-up cannot occur without certain business and technical hurdles being met.

Theoretically, as a product is determined to be viable for a given application, and a process flowsheet has been proposed, the project will move into pilot scale testing. At the pilot scale, operations are often conducted around the clock by a team of operators, instead of a single chemist or technician. From a process safety perspective, the pilot plant operation is treated as a manufacturing unit instead of lab facility. The goal of the pilot scale is typically to demonstrate feasibility and to further refine operating parameters.

The timeline to launch new specialty products is typically very quick, often a race to deliver a novel idea to the marketplace. Also, project timelines may be compressed by customer-driven or market-cycle timelines to meet the customer's product launch dates or production cycles. Due to these demanding timelines, the phases of the scale-up process may become compressed with the temptation to move to pilot or demonstration scale before appropriate chemical or process information is fully developed.

An additional issue can be generation of material for customer sampling and end-use testing. While product development and feasibility testing are still occurring, there may be a need to generate a larger quantity of material than the bench-scale can provide. For example, certain specialty applications may require enough material for food/feed or drug testing or to create several full sets of tires for road testing. This may push a product into pilot scale equipment before the process to synthesize the molecule is fully refined.

As new molecules are invented, or new pathways are developed to synthesize existing chemicals, process safety risk must be managed from the initial product development through commercialization. In "Integrating Process Safety and Innovation", Lodal and Mize recommend techniques to consider potential risk at a commercial scale early in process design [2]. Beginning with the end in mind is crucial to address commercial scale process safety issues, but the project must also not lose sight of potential process safety risk during the scale-up process. Being proactive and efficient early in the product development effort with respect to mitigating chemical hazards, will save critical time and resources in the race to deliver innovative products to customers.

2 Process Safety Risk at the Bench Scale

Development of a specialty product begins at the bench scale with developmental chemistry work. New products are created and screened for customer applications, and new reaction routes for products are explored for viability. Often the focus of this work is fitness-for-use and reaction selectivity and yields. Process safety considerations often focus on protecting the chemist or technician from chemical exposure or glassware hazards in the lab environment. Applying an increased focus on chemical and reactivity hazards and their implications at larger scales while conducting bench-scale work will save significant time and resources in the process safety design activities as the project development progresses.

Understanding the properties of the newly developed molecule and any raw materials, intermediates, and byproducts is important for both the management of risk through the scale-up effort and the overall economic viability of a project. This includes determining basic physical properties, chemical hazards such as flammability and toxicity, and the intended and unintended chemistry that might occur with these chemicals. The rapid development of specialty products necessitates a certain level of in-house chemical property testing capability to quickly screen components for potential hazards.

By understanding and recognizing the hazards of the proposed chemistry early in the effort, it is possible to employ inherently safer design principles early in process development. However, this can only be done if the chemists and engineers involved have been trained in hazard recognition and devote the necessary time to evaluating chemical hazards at the bench scale.

2.1 Example: Energetic Product

A product under development exhibits strongly unstable and energetic behavior based on results of initial differential scanning calorimetry (DSC) testing. Based on the proposed temperatures and pressures for the process, the unstable behavior is thought to be credible during normal operation and would likely make shipment of the product difficult due to Department of Transportation (DOT) requirements. However, thermal stability testing was not performed on the product until the process was proposed for scale up in the pilot facility. Significant project time was lost while further thermal stability testing was conducted to confirm the instability, and subsequent development work was conducted to substitute the molecule to one with improved stability. If the energetic behavior of the product and its implications on process risk had been understood earlier in the product development phase of the project, then the lab work could have been refocused on more viable molecules that provided acceptable performance without the added cost of mitigating the instability risk.

2.2 Example: Understanding Reactivity Testing

A new project was brought to the pilot plant facility from the research lab. Engineers in the pilot plant worked with the chemists to prepare the process for piloting, including equipment design and installation and development of operating procedures. As they prepared for start-up, the required hazard assessment was conducted by pilot plant personnel and company process safety professionals. The review revealed a significant concern with thermal stability that had not been understood by the team working on the project. The company reactive subject matter expert provided guidance on how to address the concerns, resulting in the need for additional safety systems. The modifications delayed the piloting work by over a month.

For chemists or engineers who are inexperienced in interpreting thermal stability testing results, confusion can exist on the criteria for follow-up testing or additional safety controls. This can lead to delays in project development work as further consultation is required to determine the next steps. Guidelines for reactive chemistry testing and evaluation at the bench scale and the pilot scale should be established and documented. A qualitative reactivity screening program can be found in Chapter 3 of Essential Practices for Managing Chemical Reactivity Hazards [3]. Additionally, guidance for interpreting the results of thermal stability testing such as DSC, accelerating rate calorimetry (ARC), or other methods can be helpful for engineers and chemists involved in the scale-up effort. This guidance can be based on in-house process safety knowledge and experience or industry literature [4] and expertise. Once guidelines are established, screening tools can be built to quickly assess the hazard and specify requirements for additional testing or controls. See Figure 1 for a proposed spreadsheet tool for thermal stability hazards.

Figure 1: Dashboard from Thermal Stability Screening Spreadsheet Tool

	Chemical Name(s)	Product 1	Reagent 1	Intermediate 1
	DSC Onset T C	179	246	276
	Energy J/g	1739	477	49
	Energy Cal / g	415	114	12
Does your compound have	C-C unsaturation	Yes	No	Yes
Any of these groups?	C-metal N-metal	No	No	No
See Reference Tab	Contiguous nitrogen atoms	No	No	No
Check example compounds	Contiguous oxygen atoms	No	No	No
	Nitrogen - oxygen	No	No	No
	Nitrogen - halogen	No	No	No
	Oxygen -halogen	No	No	No
	ARC Onset T C	105.7		
	Energy J/g	429		
Reactive Chemical Program	ARC Isoage T C			
Company Reactivity Program Guidelines	DSC TLP Process Solution	70.3	117.20	138.2
TLP = Temperature of low probability	DSC TLP Neat Material	50.3	97.20	118.2
This applies to both process	ARC TLP	85.7		
& utility temperatures.	ARC Isoage TLP			
	Here are your action items:	Reactive Chemicals Program applies, get safety help	Reactive Chemicals Program applies, get safety help	
		TLP guidelines are calced above	TLP guidelines are calced above	
		Run ARC		
		Evaluate for UN/DOT Class 1 testing review		

3 Ground Rules for Scale-Up

Another mechanism for improving efficiency in the scale-up process is the use of ground rules for the scale-up process or the facility chosen for scale-up. The concept of process safety ground rules or cardinal rules was introduced in a paper entitled "Reducing Process Safety Events: An Approach Proven by Sustainable Results." These are intended to be simple, easy to remember rules that are always followed with the goal of preventing process safety incidents in a facility. [5]

Introducing rules or boundaries to an organization tasked with innovating and developing new product may seem counterintuitive. How can true innovation occur in a world with limitations? The key is that this innovation will have to be scaled up in a real, functioning facility where process safety risk has cost in time and money. Developing ground rules for a given facility or for the overall scale-up process can guide project decisions in a way that prevents rework and redesign of the process, leading to a more efficient use of innovation resources and getting products to market faster.

3.1 Facility Limitations

If a pilot plant or demonstration facility has a physical limitation such as the electrical classification or a facility siting limitation on the use of certain highly toxic or high odor chemicals, this could be applied as a ground rule. If a project is being developed with a scale-up plan for a specific facility, knowledge of these limitations should guide the development of the chemical process. Modifying facilities to overcome such limitations could be very expensive, impractical, or time consuming, thus jeopardizing the success of the new product. The chemists or scale-up engineers should not wait until the safety review at the pilot scale to evaluate alternative raw materials or chemistry that would be more feasible in the desired facilities.

3.2 Process Safety Information

To prevent safety incidents as the product is scaled up, it is critical that sufficient information be available to make risk decisions at each point in the project. Many pilot scale facilities may fall below threshold quantities of chemicals for process safety management (PSM) regulations, which specify required process safety information (PSI) elements. While all elements of PSI required for a full-scale plant may not be necessary for each transient scale-up operation, at a minimum, chemical hazard information, reactivity information, and a proposed procedure should be available for each process. A minimum required PSI could be applied as a ground rule for the scale-up process no matter where the project plan for a given new product may lead.

3.3 Inherently Safer Design - Equipment

Inherently safer design (ISD) principles can be applied to the design, selection, and care of pilot plant equipment. Intentionally purchasing equipment with robust pressure and temperature ratings and materials of construction can allow much greater flexibility in the scale-up of a variety of specialty chemical products. Additionally, certain investments in automation or other equipment can reduce the risk of operator exposure to process chemicals or other hazards. Ground rules could be applied to the design processes for new or modified pilot or demonstration scale equipment. It should be noted that equipment and systems must be maintained in a way that preserves the original rating and design throughout its life cycle. Pilot plant systems should be included in a comprehensive asset integrity program to ensure adequate performance throughout the equipment life cycle.

3.4 Inherently Safer Design – Scale-up Process

The early product and process development is precisely when the concepts of ISD should be applied to maximize its benefits and prevent either the need for rework in the process development later or expensive mitigation techniques in full-scale process engineering. Specifically, the Substitute, Moderate, and Simplify options can be explored as a part of the early lab work to find design options that reduce overall process safety risk and the subsequent costs to mitigate such risk as the product is brought to market [6]. Figure 2 contains examples of the types of questions that should be asked early in the development process. The scale-up process could be set-up to require the evaluation of a proposed process with an ISD checklist to identify and implement potential improvements in chemistry or process conditions prior to proceeding with scale-up beyond the lab scale.

Figure 2: Applying Inherently Safer Design in early process development

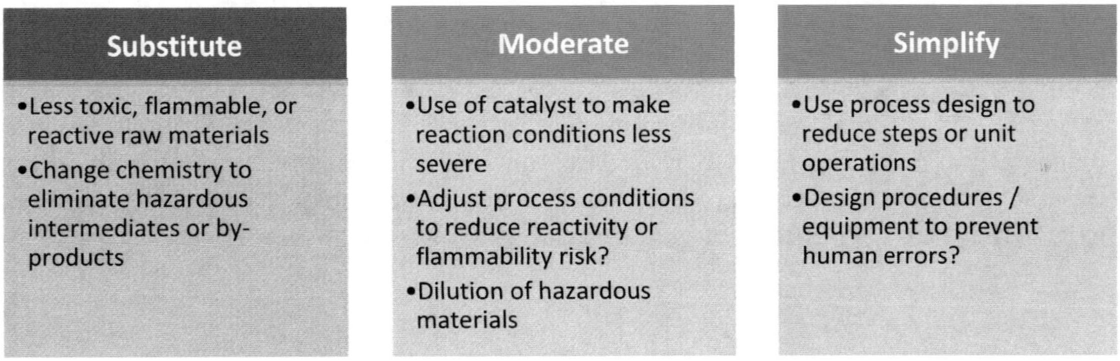

4 Risk Assessment Techniques

As mentioned previously, processes at a pilot or demonstration scale may not meet threshold quantities of hazardous chemicals to be subject to regulatory requirements for a process hazard analysis. However, hazard identification and risk assessment are critically important at each iteration of the scale-up process. As equipment size and chemical amounts are increased and operations are transitioned from laboratories to 24-7 operations, the potential consequence of the hazard may also increase, and new hazards may present themselves.

The hazard identification and risk assessment processes within the innovation process must be flexible enough to allow process safety professionals to tailor the hazard evaluation technique to the needs of the process, based upon some initial screening or hazard criteria. This can allow for speed and agility in the scale-up effort when appropriate or provide for a pause to consider potentially high-risk processes more carefully. "A successful hazard evaluation can be defined as one in which (1) the need for risk information has been met, (2) the results are of high quality and are easy for decision makers to use, and (3) the study has been performed with the minimum resources needed to get the job done." [7]

For a given process entering pilot scale testing, three different types of risk assessment could be considered.

4.1 Peer Review

For processes involving no intentional chemistry, minimal risk of unintended chemistry, and low flammability or toxicity hazards, a peer review among project engineers or chemists, pilot plant engineering and operations staff, and area management may be sufficient to identify potential hazards and determine appropriate mitigations. This type of review may have a defined checklist for review, and consider chemical handling, flammability, reactivity, process excursions, ergonomics, and perhaps even environmental issues. Figure 3 shows the suggested range of questions for this type of review.

4.2 Peer-Management Review

For processes involving well-understood chemistry, minimal risk of unintended chemistry, and medium-level flammability and toxicity hazards, a higher level of department safety review can be used. This review could have a more extensive list of required attendees, such as management personnel responsible for risk decisions in the facility, process safety expertise, industrial hygiene, and/or engineering representatives, in addition to the pilot plant staff involved in a Peer Review. The same checklist may be used for both a Peer Review or a Peer-Management Review, however more extensive discussion would be expected due to the higher level of hazards in these processes.

Figure 3: Peer Review Checklist Example Questions

Product Safety, Toxicology, Industrial Hygiene	•Hazard information available to personnel? •Any special handling requirements? •Recommended PPE?
Chemical Specific Hazards	•Respiratory hazards? Skin / eye hazards? •Any known sensitizers? •Carcinogen or reproductive hazards?
Fire and Explosion Hazards	•Required electrical classification? •Unusual flammability/explosivity hazards? •Dust explosion potential?
Reactivity Hazards	•Expected heat of reaction? •Thermal stability issues? •Potential chemical incompatibility? (In-process or nearby?)
Process Specific Hazards	•Impact of process variable upsets? Temperature, pressure, flow, level? •Impact of order of operations or duration of steps?
Equipment Specific Hazards	•Impact of utility losses or loss of agitation? •Need for interlocks? Interlocks tested? •Pressure relief design, fire protection adequate?
Procedures and Training	•Procedures developed and approved? •Safe operating limits documented? •Need for formalized training?
Asset Integrity	•Equipment inspections completed? •Interlock testing completed? •Additional inspection/testing required?
Environmental	•All waste streams documented and approved? •Any low reportable quantity chemicals? •Any safety concerns with waste disposal paths?
Miscellaneous Concerns	•Prior events or near misses in equipment / process? •Any potential significant human errors? •Need for higher level or additional safety reviews?

4.3 Process Hazard Analysis

At times, a new process will be developed with significant potential flammability, toxicity, or reactivity hazards. For these, even at a small scale, a full process hazard analysis (PHA) using a HAZOP, What-If, or other applicable hazard analysis methodologies may be warranted. These studies should be led by a trained PHA facilitator using an approved PHA methodology and team make-up. The resource investment for this type of study is much higher due to the time commitment involved as well as the higher level of process safety information needed by the team. Additionally, the process safety organization supporting the research and development groups must be flexible enough to support the very quick timelines often required in specialty product development.

A documented process should be developed by the research and pilot plant organizations for selecting a risk assessment technique and should specify required team make-up for each technique and requirements for minimum available PSI. The risk assessment process should specify the qualified technical, management, or process safety resources responsible for determining the appropriate technique based on knowledge of the proposed process design and procedure. The procedure should also specify any predefined checklist to be used in the Peer Review or Peer-Management Reviews. One important question should always be asked at the end of this checklist, "Are additional safety reviews needed for this process?" This allows the team to consider if a higher level of review or management approval is needed based on the review discussion or if a different type of review (e.g. personal protective equipment assessment or task safety review) is needed to effectively mitigate the risks associated with the proposed process.

Figure 4: Examples of Process Safety Ground Rules for Scale-Up Processes

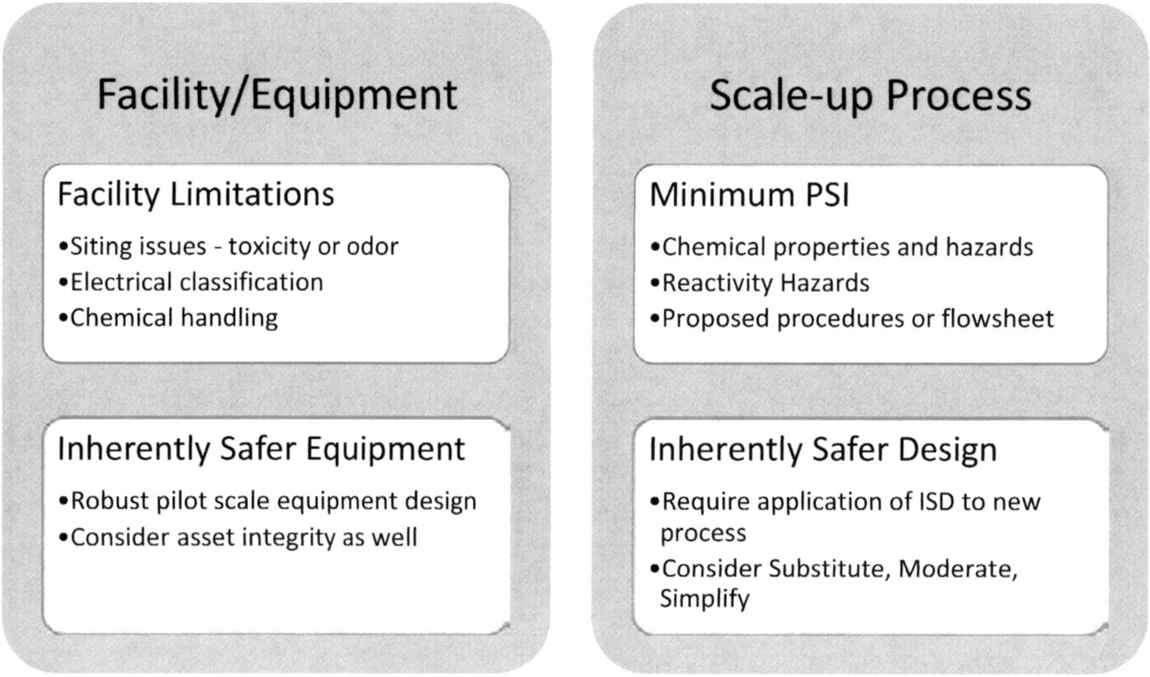

5 Conclusion

Becoming a leader in specialty materials innovation requires not only talented chemists and engineers and customer-focused marketing, but also a thoughtful approach to managing the process safety risk associated with new chemical products from the bench scale through commercialization. Novel specialty products and the processes to synthesize them must be delivered to potential customers quickly to outpace competitors and meet aggressive customer launch deadlines. To prevent process safety events during scale-up and unexpected project delays or costs, the scale-up process must be designed to consider process safety risk early and appropriately throughout the development of the product. Building process safety hazard recognition into the early bench scale work processes can prevent process safety incidents at the small scale and facilitate process and chemistry improvements to reduce risk as the project proceeds towards commercial scale. Education about hazards such as reactivity and thermal stability and the development of screening tools or processes can help to catch such issues early. Developing process safety ground rules for both the facilities used for scale-up and the scale-up process can reduce rework in projects by providing a framework for acceptable process safety risk in the piloting effort. A summary of ideas for process scale-up ground rules is given above in Figure 4. Finally, allowing flexibility in hazard evaluation techniques will provide the agility required for specialty chemical development while still managing process safety risk at a level appropriate for the process hazards. Improvements in each of these areas allows development projects to proceed smoothly and to safely and quickly bring new products to market.

6 References

[1] S. J. Edgett, "The Stage-Gate Model: An Overview", available at https://www.stage-gate.com/new-product-development-process/, accessed on January 13, 2019.

[2] P. N. Lodal and J. F. Mize, "Integrating Process Safety and Innovation", Chemical Engineering Progress, October 2016, 36-40.

[3] R. W. Johnson, S. W. Rudy, and S. D. Unwin, Essential Practices for Managing Chemical Reactivity Hazards, Center for Chemical Process Safety, American Institute of Chemical Engineers, New York, NY, 2003.

[4] M. S. Mannan, W. J. Rogers, and A. Aldeeb, "A Systematic Approach to Reactive Chemicals Analysis", Institution of Chemical Engineers Symposium Series, Volume 148 (2001), 41-58.

[5] J. Champion, S. Van Geffen, and L. Borrousch, "Reducing Process Safety Events: An Approach Proven by Sustainable Results", Process Safety Progress 36 (2017), 323-337.

[6] D. C. Hendershot, "Inherently Safer Design: The Fundamentals", Chemical Engineering Progress, January 2012, 40-42.

[7] Center for Chemical Process Safety, Guidelines for Hazard Evaluation Procedures (3rd edition), American Institute of Chemical Engineers, John Wiley and Sons, Hoboken, NJ, 2008.

Guidelines for Integrating Process Safety into Engineering Projects

Michael P. Broadribb
Baker Engineering and Risk Consultants, Inc.
3330 Oakwell Court #100, San Antonio, TX 78218
mbroadribb@bakerrisk.com

Prepared for Presentation at
American Institute of Chemical Engineers
2019 Spring Meeting and 15th Global Congress on Process Safety
New Orleans, LA
March 31 – April 3, 2019

Guidelines for Integrating Process Safety into Engineering Projects

Michael P. Broadribb
Baker Engineering and Risk Consultants, Inc.
3330 Oakwell Court #100, San Antonio, TX 78218
mbroadribb@bakerrisk.com

Keywords: engineering, project, process safety

Abstract

This paper describes a new guideline book that CCPS has published on integrating process safety into engineering projects. Engineering projects in the process industry are temporary endeavors generally undertaken to create a new or modified facility, and vary widely in scope and size ranging from large capital projects to small management of change (MOC) works.

The main focus of the guidelines is on *proactively* implementing process safety activities at the optimum timeframe, while also *reactively* conducting "cold eyes" reviews to provide assurance that nothing significant has been missed. This approach ensures that, if the right process safety activities are conducted at the right time, project leadership will have the right (process safety) information in order to be able to make the right risk management decisions regarding safety.

Process safety in engineering projects involves leadership, managers, engineers, operating and maintenance personnel, contractors, vendors, suppliers, and support staff. Therefore, these guidelines were prepared for a wide audience.

1 Introduction

CCPS has published a new book titled *Guidelines for Integrating Process Safety into Engineering Projects.* While there is much industry guidance on implementing engineering projects and a similar amount of guidance on Process Safety Management (PSM), this book fills a gap by addressing the transfer of key deliverables from Engineering to Operations, where PSM is implemented. The book provides the engineering and process safety deliverables for each stage of the project life cycle from concept appraisal through to decommissioning at end of life. Impacts to the project budget, timeline, and the safety and operability of the delivered equipment are also discussed.

Engineering projects in the process industry are temporary endeavors generally undertaken to create a new or modified facility. Historically, project managers have focused on managing the risks and performance indicators related to costs, schedules and technological risks, such as whether the facility can meet production and quality targets. From a project manager's perspective, safety concerns were often focused primarily on safety during the construction stage and the occupational safety of a contractor's workforce. Increasingly, however, major operating companies recognize the need to more comprehensively address process safety in their projects as a means of optimizing the residual risk that operations teams are required to manage for the life of the facilities.

Because engineering projects vary so widely in scope and size, these guidelines discuss effective approaches for integrating process safety into both large and small projects, including management of change (MOC) works. Projects also cover a broad range of facilities as illustrated in Table 1. Only variations in project scope determine whether certain process safety activities are applicable for a specific project.

Table 1. Types of Projects Covered by these Guidelines

Types of Projects
Greenfield and Brownfield
Onshore and Offshore
Continuous and Batch Operations
Indoors and Outdoors
Modifications (covered by MOC)
Modular and Stick-built
Pilot Plants and Full-scale Process Units
Chemical Complexes and Refineries
Fixed and Semi-Submersible Production Platforms
Drilling Rigs and MODUs
Debottlenecking
Control Systems (DCS, SCADA, SIS, HIPS, etc.)
Tankage and Storage
Utility Systems (Electrical Power, Fuel Gas, Cooling Water, Nitrogen, Compressed Air, etc.)
Buildings (Control Rooms, Offices, Workshops, Warehouses, etc.)
Loading and Offloading Systems (Road, Rail, Marine)
Pipelines (Cross-Country, Intra-Plant, Subsea)
Other Infrastructure

The intent of these guidelines is not to describe in detail ***how*** to perform specific process safety activities, but rather to identify ***what*** needs to be addressed at each stage of a project. Other CCPS publications and industry codes and standards describe methods for specific process safety activities and are extensively referenced throughout the book.

No matter how good the process safety input is into any engineering project, the newly installed and commissioned facility has a residual risk that the Operations team must manage through an effective process safety management system for the life of the facility. This is true for all projects. Therefore, one of the main benefits of successfully integrating process safety into a project is to help reduce this residual risk. The guidelines discuss the use of an inherently safer design (ISD) approach to meet this goal that should place fewer demands on Operations personnel, while also limiting the potential for major incidents.

Process safety in engineering projects involves leadership, managers, engineers, operating and maintenance personnel, contractors, licensors, vendors, suppliers, and support staff. Therefore, these guidelines have been prepared for a wide audience, who it is anticipated will gain understanding and enhance their projects by applying the guidance to meet a range of pertinent process safety challenges.

2 Project Management Concepts

While it is not the intent to present an in-depth guide to project management, the book discusses some general project concepts, project management, and underlying principles of structure and execution in Chapter 2. It also introduces terminology that is common to most projects, so that process safety engineers can effectively communicate with project personnel, ensuring that process safety issues are fully considered.

This chapter of the book covers an introduction to types of projects, key documents (e.g., Statement of Requirements, Basis of Design), project governance, project controls, typical project organizations, contracting strategies for implementation, and risk management. It should be recognized that project risk is different to process safety risk, and focuses on any set of circumstances that have a positive or negative effect on project value, such as cost, schedule and technology.

3 Project Life Cycle

The book devotes a chapter to the process safety studies and activities that may be relevant for each stage of a project. Figure 1 illustrates the typical stages of a capital project. These stages may be combined for smaller projects, although the overall range of process safety studies and activities may be appropriate depending upon the scope of the project.

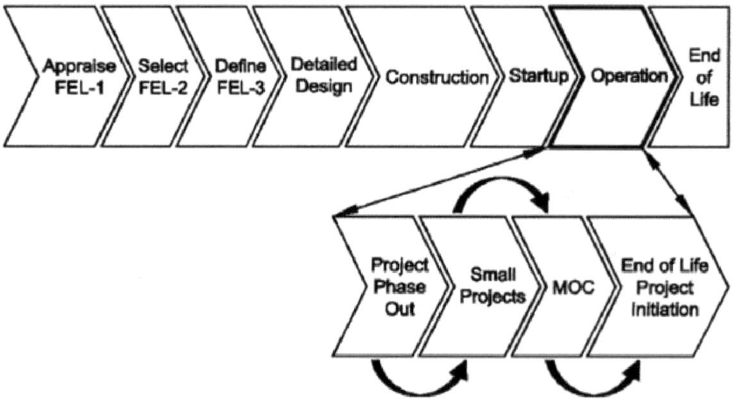

Figure 3. Typical Stages of a Capital Project
(FEL = Front End Loading)

The objectives of each stage from a business and project management perspective are as follows:

Appraise (FEL-1)

A broad range of development options is identified, and the commercial viability of the project is evaluated. A technical and commercially viable case plus alternatives should be identified for the project to proceed.

Select (FEL-2)

The alternative concept options are evaluated seeking to identify the optimum project by maximizing opportunities, while reducing threats and uncertainties to an acceptable level. Upon completion of technical and commercial studies, a single concept is selected.

Define (FEL Error! Bookmark not defined.**-3)**

The technical definition and execution plan for the project are improved to confirm the conceptual design, cost and schedule. A basic design is developed with plot plan, preliminary process flow diagrams, material and energy balances, and equipment data sheets. Timing varies between companies/projects, but sanction for financial investment usually occurs at the end of this stage, if sufficient confidence in the project is achieved.

Detailed Design

Detailed engineering of the defined scope from the front end loading (FEL) process is completed, scope changes managed, and materials and equipment procured.

ConstructionError! Bookmark not defined.

Fabrication, construction, installation, quality management, and pre-commissioning activities are completed. Operational readiness activities are performed in preparation for commissioning, startup and operation.

Commissioning and Startup

The project is commissioned, and the facility and documentation handed over to the operations team for normal operation.

Operation

Test runs may be required to confirm that performance specifications are met before the project is closed. The project team may conduct a lessons learned review to aid future projects. At this point the facility is handed over completely to the client Operations team, the project team phases out, and the project is closed.

End of Life Error! Bookmark not defined.

When a business decision is taken to cease operations, the facility is de-commissioned. Depending upon local circumstances and regulations, the facility may be dismantled, disposed and/or demolished, or modified for future use. End of facility life typically involves a new project.

Each stage of a project has specific process safety activities in support of the overall project management objectives. These process safety activities are summarized below.

3.1 Appraise (FEL-1)

Chapter 3 covers the feasibility of proceeding with a new project to produce a specific facility(s) in a certain location, employing various process technologies. This initial phase of Front End Loading (FEL-1) involves preliminary Hazard Identification and Risk Analysis (HIRA) of multiple development options, from which a range of viable options are identified.

The best opportunity to make a positive impact on the life-cycle of a major capital project is during the early conceptual and planning stages before capital outlay occurs. FEL-1 involves developing sufficient strategic information on multiple development options with the highest potential of meeting business objectives. From a process safety perspective, this involves understanding the hazards, risks and uncertainties of each option and location when ISD principles are applied. This understanding represents the foundation for further development and selection of a preferred option in FEL-2.

3.2 Select (FEL-2)

Chapter 4 deals with the next phase of FEL (FEL-2) where the various development options are reduced through a concept selection process involving more detailed HIRA, including offsite major accident risk. The site, process technology, facilities, and infrastructure requirements are determined using an ISD approach, and preliminary environmental, health and safety (EHS) and Process Safety plans developed.

As previously described, the best opportunity to make a positive impact on the life cycle of a major capital project is during the early conceptual stages. FEL-2 continues evaluation of the hazards, risks and uncertainties of each development option started in FEL-1 to the point that a preferred option can be selected. This option is then further developed applying ISD principles to prepare a Basic Engineering Package (BEP) containing preliminary process safety information (PSI) on materials, technology and equipment. This BEP represents the foundation for further development in FEL-3 and for the preparation of a Front End Engineering Design (FEED) package that can be given to an engineering contractor to complete the detailed engineering.

3.3 Define (FEL-3)

Chapter 5 addresses the final phase of FEL (FEL-3) during which the technical scope of a single development option is defined. Increasingly more detailed HIRA studies are used to determine the site layout, spacing, grading and other siting concerns as a result of potential fires, explosions and toxic releases. The FEED, including assumptions, philosophies, and engineering codes and standards, is completed, as well as the detailed EHS and Process Safety plans.

Once a single development option has been selected FEL-3 improves the technical definition and project execution plan. This involves the preparation of the FEED package that can be given to in-house engineers or an engineering contractor to complete the detailed engineering. This package includes refined design hazard management (DHM) with reduced risk achieved through understanding major hazards, and optimizing inherently safer design, functional safety, and other safety measures. Typical steps in DHM are illustrated in Table 2. The ultimate goal is to have confidence in the design, cost and schedule, thereby confirming the business case and receiving financial sanction from the client.

Table 1. Typical Steps in a Design Hazard Management Process

Project Stage	Step
FEL-1	• Identify major accident hazards in each development option
FEL-2	• HIRA • Optimize layout and design of facilities • Establish design safety concepts/critical design measures (including functional safety)
FEL-3	• Continue ISD optimization • Refine design safety concepts/critical design measures (including functional safety) • Set performance standards • Re-evaluate major accident risk • Finalize important design safety decisions • Finalize Basis of Design (BOD)

3.4 Detailed Engineering

Chapter 6 covers Detailed Design which is the first stage of project execution and involves completion of detailed engineering of the defined scope (FEED package) from FEL. This addresses layout and detailed engineering of individual items of equipment. Change management is introduced following the final HIRA study, and process safety information is documented and compiled. The procurement of long-lead items of equipment is also covered.

Engineering of inherently safer design, functional safety and other process safety and technical issues requires further development before the client is likely to make a final investment decision. Nevertheless, detailed design represents a significant financial commitment.

Other important project activities include management of any scope changes, planning for procurement of materials and equipment, planning for construction and commissioning, and interface management between multiple organizations including contractors, fabricators, vendors and suppliers. The ultimate goal is to resolve remaining project risks and uncertainties and complete a design package that includes all necessary information required for construction in order to stay on schedule and facilitate efficient commissioning and handover.

3.5 Construction

Chapter 7 addresses the construction phase of the project with the objective of safely building the facility in accordance with the design. This includes construction plans and management, procurement of equipment and materials, fabrication, installation, and management of engineering and integrity baseline documentation. It also involves multiple interfaces with fabricators, contractors, sub-contractors, suppliers, and vendors that require good management and regular performance reporting.

A number of process safety activities are essential to the success of this stage, including risk management, safe work practices, asset integrity management, management of change, and training/competence assurance. Achievement of these and other activities is necessary for a thorough mechanical completion, so that the project may safely proceed to commissioning and startup.

3.6 Commissioning and Startup

Chapter 9 deals with two main steps: commissioning and startup, commencing with commissioning to shakedown, check-out and resolve problems, and hand-over to Operations before proceeding with startup. The project's primary objective is turning a collection of process equipment into an operational facility meeting the client's requirements. To do so requires various process safety activities including an operational readiness review prior to commissioning to verify that the facility meets the design intent, hazards and risks are understood and actions completed, personnel are trained and competent, adequate procedures (EHS, process safety, operating, asset integrity) are in place, and the future Operator is prepared for ongoing operations. In this latter regard, successful projects have clearly defined roles and responsibilities between Project and Operator, and involve the Operator throughout, especially during Startup stage activities.

3.7 Operation

Chapter 10 addresses post-project operation, when the facility is running with acceptable product quality. Once the facility is handed over and accepted by the Operator, it is ready for commercial operation. The project has been closed out and the facility, data, and documents have been handed over to Operations.

The Operator's primary objective is achieving safe and reliable operation at the design production rate, for which a business management system comprising all the elements of *Risk Based Process Safety* is essential. Robust process safety and EHS procedures should be in place and properly implemented. Some engineering and technical support from the Project may be appropriate for early operation. Thereafter, the responsibility is on the Operator who should periodically perform various process safety and technical studies to maintain safe, efficient operation and asset integrity. Some technical, safety and debottlenecking projects may be performed periodically throughout the operational phase to ensure performance specifications are met, maximize return to shareholders, and protect license to operate.

3.8 End of Life

Chapter 11 covers decommissioning, abandonment, demolition/dismantling and other end-of-useful-life issues from a process safety perspective.

Decommissioning a process facility, structure or equipment can be a complex process, and demands a structured approach. In fact, it should be treated as a project in its own right. There are many hazards and opportunities for process safety incidents and regulatory non-compliance. To complete a decommissioning project without incident requires the right culture and commitment of the site, project and contractor's workforce, detailed planning, rigorous hazard identification and risk analysis, competent contractors, and disciplined implementation of detailed deconstruction and demolition procedures and practices. Each facility to be demolished may have unique characteristics that require specific procedures that are likely to be much more complex than simply 'knocking it down'.

Two other chapters discuss quality management and documentation in detail.

3.9 Quality Management

Chapter 8 covers quality management activities to ensure that the new facilities are designed, procured, fabricated and installed according to the technical specifications.

Although generally safety is viewed as a 'value' and 'priority', and quality as a 'standard', there is an interdependency between quality management and process safety. For safe, reliable and operable facilities, it is necessary for project teams to implement **both** quality and process safety at a high level. In constructing new plants and equipment, it is important that equipment as it is designed and fabricated is suitable for its process application. Appropriate quality checks and inspections must be performed to ensure that equipment is installed properly and is consistent with design specifications and manufacturer's instructions. Later in the lifecycle, equipment deficiencies outside acceptable process safety limits must be identified and corrected, and maintenance materials and spare parts must be suitable for their process application.

3.10 Documentation

Chapter 12 reviews the essential design files and process safety information (PSI) that must be compiled by the project team for hand-over to Operations. Information is the life blood of projects, which use and generate large quantities of documentation. A significant proportion of these documents are process safety information that is critical to the design and residual risk of the completed project.

In liaison with the future Operator, Project should define *as early as possible* the documentation to be retained, and that to be handed over to the Operator. When seeking tenders, contracts should define the information, including formatting and coding, to be produced by contractors. Above all else, the timing of availability of certain documents is critical for risk management decision-making and meeting the project schedule. An appendix provides an extensive list of documentation that may be appropriate at each stage of the project life cycle.

4 Process Safety Plan

The main focus of the guidelines is on proactively implementing process safety activities at the optimum timeframe, while also reactively conducting "cold eyes" reviews to provide assurance that nothing significant has been missed. This approach ensures that, if the right process safety activities are conducted at the right time, project leadership will have the right (process safety) information in order to be able to make the right risk management decisions regarding safety. It is therefore essential that projects have a *living* Process Safety Plan that evolves as the project scope and design develops.

The Process Safety Plan should describe all the process safety activities and their timing necessary to deliver a safe and reliable operating facility. For a large project these activities comprise a plethora of studies, assessments, competencies and training, documentation, reviews and inspections covering all the elements of CCPS Risk Based Process Safety, plus Inherently Safer Design and Design Hazard Management. Some companies combine the process safety plan with the Project EHS Plan, while others have separate plans.

The generic content of a typical project process safety plan is described in an appendix to the book, and is included in Table 2 below. The list is not meant to be all inclusive nor to imply that every item should be included for every project.

Table 2. Generic Content of a Typical Project Process Safety Plan

Content
Project description
Key project milestones and timing of key process safety activities
Roles and Responsibilities for managing process safety activities
Communication, meetings, etc. for integrating process safety into overall project plans, and for promoting safety
Process Safety Culture (activities to promote positive culture)
Compliance with Standards (regulations, engineering codes/standards, variances, etc.)
Workforce Involvement (activities to involve the workforce (i.e. Project, contractors, Operator, etc.) in process safety and EHS)
Stakeholder Outreach (regulator, emergency services, local community, non-governmental organizations (NGO's), etc.)
Process Knowledge Management (chemicals/materials, process technology, process equipment, etc.)
Hazard Identification and Risk Analysis (HIRA) o Hazard identification (HAZID), hazard and operability analysis (HAZOP), What If, etc. o Consequence analysis (fire, explosion, toxicity, reactivity, equipment vulnerability, etc.) o Risk matrix, layer of protection analysis (LOPA), quantitative risk analysis (QRA), o Facility Siting (location, occupied buildings, off-site impacts, etc.) o Human Factors (ergonomics, human performance, human error, etc.)
Inherently Safer Design (ISD) and Design Hazard Management (DHM) o Design strategies and philosophies, o Technical studies (process safety input on siting and layout, corrosion, geotechnical, resistivity, reliability, electrical, constructability, etc.), o Risk reduction measures (prevent/detect/control/mitigate devices), o Performance standards for risk reduction measures.
Operating Procedures (normal operation, startup, shutdown, operating limits, etc.)
Conduct of Operations (standing instructions, routines, shift handover, inhibit/override/bypass, etc.)
Safe Work Practices (for office, fabrication, installation/construction, commissioning, startup, operation, decommissioning, simultaneous operations (SIMOPS), etc.)
Asset Integrity (inspection, testing and preventive maintenance (ITPM) tasks, safety critical equipment (SCE), reliability/availability, quality management, condition monitoring, corrosion protection, etc.)
Process Safety Competency and Training & Performance Assurance (process safety & EHS, task requirements, assessment, training, subject matter experts (SME's), etc.)
Management of Change (hazard/technical reviews, approval, design change notice (DCN), late design changes, organization changes, etc.)
Contractor Management (pre-qualification, evaluation, oversight, interface management, etc.)
Operational Readiness (pre-startup reviews, go/no go decision-making, etc.)
Auditing (stage gate reviews, contractor performance, work permits, etc.)
Metrics and Measurement (injury/illness, spills, releases, etc.)
Emergency Plan (procedures, alarms, crisis management, evacuation, shelter-in-place, drills, etc.)
Accident/Incident Reporting and Investigation (injury, loss of containment, property damage, near-miss, high potential, etc.)
Management Review & Continuous Improvement (regular performance reviews by Project and/or Operator management)
Action tracking and resolution (all studies, reviews, investigations, etc.)
Risk Register
Documentation (for all process safety elements, archive and field management, obsolete/superseded documents, handover package, etc.)

5 Stage Gate Reviews[1]

Many operating companies within the process industries conduct reviews at key milestones during the life cycle of capital projects. These reviews are variously known as stage gate reviews, 'cold eyes' reviews, peer reviews, project EHS reviews, etc., and are normally conducted by an independent and experienced multi-discipline team familiar with the relevant facility/process and technology.

The objectives, scope and extent of the reviews vary between companies, but typically focus on technical and EHS issues, relevant to the stage of the project development, with a strong emphasis on process safety. One of the primary goals is to assess whether the project management team (PMT) has adequately identified and evaluated the hazards and their associated risks inherent in the project, and developed (or is capable of developing) plans to effectively manage those risks. The Project's process safety and EHS plans are a major input to the stage gate reviews. The stage gate review team may use a protocol and/or checklist to guide their assessment. The book includes an appendix with examples of protocols for process safety at each stage of the project life cycle.

If the review team identifies that the PMT has not completely met their technical, EHS and process safety objectives (activities and deliverables) for the stage, they will make recommendations for any improvements needed. Finally, the review team may recommend to a senior manager (sometimes known as the Gatekeeper) whether the project is ready to proceed to the next stage, although the final decision rests with the manager.

6 Appendices

Other appendices in the book address:

- Typical Process Safety Studies over Project Life Cycle
- Typical Hazard & Risk Register
- Safety Checklist for Process Plants
- Example of Site-Specific Decommissioning Checklist

The first item is a large table illustrating some of process safety studies and reviews that ***may*** be appropriate for a project to undertake in order to develop and deliver a facility that is safe and reliable to operate. A 'user beware' caution is added as no two projects are the same. They may vary in strategy, scope, complexity, location, design basis, local laws and regulations, and various other factors. When reviewing the table, the user should consider all project-specific factors before determining which process safety studies and reviews apply. A color copy of this table is available on a CCPS website to purchasers of the book.

---

[1] The term "stage gate" is being used in its common and generic form as it is used throughout industry. Stage-Gate® is also a registered trademark of the last listed owner, Stage-Gate International. CCPS and BakerRisk® disclaim any proprietary right or interest in the registration of the mark.

7 Conclusions

The book fills a gap in CCPS's library of publications that address the range of process safety challenges that companies face. As Trevor Kletz was fond of saying:

> *"... if you think safety is expensive, try an accident. Accidents cost a lot of money. And, not only in damage to plant and in claims for injury, but also in the loss of the company's reputation."*

Certainly, process safety activities can incur significant resource requirements. However, several major incidents that involved newly commissioned projects with a range of inherent weaknesses bear testimony to the need for building process safety systematically into future engineering projects. The aim of this guideline book is to raise awareness of what companies can do to integrate process safety into their projects in order to minimize incidents and ensure a safe and reliable facility.

8 Acknowledgements

As with all CCPS projects, this book was a team effort and the following contributions are acknowledged:

The Chemical Center for Process Safety (CCPS) thanks all of the members of the Integrating PS into Engineering Projects Subcommittee for providing technical guidance in the preparation of this book. CCPS also expresses its appreciation to the members of the Technical Steering Committee for their advice and support. The chairman of the Subcommittee was Eric Freiburger (Praxair). The CCPS staff consultant was Dave Belonger.

The Subcommittee had the following key contributing members: Ignacio Alonso (Consejo de Securidad de Procesos), Christopher Buehler (Exponent), Donnie Carter (Retired, formerly BP), Robert Dayton (Chevron), Dr. S. Ganeshan (adjunct Professor of Chemical Engineering, Bombay), Andrew Goddard (Arkema), Anil Gokhale (CCPS), Emmanuelle Hagey (Nova Chemicals), and Kevin Watson (Chevron).

Baker Engineering & Risk Consultants, Inc. (San Antonio, Texas) prepared the book manuscript under contract with CCPS. Michael P. Broadribb was the lead contract author, with input from Joe Zanoni and Chuck Peterson.

CCPS gratefully acknowledges the comments submitted by the following peer reviewers, whose insights and suggestions helped ensure a balanced perspective: Anne Bertelsmann (Marathon Petroleum), Denise Chastain-Knight (Exida), Marlon Harding (Merck), Patti Jones (Praxair), Pamela Nelson (Solvay), John Remy (LyondellBasell), and Steven Thomas (Chevron).

9 References

CCPS, *Guidelines for Integrating Process Safety into Engineering Projects*, Center for Chemical Process Safety, American Institute of Chemical Engineers, New York, NY, 2018.

Return to Operation-The Final Check

Commissioning of the green field project, integrated refining and petrochemical complex or even single production unit in Hydrocarbon industry poses challenges before the leadership of the organisation. Challenges arsing due to novice or unskilled workforce and unforeseen events become more serious during delayed mechanical completion of turnaround activities. These challenges demand improvements in both organisational and management effectiveness for assuring operational readiness and preventing human error before the hazardous chemicals are introduced on initial commissioning or recommissioning.

The initiative of flawless start up drives to achieve world-class commissioning, start-up and operational excellence for mega projects including the first cycle of operation. Major objective of this program is to ensure functional integrity of equipment, instruments, components so that it results in No Harm to people, environment, asset or reputation of the company. The program is designed based on risk based approach. This paper addresses the essence of flawless commissioning of the plant, i.e. introduction of Hydrocarbon after successful pre-commissioning phase.

We at Reliance Industries Limited operate 200 odd process plants at different geographical locations. All these plants have been commissioned in last 40 years and nearly half of these plants have been commissioned in last 15 years. Numerous issues like connection of utility lines to Hydrocarbon lines, commissioning the lines with no end blinds, Commissioning of steam headers without draining condensate, plugging of control valves, configuration of wrong logics, equipment box up without internals, workplace incidents due to poor access to the equipment etc. has been experienced by our teams and has been documented. This paper would focus on major problems faced and improvements done in processes and procedures to eliminate human errors and ensure safe introduction of Hydrocarbon in the equipment. The paper would address how Risk based approach helped organisation in recent commissioning of World's largest Refinery Offgas Gas Cracker (ROGC) plant at Jamanagar, India.

- **Flaws during start up**
  Major accidents that happened during ethylene plant commissioning are related to temporary modifications, taking short cuts, unhealthy safety barriers etc. Effect of these accident is either Process safety incident involving loss of containment, explosion, equipment failure (such as brittle failure etc), off-specification product generation etc ultimately affecting the start-up schedule. The flaws are caused by factors such as human error during schedule pressure, human tendency to take short cuts, poorly trained workforce etc. Trevor Kletz said that history of these process incidents shows that many incidents are repeated after few years. The people involved in older safety incidents either retire or change the employment and thus reasons for some operating instructions and design preferences are being lost.

FLAWLESS commissioning in ROGC

Design of ROGC is influenced by three key categories

1. Feedstock affecting design requirements
2. Design requirements based on RIL(Reliance Ind Ltd.) desirable features
3. Latest technological trends – with best Energy, operability and safety standards

Implemented/Executed during Construction

1. Align Construction team with plant Operation & Maintenance team

Systematic construction approach by segregating the entire plants into sections. Due to large plot size, hundreds of equipment, thousands of km of pipelines, variety of pipe specifications, thousands of km of cables, the engineering work was divided into smaller sections. This concept helped not only in getting early deliverables from DEC(Detail Engg Contractor)) to client but it helped client to align construction workforce right from engineering phase.

Dividing entire plant into smaller sections and dividing total manpower into smaller teams lead by area managers was key step to ensure critical scrutiny of all safety elements. Forming operations team section wise also helped to collaborate with construction teams to ensure operations requirement of pre-commissioning, commissioning and even steady state are properly addressed during construction phase.

Close monitoring of construction progress to ensure early pre-commissioning initiated by a core team comprising of Operation & Maintenance. Daily progress review meetings by each sections helped to stick to plan and also helped to align the plan based on recent developments. Close aligning with construction team helped to direct construction progress from the point of plant start-up sequence.

2. <u>CLEAN Built efforts and preservation of Equipment</u>:

Majority of efforts in pre-commissioning is to remove debris, water, foreign materials, oil and grease etc from pipelines and equipment. The debris typically gets deposited in opened equipment and pipes either at store or construction site. Since construction team and operating teams are typically different, history of equipment and pipeline cleaning not maintained till pre-commissioning. Large unskilled / poorly skilled workforce involved in construction typically has very little knowledge about process plant operation in which equipment or pipeline being construction are going to be used. The efforts of operation and plant maintenance people for ensuring cleanliness prior to start up can greatly reduced if apt care is taken during installation. This "Clean Build" concept was started with explaining the process sections to construction team and explaining them significance of equipment and pipelines cleanliness in operation. This was win-win for both construction and operation/maintenance team to know areas of concern from each other and work hand in hand.

3. <u>Systematic review of cleaning work</u>
   a. Cleaning of large cooling water network

   Manual cleaning of main cooling water underground heaters and Chemical treatment/Flushing. Extensive use of CW network PIPENET software simulation model to predict the required velocities in different networks during cleaning. Based on the model, systematic approach was taken to carry out CW network header cleaning in steps.
   b. Chemical cleaning of Compressor Lube Oil circuit

   Compressor circuit pressure drop and velocities were estimated at different steps of high velocity blowing and chemical cleaning. The entire progress of cleaning was monitored based on pre-ascertained conditions. This gave confidence of effective cleaning of complex system.

c. Furnace Steam and Hydrocarbon Convection banks physical checking for cleanliness and blowing critical components with air.

Based on experience from earlier commissioning it was decided to do air blowing of furnace convection section to ensure no blockage is observed in flanged and welded sections.

4. P&ID walk through in field

A multidisciplinary team did the section wise P&ID walkthrough which surfaced out many small but equally important factors

a. Direction of critical isolation valves :

Directions of critical valves such as Triple eccentric butterfly valves and Cryogenic service cavity valves are key for effective isolation. The effectiveness of these valves to hold the pressure is only ensured if they are planed in correct direction of flow. During initial installation of these valves in different sections, different methodology was used by different sub-contractor of constructions. However with detailed discussions with valve manufactures, the corrective action was initiated timely.

5. When different layers of contractors are involved

Jobs involved in multiple contractors and sub-contractors poses a chance of potential miss which could lead to process incidents. These factors typically comes out as surprises and could even impact the schedule.

a. Incorrect fabrication of distillation column internals
b. Improper support strength of de-mister pads in vessels and columns
c. Oil soaked de-mister pads in cryogenic services
d. Lube oil overhead tank incorrect elevation
e. Improper alignment in vendor skids with outside pipelines.
f. Modifying 8 leg vortex breaker to 4 leg vortex breaker to overcome flow restrictions in vessels bottom.
g. Initiative taken to check the quality of Alumina balls (by *In situ* impact test and detailed laboratory analysis). This lead to selection of high purity alumna balls from different vendor.

Despite best efforts to align O&M team with construction to ensure Clean Built, the efforts on cross –confirmation of cleaning were not reduced as the schedule impact not only delays the start-up but it also affect the morale of team after tiring efforts.

- **<u>Pre-commissioning good practices which were adopted in ROGC and giving an excellent result during start up and steady state:</u>**

1. Cleaning of large diameter / long length pipelines :
   High velocity / momentum blowing using Air Separation Unit (ASU) air compressor done. To reduce the total air blowing duration a logical approach was used to maximize the use of available air from ASU by creating parallel air blowing paths. Need of air quantity required for each of the header was estimated based on momentum which helped in effective cleaning of pipelines and also helped in drying of cryogenic sections in plant.

2. Cleaning of long pipelines connected to cryogenic columns : Approach of using column as large inventory vessels for blowing connected pipelines was used. Required cleaning velocity was achieved by pressurizing the columns and using a controlled blowing. Welding tapes in cryogenic pipelines could be effectively removed using this approach.

3. Small bore pipelines: Based on pre-identified criteria based on service, application of cleaning method for small bore pipeline were decided. Additionally control valves were provided with inlet strainers to capture commissioning mucks during start up.

4. High Pressure Leak check. No Flange leaks (in HC service)

5. Dry out loops mark-up & execution. Not a single instance of moisture ingress in cold sections. High volumetric flow to achieve (more than 30m/sec velocity) cryogenic dew points in fastest possible time.

6. Used external fuel oil burner for refractory dry out of all furnaces.

- **<u>Best Start up and steady state approaches</u>**

A. <u>Known Challenges</u>

How to get the people ready for start up :

1. With limited manpower with cracker experience and no Panel officer and field executive with cracker experience a mammoth challenge was to train operators before pre-comm and commissioning. OTS (Online Training Simulator) was developed and the young and fresh workforce was trained for all start ups and critical trouble shooting exercises in a structured manner.
OTS was utilised extensively for training for all possible emergencies. This has helped in updating SOPs (e.g. ERC, Ethylene compressor side out control scheme was modified after start up review or PSV popping of ERC side out was first experienced in ERC start up in OTS. Side out flow control is outcome of further deliberation with licensor.)

Close coordination with CCC (Compressor Control Corp.) during main machine start up Ensuring training by CCC expert and sharing of experience of other crackers with CCC.

B. Unforeseen Surprises

1. MIC (Microbes induced corrosion) related pipelines identification & replacement upfront to air blowing.

2. Chemical cleaning of Finned tube carbon steel exchangers

   Finned tubes of carbon steel exchangers were prone for corrosion despite preservation attempts from the time of manufacturing till it's taken in service. Based on experience these fixed tube sheet exchangers were cleaning by chemical cleaning method.

3. Column internals incorrect fabrication :
   Suite-at-site modifications in column internals to address incorrect fabrication
4. Compressor reverse rotation :
   Efforts were made to ensure causes of reverse rotation are eliminated and thus suitable modifications are done.
5. High vibration / noise in pipeline downstream of restriction orifices. – Provision of multi hole orifices in place of single hole helped to fulfil site requirement.
6. First start-up of front end hydrogenation reactor… Provision of CO soaking in reactor

7. Frequent cleaning requirement of strainers at the inlet of Aluminium cold box : Rectification and strengthening of all cold box strainers & blowing of upstream lines to remove foreign particles executed during opportunity.

What can we do to address human error?

Let's take a look at the environment. We cannot change human condition but we can change the conditions where humans work, so let's review our systems and operation control. Some areas to evaluate are:

1. Management systems: documentation control, investigation management, risk management and project management are important to set the bases for the rest of the operation.
2. Procedures: these need to be accurate, human-engineered, available and enforceable.
3. Human factors engineering: Work areas need to be designed with human factors and human capabilities in mind. Excessive monitoring, mental calculations, housekeeping, and work layout, among others, become the main reasons for error in this category.
4. Training: training needs to include the whys as well as the whats and hows. Also, on the job training and qualification (OJT) is necessary, especially for critical tasks and activities.
5. Immediate supervision: pre-job briefs, walkthroughs, presence and instructions to workers are necessary. We need to have supervisors on the floor, not in the office.
6. Communication: between groups, shifts, radio communication rules and training. Employees should know what needs to be achieved daily and the proper way to do it.
7. Individual Performance through CAS (Competency Assessment System): Evaluate conditions that could potentially create cognitive overload that creates attention and memory failures. Some of these conditions include available time for the job, fitness for duty or fatigue management, and complexity and task design.

Prepared by Dipankar Bandyopadhyay
Sr.Vice President
Head Operation C2 Complex.
Reliance Industries Ltd.
Jamnagar . Gujrat
India.361140
Mail Id: Dipankar.bandyopadhyay@ril.com
Mobile: +91 9970050031
Landline:+912884035017

Study of the Initiator Mischarging Effect on Runaway Reaction of Styrene Polymerization

Lin Zhao
Artie McFerrin Department of Chemical Engineering, Texas A&M University, 3122 TAMU, College Station, Texas 77843-3122, USA
Mary Kay O'Connor Process Safety Center, Texas A&M University, 3122 TAMU, College Station, Texas 77843-3122, USA
linzhao@tamu.edu

Maria I. Papadaki
Mary Kay O'Connor Process Safety Center, Texas A&M University, 3122 TAMU, College Station, Texas 77843-3122, USA
Department of Environmental and Natural Resources Management, School of Engineering, University of Patras, Seferi 2, Agrinio 30100, Greece
marpapadaki@upatras.gr

M. Sam Mannan
Artie McFerrin Department of Chemical Engineering, Texas A&M University, 3122 TAMU, College Station, Texas 77843-3122, USA
Mary Kay O'Connor Process Safety Center, Texas A&M University, 3122 TAMU, College Station, Texas 77843-3122, USA

Mustafa Akbulut
Artie McFerrin Department of Chemical Engineering, Texas A&M University, 3122 TAMU, College Station, Texas 77843-3122, USA
makbulut@tamu.edu

Prepared for Presentation at
American Institute of Chemical Engineers
2019 Spring Meeting and 15th Global Congress on Process Safety
New Orleans, LA
March 31 – April 3, 2019

Study of the Initiator Mischarging Effect on Runaway Reaction of Styrene Polymerization

Lin Zhao
Artie McFerrin Department of Chemical Engineering, Texas A&M University, 3122 TAMU, College Station, Texas 77843-3122, USA
Mary Kay O'Connor Process Safety Center, Texas A&M University, 3122 TAMU, College Station, Texas 77843-3122, USA
linzhao@tamu.edu

Maria I. Papadaki
Mary Kay O'Connor Process Safety Center, Texas A&M University
Department of Environmental and Natural Resources Management, School of Engineering, University of Patras

M. Sam Mannan
Artie McFerrin Department of Chemical Engineering, Texas A&M University
Mary Kay O'Connor Process Safety Center, Texas A&M University

Mustafa Akbulut
Artie McFerrin Department of Chemical Engineering, Texas A&M University

Keywords: Reactive chemical, Calorimeter, Polymerization, Thermal runaway

Abstract

Polymerization reactions are prone to runaway risks due to the unstable nature of monomers and complex exothermic interactions between reactants. Deviation from a normal recipe or designed operation condition may lead to auto-accelerated temperature and pressure rise, followed by rupture of reaction vessels, fire, and explosion. Mischarging of initiators (radical generators) is one of the most common credible mal-operation cases, which has only been studied to a limited extend. In this study, benzoyl peroxide (BPO) and azobisisobutyronitrile (AIBN) were used as initiators for styrene polymerization as a function of high concentrations to mimic the mischarging scenarios. Both screening and adiabatic tests were performed to obtain the most critical safety parameters such as onset temperature, adiabatic temperature rise, and time to maximum rate. It was found that there exists a significant difference in reaction pathways and safety indicators with respect to initiator type and concentration. Uncontrolled initiation reactions not only lead to subsequent faster polymerization but also pose threats to relief valve design due to the generation of non-condensable gas. We anticipate that this study will provide beneficial guidelines for polymer manufacturers on the initiator selection and also thermal runaway prevention.

1 Introduction

Styrene monomer is a thermally unstable vinyl type monomer with the potential of self-polymerization and triggering uncontrolled thermal runaway reaction during handling, storage, and polymer production. The polymerization reaction is highly exothermic and auto-accelerates due to the drop of termination rate hampered by the long polymer chains, which may result in unexpected temperature increase beyond design cooling capacity, higher pressure building-up in vessels, loss of containment, fire, and explosion. According to the statistics, 48% of the runaway incidents in the UK during 1962-1987 were polymerization reactions.[1] A more recent study from the Chemical Safety Board (CSB)[2] revealed that polymerization thermal runaways contributed to 15% of total runaway incidents in the US between 1980-2001. Uncontrolled polymerization of styrene kept reoccurring all over the world, resulting in fatalities and asset loss. Thus, understanding the thermal runaway characteristic of the styrene polymerization reaction is vitally important regarding estimating the risks of runaway reactions and designing proper protection measures.

Thermal runaway of reactions can be caused by various types of deviations from the normal operation condition, and the mischarging of reactants has been identified to be one of the most common scenarios.[3-4] For a typical polymerization process of styrene, thermally unstable initiators are used to generate free radicals and trigger the chain-addition polymerization of styrene in the presence of a small portion of diluent (solvent). The type and the initial concentration of initiators strongly affect the initiation process and also the overall polymerization behavior. Maschio[5] investigated the free-radical polymerization of methyl methacrylate (MMA) using an adiabatic calorimeter and identified that the celling temperature, and initial polymerization rate varied with initiator type. The study of Theis[6] on copolymerization of MMA and styrene with a pseudo-adiabatic calorimeter showed an increase of over 100 °C/min on self-heating rate when the initiator concentration was doubled. A more recent study on the polymerization of styrene with different benzoyl peroxide (BPO) concentration[7] identified that higher BPO dosage affected adiabatic temperature rise of the runaway reaction.

In this work, a comprehensive study on the runaway behavior of styrene polymerization in the solvent was conducted with two different types of initiators. Benzoyl peroxide (BPO) and azobisisobutyronitrile (AIBN) were used at a series of elevated concentrations to mimic the initiator mischarging scenarios. Both screening and adiabatic calorimeter test were performed to understand the thermal activity at small and bulk scale. Important safety indicators such as "onset" temperature of polymerization, the time to maximum rate, the heat of reaction, as well as pressure increase during the exothermic were reported.

2 Experimental

2.1 Materials

In this paper, the two initiators, benzoyl peroxide (Sigma-Aldrich, 98%) and 2,2'-Azobis(2-methylpropionitrile) (Sigma-Aldrich, 98%) were used without any further purification. Ethylbenzene (Sigma-Aldrich, ReagentPlus, 99%) was used as a diluent to reduce the viscosity of the system and mitigate the gel effect. Styrene (Sigma-Aldrich, ReagentPlus, 99%) monomer stabilized with 15 ppm of 4-tert-butylcatechol (TBC) was purified with a packed column filled with alumina inhibitor remover (Sigma-Aldrich) powders. The inhibitor-free styrene monomer was mixed with ethylbenzene and initiators to prepare a serious of reaction solutions and was used immediately. The mass fraction of ethylbenzene was 15% for all polymerization samples, and the initiator mass fraction varied from 500 ppm to 10000 ppm based on the total styrene mass. Initiators were also dissolved into ethylbenzene at 5% to investigate their thermal decomposition in solvent.

2.2 Differential Scanning Calorimeter (DSC)

The test samples were first measured in a Q20 differential scanning calorimeter (TA Instruments, New Castle, DE) to obtain preliminary thermal activity results during the reaction. Approximately 5 mg of test samples were sealed in a high-pressure (15 MPa), stainless-steel sample capsule and tested between 30 °C and 300 °C in a modulated DSC (MDSC®) mode. The MDSC operates differently from the traditional DSC that a sinusoidal temperature change which oscillates around the average heating rate is applied, and the total modulated heat flow signal obtained can be separated into two components based on whether it responds to the modulated heating rate.[8] The reversible heat flow represents the heat capacity related thermal activity, i.e., heat capacity change, glass transition, and melting. The non-reversible heat flow stands for kinetic involved thermal behaviors such as reaction, decomposition, curing, etc. The monomer-contained samples and pure initiators were heated up at a fixed average heating rate of 4 °C/min oscillating at 0.6 °C per 60 s. In addition, the 5% initiator in ethylbenzene samples were tested at four different average heating rates of 2, 4, 8, 10 °C /min to obtain the kinetic information.

2.3 Automatic Pressure Tracking Adiabatic Calorimeter (APTAC™)

Adiabatic calorimeter tests using APTAC™ were performed on the styrene-ethylbenzene-initiator system for different initiator dosages to investigate the runaway behavior of bulk samples. Purified styrene was mixed with ethylbenzene at a fixed monomer mass fraction of 85%. The initiators were then dissolved in the monomer-solvent mixture and used immediately. The initiator ratios used in adiabatic tests were 500 ppm, 2500 ppm, and 10000 ppm based on the mass of the monomer for both BPO and AIBN. In this work, *ca.* 10 g of mixture sample was loaded into a glass test-cell, and the corresponding thermal inertia factor (φ-factor) was around 1.26. An initial nitrogen pad pressure of 1000 kPa (147 psi) was used to suppress the boiling of the reactant. During the heat-wait-search operating, a temperature increment of 5 °C, a heating rate of 5 °C/min, and a 30 min waiting period between each temperature step were used. The "onset" of exothermic activity was defined as when the self-heating rate exceeded 0.1 ℃ /min. The temperature and pressure during the exothermic were recorded, and the rate of self-heating and pressure-rise were calculated.

3 Results and Discussion

3.1 DSC results of initiators

The thermal decomposition of the two solid initiators was first examined by MDSC, and the results are summarized in Figure 1. The non-reversible heat flow curve denoted one single exothermic peak, which was the thermal decomposition process. The reversible heat flow curve showed a single endothermic peak, overlapped with the non-reversible heat flow, which was identified as the melting process of the initiator from the solid state to the liquid state. This coupled phenomena of heat absorption and generation has been observed by many other researchers, and efforts have been made to decouple the endothermic peak from the decomposition peak.[9] As it is clearly shown in Figure 1, MDSC provided a convenient way to separate overlapping thermal activities and rule out the phase transition from the hazardous exothermic behavior. As a result, the determination of the "onset" temperature of the thermal decomposition became straightforward. The non-reversible heat flow denoted the "onset" temperature of solid BPO and AIBN was 103 °C and 100 °C, respectively. The heat of decomposition integrated from the non-reversible heat flow was 1352 J/g for BPO, and 1140 J/g for AIBN, which were slightly higher than the heat of decomposition obtained from the traditional method,[10-11] implying the MDSC decoupling method provided a more conservative estimate of the reaction heat and thermal hazard.

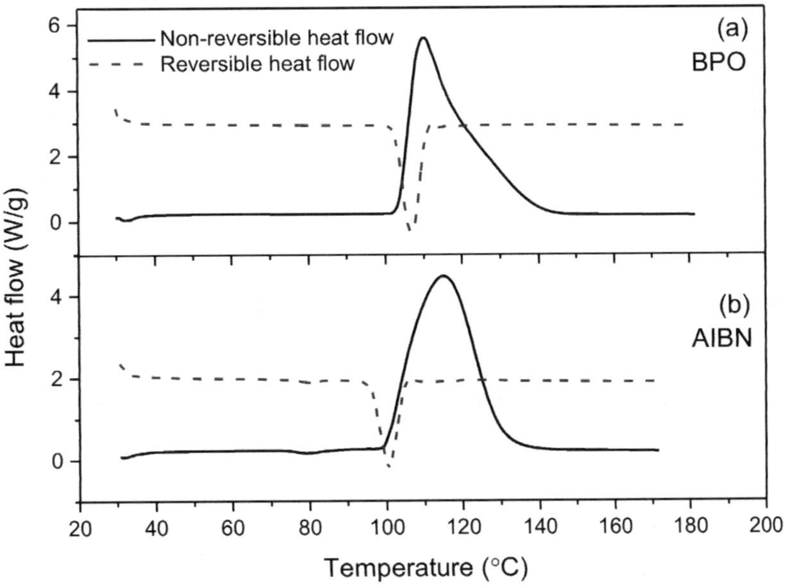

Figure 1. Non-reversible heat flow (——) and reversible heat flow (– –) of the two solid initiators during thermal decomposition from the MDSC experiments. (a) BPO. (b) AIBN.

In general, industrial practices utilizes initiators which dissolved in the solvent to trigger the polymerization reaction, and the thermal decomposition of the two initiators was examined in ethylbenzene following the ASTM E698 standard (Ozawa method).[12] Four different scanning rates were applied to the 5% initiator-ethylbenzene mixture as described in the *Experimental* section. The corresponding results were summarized in Table 2.

Table 2. The peak temperature and the heat of reaction of the 5% initiator-ethylbenzene solution at different heating rates in MDSC

β (℃ /min)	*BPO*		*AIBN*	
	T_{max} (℃)	ΔH (J/g)	T_{max} (℃)	ΔH (J/g)
2	114.9	1341	111.5	1394
4	125.9	1320	118.5	1478
8	138.1	1319	127.1	1477
10	140.5	1257	129.5	1304

In Table 2, the peak temperature (T_{max}) corresponding to the maximum heat generation was recorded for each heating ramp, and the activation energy (E_a) of thermal decomposition in the solvent was obtained by using Equation (1), where E_a is the activation energy of thermal decomposition, R is the gas constant, β is the heating rate in ℃ /min, and T is the peak temperature in Kelvin.

$$E_a \cong -2.19R[\mathrm{d}\ \mathrm{Log}_{10}\beta/\mathrm{d}(1/T)] \quad (1)$$

When initiators were dissolved in ethylbenzene, the phase transition was eliminated, and the reversible heat flow became a trivial part. The overall heat flow obtained was identical to the non-reversible heat flow through the process of thermal decomposition. As shown in Figure 2, when initiators were dissolved in ethylbenzene, the thermal decomposition became detectable at a significantly lower temperature when compared to the concentrated solid form. For the heating rate of 4 ℃ /min, 5% BPO in ethylbenzene "onset" at around 84.8 °C, and the 5% AIBN in ethylbenzene "onset" at around 70.7 ℃ . The maximum temperature of the exothermic peak increased with a higher heating rate. The activation energy E_a obtained of 5% initiators in ethylbenzene was 79.7 kJ/mol for BPO and 108.1 kJ/mol for AIBN. The experimental results in the solvent denoted that AIBN was able to thermally decompose and trigger polymerization at a lower temperature when compared to BPO which usually had a higher activation energy and was more sensitive to temperature.

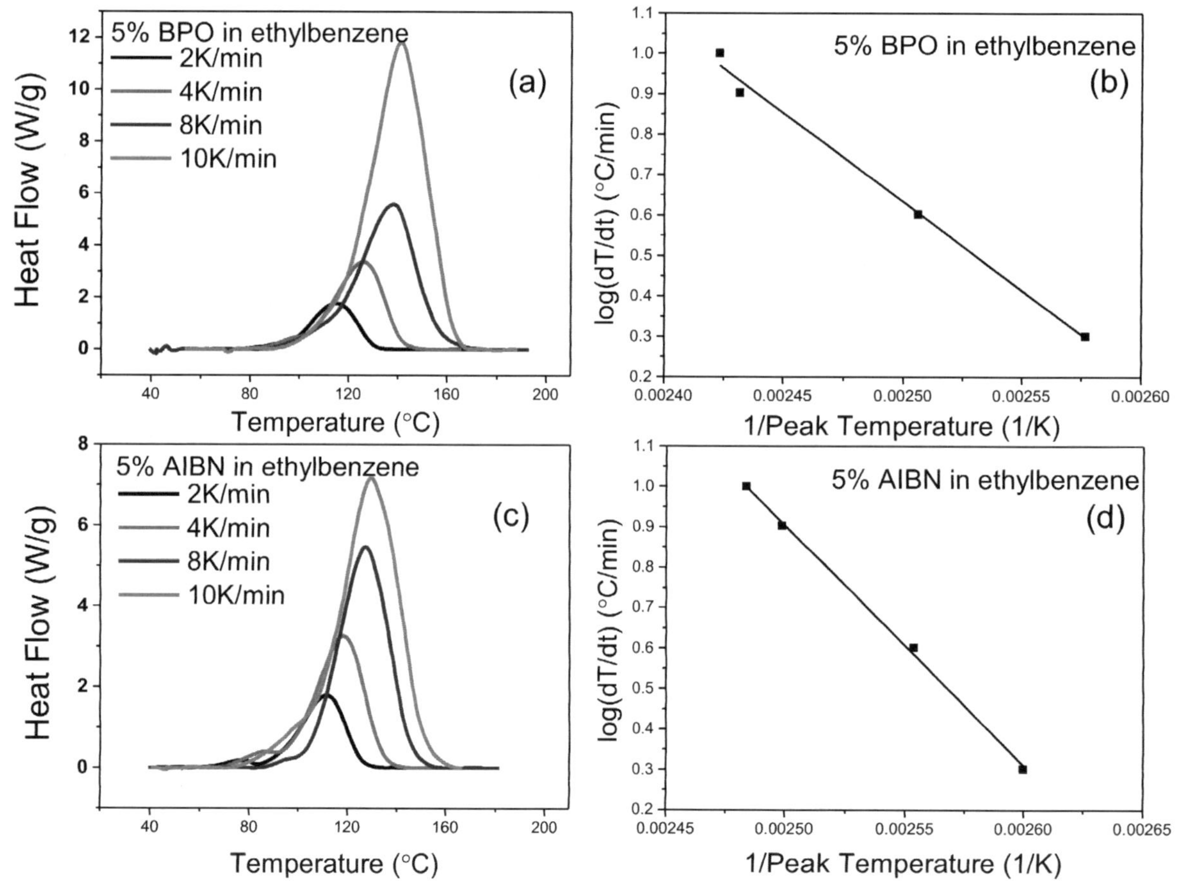

Figure 2. Heat flow of the 5% initiator-ethylbenzene solution at different heating rates in the MDSC and the kinetic fitting. (a), (b) BPO. (c), (d) AIBN.

3.2 *DSC results of polymerization systems*

After the thermal hazard of the initiators was assessed in the solvent, the initiators were then mixed with 15% styrene-ethylbenzene at a series of initiator concentrations and tested with MDSC. The heat flow results are depicted in Figure 3 and Figure 4 for BPO and AIBN, respectively. For free-radical polymerization triggered by both initiators, the total heat flow was equal to the non-reversible flow, and the kinetic related thermal activity dominated the whole reaction.

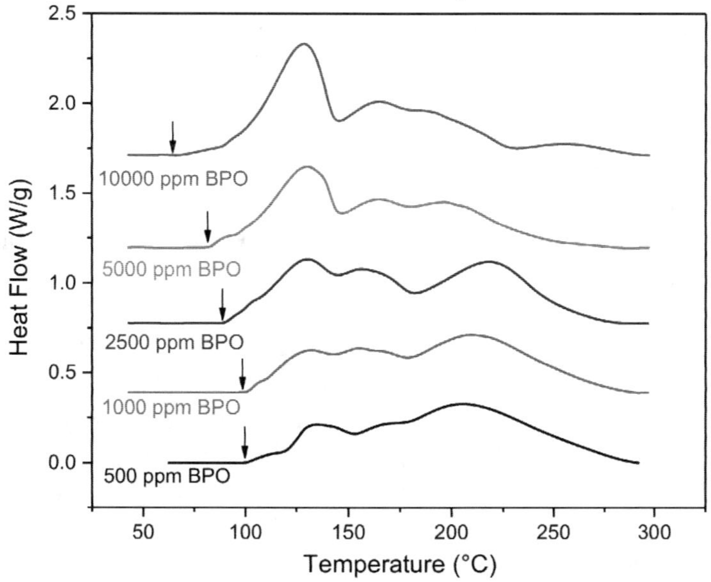

Figure 3. Heat flow of the polymerization of 85% styrene-ethylbenzene with different BPO concentrations in MDSC at 4 °C/min

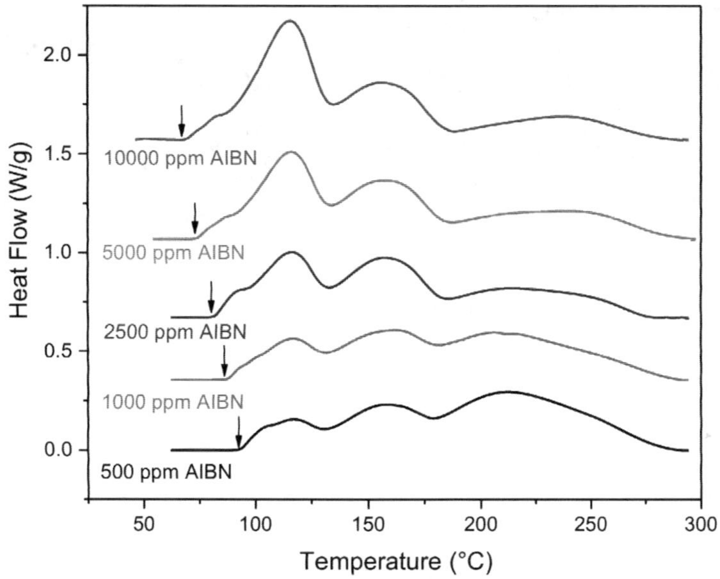

Figure 4. Heat flow of the polymerization of 85% styrene-ethylbenzene with different AIBN concentrations in MDSC at 4 °C/min

The polymerization with the two initiators expressed some common trends. First, more than one exothermic peak emerged upon heating, denoting a complex reaction pathway including initiation, chain propagation, chain-transfer, and termination. Compared to the thermal polymerization of styrene without an initiator, Liao[13] identified two major peaks to represent two different types of reaction patterns. However, when the initiators were present, peak numbers increased, and no clear conclusion can be drawn whether the peaks were associated with a specific reaction step as they highly overlapped with each other. A significant difference regarding the "onset" temperature and the relative peak area was noticed under different initiator dosages. For both BPO and AIBN, the free-radical polymerization reaction started "onset" at a lower temperature when the initiator concentration was increased. The lowest "onset" temperature for AIBN initiated polymerization was 71.3 ℃ , which was identical to the starting temperature of the thermal decomposition of AIBN in ethylbenzene at 5%. For BPO initiated system, the lowest "onset" temperature obtained at 10000 ppm BPO in styrene-ethylbenzene was 70.1 °C, which was even lower than the thermal decomposition of BPO in ethylbenzene (84.8 °C). This finding implied that the presence of monomer further promoted the decomposition of BPO.

Another difference noticed from the DSC results was that the main exothermic peak shifted to a lower temperature upon the addition of the initiators. The area under the heat flow curve represents the heat generated during the reaction, and thus the relative size and position of the peak can be used to estimate the heat release and at different temperatures, in other words, assess the severity of the thermal hazards. When BPO and AIBN dosage was low (500 ppm), the major exotherm took place at the late stage of the reaction at around 210 °C. With the addition of the initiators, the main heat release was observed at a much-lowered temperature. For systems with 10000 ppm of BPO or AIBN, the maximum heat generation took place at around 120 °C; and the sharper, narrower peak denoted a sudden heat release during a short period.

The overall heat of reaction was calculated for all test samples and summarized in Table 3 together with the "onset" temperature. The change of these important thermal hazard indicators obtained from the MDSC study with respect to initiator dosage was depicted in Figure 5.

Table 3. Thermal kinetic data for the polymerization of 85% styrene-ethylbenzene with different initiator concentrations in MDSC at 4 °C/min

Initiator concentration (ppm)	*BPO*		*AIBN*	
	T_{on} (℃)	ΔH (J/g)	T_{on} (℃)	ΔH (J/g)
500	102.0	478.0	94.1	503.7
1000	100.0	531.8	87.5	498.9
2500	91.1	558.0	80.9	499.9
5000	83.5	584.7	75.1	565.3
10000	70.1	596.1	71.3	637.2

As shown in Figure 5(a), when BPO was used to trigger the polymerization of styrene in ethylbenzene, the reaction heat increased monotonically upon BPO addition, and this increase was most significant when the BPO concentration was below 5000 ppm. At the same time, adding BPO to the monomer-solvent system also lowered "onset" temperature, and the lowest "onset" temperature obtained was around 15 °C lower than the initial thermal decomposition temperature of BPO itself when dissolved in ethylbenzene at 5%. When compared to the case of AIBN in figure 5 (c) and (d), it was noticed that the reaction heat roughly remained constant when AIBN concentration was below 2500 ppm, and the further charging of AIBN resulted in a sharp increase of reaction heat. The "onset" temperature of the polymerization system with different AIBN dosage showed a similar declining trend as what was observed for the case of BPO. However, the lowest "onset" temperature was not affected by the monomer.

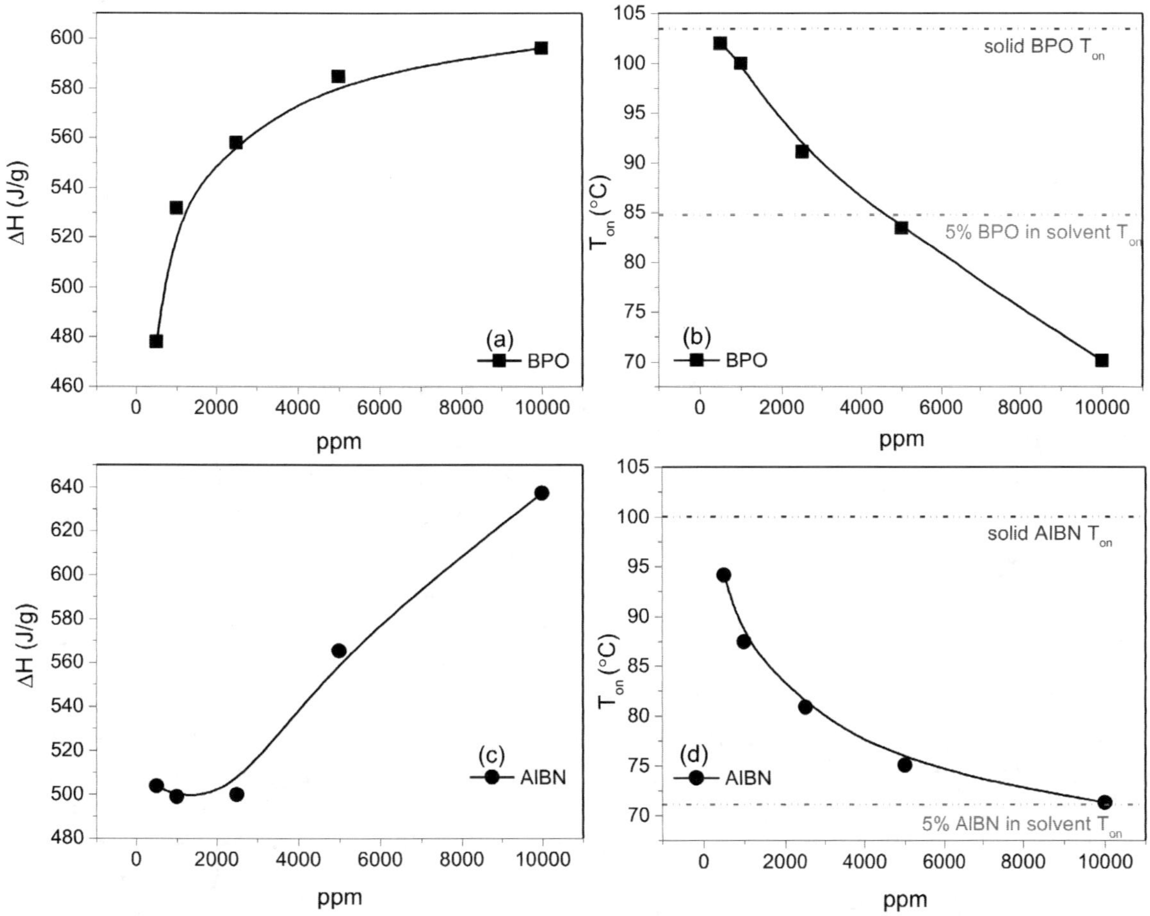

Figure 5. The "onset" temperature and heat of reaction for the polymerization of 85% styrene-ethylbenzene with different initiator concentrations in MDSC at 4 °C/min. (a), (b) BPO in styrene-ethylbenzene. (c), (d) AIBN in styrene-ethylbenzene

3.2 Adiabatic calorimeter results of polymerization systems

The DSC provided some preliminary heat generation information of the samples at a small scale, where the heat and mass transfer did not play an important role in the overall reaction. The thermal runaway behavior of the polymerization process with various initiators dosages was further investigated in APTAC to obtain bulk temperature and pressure data under adiabatic condition. BPO and AIBN were added into the 85% styrene-ethylbenzene system at 500 ppm, 2500 ppm, and 10000 ppm to represent the normal operation and two initiator mischarging scenarios. The adiabatic temperature rise with respect to time and the self-heating rate with temperature was shown in Figure 6. The detected "onset" temperature, time to maximum rate under adiabatic conditions (TMR), the maximum self-heating rate (dT/dt_{max}) and the maximum pressure-rising rate (dP/dt_{max}) were calculated and summarized in Table 4.

Table 4. The adiabatic thermal kinetic data for the polymerization of 85% styrene-ethylbenzene with different initiator concentrations

Initiator (ppm)	BPO				AIBN			
	T_{on} (°C)	TMR (min)	dT/dt_{max} (°C/min)	dP/dt_{max} (kPa/min)	T_{on} (°C)	TMR (min)	dT/dt_{max} (°C/min)	dP/dt_{max} (kPa/min)
500	83.8	206.7	3.8	22.8	71.4	236.8	2.8	17.7
2500	73.4	78.3	3.6	17.5	62.8	68.6	3.9	18.4
10000	62.7	70.2	37	105.5	55.1	73.4	25.0	93.4

The adiabatic "onset" temperature monotonically decreased with initiator dosage for both BPO and AIBN, which was consistent with the observations in MDSC tests. It is also noteworthy that the APTAC identified the thermal activity at a lower starting temperature when compared to DSC, highlighting that the "onset" threshold value strongly depends on the sensitivity of the instrument, and can only be used as a rough estimate of the thermal hazard. At the same time, the results emphasized the necessity of employing adiabatic calorimeter in thermal hazard assessment, as more conservative and realistic value can be obtained. The lowest "onset" temperature obtained in this study was 55.1 °C when 10000 ppm of AIBN was charged into the 85% styrene-ethylbenzene system. This value was around 50 °C below the regular recognized thermal decomposition temperature of pure AIBN,[10, 14-15] and also much lower than the "onset" of thermal polymerization temperature of styrene of around 100 ℃ .[16] Mischarging of a small quantity of AIBN into the system may result in an unexpected thermal activity at a very low temperature.

At the same time, a significant decrease in TMR was noticed when the initiator concentration was higher than 2500 ppm for both BPO and AIBN. This can be observed in Figure 6 (a) and Figure 6 (c) that as BPO and AIBN was 500 ppm in 85% styrene-ethylbenzene, the polymerization reaction activity became notable between 70-80 °C, and the temperature building-up was relatively slow as only around 20 °C increase in temperature was observed in the first 2 hrs. The main temperature increase was noticed after 200 min of the "onset", where a temperature increment of over 100 °C was achieved within less than an hour. The time used for the system to reach its maximum self-heating rate (TMR) was 206.7 min for BPO and 236.8 min for AIBN, respectively. When initiators were mischarged into the polymerization system at 2500 ppm, the TMR was shortened to around 80 min for both initiators; a steep temperature increment was observed, denoting that the runaway behavior changed to almost explosive type upon initiator mischarging. The corresponding massive heat release in such a short time may result in a huge increase in cooling demand which might not be able to meet by equipment design.

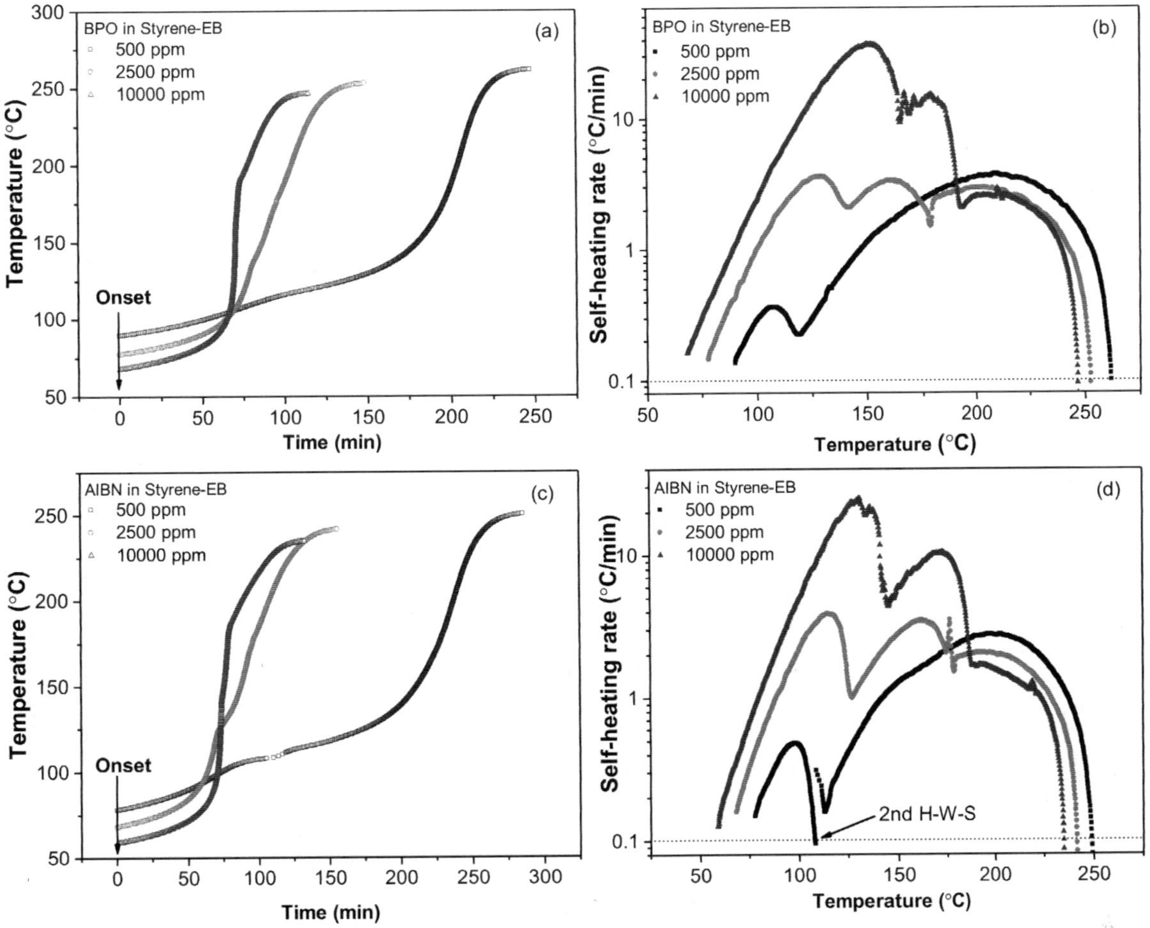

Figure 6. The adiabatic runaway behavior of 85% styrene-ethylbenzene system with various initiator charges. (a), (b) BPO. (c), (d) AIBN.

The energy release, or the heat generation, that needs to be taken care of by proper cooling is expressed as the self-heating rate versus temperature in Figure 6 (b) and Figure 6 (d). The polymerization of 85% styrene in ethylbenzene initiated by BPO and AIBN went through multiple exothermic steps, which resulted in several peaks on the curve. When BPO dosage was 500 ppm, the self-heating rate exceeded 0.1 °C/min at 83.8 °C and resulted in a mild exothermic peak with a maximum self-heating rate of 0.2 °C/min before 125 °C. This peak was followed by a major exotherm with a peak temperature at over 200 °C. The latter reaction step also possessed a characteristic of auto-acceleration as the self-heating rate was first linearly related to the reciprocal of temperature (pseudo-first-order) and then had a negative deviation (higher than first order) after around 145 °C. This unique heat generation pattern represented a typical free-radical polymerization process with a gel-effect[17]. The gel-effect poses a great threat to the polymer production as the polymerization rate, and the heat production rate suddenly increases due to the reduced mobility of the long polymer chains after the conversion achieved a certain critical point[18]. For the case of 500 ppm in AIBN, similar temperature pattern was observed only that the initial exothermic step was not able to generate enough heat to sustain the following reaction at around 110 °C. The APTAC entered heat-wait-search (H-W-S) mode and then heated the system up again to 120 °C; the following reaction developed to complete after that. The temperature rate curve confirmed that when BPO and AIBN dosage was at 500 ppm, the thermal risk associated with the polymerization reaction was due to the sudden heat release affected by the gel-effect at high temperature, and the maximum self-heating rate was below 4 °C/min. When initiator dosage was increased to 2500 ppm, three exothermic peaks were observed for both BPO and AIBN. The maximum self-heating rate of the reaction was achieved in the first reaction step at a lowered temperature of around 125 °C for BPO and 114 °C for AIBN, and the value of this peak self-heating rate had no obvious change when compared to the case of 500 ppm initiator. Further increasing of initiator concentration to 10000 ppm resulted in a remarkable change of maximum self-heating rate which was almost ten times higher than that of the 500-ppm case. The heat release rate of the first reaction step overwhelmed the following reactions.

The change of the maximum pressure-rising rate with respect to initiator concentration also followed the same trend. As summarized in Table 4, there is no significant change in the peak pressure-building up rate until the initiator concentration increased from 2500 ppm to 10000 ppm. The polymerization of the styrene-ethylbenzene system was a *vapor system* according to the Design Institute for Emergency Relief Systems (DIERS) classification, as such the pressure increase was solely related to the vaporization of the volatile species. When initiators were present, the decomposition of the BPO resulted in the release of oxygen, while the AIBN generated nitrogen upon heating, which also contributed to the total pressure of non-condensable gases. However, in this study, the initiator dosage was remained at a very low concentration (less than 1%), and the contribution of oxygen/nitrogen generation was trivial. The pressure building-up was related to the vapor pressure change caused by the temperature rise.

4 Conclusion

In this study, the runaway behavior and thermal hazard of mischarging of two initiators (BPO and AIBN) into the styrene-ethylbenzene polymerization system were investigated using screening and adiabatic calorimeters. The thermal decomposition of solid initiators was first characterized by a novel MDSC method to decouple the endothermic phase change and the exothermic decomposition reaction. The "onset" temperature and the activation energy of the thermal decomposition of initiators in ethylbenzene were investigated by MDSC, revealing that when dissolved in ethylbenzene, the thermal decomposition of initiators became notable at a much-lowered temperature when compared to their solid state.

MDSC studies on the runaway behavior of styrene-ethylbenzene with different initiator charges discovered that the thermal hazard of free-radical polymerization was strongly related to the initiator dosage regardless of the initiator type. When the BPO and AIBN were mischarged into the system at an elevated concentration, the exothermic onset temperature dropped 30 °C when compared to the thermal polymerization temperature of styrene and also the thermal decomposition temperature of the initiators themselves. Furthermore, the total heat generated by the system with over-dosed initiators increased monotonically with the increase of BPO concentration and had an approximate linear increment after AIBN concentration was over 2500 ppm. The overcharging of both initiators also resulted in a shift of main exothermic activity temperature from around 200 °C to around 120 °C.

The adiabatic runaway test of bulk samples also demonstrated a similar trend that the free-radical polymerization started "onset" at lower temperature upon initiator dosage, and the time required for the reaction to develop to its maximum self-heating was significantly reduced. With an elevated initiator dosage of 10,000 ppm, increased peak heat-release was observed at a lowered temperature associated with higher pressure building-up. The adiabatic data provided extremely useful information regarding thermal behavior and pressure profile for large-scale samples regarding scenarios of mischarging initiators into the polymerization system.

5 Reference

1) Barton, J.; Nolan, P., Incidents in the chemical industry due to thermal runaway chemical reactions. *Hazards X: Process Safety in Fine and Speciality Chemical Plants* **1989,** (115), 3-18.
2) CSB, Improving Reactive Hazard Management. *US Chemical Safety and Hazard Investigation Board (CSB)* **2002**.
3) Saada, R.; Patel, D.; Saha, B., Causes and consequences of thermal runaway incidents—will they ever be avoided? *Process Saf. Environ. Prot.* **2015,** *97*, 109-115.
4) Sales, J.; Mushtaq, F.; Christou, M.; Nomen, R., Study of major accidents involving chemical reactive substances: analysis and lessons learned. *Process Saf. Environ. Prot.* **2007,** *85* (2), 117-124.
5) Maschio, G.; Feliu, J.; Ligthart, J.; Ferrara, I.; Bassani, C., The use of adiabatic calorimetry for the process analysis and safety evaluation in free radical polymerization. *J. Therm. Anal. Calorim.* **1999,** *58* (1), 201-214.
6) Theis, A. E.; Burelbach, J. P.; Askonas, C. F., Safely scale-up processes and accommodate recipe changes. *Process Safety Progress* **2009,** *28* (2), 135-140.
7) Wang, W.; Fang, J.; Pan, X.; Hua, M.; Jiang, J.; Ni, L.; Jiang, J., Thermal research on the uncontrolled behavior of styrene bulk polymerization. *J. Loss. Prevent. Proc.* **2019,** *57*, 239-244.
8) Thomas, L. C., Modulated DSC® Paper# 1 Why Modulated DSC®?; An Overview and Summary of Advantages and Disadvantages Relative to Traditional DSC. *TA Instruments Technol. Pap* **2005**.
9) Zhang, C.-X.; Lu, G.-B.; Chen, L.-P.; Chen, W.-H.; Peng, M.-J.; Lv, J.-Y., Two decoupling methods for non-isothermal DSC results of AIBN decomposition. *J. Hazard. Mater.* **2015,** *285*, 61-68.
10) Liu, S.-H.; Lin, W.-C.; Hou, H.-Y.; Shu, C.-M., Comprehensive runaway kinetic analysis and validation of three azo compounds using calorimetric approach and simulation *J. Loss. Prevent. Proc.* **2017**.
11) Tsai, L.-C.; Chen, J.-W.; Hou, H.-Y.; Liu, S.-H.; Shu, C.-M., Exothermic behaviors in decomposition of three solid organic peroxides by DSC and VSP2. *J. Therm. Anal. Calorim.* **2012,** *109* (3), 1303-1309.
12) ASTM., Standard test method for Arrhenius kinetic constants for thermally unstable materials using differential scanning calorimetry and the Flynn/Wall/Ozawa method. West Conshohocken New York (NY): 2005.
13) Liao, C.; Wu, S.; Su, T.; Shyu, M.; Shu, C., Thermokinetics evaluation and simulations for the polymerization of styrene in the presence of various inhibitor concentrations. *J. Therm. Anal. Calorim.* **2006,** *85* (1), 65-71.
14) Liu, S.-H.; Shu, C.-M.; Hou, H.-Y., Applications of thermal hazard analyses on process safety assessments. *J. Loss. Prevent. Proc.* **2015,** *33*, 59-69.

15) Guo, S.; Wan, W.; Chen, C.; Chen, W., Thermal decomposition kinetic evaluation and its thermal hazards prediction of AIBN. *J. Therm. Anal. Calorim.* **2013,** *113* (3), 1169-1176.
16) Chen, C.-C.; Duh, Y.-S.; Shu, C.-M., Thermal polymerization of uninhibited styrene investigated by using microcalorimetry. *J. Hazard. Mater.* **2009,** *163* (2), 1385-1390.
17) Leung, J. C.; Fauske, H.; Fisher, H., Thermal runaway reactions in a low thermal inertia apparatus. *Thermochim. Acta* **1986,** *104*, 13-29.
18) Achilias, D. S., A review of modeling of diffusion controlled polymerization reactions. *Macromol. Theory Simul.* **2007,** *16* (4), 319-347.

A semi-quantitative tool to estimate risks of combustible dust fires and explosions from process equipment

Emmanuel Kwasi Addai
DEKRA Process Safety
113 Campus Dr
Princeton, NJ, USA
emmanuel.addai@dekra.com

Jeff Foisel
DEKRA Process Safety
113 Campus Dr
Princeton, NJ, USA
jeff.foisel@dekra.com

John Wincek
DEKRA Process Safety
113 Campus Dr
Princeton, NJ, USA
john.wincek@dekra.com

Christopher Headen
Cargill, Inc.
15407 McGinty Road West
Wayzata, MN, USA
Christopher_Headen@cargill.com

Prepared for Presentation at
American Institute of Chemical Engineers
2019 Spring Meeting and 15th Global Congress on Process Safety
New Orleans
March 31 - April 4, 2019

UNPUBLISHED

A semi-quantitative tool to estimate risks of dust fires and explosions from process equipment

Emmanuel Kwasi Addai
DEKRA Process Safety
113 Campus Dr
Princeton, NJ, USA
emmanuel.addai@dekra.com

Jeff Foisel
DEKRA Process Safety
113 Campus Dr
Princeton, NJ, USA
jeff.foisel@dekra.com

John Wincek
DEKRA Process Safety
113 Campus Dr
Princeton, NJ, USA
john.wincek@dekra.com

Christopher Headen
Cargill, Inc.
15407 McGinty Road West
Wayzata, MN, USA
Christopher_Headen@cargill.com

Abstract

Consistently assessing risks associated with a particular process or piece of equipment can be difficult, even for very low risk operations. Furthermore, when utilizing industry experts to complete risk assessments, inconsistencies remain in results due to the still highly qualitative nature of combustible dust risk assessments. In response, the industry is advancing efforts to bring a higher level of consistency in evaluating risks for combustible dust processes [1]. In line with the industry focus on improving combustible dust risk assessments, this work presents a semi-quantitative tool to consistently and effectively estimate risks of combustible dust fires and explosions in non-complex unit operations. This tool has been piloted for bucket elevators in the grain industry. Additional tool modules are in development and validation phases that are designed to analyze unit operations such as: grain dryers, dust collection systems, hammer mills and belt conveyors. The function of this tool is to identify and evaluate the most common preventive, protective and mitigative measures used in industry to prevent dust fires and explosions. When this measures evaluation is combined with combustible dust test data (such as explosibility properties of the material; minimum ignition energy, minimum ignition temperature, deflagration index, maximum explosion pressure, self-ignition temperature, limiting oxygen concentration, minimum explosible concentration etc.), a more complete picture of the risk is obtained. In addition to providing an estimated risk level, the tool identifies potential safeguarding opportunities and thus providing guidance for improvement.

Keywords: dust explosion, risk assessment, semi-quantitative risk tool, bucket elevator

1. Introduction

Dust explosions present serious and widespread hazards in processing industries such as the grain industry [2, 3]. Incidents involving dust explosions often lead to equipment failures, injuries to people, negative impacts to the surrounding environment, plant shut-downs and sometimes destruction of buildings resulting in fatalities and enormous financial losses. For many years in the processing industry, efforts have been in place to develop and improve measures intended to prevent or mitigate combustible dust explosions [4, 5, 6]. Despite extensive research, increasing regulations, standardization and advancements in technologies surrounding combustible dust hazard prevention and mitigation, these types of incidents still occur. In 2006, the US Chemical Safety Board (CSB) [7] conducted a study that reported 281 dust explosions occurred between 1980 and 2005 in the USA. As a result of these dust explosions, 119 workers were fatally injured while 718 experienced life-altering injuries. Since publication of the 2006 study, the CSB has confirmed an additional 105 combustible dust incidents [8]. The CSB conducted in-depth investigations of 5 of these high-profile incidents, including most recently the Didion Milling dust explosion in Cambria, Wisconsin. 27 lives were lost and dozens more injured as a result of these 5 incidents investigated by the CSB. In the processing industry, approximately 70% of dusts handled are classified as combustible. Most reported dust explosions involve organic materials from industries such as agricultural, food, fuel and pharmaceuticals.

Preventing these catastrophic dust explosions requires a comprehensive assessment of materials and processes to identify operations posing risks. One of the many challenges for organizations is determining how to consistently assess the risks for even common low risk operations. A qualitative dust hazard analysis relies heavily on expertise of the assessment team and facilitator. Experience shows a high level of variation in qualitative risk assessment results due to bias and a lack of industry accepted combustible dust data. In order to obtain an overview of the risk of fires and explosions in a process, this work presents a semi-quantitative tool to consistently estimate the risks of combustible dust fires and explosions in a unit operation.

This tool has been piloted for bucket elevators in the grain industry. Tool modules are in development and validation phases and are designed to analyze unit operations, such as grain dryers, dust collection systems, hammer mills and belt conveyors. The tool is designed for use by on-site supervisors and local Subject Matter Experts (SMEs) trained in the hazards of combustible dust and use of the tool. It uses multiple-choice and yes/no questions to assess both likelihood and severity of risk. Values are assigned to each response, providing the opportunity to weigh the relative importance of each question. The values for Severity and Likelihood are used to display the mitigated risk in real-time, as questions are answered.

The function of this tool is to establish a process baseline using material properties of the dust (such as explosibility properties of the material; minimum ignition energy, minimum ignition temperature, deflagration index, maximum explosion pressure, self-ignition temperature, limiting oxygen concentration, minimum explosible concentration etc.), the location or siting of equipment and management system elements (i.e. Asset Integrity) in place. The tool allows users to identify and evaluate the most common preventive, protective and mitigative measures taken to address combustible dust fire and explosion risks. As a result, a more complete picture of the risk is obtained. In addition to providing an estimated risk level, the tool identifies potential safeguarding opportunities and the impacts safeguards on level of risk.

2. Tool Creation

The key deliverables for this tool were to provide:

1. Consistent and effective risk identification
2. Consistent assessment of risk based on data
3. Customized risk reduction recommendations based on facility data
4. Data driven logic
5. Ease of use

Microsoft Excel was chosen as a development platform for the tool because it provides A) a familiar platform, B) programming capability, and C) a straightforward method to perform calculations associated with determination of mitigated risk.

The first step in tool creation involved applying core concepts of risk management, operating knowledge and experience, industry data, expert opinion and incident case studies in designing a background worksheet (see Table 1) that includes factors affecting frequency and consequence. These sheets are integral to tool functionality and allow for:

A. Questions compilation – All questions are compiled allowing experts to review progress as the tool is developed. Capturing the right questions in ways that are clear to the user is critical. The questions are limited to true / false, or a selection from a short list of options to further aid the user in responding.
B. Maximum scoring – This is the percentage of overall risk factor assigned to a specific question/response. This value is used to determine risk increase based on a given response. For example, though both are important and contribute to the overall risk factor, locating the bucket elevator outside and limiting people near it has a much greater impact on the consequence than having the explosion suppression device on a preventive maintenance program.
C. User Value Entry Sheet – Values returned from a form control in an Excel worksheet are connected to a cell. By placing the return value into the Background Sheet, it keeps the response directly connected to the other information associated with that particular question.
A. Scoring – This is the value from the form control that is converted to the score for a given question on the basis of the maximum score. This could also be called the weighted score for the question. The weighted score is foundational to tool functionality and, as aforementioned, is built upon operating knowledge and experience, industry data, expert opinion and incident case studies.
B. Logic – Several logic steps are used to evaluate relative sections, conditional scoring limits, and even tool use boundaries communicated directly to the user. For example, if the parameters selected in response to certain questions place results outside the intended tool scope, the user is notified by a prompt box (i.e. low MIE powders, self-heating materials, etc.)
C. Additional information, such as considerations to decrease the risk.

Table 1: Typical Layout of Background Sheet

Type of Text	Text / Question	Yes/Option Buttion Return	No	Index for Option Button Lookup	Score	Max Response Score	Max Question Score	Section Score	Section Max Score	Calculation Logic	Stop Logic	Considerations
H	1.0 Minimum Ignition Energy							40	45			
TF	Is the Minimum Ignition Energy <30 mJ?	FALSE	TRUE		0						Stop if True	*** Tool Not Valid for
SH	1.1. Electrical and Electrostatic Discharge	FALSE	TRUE									
TF	Is the belt constructed with conductive or static di	TRUE	FALSE		5	5	5					Consider using a belt
TF	Are the buckets made of non metal or other non-c	TRUE	FALSE		5	5	5					Consider using bucket
SH	1.2. Hazard Area Classification											
TF	Has the location (bucket elevator) completed a su	TRUE	FALSE		5	5	5					Consider completing a
TF	If an area was found to be at electrical hazard loca	TRUE	FALSE		5	5	5					If an area was found t
TF	Is there routine inspection of electrical equipmen	TRUE	FALSE		5	5	5					Ensure routine inspec

The second step in tool creation involved developing worksheets for user interface (see Figure 1 – Example User Entry **Sheet**). The user interface is built with a series of standard Excel controls such as check boxes and multi-button or option button controls. The result is a clean user entry sheet that enables the user to focus on answering questions presented via the form tools. As responses are entered, resulting values are linked to cells on the Background Sheet. The responses are then converted from Excel form tool responses into the scored values for the questions.

Questions from the User Entry Sheets are designed for the appropriate level of technical information to be entered. Most questions are simple yes/no type and describe physical properties of materials being handled, bucket elevator configuration, existing protections in place, and operational controls used to prevent incidents. The chosen format is helpful to the user by allowing a simple click rather than typing lengthy responses, and further helps reduce errors by displaying the responses selected.

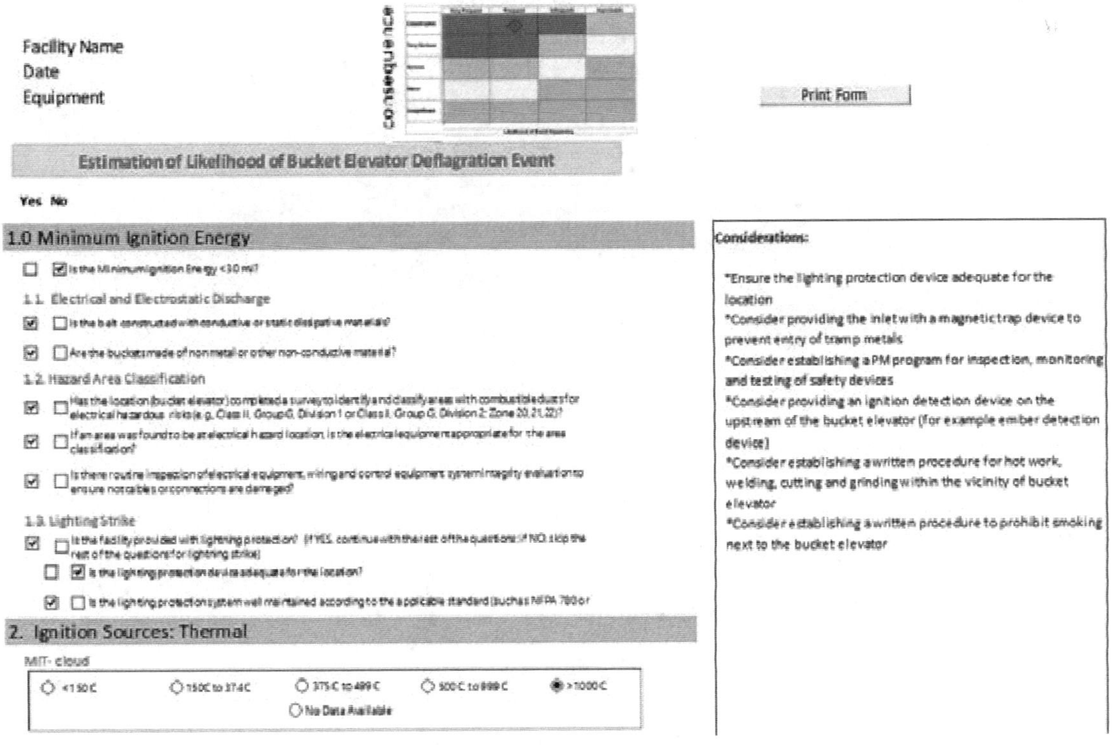

Figure 1 – Example User Entry Sheet

The user interface page is enhanced with a mitigated risk graph depicting risk relative to an organization's risk tolerance levels. A digital readout of the values shows the overall change as data is entered. Lastly, there is a place where the user can view suggestions for reducing overall risk.

3. Risk Concepts

According to Risk Based Process Safety concepts, there are three main questions associated with risk. They include [9]:

Hazard - What can go wrong?

Consequences - How bad could it be?

Likelihood - How often might it happen?

These questions are fundamental to adequately understand the risk at hand. When we ask these questions, the objective is to perform an analysis sufficient to understand the level of risk and to determine if existing levels of management system elements and safeguards employed align with the organization's risk tolerance [1].

Each person and organization have different risk mitigation approaches and level of risk tolerance. The organization's risk acceptance criterion is defined in the tool as a matrix (see Figure 2) displaying the range from broadly acceptable risks through risks that are completely intolerable. Though this is not an extremely precise value, it provides a visual representation of risk level based on a more quantitative approach. A more quantitative approach reduces personal biases in interpreting risk and aligns results to an organization's expectations, consistently.

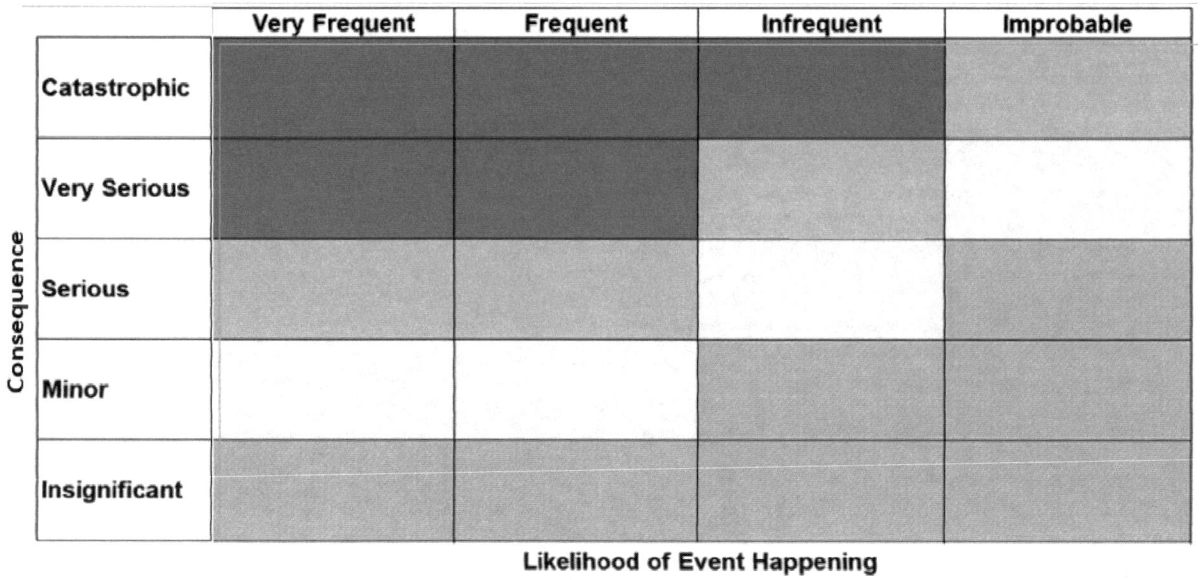

Figure 2: Example Risk Map

Green – **Low risk:** risks here are insignificant with no further actions required

– **Medium:** risks ranked here require reasonable steps to reduce; not the most urgent priority; action plan

Orange – **High risk:** risks here are high priority and call for immediate action; action plan

Red – **Extreme:** risks here are critical and the highest priority; immediate action; action plan

Likelihood and Consequence factors are used to distill conditions and controls associated with a bucket elevator into values used to prioritize implementation of potential solutions. For the tool, these factors are built into a summary form on a sheet and into the headers of the User Entry sheets for both the likelihood and consequence. This approach to risk, when combined with the quantified questions used on the User Entry Sheets, provides users with a mitigated risk aligned with an organization's risk tolerance.

4. Results and Discussions

As a result of this tool and associated work, an end user is able to bring a higher level of consistency to the analysis. The user is also able to be more quantitative when identifying dust hazards. The tool takes into consideration the various factors which could cause combustible dust explosions. As generally known, three elements are required for a fire (1) a fuel; (2) an oxidant, typically the oxygen in air; and (3) a sufficiently energetic ignition source. If any one of these three elements are removed, a fire cannot be initiated. The first two of these elements (fuel and oxidant) when in an appropriate ratio, are referred to as the flammable atmosphere. For an explosion involving a combustible dust to occur, two additional requirements are necessary: (4) suspension or mixing of the combustible dust in air at or above the lower explosible concentration, and (5) confinement. When a combustible dust cloud is ignited, a flash fire (flame propagation without significant pressure effects) will ensue and the developing fireball can cause serious injuries to people and damage to process equipment. If the dust-air mixture is confined in a vessel or building, the expansion of the hot combustion materials will be impeded while the pressure increases, leading to an explosion. Due to the increase in pressure, equipment and building damages are likely.

This tool also takes into consideration the possibility of fuel generation, ignition sources consequences. In the case of event consequences, various parameters such as construction and design, equipment location, personnel and public risk, and mitigating measures such as explosion protection (relief venting or suppression and isolation) are evaluated. Moreover, housekeeping in the process area being assessed is considered. Detailed questionnaires based on operation experience, industrial data, operational principles of the equipment are developed for each parameter mentioned above and values are assigned for likelihood and consequence. The values generated from each questionnaire are summed and plotted in the risk matrix as a mitigated risk. In order to test the tool, five grain elevator scenarios were used as presented in the Figures 3 to 7.

Figure 3 displays the mitigated risk map for scenario A. In this example, the bucket elevator is well designed and located outdoors with personnel egress restrictions. The bucket elevator is provided with explosion relief venting according to NFPA 68 and equipped with upstream and downstream isolation. It has all recommended safety devices and a monthly preventive maintenance program for inspection that includes monitoring and testing. As detailed above, personnel restrictions are in place and operators work around the bucket elevator once a week. The mitigated risk of an explosion is at the intersection of "infrequent" and "minor". No additional requirements are required for this bucket elevator.

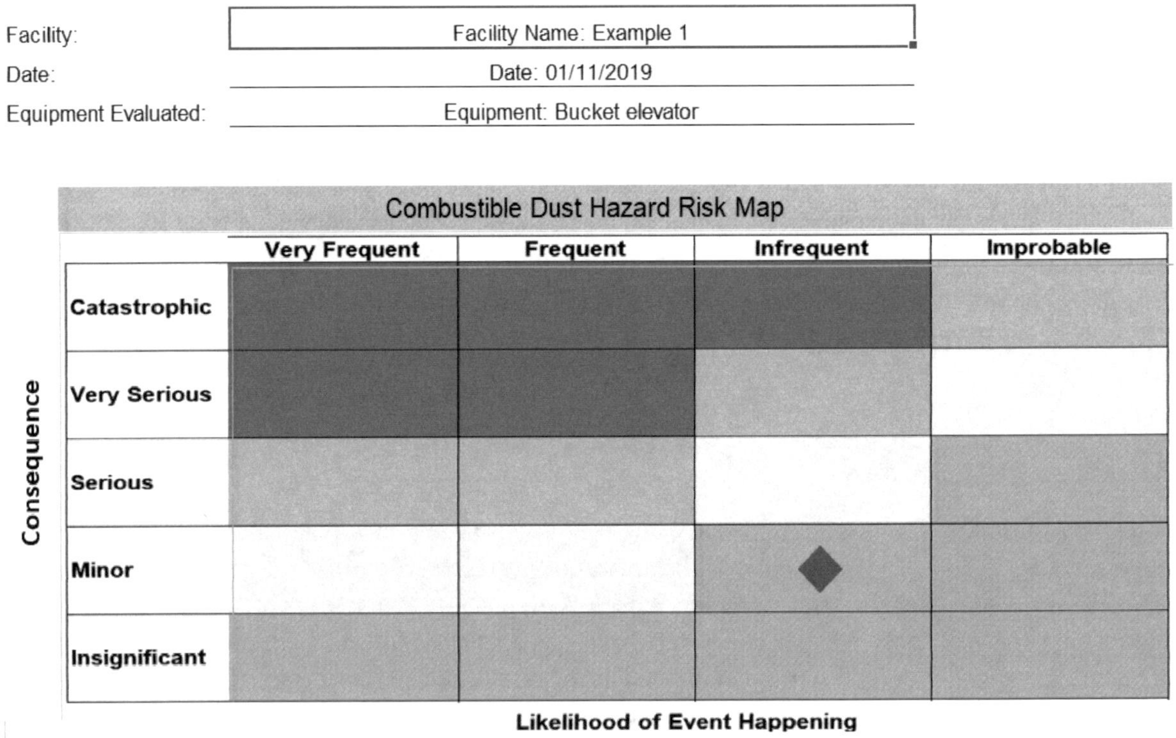

Figure 3: Example Bucket Elevator Scenario A

Figure 4 presents the mitigated risk map for scenario B. In this example, the bucket elevator is designed with safeguards and located indoors with no personnel egress restrictions (that is personnel work 25 feet around it while the bucket elevator is running). The elevator leg is provided with explosion relief venting ducted outdoors and equipped with explosion isolation devices. It has all required safety devices and a monthly preventive maintenance program for inspection, monitoring and testing. In this case, the bucket elevator is located indoors with evidence of housekeeping issues. There is an established housekeeping program, but it is not strictly followed. The likelihood of experiencing an explosion inside the bucket elevator is "infrequent" and the consequence of explosion is defined as "serious".

With regards to scenario C represented in Figure 5, the bucket elevator is located outdoors. It is equipped with explosion relief venting, but no isolation. The bucket elevator has all the required safety devices including a monthly preventive maintenance program for inspections, monitoring and testing. The likelihood of experiencing an explosion inside the bucket elevator is "infrequent" and the consequence of an explosion in the bucket elevator is categorized as "very serious". Any deflagration event inside the bucket elevator could propagate to upstream and downstream operations, thereby potentially harming the operators in these locations.

Facility:	Facility Name: Example 2
Date:	Date: 01/11/2019
Equipment Evaluated:	Equipment: Bucket elevator

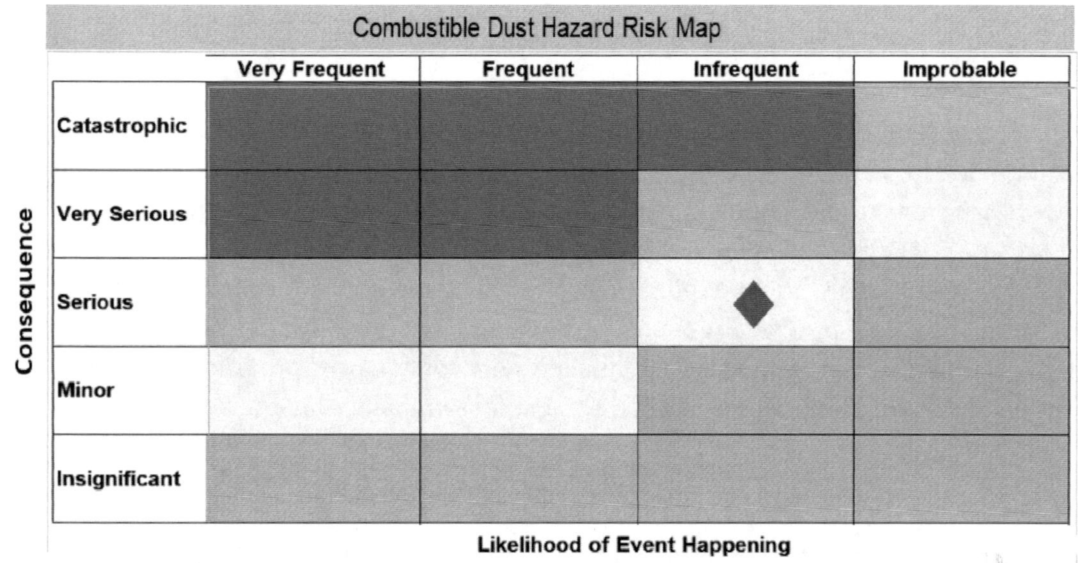

Figure 4: Example Bucket Elevator Scenario B

Facility:	Facility Name: Example 3
Date:	Date: 01/11/2019
Equipment Evaluated:	Equipment: Bucket elevator

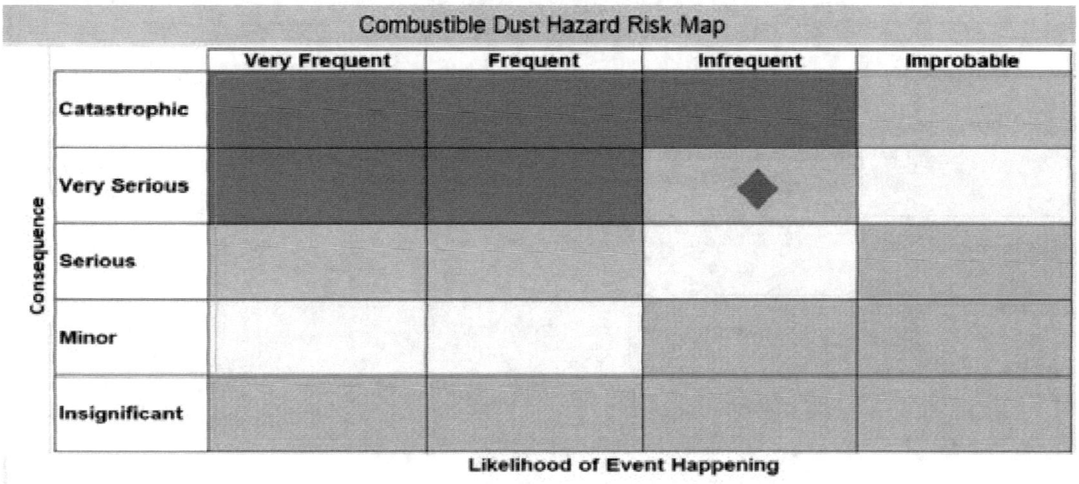

Figure 5: Example Bucket Elevator Scenario C

Figure 6 presents the mitigated risk map for scenario D. In this example, the bucket elevator is located indoors where personnel work within 25 feet of the elevator while it is running. The elevator is provided with explosion relief vents routed outdoors without explosion isolation. The consequence of an explosion in the bucket elevator is categorized as "very serious". Any deflagration event inside the bucket elevator has potential to propagate upstream and downstream to adjacent operations, with the potential to harm operators in these locations. This example simulates a facility that has limited explosion prevention measures. For example, there are no belt and pulley alignment sensors to detect misalignment and there is no preventive maintenance program to inspect, monitor and test safety devices. The likelihood of fire or explosion inside this bucket elevator is "frequent".

Finally, Figure 7 depicts the mitigated risk map for example scenario E. The bucket elevator is located indoors with personnel working within 25 feet while the bucket elevator is running. It is neither provided with explosion relief venting nor explosion isolation. There is no housekeeping program in place to maintain accumulation levels below acceptable limits. Dust accumulations in this process area create a secondary explosion hazard and the consequence of explosion in the bucket elevator is categorized as "catastrophic". Moreover, this example was populated to simulate the impacts of operations without explosion protection measures in place. Electrical equipment are not classified, belt and pulley alignment sensors are not installed to detect misalignment, no preventive maintenance program, no established hot work permit program, no upstream ignition detection devices, etc. The likelihood of fire or explosion inside this bucket elevator is considered "very frequent"

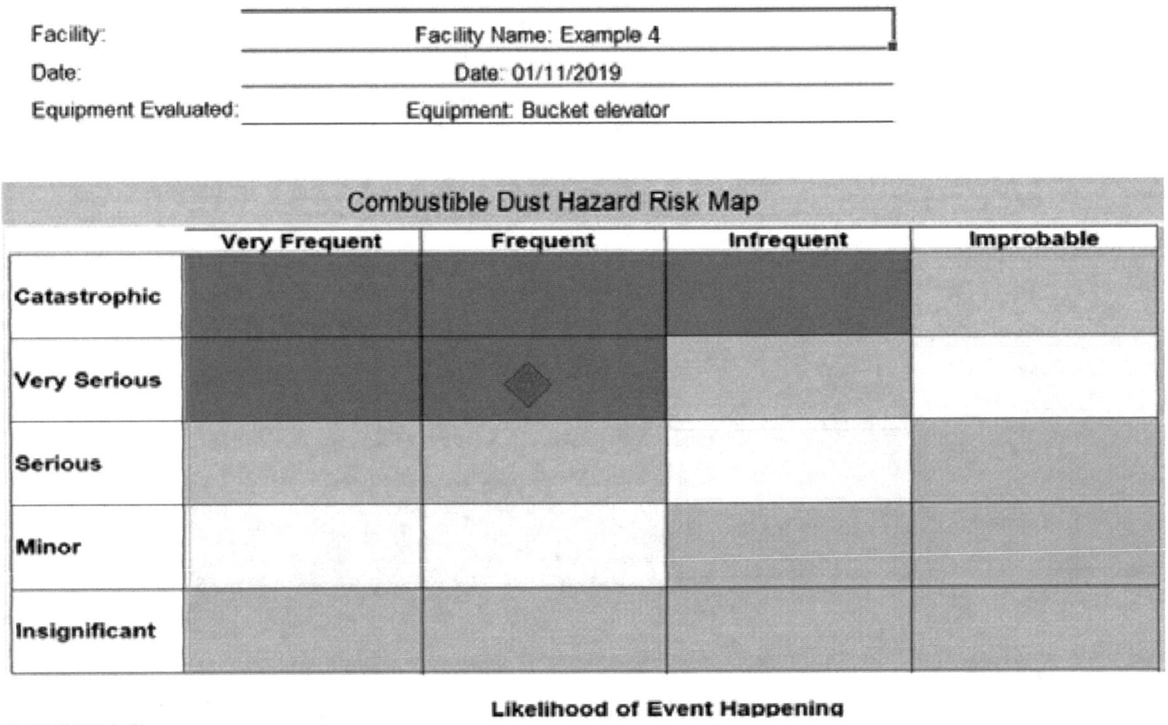

Figure 6: Example Bucket Elevator Scenario D

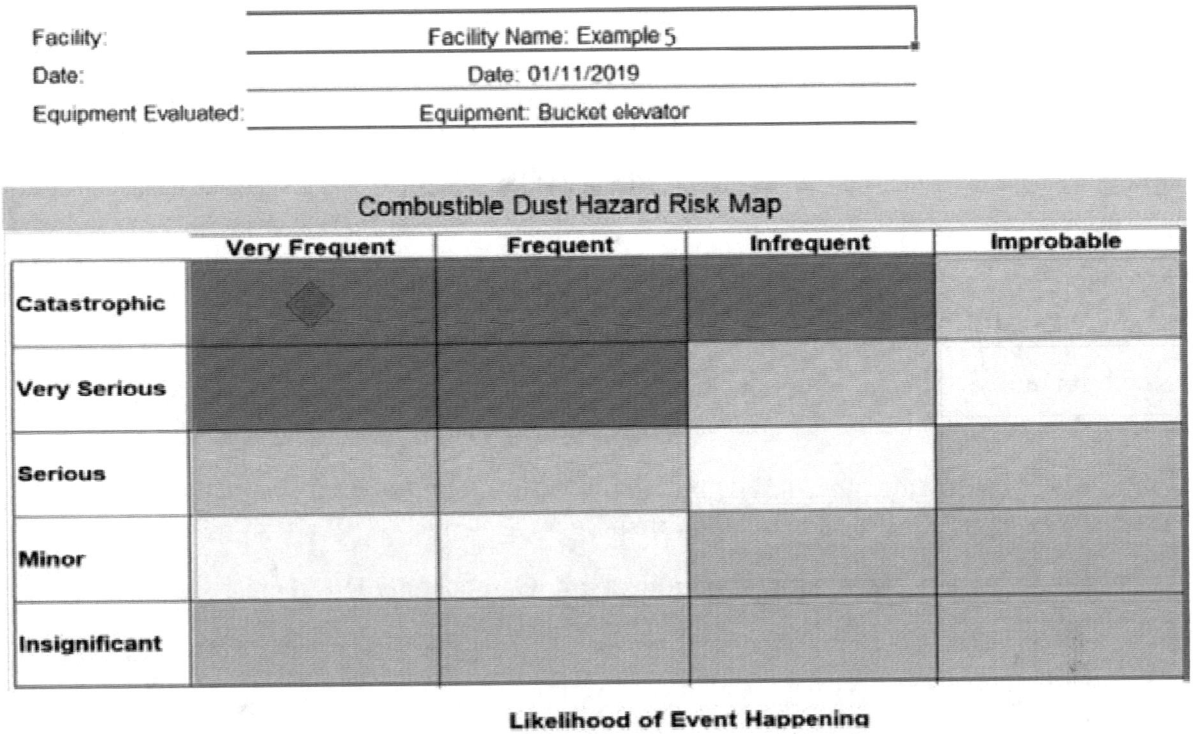

Figure 7: Example Bucket Elevator Scenario E

5. Future actions

Future actions include applying similar methods in developing a suite of customized tools designed to address common solids handing unit operations such as hammer mills, belt conveyors, silos, pneumatic conveying, dust collection system, etc. Future work plans include continued validation of tool modules by applying actual facility data and benchmarking with incident histories. These iterative steps allow for further calibration of the tool modules.

6. Conclusion

A semi-quantitative risk-based screening tool has been developed to consistently and effectively evaluate dust explosion risks for bucket elevators with future tool modules in development. This tool is designed for use by on-site supervisors and local subject matter experts (SMEs) trained in combustible dust hazards and the tool. The tool uses multiple-choice and yes/no questions to assess both likelihood and severity of risk. Values are assigned to each response, providing the opportunity to weigh relative importance of each question, in real-time. Values for Severity and Likelihood are used to automatically show the mitigated risk plotted on a Risk Matrix. The function of this tool is to identify and evaluate the most common preventive, protective and mitigative measures an organization has taken to prevent combustible dust related fires and explosions. When this measures evaluation is combined with combustible dust test data (such as explosibility properties of the material; minimum ignition energy, minimum ignition temperature, deflagration index, maximum explosion pressure, self-ignition temperature, limiting oxygen concentration, minimum explosible concentration etc.), a visual representation of risk is obtained. In addition to providing an estimated risk level, the tool identifies potential safeguarding opportunities and thus providing guidance for improvement.

References

[1] CCPS Guidelines for Combustible Dust Hazard Analysis, 2017

[2] NFPA 61 Standard for the Prevention of Fires and Dust Explosions in Agricultural and Food Processing Facilities, 2019.

[3] Mannan, S., Lees' Process Safety Essentials: Hazard Identification, Assessment and Control. 2013: Butterworth-Heinemann.

[4] Khan, F.I. and S. Abbasi, Major accidents in process industries and an analysis of causes and consequences. Journal of Loss Prevention in the process Industries, 1999. 12(5): p. 361-378.

[5] Eckhoff, R., Dust explosions in the process industries: identification, assessment and control of dust hazards. 2003: Gulf professional publishing.

[6] Chemical Safety and Hazard Investigation Board, Combustible Dust Hazard Study, 2006-H-1, November 2006,

[7] CSB incident database, as of 4/23/2018.

[8] There were 14 fatalities and 36 injured at Imperial Sugar (2009), three fatalities and one injured at AL Solutions (2014), five fatalities and three injured at Hoeganaes (2012), seven injured at US Ink (2012), and five fatalities and 14 injured at Didion Milling (2017). The CSB also investigated West Pharmaceutical Services (2004), CTA Acoustics (2005), and Hayes Lemmerz (2005). In total, these eight incidents cost the lives of 41 individuals and injured 142 others.

[9] CCPS Guidelines for Risk Based Process Safety, 2007

19 AIChE Spring Meeting
15TH GLOBAL CONGRESS ON PROCESS SAFETY
NEW ORLEANS • 2019

Two Plants, One Fence Line: Understanding Process Safety Management Responsibilities in Shared Facilities

Matthew S. Walters
Exponent, Inc.
Warrenville, IL 60555
mwalters@exponent.com

Sean J. Dee, Russel A. Ogle
Exponent, Inc.
Warrenville, IL 60555

Prepared for Presentation at
American Institute of Chemical Engineers
2019 Spring Meeting and 15th Global Congress on Process Safety
New Orleans, LA
March 31 – April 3, 2019

Two Plants, One Fence Line: Understanding Process Safety Management Responsibilities in Shared Facilities

Matthew S. Walters
Exponent, Inc.
Warrenville, IL 60555
mwalters@exponent.com

Sean J. Dee, Russel A. Ogle
Exponent, Inc.
Warrenville, IL 60555

Keywords: shared facilities, Risk Based Process Safety, emergency response

Abstract

Multiple companies operating process plants at a single site has become increasingly common as aging facilities are sold and re-purposed, but this scenario creates unique challenges for effective Process Safety Management (PSM). The PSM responsibilities of the owner/operator of a plant located within a shared facility will be explored using a case study of a fertilizer company and a specialty chemicals company operating at a shared site that started as a single plant over half a century ago. This paper will examine the responsibilities of the owner/operator of each facility at the site in the context of the Risk Based Process Safety (RBPS) approach to PSM developed by CCPS. Specifically, the RBPS elements of "Stakeholder Outreach" and "Conduct of Operations" will be examined in detail, emphasizing the concerns that may arise when more than one entity is responsible for the operation of processes within the fence line. Additionally, the element of "Emergency Response" will be analyzed through discussion of an incident in the case study where a contract employee working on behalf of the specialty chemical company was allegedly exposed to a hazardous chemical originating from the fertilizer plant. This case study will provide guidance for effective PSM practices at process facilities that contain more than one owner/operator.

1 Introduction

The end of World War II brought an excess of anhydrous ammonia production capacity to the United States [1]. With drying up federal subsidies and a dwindling demand for explosives, anhydrous ammonia (NH_3) producers were forced to find a new market or face extinction. As a result of extensive experience gained from ammonium nitrate (NH_4NO_3) explosive manufacturing, improved packaging techniques for safe transportation, and increased nitrogen demand in the agricultural industry, ammonia plants expanded their capabilities to generate valuable nitrogen fertilizer products.

At one such plant, Company X added a number of production units to manufacture nitric acid (HNO_3, a precursor to NH_4NO_3) and NH_4NO_3 itself. However, the 1940s-era process of NH_3 production through coal gasification operated by Company X became antiquated as modern, large-scale NH_3 plants using natural gas or naphtha feedstock and efficient steam-driven centrifugal compressors came online. Company X ultimately decided to sell off its nitrogen products business to Company A and shut down the anhydrous NH_3 process. Company A now receives anhydrous NH_3 in a more economical fashion via pipeline. The shuttered NH_3 production equipment was eventually acquired by Company B and re-purposed for the manufacture of specialty chemicals. Companies A and B now operate two independent processes within a once-integrated site that still shares utilities, office space, and a common fence line. A simplified layout of the site shared by Companies A and B is shown in Figure 1.

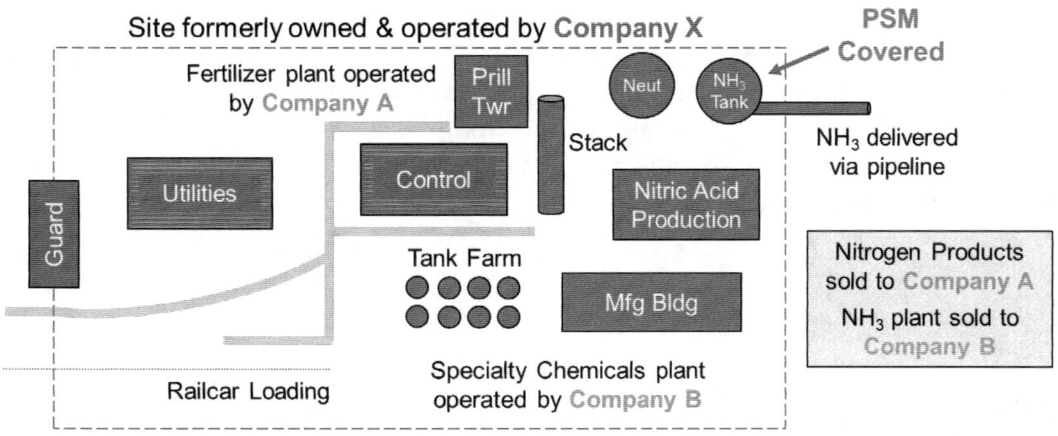

Figure 1. Overview of the former ammonia plant site now shared by Companies A and B.

Two independent companies sharing a plant site is not unique to former ammonia plants; chemical manufacturers are frequently forced to sell or re-purpose aging facilities in an increasingly competitive market [2]. There are additional situations where independent operators exist within a single facility and operate separate portions of one process, such as an industrial gas plant within a refinery. This paper focuses on an example where there is no material flow between separate entities within the site. However, the learnings and guidance highlighted in this paper may be relevant and applicable to similar facilities where materials may flow into a shared production process. While safety guidelines or requirements for multi-employer worksites exist, they are often focused on the relationship between contractors and the site owner/operator [3]. Using Companies A and B as a case study, the Process Safety Management (PSM) responsibilities when two chemical manufacturers are *operating independently* at the same site will be explored in the context of the Risk Based Process Safety (RBPS) guidelines developed by the Center for Chemical Process Safety (CCPS) [4].

2 Case Study Overview

Company A uses the Ostwald process to manufacture HNO_3 (see Figure 2) [5]. In the Ostwald process, a catalytic reactor converts a mixture of NH_3 and air to nitric oxide (NO). This highly exothermic reaction is used to produce steam in a waste heat boiler, and a portion of the steam is sold to Company B. As the NO is cooled, it is mixed with additional air for further oxidation to nitrogen dioxide (NO_2). The NO_2 is subsequently absorbed in a pressurized column to form the weak HNO_3 product. The HNO_3 is then either concentrated for sale or neutralized with NH_3 to form NH_4NO_3. In contrast, Company B handles malodorous but generally nonhazardous chemicals in its own proprietary specialty chemical production processes.

Although Companies A and B have a good working relationship, there is limited process knowledge cross-over or communication between engineers and operators of the two organizations due to the dissimilarity of their respective processes. Furthermore, both companies regularly hire contract employees to perform construction or maintenance activities in their respective portions of the site. Company A is covered by the OSHA PSM standard due to the amount of anhydrous NH_3 stored onsite and therefore has specific requirements for managing the hazards associated with NH_3 [6]. Company B is not covered by the PSM standard, but is nevertheless still responsible for ensuring that the workplace is safe for work given the recognized hazards [7]. In the following sections, the roles and responsibilities of the two cohabitating companies will be analyzed based on RBPS best practices.

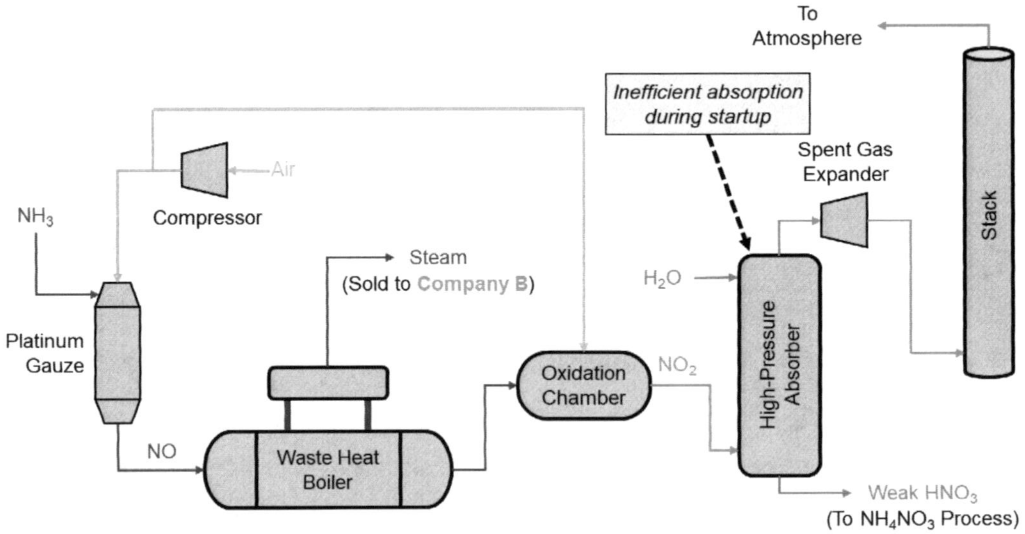

Figure 2. General process flow diagram for the Ostwald process.

3 PSM at Multi-Employer Worksites

Assigning appropriate RBPS responsibilities at multi-employer worksites requires clearly designated roles for all parties. Using definitions set by the local authority having jurisdiction for different categories of employers that exist at a shared worksite can be a convenient tool for understanding relationships within the site. As an illustrative example to define the potential roles and responsibilities of the companies involved in the case study, this paper will use the classifications provided by the United States Occupational Safety and Health Administration (OSHA) as a frame of reference.

3.1 OSHA Multi-Employer Classifications

When a hazardous condition occurs at a multi-party site within the United States, OSHA uses the following employer classifications: *Creating*, *Exposing*, *Correcting*, and/or *Controlling* [3]. A definition for each classification is provided in Figure 3.These classifications are neither mutually exclusive nor static; the classification for the individual employers will change depending on the hazard scenario. OSHA's Multi-Employer Citation Policy will not be addressed in this paper or case study, but the classifications will be used as a framework for defining roles within a site.

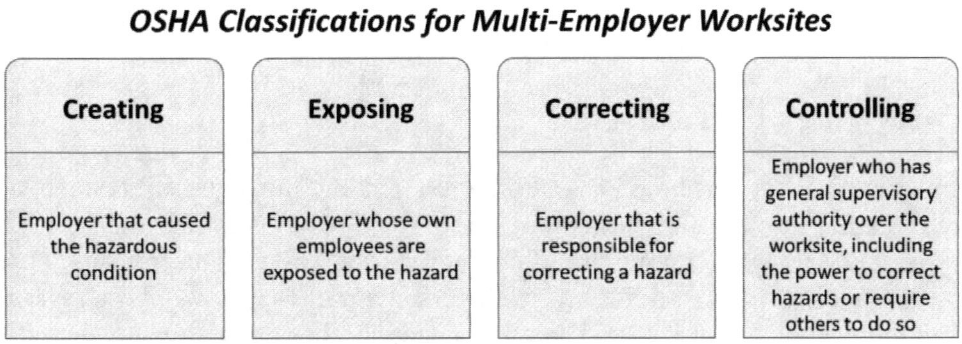

Figure 3. Definitions for employers at a multi-employer worksite.

3.2 RBPS Approach

The RBPS approach developed by CCPS provides guidelines for designing and implementing effective PSM systems [4]. Three elements of RBPS that present unique challenges to multi-employer worksites and will be considered in this case study are Conduct of Operations, Stakeholder Outreach, and Emergency Management, as outlined in Figure 4. Note that this paper does not address legal obligations for multi-employer worksites under OSHA PSM or any other regulation, but rather discusses best practices based on RBPS guidelines.

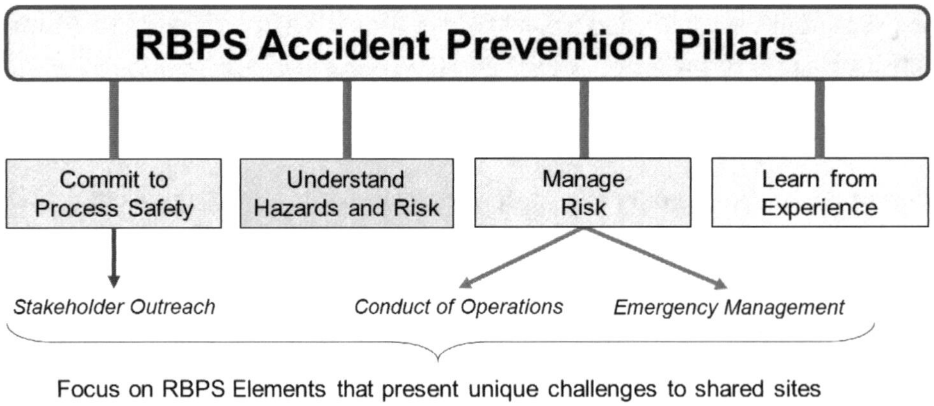

Figure 4. Overview of RBPS, with the elements that will be considered in this case study for multi-employer worksites emphasized.

3.2.1 Conduct of Operations

Conduct of operations is the execution of both operational and management tasks in a deliberate and structured manner. A clear chain of command, defined responsibilities, and lines of communication are an integral component of conduct of operations. This not only applies within one organization, but should exist between employers at a multi-party worksite as well. This is because the actions of one company can directly impact another company at a shared facility. For example, an effective line of communication must exist between the Controlling Employer and Creating Employer in the event that they are different companies at a multi-employer site.

3.2.2 Stakeholder Outreach

Relevant stakeholders for a company at a shared facility invariably include the other employers (and their respective employees) at the site. Stakeholder Outreach can be further complicated when accounting for contractor relationships given the separation between the entity that has hired the contractor and the other entities on the site. Clear lines of communication to contractors can play a key role in incident response, as discussed in the case study. As part of the RBPS element of Stakeholder Outreach, neighboring employers should openly communicate potential hazards with one another. This information should also have a clear line of communication to third-party contractors onsite. Additionally, routine activities that may impact the operations of other stakeholders at a site (e.g. tanker truck delivery or utilities interruption) should also be communicated. Simply put, competent personnel from each respective employer at a multi-employer site should have a general understanding of the activities occurring around them. However, engineers from one organization do not need a working knowledge of their neighbor's process; rather, all persons on site need awareness of the recognized hazards that they may encounter and the appropriate response. Stakeholder Outreach does not require the sharing of confidential business information, although certain individuals may need detailed knowledge of process hazards if that information is necessary for effective emergency response.

3.2.3 Emergency Management

The RBPS element of Emergency Management encompasses all aspects of preparation for and response to emergency situations, including planning, providing resources, practicing and improvement, training, and effective communication with stakeholders during an incident. At a multi-employer site, there can be more than one Controlling employer. Therefore, emergency plans should clearly define an incident command structure, including channels of communication for Controlling employers.

4 Case Study Involving Alleged Hazardous Release at Shared Facility

4.1 The Situation

Company B hired Contractor C to complete construction work at the shared site. Contractor C subsequently hired Subcontractor S. Neither Contractor C nor Subcontractor S have any relationship with Company A. Both Contractor C and Subcontractor S were in the field when Company A began a planned startup of the HNO_3 production process. During startup, increased NO_2 emissions are known to occur as the absorption column comes up to operating pressure. NO_2 gas can appear yellow, orange, brown, or red, so the NO_2 emissions during startup are visible from Company A's stack [8]. Although Company A regularly conducts startups of its HNO_3 process as a result of planned maintenance or unscheduled outages, Subcontractor S had never been onsite during a startup occurrence. Upon observing the reddish-brown cloud of NO_2 emissions from the stack, employees for S believed that an uncontrolled chemical release was underway and that they may have potentially been exposed to a hazardous chemical.

In this situation, the roles of each employer using OSHA definitions as an illustrative guideline are as follows (also summarized in Figure 5):

4.1.1 Creating Employer

Company A allegedly released a hazardous chemical from its process, so A is the alleged Creating Employer.

4.1.2 Exposing Employer

Employees from Subcontractor S believed that they were exposed to a hazard, making S the alleged Exposing Employer.

4.1.3 Correcting Employer

In addition to being the Creating Employer, Company A is also the Correcting Employer. Only Company A has the ability to correct a hazard arising from the HNO_3 production process.

4.1.4 Controlling Employer

The Controlling Employer in this situation is not obvious, and the examples of Controlling employers provided in the OSHA guidelines and interpretations do not address this specific scenario [3]. As defined in Figure 3, the Controlling Employer has general supervisory authority over a worksite. This definition would likely preclude Company A because it has no relationship with Contractor C or Subcontractor S. Contractor C conducts periodic safety inspections for subcontractors and holds daily safety meetings with all contract and subcontract employees. Additionally, Contactor C has stop work authority over Subcontractor S. In the case of a hazard arising from construction activity (e.g. broken handrail on scaffolding), Contractor C would likely be considered the Controlling Employer. However, Contractor C has no power to correct an alleged safety issue arising from Company A, so C would likely not be considered the Controlling Employer in this situation. Therefore, Company B is the most likely candidate for the Controlling Employer. Only Company B has authority over both Contractor C and Subcontractor S, as well as the power to correct the hazard through the site emergency response plan developed in conjunction with Company A.

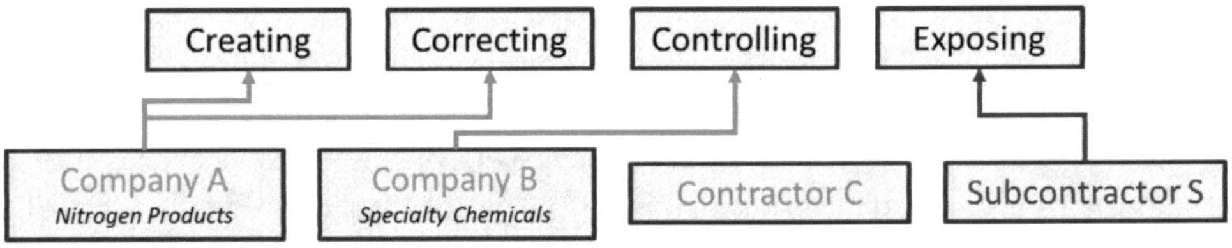

Figure 5. Classification of employers in the case study.

4.2 The Incident Response

Upon learning of the alleged release, a supervisor from Subcontractor S instructed his employees to stop work and assemble at the designated safety area. He also informed the safety director of Contractor C of his intention to stop work and recommended that Contractor C instruct its employees to move to a safe location as well. The safety director for Contractor C did not believe that a hazard existed and ordered Subcontractor S to return to work. The conflicting directions from representatives for Subcontractor S and Contractor C potentially created confusion for the contract and subcontract employees, with some employees for both C and S assembling at designated safety areas while others stayed in the field.

Despite not believing there was a hazard, the safety director for Contractor C called the site superintendent for Company B to report the alleged exposure. The site superintendent subsequently called the EHS director for Company A. The EHS director for Company A was able to quickly determine that the HNO_3 process was undergoing a startup and that no abnormal conditions had been encountered. The EHS director for Company A relayed this information to his counterpart at Company B, and then went into the field to further assess the situation. It is unclear what information was ultimately passed along to the Subcontractor S employees, but some Subcontractor S employees came to believe that they had been exposed to HNO_3 (not NO_2, which is what was visually observed being emitted from the stack). Several Subcontractor S employees then decided to seek medical attention after complaining of headaches and nausea.

4.3 Good Emergency Management Practices

The RBPS element of Conduct of Operations can be applied to a situation where an employee believes a hazard exists. A clear organizational structure and chain of command should be in place to respond, especially when a process hazard could potentially arise from two independent companies. The chain of command in this case study was reasonably well-defined. The alleged Exposing employer (Subcontractor S) was able to communicate the perceived emergency to the Controlling employer (Company B) via Contractor C. The Controlling employer (Company B) then determined that the worksite was safe by contacting the appropriate individual at the alleged Creating employer (Company A). It appears that personnel at each of the four entities at the site were aware of the appropriate reporting structure, as shown in Figure 6 (a).

Good Conduct of Operations would also dictate that emergency response procedures be followed until a potential hazard has been addressed. Once the supervisor for Subcontractor S gave the stop work order, the Controlling employer (Company B) should have initiated its emergency response protocol once it was informed and taken control of operations for all contract and subcontract employees. Because Contractor C does not have the competency to assess a process hazard, Contractor C should not have had the authority to order any workers on site to return to the field. The actions taken (or not taken) by Company B and Contractor C after a reported hazard did not demonstrate good Conduct of Operations.

Stakeholder Outreach is its own RBPS element, as well as a component of the element of Emergency Management. Addressing both real and perceived process safety concerns is therefore good practice for maintaining a safe worksite and a well-functioning relationship with other parties. Company A has a responsibility to openly communicate hazards, which it sufficiently did with respect to Company B. Company B must then enable the transfer of that knowledge to its own employees, as well as to employees at any employer it controls. It may be appropriate for Company B to facilitate direct communication from Company A to either Contractor C or Subcontractor S. In this case study, Company B could have asked for a representative from Company A to provide a portable gas detector and address the concerns raised by Subcontractor S. The flow of information that should have occurred is shown in Figure 6 (b).

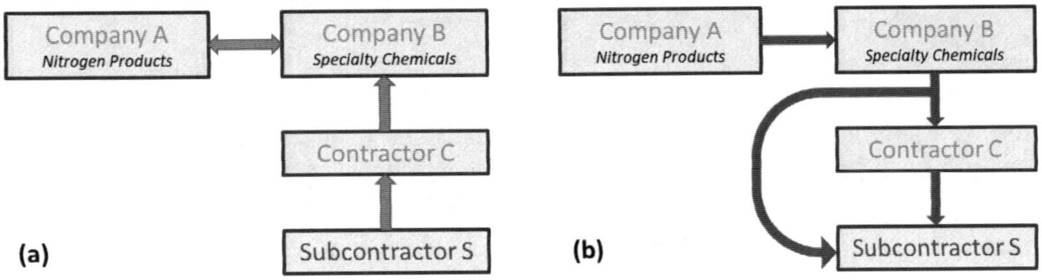

Figure 6. Good practice for (a) incident reporting and (b) information sharing in the event of a potential process hazard created by Company A.

4.4 Discussion

The false alarm described in this case study exposed RBPS deficiencies at the former NH_3 plant now shared by a nitrogen products company and a specialty chemicals company. Company B had a responsibility to manage process safety issues that may impact its regular or contract employees, regardless of whether or not Company B was the Creating employer. However, the inadequacies discussed in this section with regards to Conduct of Operations, Stakeholder

Outreach, and Emergency Management demonstrated that the site was not equipped to handle a process safety incident at Company A that could potentially effect contractors working for Company B. Furthermore, Subcontractor S employees onsite came to believe that they were exposed to a hazardous chemical because of these inadequacies. The Lessons Learned below provides guidance for preventing breakdowns in good RBPS practices at shared process facilities based on this case study.

5 Lessons Learned

5.1 Routine activities should be communicated at shared sites

Operators for Company A have direct radio access to the control room of Company B. It is regular practice to alert Company B that a site-wide steam interruption will occur during the HNO_3 process startup. This communication of routine activity that may affect another company onsite demonstrates good Conduct of Operations. However, this case study showed that it is worthwhile for all employees onsite to have a general understanding of what they should and should not see, hear, or smell in the workplace. For example, it would be appropriate to tell contract employees at their daily safety meeting that they may see a yellow, orange, brown, or red cloud emitted at some point during the day when an HNO_3 startup is scheduled to occur. Knowing what is "normal" will aid employees in reporting potential safety hazards and reduce false alarms. Additionally, defining appropriate response to process safety concerns can be beneficial during incident response. As an example, the safety director at Contractor C might have asked for an NO_2 concentration field survey to be performed to validate the safe for work conditions on site prior to instructing employees from both Contractor C and Subcontractor S to return to work. This response would have provided all the employees with information about the hazard and reassured them of the safe work environment onsite.

5.2 Process safety incidents require special reporting structures at shared sites

Process hazards are unique compared to other safety issues at multi-employer worksites. It is unlikely that Company B would be directly involved with construction or occupational hazards encountered by Subcontractor S. Instead, the responsibility for these types of hazards would be with the Contractor C. Process hazards, however, require involvement from the process operator (the Creating and Correcting employer) regardless of whether or not the operator is the Exposing or Controlling employer. This means special consideration should be given when deciding which entity should be the Controlling employer in the emergency plan.

5.3 The Controlling employer can view Stakeholder Outreach as a means to educate other parties onsite of the hazards to the process

Relevant stakeholders typically include groups such as neighbors, local authorities, or community organizations. At multi-employer sites, other employers onsite should also be considered as part of Stakeholder Outreach. The case study demonstrates that Stakeholder Outreach within the fence line is just as important at a shared site as outreach to external parties.

5.4 Employees from the Exposing employer need information to understand the hazard

The employees from Subcontractor S who decided to seek off-site medical attention did so without any of the pertinent information about the hazard. Not only did the Subcontractor S employees arrive at the emergency room without Safety Data Sheets or treatment instructions, but they incorrectly believed the emissions to be HNO_3. Had an actual exposure to a hazardous substance occurred, the employees may have received delayed or improper treatment and potentially put others in the community at risk. The Exposing employer needs timely access to relevant information about the potential hazard, which Company B should have either provided directly or compelled Company A to provide.

6 Conclusions

Even though it was ultimately shown that no hazard existed as a result of the routine activity of Company A, all employees onsite have the right to raise a safety concern and have that concern addressed. This case study demonstrated that shared facilities present unique challenges when responding to potential process safety issues. A legitimate process safety concern was not adequately addressed at the shared site as a result of these challenges and ineffective channels of communication. Care must be taken at shared worksites when applying the RBPS approach to ensure preparedness for process safety incidents, with special attention to Conduct of Operations, Stakeholder Outreach, and Emergency Management.

7 References.

[1] F.T. Nielsson (Ed.), *Manual of Fertilizer Processing*, Fertilizer Science and Technology Series – Vol 5, CRC Press, Boca Raton, FL, 1987. pp. 35–53.

[2] CCPS, Dealing with Aging Process Facilities and Infrastructure, American Institute of Chemical Engineers, Center for Chemical Process Safety, John Wiley & Sons, 2018.

[3] Multi-Employer Citation Policy (CPL 02-00-124), U.S. Occupational Safety and Health Administration, December 1999. www.osha.gov.

[4] CCPS, *Guidelines for Risk Based Process Safety*, American Institute of Chemical Engineers, Center for Chemical Process Safety, John Wiley & Sons, 2007.

[5] M. Thiemann, E. Scheibler, and K.W. Wiegand. "Nitric Acid, Nitrous Acid, and Nitrogen Oxides," *Ullmann's Encyclopedia of Industrial Chemistry*, Vol. 24, pp. 177–225.

[6] Process Safety Management of Highly Hazardous Chemicals (29 CFR 1910.119), U.S. Occupational Safety and Health Administration, May 1992. www.osha.gov.

[7] Section 5(a)(1) – General Duty Clause, Occupational Safety and Health Act of 1970, Public Law 91-596, 29 USC 654, December 29, 1970. www.osha.gov.

[8] Safety Data Sheet for Nitrogen Dioxide, Airgas, 2/14/2018, Version 1, Accessed 1/14/2019. www.airgas.com/msds/001041.pdf.

19 AIChE Spring Meeting
15TH GLOBAL CONGRESS ON PROCESS SAFETY
NEW ORLEANS • 2019

Strathcona County's Industrial Engagement Program: Leading the Way Using the MIACC Model

Modusser Tufail, P. Eng., M. Eng.
Fire Protection Engineer

Strathcona County Emergency Services
915 Bison Way,
Sherwood Park, Alberta T8H 1S9
Modusser.tufail@strathcona.ca

Prepared for Presentation at
American Institute of Chemical Engineers
2019 Spring Meeting and 15th Global Congress on Process Safety
New Orleans, LA
March 31 – April 3, 2019

Strathcona County's Industrial Engagement Program: Leading the Way Using the MIACC Model

Modusser Tufail, P. Eng., M. Eng.
Fire Protection Engineer

Strathcona County Emergency Services
915 Bison Way, Sherwood Park Alberta T8H 1S9
Modusser.tufail@strathcona.ca

Keywords: land buffering, risk based land use, process safety management, fire inspections

Abstract

Strathcona County is in the heart of Alberta's petrochemical district with oil refineries, chemical plants, oil and gas pipelines, and dangerous goods trains bringing unique challenges for public safety. Safe and sustainable industry is vital for the community to maintain lifestyle and stay as a 'World leader in petrochemical cluster'. The municipality's guiding document, Strathcona County Vision 2030 – Powering our New Tomorrow [1] incorporates this important strategic goal.

Therefore there is, and there will be a continuous presence of hazardous materials for storage, use, handling, transportation and similar operations. These materials and their operational situations are prone to industrial accidents, and there should be adequate protection and effective risk reduction programs in place to address them.

The industrial sector has been expanding for the last 50 to 60 years, and with every growth, the risk grows as well. Strathcona County has taken the initiative of addressing industrial risk by utilizing guidelines established by MIACC (Major Industrial Accident Council of Canada) and published by Canadian Society for Chemical Engineering Process Safety Management Division (CSChE-PSM). These guidelines provide a 'risk acceptability criteria' and land use solution to control risk on existing and new industrial sites. It also gives recommendations on how to develop the nearby lands to avoid encroachments and unwanted exposure of risk to the public. The municipal land use bylaw restricts the residential development up to 3 km from the land zoned 'heavy industrial', and at a regular interval, Strathcona County measures its cumulative risks exposed to the adjacent residential lands as part of its risk control activities.

Strathcona County Emergency Services (SCES) is part of the development review team and has now established a MIACC-based industrial follow up program called the Industrial Engagement Program, the first in Canada which focusses on risk control inspections and PSM education based on PSM Standard CSA Z767-17.

Strathcona County earned credibility for its leadership in using a fixed approach on land buffering in the MIACC guidelines. With the Industrial Engagement Program going forward, Strathcona County will continue its role as a leader in Public Safety in Canada which is in line with SCES's vision, "We will lead our community to be the safest in Canada" while fostering economic activities.

1 Background

In Canada and internationally, the Major Industrial Accident Council of Canada (MIACC) guidelines are known as a best practice to prevent major accidents by using Process Safety Management principles in land use planning in municipalities. It provides a risk acceptability criteria which helps to establish land buffers to maintain a sustainable infrastructure for public safety and municipal planning. Major industrial accidents may happen in heavy industrial occupancies such as crude oil refineries, chemical storage facilities, oil and gas pipelines, chemical and manufacturing plants, dangerous-goods trains carrying different hazardous products, etc.

MIACC guidelines principally deal with the hazardous substances manufactured, handled, stored, or transported to, from or through a community. It helps to identify what situations, materials or processes can go wrong, evaluating the potential consequences of those possible events and estimating how likely it is going to happen based on available scientific information.

Currently, the Canadian Society for Chemical Engineering Process Safety Management Division (CSChE-PSM) [2] are managing the MIACC guidelines. These guidelines are based on the principles of Process Safety Management (PSM), and that is why the following CSChE-PSM publications are directly related to its implementation:

1. Risk-Based Land-use Planning Guidelines
2. Hazardous Substances Risk Assessment: A MiniGuide for Municipalities and Industry.
3. Risk Assessment Guidelines for Municipalities and Industries - An Initial Screening Tool (1997)
4. Risk Assessment – Recommended Practice for Municipalities and Industry (2004)
5. Process Safety Management Standard 1st Edition
6. Process Safety Management Guide – 4th Edition
7. Process Safety Management Audit Protocol Ver.1.01
8. Guidelines for Site Risk Communication
9. Site Self Assessment Tool

Risk Acceptability Criteria of MIACC Guidelines

MIACC risk acceptability is based on the concept of location-specific annual individual fatality risk also called annual location risk (chance of fatality per year)

The maximum tolerable level for individual risk is that which increases the risk of death by all involuntary exposures caused by one order magnitude concerning everyday risk. The maximum acceptable mortality risk to human beings from all risk sources (major accidents, exposure to substances, radiation etc.) is defined such that the combined probability of mortality for these hazards should not exceed 1×10^{-6}. As it would be difficult to have an enumeration of all possible sources, for a single risk source the value of 1×10^{-6} per year is adopted as the maximum acceptable individual risk of death. Below this value, the tendency is to adopt an ALARP principle (As Low As Reasonably Practicable).

Figure 1 below represents the criteria for allowable land uses. The objective is to determine risk contours and do a 'risk based land use planning' allowing certain similar developments in an area and restricting the incompatible ones providing a land buffer for effective emergency response planning.

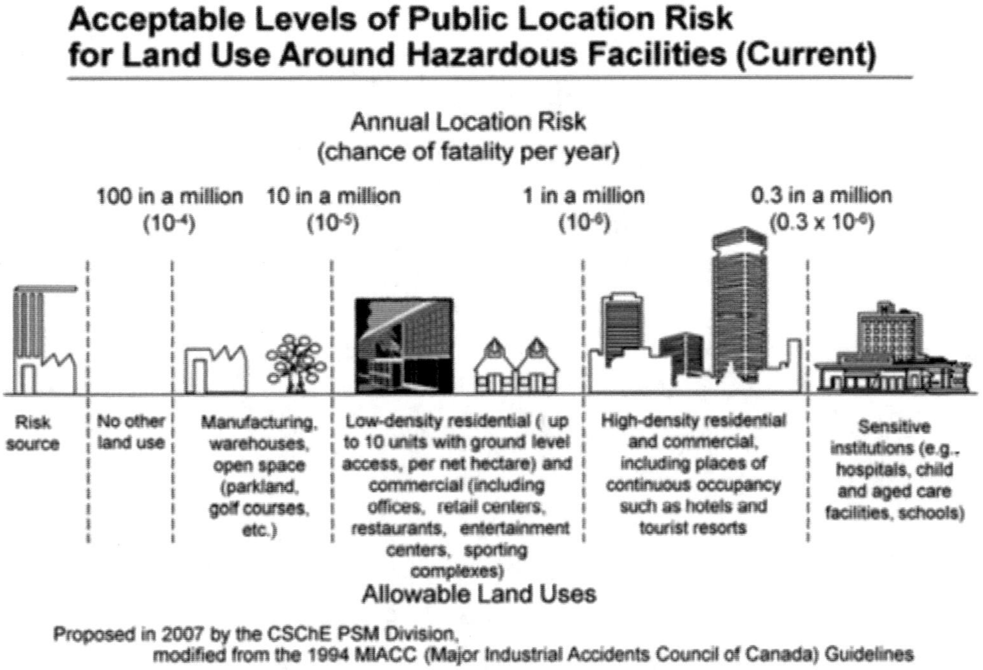

Figure 1: MIACC Criteria for Allowable Land Uses New Version 2007 [3]

From risk source to 1 $x10^{-4}$ risk contour: no other land uses except the source facility, pipeline or corridor.

From 1 $x10^{-4}$ to 1 $x10^{-5}$ risk contours: uses involving continuous access and the presence of limited numbers of people but easy evacuation, e.g. open space (parks, golf courses, conservation areas, trails, excluding recreation facilities such as arenas), warehouses, manufacturing plants.

From 1×10^{-5} to 1×10^{-6} risk contours: uses involving continuous access but easy evacuation, e.g., commercial uses, low-density residential areas, offices.

From 1×10^{-6} to $0.3 x10^{-6}$ risk contours: all other land uses including institutional uses, high-density residential areas, etc., except for sensitive receptors, such as schools, hospitals, elderly and child care facilities.

Beyond the $0.3x10^{-6}$ risk contour: all land uses without restrictions [3].

According to Alp Ertugrul [4], land use planning can be viewed as another layer of protection used in PSM, demonstrated by Figure 2 which is true from a municipality perspective. MIACC Risk Acceptability Criteria can also help in evaluating the risk acceptability in all of these layers of protection:

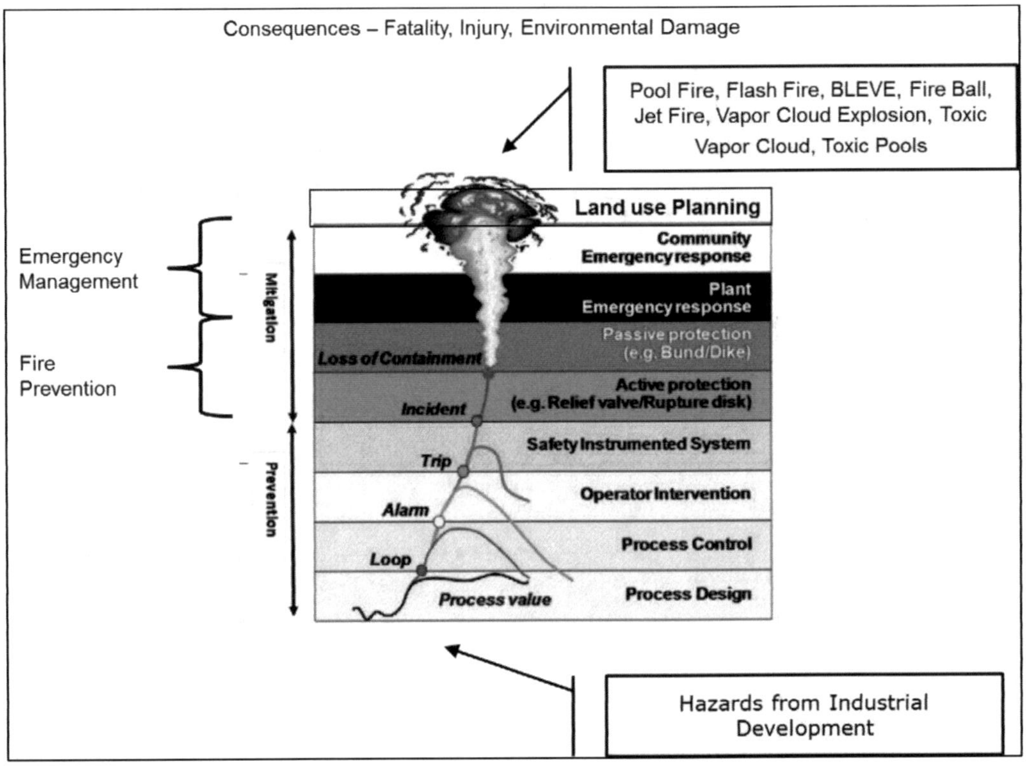

Figure 2: Modified from Alp Ertugrul [4] and ISA [5]

Implementation of MIACC guidelines:

Implementation of MIACC guidelines in any municipality requires that the representatives of the local government engage with the industry for the follow-up activities as indicated by Figure 3. This engagement must have the following prevention tasks to meet MIACC risk acceptability criteria:

a. Document industrial accidents
b. Process safety
c. Land use planning and control

The following diagram from MIACC highlights the tasks that a municipality must perform.

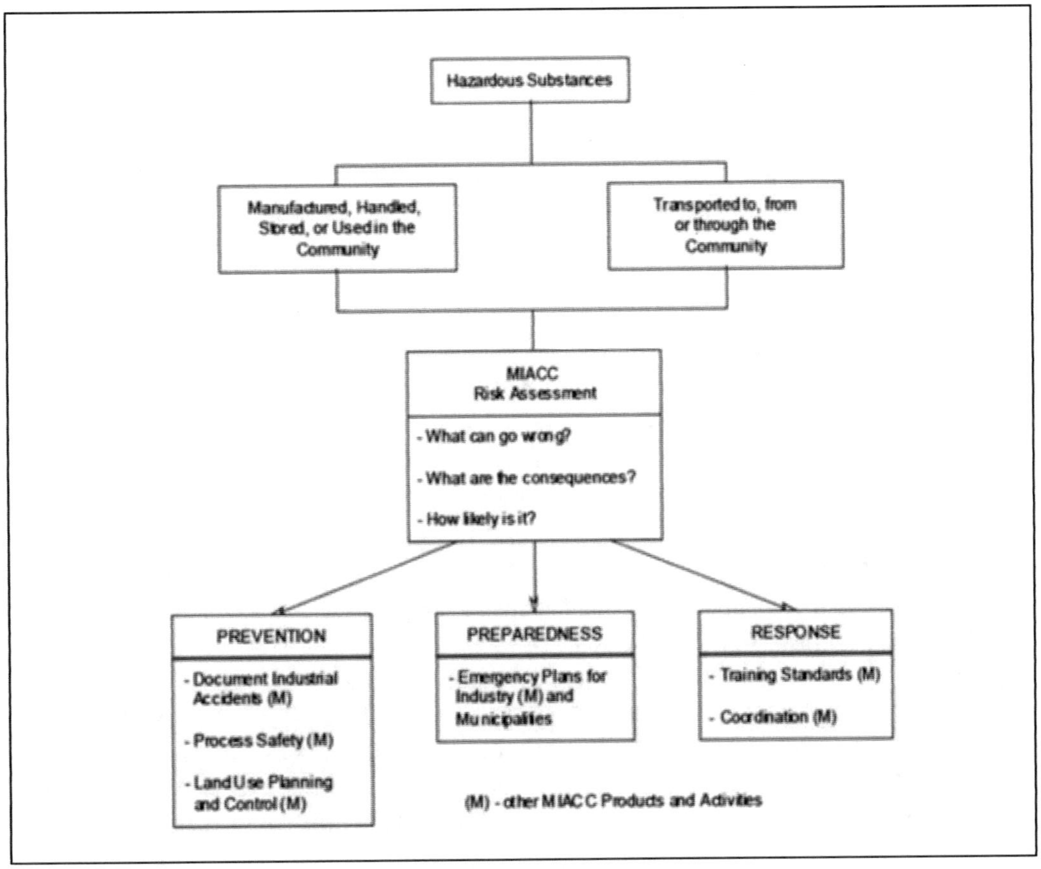

Figure 3: The Process of Risk Assessment in the Community [6]

In the diagram above, the process of risk assessment in the community has certain action items as a follow up in prevention, preparedness, and response that correspond to MIACC publications vital to ensure that safety measures are effectively incorporated.

The majority of the activities associated with prevention are for the validation of the risk assessment and the treatment of residual risks. Hence the objective of public safety and sustainable development is attained throughout the lifecycle of the facility or development.

a) Document industrial accidents

This prevention activity is important to gather data on failures, root causes, and similar statistical information to analyze trends in industrial events. There can be certain companies or facilities which could be facing frequent accidental releases; for example, some could have depressurization issues with their plants and are frequently flaring, etc. All these could be taken as pre-cursors to major events and are important to evaluate the performance of prevailing risk management programs.

b) Process safety

Process safety (M) refers to the Process Safety Management Standard, which is a product of MIACC, published by CSChE-PSM and employed for the Process Safety Evaluation and validation of risk assessments. There is assurance that the risks are kept within MIACC criteria and risk reduction or risk management programs are adequate to meet MIACC criteria.

The MIACC-based Process Safety Management Standard should be used to assess facilities to ensure MIACC guidelines are followed and risks are managed within the allowable limits. The action of doing this inspection is referred to as a Process Safety Management Audit. There are certain guidelines and standards published by MIACC or CSChE-PSM publications to provide direction on how to comply with this requirement.

c) Land use planning and control

Land use planning refers to policies and land use bylaws in a community to address public safety around an industrial facility, dangerous goods trains, and pipelines containing hazardous materials. Zoning certain areas for industry assures there is enough spatial separation for public safety. Zoning helps to decrease public encroachment towards industry and hence helps to reduce risk. This action also reduces risk within emergency planning zones. These zones are the areas adjacent to industrial locations which could get the impact of any major event (fire, explosion, toxic spill). Historically, major accidents are followed by possible evacuations or shelter-in-place protocols in these emergency planning zones.

After a policy is adopted by the municipality, every land use request at an industrial site should have a MIACC-based risk assessment conducted at a minimum every five years (PSM Standard/CSA Z767) for the project to analyze if MIACC criteria are met. All accepted developments tend to have residual risks which require treatment time after time depending on the successive changes which are eminent in every company. These residual risks grow over time. The following diagram demonstrates this effect:

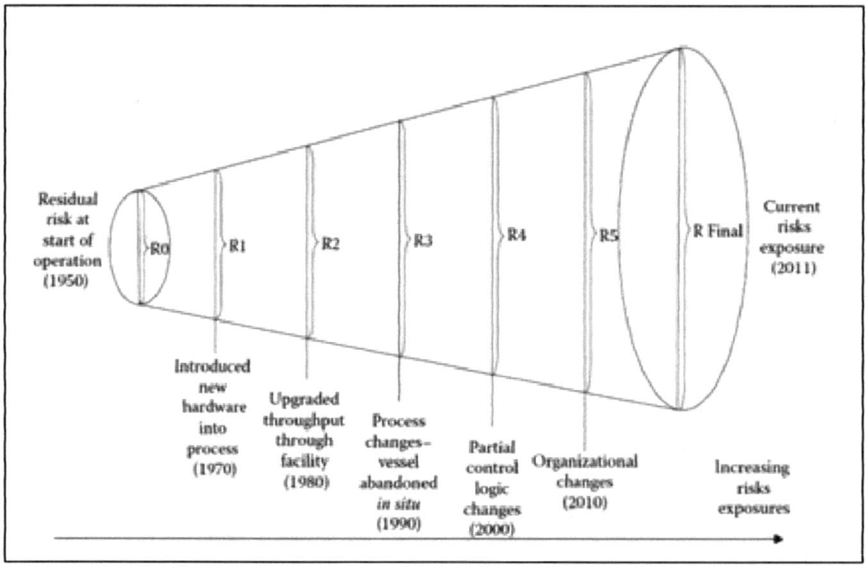

Figure 4: Increasing risk exposures with successive changes [7]

For an industrial site, the residual risk is the risk accepted by the organization to handle or operate a material throughout its life cycle. The changes in processes, material combinations, operating conditions, people handling the material, life of equipment dealing with the material, new processes attached with the material, and similar factors introduce more risks throughout the life of the facility. The observation is the residual risk at the start of the facility or year 1 of the operations of the facility continues to grow throughout its life.

Treatment of residual risk is as important as conducting a risk assessment for the first time. Safety audits and similar actions are helpful to assist the risk control programs. In technical terms, Process Safety Management Audits should be conducted after certain durations to manage the residual risks.

Collectively all these prevention activities help to maintain the site risks within the acceptability criteria. Preparedness to handle situations through drills and exercises is also important, however the majority of the efforts have to be concentrated in the prevention side of the bow tie.

2 MIACC Guidelines and Strathcona County

Initially Strathcona County was using the first version of the MIACC guidelines published in 1995 wherein there was no separate risk contour for sensitive and vulnerable developments. Both residential and institutional developments were accepted beyond the risk contour of $1x10^{-6}$. The new version of the guidelines published in 2007 has introduced another contour of 0.3×10^{-6} for sensitive and vulnerable developments (Figure 1). In The Netherlands, the comparable level is the risk contour of 1×10^{-8}.

Strathcona County has adopted the new version and adjusted their criteria for sensitive land use to be 0.3×10^{-6}. The rest remained unchanged; that is, less vulnerable developments are allowed in grey areas between 1×10^{-5} and 1×10^{-6}, and all residential developments are accepted beyond 1×10^{-6}.

Figure 5 compares other risk based land use planning thresholds practiced around the world. The Netherlands has different minimum acceptances for new facilities and existing facilities. It relaxes new developments in existing facilities, and at the same time provides enough cushion to avoid extensive risk accumulation. Strathcona County is in line with the rest of the world and is using a risk contour of 1×10^{-5} as the probability of a fatality from combined sources of accidents for industrial zones, following the MIACC criteria. For sensitive institutions and vulnerable developments a risk contour of 0.3×10^{-6} is accepted.

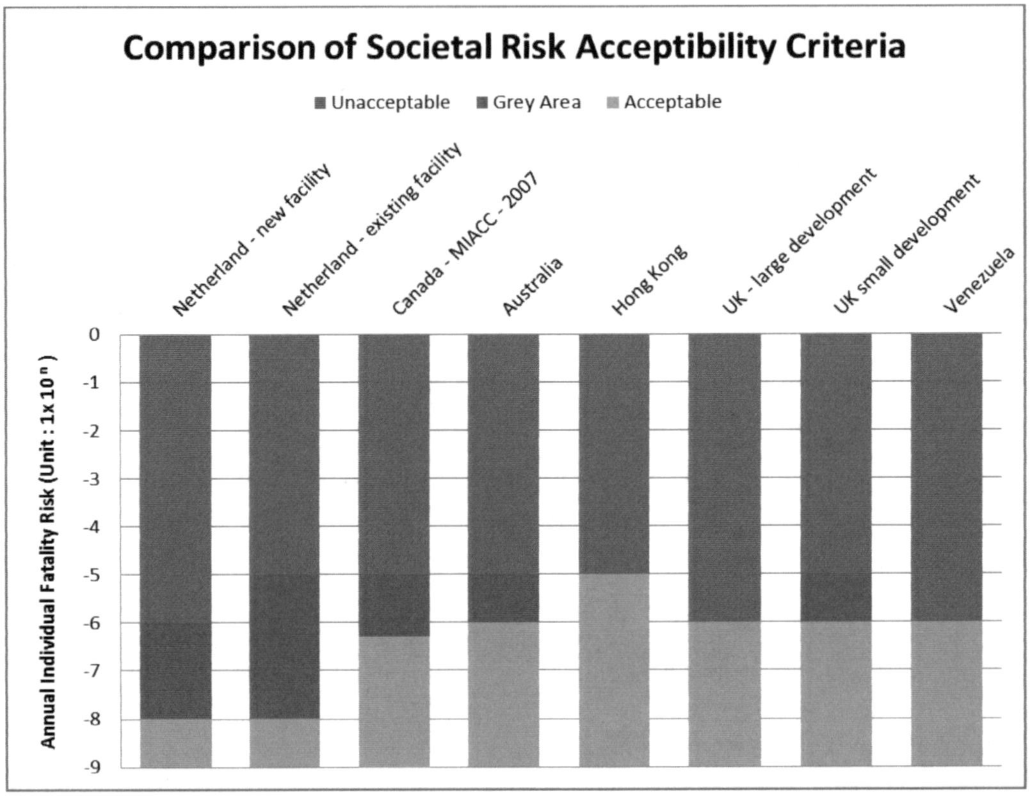

Figure 5: Individual Risk Criteria of Acceptance around the globe [8]

Strathcona County created buffer zones between land designated for industrial and residential developments by introducing a 'heavy industrial' zone and a 'heavy industrial overlay.' The heavy industrial land is the dedicated area for industrial occupancies like chemical plants, storage tanks, refineries, etc. The heavy industrial overlay is to allow medium and supportive industrial developments through a transition of risk exposure following MIACC guidelines.

Since 1978, Strathcona County's approach has been to keep a 3 km separation distance between the existing heavy industrial and residential areas. This approach was incorporated into Strathcona County's General Municipal Plan in 1986 and was further refined in 1990. In 1992, the Strathcona Industrial Association worked with Strathcona County to develop transition zones, and in 1993, the County passed the two transition zones [9] illustrated in Figure 6 below [10]. This became part of Strathcona County land use bylaw in 2002. This bylaw amendment also provides the basis to start an Industrial Engagement Program.

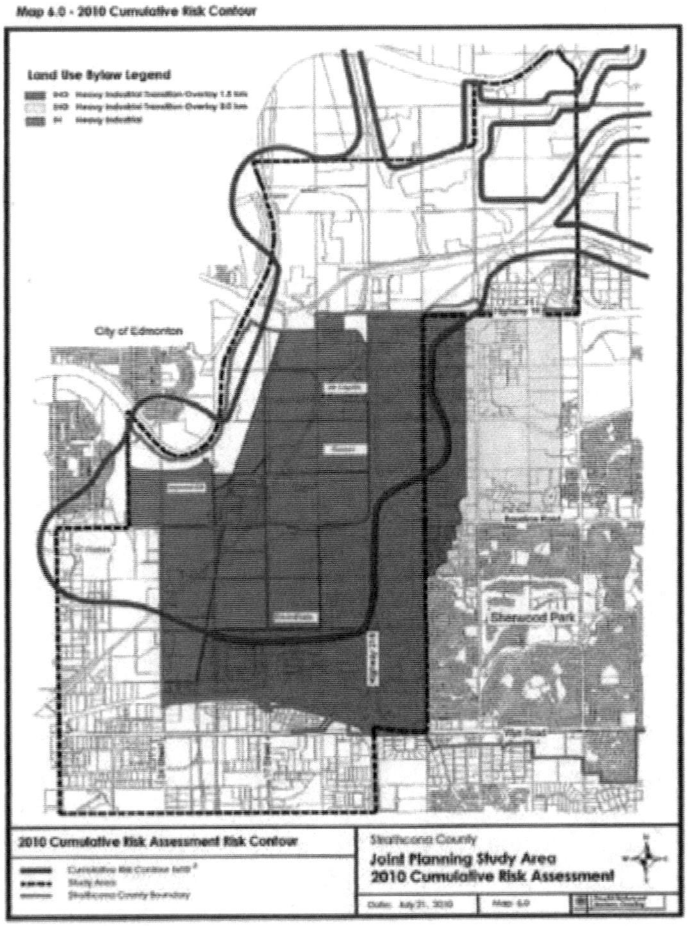

Figure 6: Cumulative Risk Contour (10^{-5}) of Industrial Area-Strathcona County [10]

Figure 6 is from the cumulative risk assessment conducted by Doug McCutcheon [10]. The land designated for Heavy Industrial occupancies is illustrated in purple. The area east of the heavy industrial land up to 1.5 kilometres has been designated for medium industrial uses is highlighted in red, and a line was created to separate these zones in 1980. The yellow blocks are another 1.5 km away from heavy industrial lands, which allows commercial and light industrial occupancies to be built in this zone. This second line of 3 kilometres from the industrial areas was introduced in 1989. Initially the 1980 and 1989 distances were arbitrarily set. The blue, red and yellow zones were later determined in 2002 as an update in for the risk based land use planning.

All types of residential developments are accepted beyond the yellow line. For the ease of planning, fixed infrastructure like roads and property lines were used to introduce these fixed lines. The blue lines are the risk contours developed based on worst case scenarios and major hazards in the area.

Both of these initiatives were taken before the publication of the MIACC guidelines. This land use has been presented as an example of fixed approach in MIACC's *Risk-Based Land Use Planning* document [2].

Figure 7 represents the two transition zones of Strathcona County mentioned in MIACC documents as an example of fixed approach in land buffering

Figure 7: Strathcona County's fixed approach to land buffering [11]

Figure 8 below represents the Cumulative Risk Contours from the risk assessment conducted in 2017. The purple region demonstrates the area between the risk levels of 1 X 10^{-4} to 1 x 10^{-5}, the blue area represents the region between 1 x 10^{-5} to 1 x 10^{-6} and the yellow thin line represents the region between 1 x 10^{-6} and 0.3x 10^{-6}[12].

To evaluate the risk level for sensitive institutions, the risk assessment determined the risk contour of 1 x 10^{-6} and 0.3 x 10^{-6} to be at 2.274 km and 2.740 km away from heavy industrial properties respectively. At present Strathcona County has not determined a fixed line for the third risk contour corresponding to the risk level of 0.3 x 10^{-6} for sensitive institutions. The recommended distance is another 100 meters away from the industrial properties applied to the area north of Highway 16 marked in blue. These risk contours are monitored with a cumulative risk assessment at a frequency of once in every 5 years.

It is not uncommon to observe flaring from the industrial sites which generate odour complaints. A study was conducted by Doug McCutcheon [13], which evaluated the nuisance reports in the area represented by Table 1 and 2, reported to Strathcona Industrial Association and the Alberta Energy Regulator (previously known as the Alberta Energy and Utilities Board). It was observed that the majority of the odour (90%) complaints were reported within the 3-kilometre separation distance from the industrial lands.

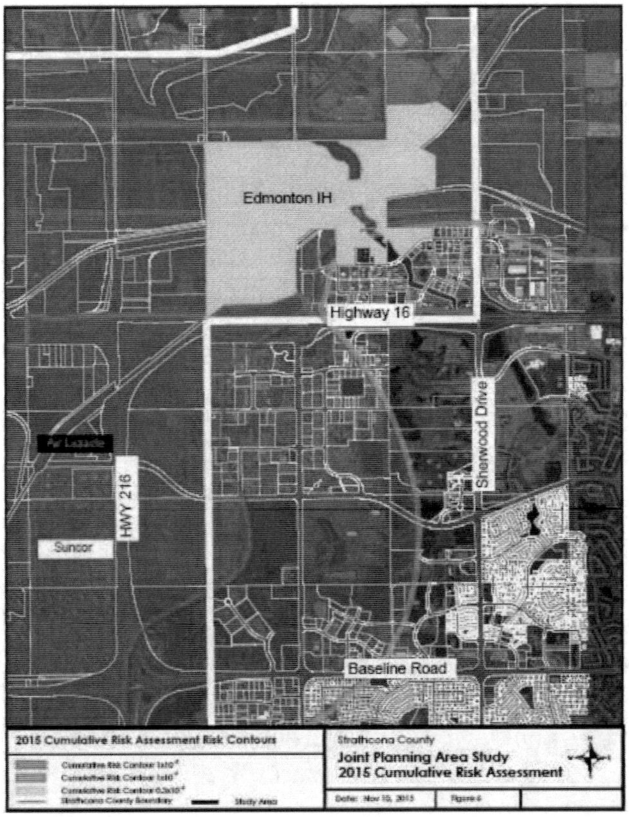

Figure 8: Cumulative Risk Contours of Strathcona County 10^{-5}, 10^{-6}, 0.3 X 10^{-6}

Table 1: Nuisance complaints reported to Strathcona Industrial Association [13]

Nuisance Reports -Strathcona Industrial Association, 1995-2000			
Nuisance Concern	Number of Complaints	Percentages	Distance Range (distance from heavy industrial zone)
Odour	74	66	90% within 3 km
Noise	14	12	70 % within 3 km
Sight	6	5	-
Alarms	8	7	-
Explosions	6	5	-
Flare	2	2	-
Other	4	3	-
Total	114	100	-
Spring	26	22	Not Applicable
Summer	16	14	-
Fall	31	27	-
Winter	41	36	-
Total	114	100	-

Table 2: Nuisance Complaints reported to the Alberta Energy Regulator [13]

Nuisance Reports - Alberta Energy and Utilities Board (now called Alberta Energy Regulator), 1996-1999		
Nuisance Concern	Number of Complaints (approx.)	Percentages
Odour	1380	41
Noise	330	10
Flare/smoke	445	13
Public Health and Safety	415	12
Lease Management	490	14
Spills	325	10
Total	3385	100

One of the limitations of the MIACC guidelines is that the risk-based separation distance does not address the nuisance and health impact on the residents. To accommodate these factors in a land buffer, additional separation distance needs to be considered. These numbers are mostly different for different municipalities and risk assessors. The conflict arises at the borders where two adjacent municipalities might have two different distances for land buffering. In other words, the performance-based design might be working for one municipality to accept more industrial developments due to increased separation distance, however if the adjacent municipality has not considered reasonable buffering, this may conflict with the zoning and discretionary uses accepted in their land use, ultimately introducing residents to increased involuntary risks. This can potentially create a non-compliance with the risk acceptability criteria of MIACC guidelines.

In short, the safe distances from risk assessments might not necessarily provide a reasonable land buffer to avoid impact on health or nuisance complaints. A study was performed in this region to evaluate the health impact of the residents, which found a possible connection between the industrial air pollution of the area to increased health concerns of asthma patients in neighbourhood. It states "*it is possible that industrial sources of NO_2 in the Edmonton area contribute to the increased risk of asthma visits.*" [14]

With these perspectives in hand, Strathcona County made a policy to conduct a cumulative risk assessment every five years. The first one was conducted in the year 2010. One of the recommendations in the risk assessment was for Strathcona County to validate the site-specific information at industrial sites and ensure that the residual risks are managed appropriately with adequate risk reduction programs. It was another factor which influenced the Emergency Services department in launching an Industrial Engagement Program.

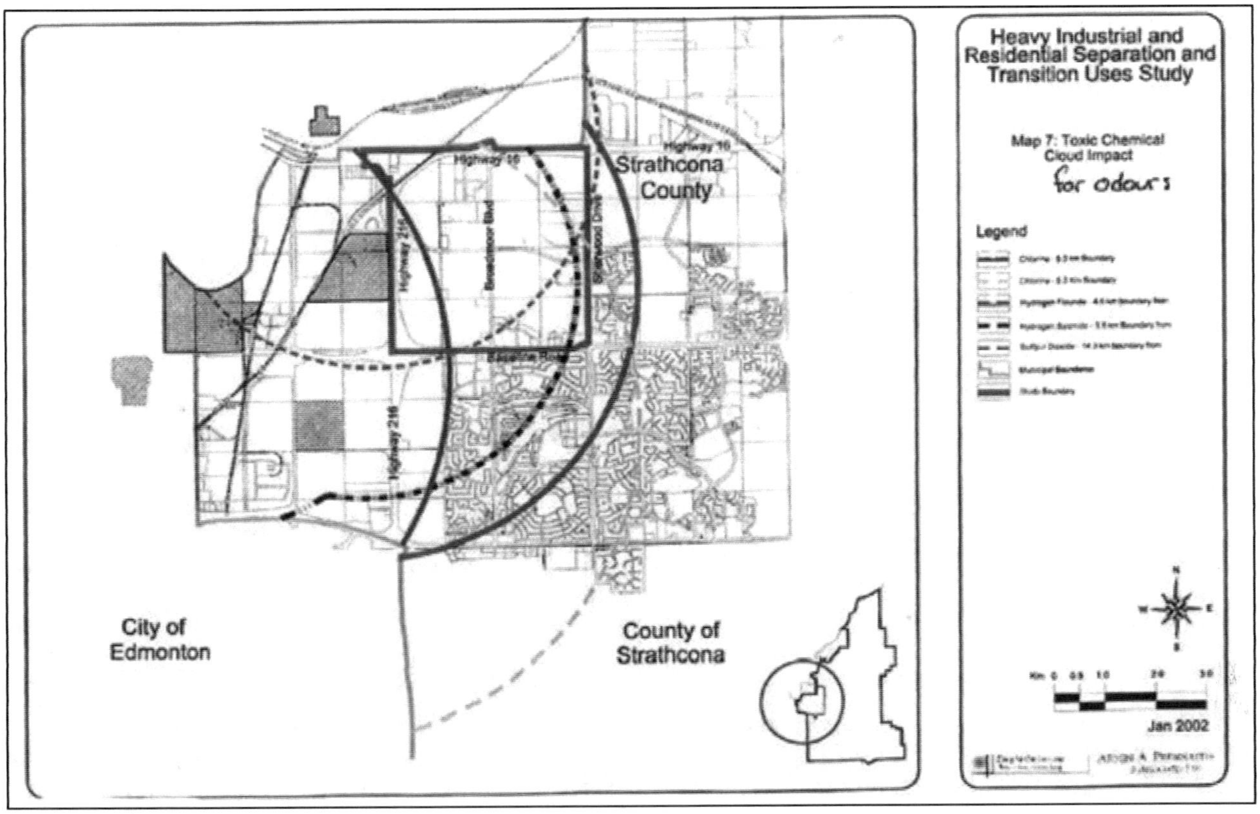

Figure 9: Toxic Cloud Dispersion Distance – Analyzing odour in the area [13]

Figure 9 above represents a consequence analysis performed for the odours in the area. It reaches close to 5.3 kilometres for a likely scenario of a one-ton liquid chlorine spill generating bad odour. Along with the nuisance factor, the emergency response time was considered as a factor in deciding for a fixed separation distance. It was determined that keeping a 3 km separation distance will have a positive effect on helping the public to be safe as long as Emergency Services has 20 minutes to react.

It affirms that using a 3 km buffer separation not only helps in minimizing impact to public safety and public health but also offers a cushion for emergency response. This goes above and beyond the acute risk level and rightly justifies the land buffering separation distance, meeting MIACC guidelines and the SCES vision of becoming the safest community in Canada.

3 Strathcona County and other Canadian Municipalities

In 2013, one of the past Fire Chiefs of Strathcona County, Iain Bushel [9] researched surveys and communications to fire departments and planning officers on community risk reduction specifically collecting information on initiatives other municipalities with heavy industries are taking regarding land buffering.

The following table sums up the findings of the research [9]:

Table 3: Canadian Municipalities with Heavy Industrial Complexes with land buffers [9]

Municipality	Refinery Operator	Capacity (bbls per day)	Buffers
Burnaby, BC	Parkland Burnaby Refinery	55,000	61 m
Come by Chance, NL	North Atlantic Refining Limited	130,000	No land buffer
Dartmouth, Regional Municipality of Halifax, NS	Imperial Oil Limited (closed)	88,000	100 ft. and no land buffer
Levis, QC	Valero	265,000	No information available
Montreal QC	Suncor Energy Incorporated	137,000	Not available, likely there is no separation distance
Nanticoke, ON	Imperial Oil Limited	118,000	Industrial influence areas of 3 km
Prince George, BC	Husky Prince George Refinery	12,000	No buffer
Regina, SK	Co-op Refinery Complex	130,000	No buffer zones, 1 km separation distance for one metal recycling plant
Saint John, NB	Irving Oil Limited	300,000	No minimum distance
Sarnia , ON	Imperial Oil Limited	124,000	No buffer or separation distance
Sarnia, ON	Shell Canada Limited	72,000	
Sarnia, ON	Suncor Energy Incorporated	85,000	
Strathcona County, AB	Shell Canada Limited	100,000	3 km
Strathcona County, AB	Shell Canada Limited	Upgrader	3 km
Strathcona County, AB	Imperial Oil Limited	187,000	3 km
Strathcona County, AB	Suncor Energy Incorporated	135,000	3 km
Sturgeon County, AB	NWR Sturgeon Bitumen Refinery	150,000	1.6 km, can be increased if indicated by a risk assessment
Lloydminster, SK	Husky Lloydminster Heavy Oil Upgrader	25,000	1.6 km or 1 mile
Wood Buffalo, AB	CNRL – Horizon	211,000	These four upgraders re remotely located and are far from the population.
Wood Buffalo, AB	Nexen – Long Lake	58,500	
Wood Buffalo, AB	Suncor Base and Millennium	440,000	
Wood Buffalo, AB	Syncrude – Mildred Lake	407,000	

As per the survey and research, the fixed buffer zones and setbacks ranged between 30 meters (100 ft.) to 3.0 km. Many municipalities are struggling with maintaining a reasonable land buffer. In case of Strathcona County, no residential development grandfathered in this policy area is closer than approximately 900 meters to any heavy industrial site.

So far this risk-based land use planning has encouraged industry to expand and new projects have been developed in the heavy industrial zones. Some commercial businesses were grandfathered in the transition zone. These businesses continue to operate, but they are not allowed to expand further. In other words, the land buffering has received appreciation and acceptance from Strathcona County's industrial partners. Because of increased risk awareness in both municipal and industry stakeholders, the land buffering has been maintained to date.

Clearly Strathcona County is the leading municipality in maintaining a land buffer which not only provides safety from industrial sites but also reasonably addresses the nuisance factors and provides a buffer for emergency response. This also partially helps in minimizing the health effects on the general public. With the launch of their Industrial Engagement Program, Strathcona County has established a leading role in industrial risk management among other Canadian municipalities.

4 Challenges in Risk-Based Land Use

History reveals problems in complying with the stringent Dutch criteria, especially concerning societal risk [15].

As reported by the CCPS *Guidelines for Developing Quantitative Safety Risk Criteria* [15], in the early 1970s, the mayor of Dordrecht/Zwijndrecht petitioned Parliament to relax the societal risk criteria to allow large scale development of offices near railway stations. Here, the goal of facilitating mass transit commuting conflicted with the risks associated with the adjacent large scale transport of hazardous materials.

The second example involved the proposed expansion of the Schiphol Airport. The expansion was deemed essential to the Dutch economy but would have resulted in a societal risk more than 10 times the total risk of the combined SEVESO sites in the Netherlands.

Studies following the disastrous fireworks warehouse explosion in Enschede revealed some circumstances in which risk reduction measures should have been implemented previously in accordance with the Dutch regulations. These measures, which had been postponed because of cost considerations, included the relocation of 600 LPG fueling stations (for storage and sale of LPG as automotive fuel to consumers) that were too close to residences.

The Schiphol issue helped precipitate risk criteria revisions in 1993, which declared the criteria for maximum tolerable societal risk to be advisory rather than mandatory. As a consequence of this change, authorities could now tolerate exceeding the societal risk criteria if a compelling rationale (e.g. economic need) could be provided for the specific case. To offset the less stringent approach to societal risk, the concept of negligible risk was abandoned for both individual risk and societal risk.

Learning from these situations, Strathcona County has determined 'permitted uses' and 'discretionary uses' to help implement regulations through the Land Use Bylaw [16].

Hence, Dutch societal risk criteria are numerically more restrictive and are advisory rather than mandatory. They have been made advisory with the expressed intention of allowing authorities, with justification, to waive them.

Some of the observations about the implementation of Dutch risk management initiatives are made by Claudia Basta [17]:

1. Risk generators are not provided with enough incentives to limit risks. In practice, the ALARA principle is often not effectively implemented, with some risk generators trying to maintain risk just below the risk criteria rather than trying to make further risk reductions.
2. The concept of group risk (societal risk) was originally intended to apply to single stationary facilities, such as LPG fueling stations. However, the societal risk criteria are now applied to markedly different types and scales of facilities (such as large industrial facilities, railroad yards, and airports), complicating compliance with the criteria.
3. Local authorities are often unwilling to enforce the external safety objectives. There is a gap between the external safety level that the Dutch government wishes to offer its citizens and the effort it is willing to put into this. As a result, the aim of creating a living environment with limited external safety risks is often not achieved under the present system of standards.

Strathcona County has also experienced similar situations in the past where requests have been made to Development Officers regarding variances in Land Use Bylaw to seek relaxation in boundaries as deemed appropriate due to the existing development pattern, for daycares, medical centers and similar developments within the vicinity of industrial developments. These variances were turned down for the reason that the developments were identified as incompatible with the zones which negates the MIACC guideline. There must be compliance with these separation distances, especially when adopted as a best practice to follow. These proposals were in conflict with the values of Strathcona County as a municipality and the vision and mission of SCES as a department.

A fixed approach of land buffering and monitoring of risk contours is necessary to control the risk in the area. Maintaining these buffer zones improves public safety; however the policies of an adjacent municipality can sometimes conflict with one another. It is quite possible that one municipality is inclined to attract more industry whereas the adjacent municipality could have accepted residential developments near to that area.

Strathcona County has faced similar concerns from a neighbouring municipality, City of Edmonton, where the prevailing policies are not as stringent regarding land buffering (e.g. to control nuisance) compared to Strathcona County. This became a concern from City of Edmonton which was later discussed in the Capital Region Board, a collaborated body of regional municipalities with a quasi-judicial role. The land use planning of Strathcona County was analyzed in depth to evaluate negligence or to point out improvements. The proceedings went in the favor of Strathcona County and the objection was overruled. This session endorsed

MIACC guidelines as a solution and critically examined the benefits and limitations of MIACC guidelines and its risk acceptability criteria.

A decision was made and some of the adjoining area was declared as a joint planning study area. [19]. Technically both sides comply with MIACC guidelines, however due to closeness to industry the buffer from City of Edmonton is limited to provide a buffer to address nuisance however that is not the case with Strathcona County. This indicates meeting MIACC guidelines is a minimum a municipality should do however it is not enough, extra buffer should be allocated to address nuisance. Also the emphasis should be to include some buffer for an emergency response. Therefore there should be long term planning to avoid compromise on nuisance and quality of life or any conflicts with the regional land use planning.

These sessions tested the application of MIACC guidelines and risk acceptability criteria in a legal environment, making it a 'tried and tested' model for any Canadian municipality to use.

5 Strathcona County Industrial Engagement Program

SCES was established as a department with the responsibility to prevent, combat and control incidents, carrying out preventable patrols, and authorized to control and mitigate incidents involving dangerous goods through Strathcona County's Emergency Services Bylaw [20].

To fully comply with MIACC guidelines and to monitor and control the residual risks identified by risk assessments, a collaborative Industrial Engagement Program was deemed necessary by SCES.

The program focusses on building relationships with industrial stakeholders. Up until 2019, SCES has been engaged with industry during the development permit application, construction phase and at start-up or occupancy phases only leaving behind the follow-up on risk assessment and residual risk management. The Industrial Engagement Program will allow SCES to continue to work with industry to monitor risks. A majority of heavy industrial business have realized the benefits of Process Safety Management (PSM) and practice its application, however many medium size businesses are not yet fully on board with PSM. This program will educate them on ways to include PSM within their facilities and operations.

A gap analysis was performed wherein the PSM Standard CSA Z767 was used to compare the practices and come up with shortcomings which needed attention in the program. The inspiration was from the methodology used by the Canadian Association of Petroleum Producers (CAPP) in the regulatory scan publication [22].

In connection to industrial accidents, in the recent past there have been many cases in North America where events at industrial sites have caused injuries and fatalities of firefighters. The most frequent contributing factor is non-recognition of the high hazard or high-risk structures and occupancies. The National Institute of Occupational and Health (NIOSH) recommends that a fire department with an industrial presence in the jurisdiction should have a written risk management plan and use risk management principles at all structure fires and especially at incidents involving high-risk hazards [21].

SCES recognizes that one of the most visible risks to Strathcona County is from the major hazards at heavy industrial sites in Strathcona District, that is, the area in between Sherwood Park and the City of Edmonton along the North Saskatchewan River.

A major hazard is defined as the hazard which can cause impact off-site from the facility and has the potential to cause injuries, fatalities, loss to infrastructure, environment (short term and long term) and disruption to business operations as well as municipal operations. The majority of these hazards are human-made or are related to loss of containment activities leading to leaks, spills, fires and explosions.

One of the planning review documents used in the quasi-judicial sessions of the Capital Region Board by IPS Consulting Inc. under the section "Responsibilities for and during Emergencies" states [23]:

"Recent trends indicate that society is holding industry and government more and more responsible for emergency responses in the event of an industrial accident. The responsibility for evacuations or other measures to protect or rescue people located on adjacent or nearby land uses to an industrial accident lies almost exclusively with the host municipality".

It emphasizes the need and importance of accident prevention programs in a municipality and provides a solid ground to launch an Industrial Engagement Program (IEP).

After analyzing the Federal, Provincial and Municipal laws and reviewing the industrial regulatory scans such as CAPP PSM Regulatory Scan [22], SCES has realized its role in creating awareness of PSM. SCES understands that PSM not only keeps the business operations sustainable for an extended period of time but also prevents accidents and injuries to occupants and emergency responders. In Canada there is no direct enforcement of PSM from any other body and to truly prevent accidents, the municipality has to play its role.

Strathcona County Emergency Services is continuing the journey towards accident prevention through its IEP. The following focus areas are going to be covered in a phased approach:

1. Identify and minimize the risk of accidents from hazardous substances and prevent recurrences of incidents.
2. Collect incident notifications, investigation reports, do trend analysis of incidents, and maintain an accessible database for monitoring and controlling of risk reduction programs.
3. Make arrangements to improve the accountability gap by making follow-up policies on industrial developments.
4. Fire inspections to include industrial occupancies, continue emergency preparedness and response planning, and governance of safety programs, maintaining industrial partner support.
5. Develop initiatives for PSM education.
6. Review PSM and risk reduction programs within the municipal jurisdiction.

The Program is planned to carry out the following tasks in phases:

Phase 1 (Started in Jan 2019)	Documenting Release Incidents
	Completing Community Emergency Response Plans, called Industrial Response Worksheets (IRW)
	Carry out Fire Inspections
Phase 2 (in the planning phase)	Industrial Engagement to verify site hazards for Cumulative Risk Assessments
	PSM Education Program to encourage businesses to incorporate PSM in Safety Management Systems
Phase 3 (to be determined)	Reviewing PSM and Residual Risk Management Programs

Figure 10 below represents the projected decrease in PSM incidents measured by process safety information and data collection. A trend similar to CCPS Process Safety Journey to Excellence is anticipated over the period of time with the implementation of the IEP.

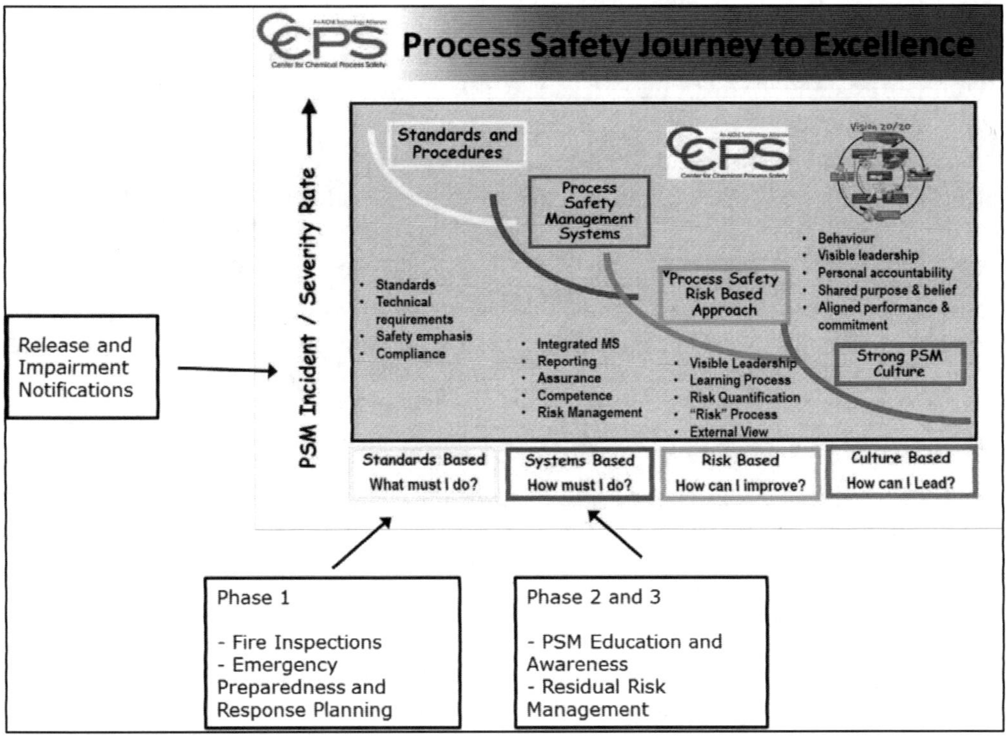

Figure 10: IEP and CCPS Vision 2020 – Process Safety Journey to Excellence

In 2018, the program was introduced to industrial partners through the mutual aid groups Northeast Region Community Awareness Emergency Response (NRCAER) and Strathcona District Mutual Aid Partnership (SDMAP).

Initially there was hesitation from the industrial partners, especially on the release reporting initiative. With further discussions on the topic, industry has agreed to provide the release data.

The Industrial Response Worksheet (IRW) was greatly appreciated. The initiative involved industry partners generating GIS based site maps with information about fire protection resources and staging areas. The intent is to provide emergency responders with site-specific information on the hazards, resources and instructions for the first 20 minutes of the response call. Another benefit of IRW is it helps in practicing emergency preparedness exercises.

In January 2019, SCES began phase 1 of the program by conducting site visits to these industrial businesses and reviewed the maintenance documents of fire protection systems. In this first phase, SCES will be establishing working relationships with industry and progressively work on educating about PSM and residual risk management programs.

This is first time a municipal fire department in Canada will be engaging and making site visits to its industrial partners to educate on PSM standards. SCES understands larger businesses have already realized the importance of PSM and have developed working PSM systems. More effort will be required to educate medium and small industries handling hazardous substances. The majority of these businesses rely on the local fire departments to respond during an accident.

Also there were some concerns whether IEP is going to increase operational costs or reporting efforts. With communications between SCES and industrial businesses about IEP, industry gradually understands the benefits and is open to engage for IEP. In principle the fire department site visits are planned to be more education and awareness focused, rather than enforcement.

Like any other inspection or safety engagement program there is cost associated both for regulated industry and municipality. However keeping the businesses uninterrupted, safe, reliable, and sustainable saves from many other losses and expenditures which compared to the program implementation costs are not that significant.

This will potentially result in minimizing the accidents and loss of containment situations ultimately helping both the industry partners and emergency responders. This will not only bring sustainability to the business by maintaining uninterrupted production for longer periods, but also prevent occupational health and safety incidents involving employees and emergency responders and provide increased protection to assets and businesses.

If evaluated in terms of monetary impact to regulated industry, fire department engagement is likely to decrease the likelihood of businesses facing costs of emergency response, business recovery, potential damage to municipal infrastructure, environmental damage, recovery claims, etc.

6 Conclusion

In Canada, the PSM initiatives are not mandatory in federal or provincial law. That is why to provide safety to the public and a quality of life to Canadian residents, the responsibility lies on municipalities to implement PSM. It involves many actions such as conducting risk assessments of facilities in their jurisdiction, encouraging businesses to establish PSM practices, maintaining risk management programs and monitoring the PSM incidents in the area.

Strathcona County is leading the way in the implementation of PSM using the MIACC risk-based land use model. Strathcona County has the largest fixed buffer (3 km) among all the municipalities in Canada that have refineries or heavy industry. Most of the larger organizations have realized the importance of implementing PSM programs in their management systems, however many mid-level companies dealing with hazardous materials are still not on board with implementing PSM for accident prevention and need further education and awareness.

When following the MIACC model, municipalities should conduct cumulative risk assessments to evaluate 1×10^{-4}, 1×10^{-5}, 1×10^{-6} and 0.3×10^{-6} risk contours from all of the heavy industrial businesses, and also consider the unoccupied land zoned for industrial uses for risk analysis. A public survey of nearby residents should be conducted to monitor the effectiveness of risk assessments. It will result in higher quality and standard of life and partially address the nuisance factor.

A land buffer should be decided based on the combination of 1) risk contours determined by effective risk modelling, 2) consequence contours of a spill or release and 3) buffer time to react and respond to emergencies, which altogether means a separation distance more than just the risk contours alone.

Up until end of 2018 Strathcona County has been engaging with industry only during the development permit application and construction phases. The IEP will allow SCES to follow up on the development requirements [24], continue to work with industry to ensure they have maintained their fire protection systems, and educate them on ways to include Process Safety Management within their facilities and operations.

The fire department or by-law enforcement officers in collaboration with industrial partners should establish engagement programs to encourage and educate about PSM. It should be followed by periodic PSM audits or safety reviews for compliance to MIACC guidelines risk acceptability criteria. These risk reduction initiatives will be helpful to decrease the number of incidents, bringing reliability and sustainability to businesses and establishing a better quality of life for residents. Any unwanted event at a heavy industrial site can potentially impact the municipal tax collection, which sometimes can be significant enough to cause budget issues especially if it is a smaller municipality.

It is well understood the results are not going to come overnight, but the journey has to begin. It is anticipated the trend will be based on the learnings from CCPS and is most likely going to follow the trend proposed by CCPS Vision 2020 [25].

Acknowledgement:

The author would like to acknowledge: Strathcona County, for allowing this information to be published for education and learning purposes, Doug McCutcheon P.Eng. for continuous professional guidance, coaching and mentorship, Bill Timbers P.Eng. and members of CSChE-PSM Division for support, and specifically the family of Strathcona County Emergency Services which breathes a mission 'to protect from risk the things that matter to those we serve', a vision 'to lead our community to be the safest in Canada', and strives to become Canada's most livable community [1].

7 References

[1] Strathcona County Strategic Plan, available at https://www.strathcona.ca/files/files/encl_1_2013_-_2030_strategic_plan_refined.pdf, Accessed on January 15, 2019.

[2] CSChE, PSM Publications, Available at https://www.cheminst.ca/psm/psm-publications, Accessed on January 15, 2019.

[3] TSSA - Technical Standards and Safety Authority, "Guidelines for the Implementation of the Level 2 Risk and Safety Management Plan", Dec 22, 2010

[4] A. Ertugrul, Integrated Risk Assessment and Management for Continuous Improvement - A short course with a focus on Hazardous Materials, Edmonton, 2006

[5] International Society of Automation, Available at https://www.isa.org/standards-and-publications/isa-publications/intech-magazine/2009/september/cover-story-operators-on-alert/, Accessed on January 15, 2019.

[6] MIACC, Risk Assessment Guidelines for Municipalities and Industries - An Initial Screening Tool, 1997

[7] L. Chitram, "Safety Management, A comprehensive approach to developing a sustainable system," CRC Press, 2012

[8] J. B. Cornwell and M. M. Meyer, "Risk Acceptance Criteria or How safe is safe enough?," Quest Consultants Inc.®, Oklahoma 1997, Available at https://www.questconsult.com/papers/risk-acceptance-criteria/, Accessed on January 15, 2019

[9] Iain P. Bushel, "How close is too close? Community Safety and an Industrial Buffer Zone", Available at https://nfa.usfa.fema.gov/pdf/efop/efo47523.pdf, Accessed on January 15, 2019.

[10] D. McCutcheon, "Cumulative Risk Assessment Study for Strathcona County – Final Report July 28, 2010"

[11] CSChE, "MIACC Risk-Based Land Use Planning," Available at https://www.cheminst.ca/psm/psm-publications, Accessed on January 15, 2019.

[12] D. McCutcheon, "Cumulative Risk Assessment Study for Strathcona County-Final Report, June 16, 2017

[13] D. McCutcheon, "Review of Land Use Policies and Regulations Affecting Heavy Industrial and Residential Separation and Transition Uses, 2003

[14] P. J. Villeneuve, "Outdoor air pollution and emergency department visits for asthma among children and adults: A case-crossover study in northern Alberta, Canada, Environ Health. 2007; 6:40, https://www.ncbi.nlm.nih.gov/pmc/articles/PMC2254596/ , Accessed on January 15, 2019

[15] CCPS - Center for Chemical Process Safety, "Guidelines for Developing Quantitative Safety Risk Criteria," American Institute of Chemical Engineers, John Wiley & Sons, 2009

[16] Strathcona County Land Use Bylaw 6-2015 Part 7: Urban Service Area Zoning Districts, Available at https://www.strathcona.ca/council-county/bylaws-and-policies/land-use-bylaw/ , Accessed on January 15, 2019

[17] Claudia Basta et al., "Risk-maps informing land-use planning processes. A survey on the Netherlands and the United Kingdom recent developments", Journal of Hazardous Materials 145 (2007) 241–249

[18] City of Edmonton vs Strathcona County, Board Order MGB 098/08, File 07/IMD-02, Available at http://www.municipalaffairs.alberta.ca/cfml/boardorders/pdf/MGB%20098-08.pdf, Accessed on January 15, 2019

[19] Strathcona County and City of Edmonton, Joint Planning Study Boundary Interface Protocols & Strategies", Available at https://www.edmonton.ca/city_government/documents/Edmonton_Strathcona_County_Joint_Planning_Study.pdf , Accessed on January 15, 2019

[20] Strathcona County Emergency Services Bylaw, Available at https://www.strathcona.ca/files/files/attachment-lls-bylaw-68-2000.pdf, Accessed on January 15, 2019

[21] The National Institute for Occupational Safety and Health (NIOSH), "Death in the Line of Duty…A Summary of a NIOSH fire fighter fatality investigation, Available at https://www.cdc.gov/niosh/fire/reports/face201311.html, Accessed on January 15, 2019

[22] Canadian Association of Petroleum Producers (CAPP), "Process Safety Management: Regulatory Scan," August 2014, Available at https://www.capp.ca/~/media/capp/customer-portal/publications/251323.pdf?modified=20180910121056 , Accessed on January 15, 2019.

[23] T. MacDougall, "Heavy Industrial Land Use Planning and Development MGB Hearing," IPS Consulting, Edmonton AB, November 2007

[24] Strathcona County Emergency Services Development Requirements, https://www.strathcona.ca/business-and-development/development/oil-and-gas-in-strathcona-county/industry-information/ , last accessed on March 01, 2019

[25] CCPS Vision 2020, Available at https://www.aiche.org/sites/default/files/docs/events/ccps_welcome_calgary_kadri.pptx , last accessed on January 15, 2019

548306 Beyond VCE: Addressing Fire Hazards As Part of Facility Siting

Background:

Following influential facility siting incidents such as the 2005 BP Texas City Explosion, it was recognized that vapor cloud explosions were a significant risk to facilities due to the sudden and catastrophic nature of the events. As a result, API's recommended practices for facility siting were revised to include additional guidance to ensure safe placement of portable and permanent buildings and tents in relation to process equipment.

One hazard that has similar catastrophic consequences, but is discussed in less detail is fire hazard. Sites that do not have significant congestion still have the potential to generate catastrophic flame impingement, particularly when examining "jet fire" hazards. As demonstrated by the Hickson & Welch LTD[1] incident that occurred on September 21, 1992, jet fire can have a sudden and devastating consequence to buildings and the occupants they house. In this incident, 5 building occupants lost their lives as a runaway reaction during maintenance activities ignited a powerful jet fire that destroyed a control building and caused significant damage to a nearby office building.

Abstract Proposal:

Marathon Pipe Line (MPL) and ABS Consulting would like to provide a presentation that highlights facility siting research conducted for terminals and pipeline stations and how this information helped the Marathon Logistics & Storage (L&S) organization develop internal guidance for safe placement of buildings at all supported facilities. Due to the limited congestion at these facilities, fire hazards are the most relevant and significant hazard for facility siting at these locations. As the values identified by internal data vary significantly from existing CCPS guidance, L&S developed a standard that specifically addresses fire hazards through a consistent modeling approach.

As part of this presentation, MPL and ABS will provide background information regarding the L&S Facility Siting Standard, including discussion on the Hickson & Welch jet fire incident, technical modeling information, and will also review the management of buildings and tents at facilities utilizing the Management of Change (MOC) process.

At the conclusion of the presentation, MPL and ABS welcome any further discussion and questions regarding fire hazards related to facility siting as part of an effort to increase the awareness of this significant hazard.

[1]HSE Books; *The fire at HICKSON & WELCH LTD*; *A report of the incident by the Health and Safety Executive into the fatal fire at Hickson & Welch Ltd, Castleford on 21 September 1992*

19 AIChE Spring Meeting
15TH GLOBAL CONGRESS ON PROCESS SAFETY
NEW ORLEANS • 2019

Confirming the SIS Layer of Protection

Denise Chastain-Knight, PE, CFSE, CCPSC
Exida Consulting, LLC
80 North Main Street
Sellersville, PA, USA
dchastainknight@exida.com

Prepared for Presentation at
American Institute of Chemical Engineers
2019 Spring Meeting and 15th Global Congress on Process Safety
New Orleans, LA
March 31 – April 3, 2019

Confirming the SIS Layer of Protection

Denise Chastain-Knight, PE, CFSE, CCPSC
Exida Consulting, LLC
80 North Main Street
Sellersville, PA, USA
dchastainknight@exida.com

Keywords: Functional Safety, SIF, SIS, maintenance, proof test, demand, operations, O&M, alarm management,

Abstract

The Safety Instrumented System (SIS) is a complex protection layer with high reliability requirements. Assuring the effectiveness of the SIS requires monitoring the effectiveness of the Safety Instrumented Functions (SIFs) as well as the Independent Protection Layer(s) (IPL) that reduce demand on the SIF(s). A SIF tripping more frequently than anticipated may be due to flaws in IPLs intended to reduce demand on the SIF. A slower than expected or failed SIF trip may be due to flaws in the SIF components, the proof testing program, the maintenance program, or inadequate SIF design. This paper will identify parameters that should be routinely monitored, illustrate how data can be captured and analyzed detect weaknesses in the SIS IPL, and provide recommendations for routine monitoring and audit activities.

1 Introduction

The chemical process industry relies on a multi-layer approach to Process safety. The layers build from the chemical process, to equipment design, controls, operator response to alarms, Safety Instrumented Systems (SIS), relief devices, containment and emergency response. Containment and emergency response manage the consequences of an incident usually after a loss of containment has occurred. Relief devices act to reduce the consequences of an incident by releasing the contents of (depressurizing) a system in a controlled way. Basic Process Control System (BPCS), operator response to alarms and SIS work to prevent an incident usually before any loss of containment occurs. These protection layers work in different ways and rely on different resources and practices to assure their effectiveness.

A chemical process comes with a certain amount of process risk that is dictated by the nature of the chemicals (flammable, toxic) and the conditions under which they are being processed (heat, reaction). Protection layers are applied to the process to reduce process risk to a tolerable risk level, or as low as reasonably practicable (ALARP). The Layer of Protection Analysis (LOPA), is a tool used to compare the process risk of a particular hazard scenario against the risk mitigation provided by independent protection layers (IPLs) to confirm that sufficient IPLs are in place to achieve the desired risk reduction. Figure 1 illustrates the LOPA layers of protection concept.

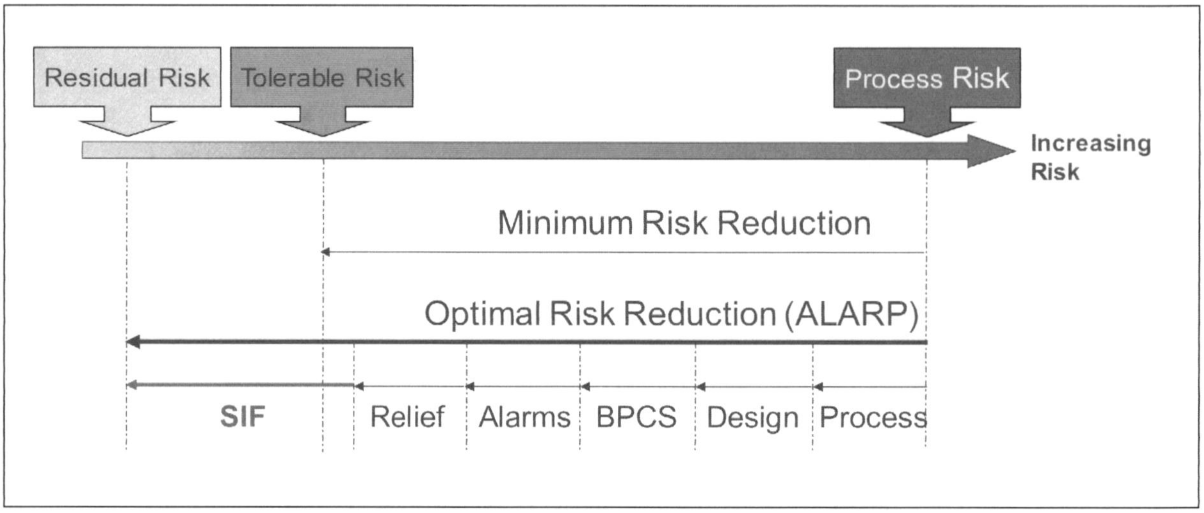

Figure 1. Layers of Protection

A primary objective of process safety is to focus resources to prevent incidents from occurring by maintaining primary containment placing a focus on prevention rather than mitigation. Because secondary containment and emergency response layers are utilized after the release from primary containment, the LOPA illustration does not consider these layers and they will not be discussed any further in this paper. IPLs for BPCS control, alarms, and mechanical relief are considered first, then a Safety Instrumented Function (SIF) is applied to close the risk gap. IPLs provide risk reduction, however, they can and do fail. In the LOPA the potential for failure is considered by assigning a probability of failure on demand (PFD) to each IPL. PFD in turn establishes the risk reduction factor (RRF) with which the IPL can be credited.

During LOPA analysis PFD values for IPLs are assigned based on broad industry experience assuming systems are managed and maintained in accordance with recognized and generally accepted good engineering practices (RAGAGEP). The IEC 61511 lifecycle includes steps to monitor the performance of the SIS IPL and to enable continuous improvement to assure that the assumed PFD is not exceeded. This paper will discuss some of the leading and lagging indicators that should be monitored for a SIS and supporting alarm IPLs.

2 LOPA Structure

Figure 2 is an example of a LOPA with multiple initiating events.

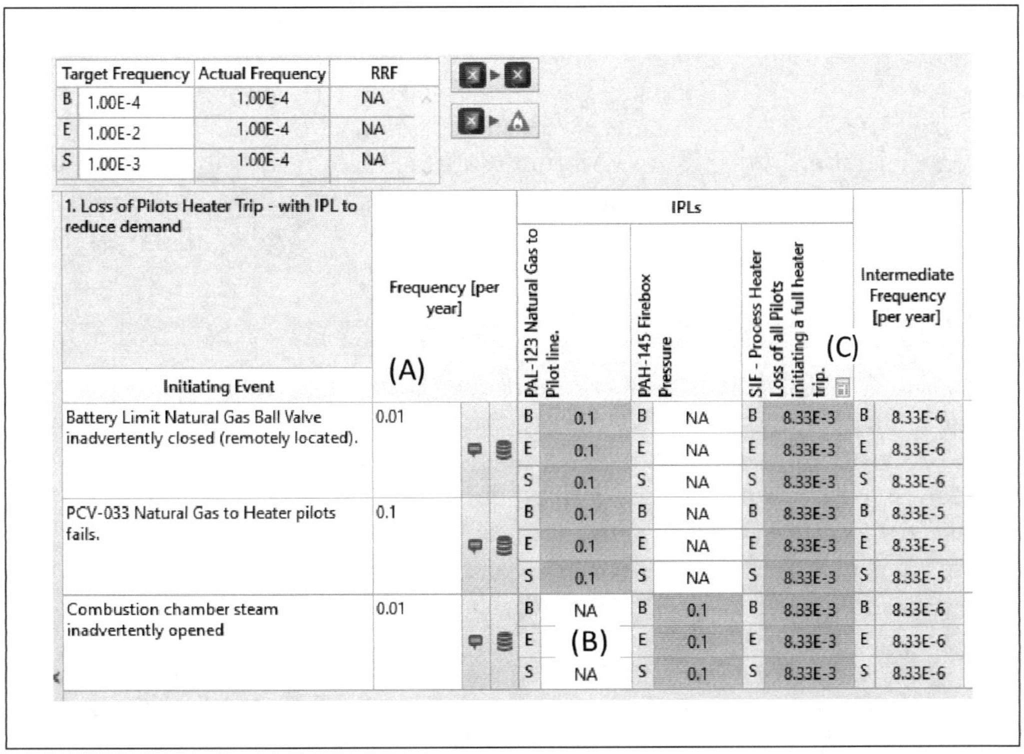

Target Frequency	Actual Frequency	RRF
B 1.00E-4	1.00E-4	NA
E 1.00E-2	1.00E-4	NA
S 1.00E-3	1.00E-4	NA

1. Loss of Pilots Heater Trip - with IPL to reduce demand Initiating Event	Frequency [per year] (A)	IPLs: PAL-123 Natural Gas to Pilot line.	IPLs: PAH-145 Firebox Pressure	IPLs: SIF - Process Heater Loss of all Pilots initiating a full heater trip. (C)	Intermediate Frequency [per year]
Battery Limit Natural Gas Ball Valve inadvertently closed (remotely located).	0.01	B 0.1	B NA	B 8.33E-3	B 8.33E-6
		E 0.1	E NA	E 8.33E-3	E 8.33E-6
		S 0.1	S NA	S 8.33E-3	S 8.33E-6
PCV-033 Natural Gas to Heater pilots fails.	0.1	B 0.1	B NA	B 8.33E-3	B 8.33E-5
		E 0.1	E NA	E 8.33E-3	E 8.33E-5
		S 0.1	S NA	S 8.33E-3	S 8.33E-5
Combustion chamber steam inadvertently opened	0.01	B NA	B 0.1	B 8.33E-3	B 8.33E-6
		E (B)	E 0.1	E 8.33E-3	E 8.33E-6
		S NA	S 0.1	S 8.33E-3	S 8.33E-6

Figure 2. Sample LOPA

This LOPA includes three initiating events for the hazard scenario, two alarm IPLs, and a safety Instrumented Function (SIF) IPL. Values for initiating event frequency (A) and alarm PFD (B) are assigned based on operating experience and industry data. The SIF PFD (C) is calculated to mitigate remaining residual risk. Each of these assumed values must be monitored and verified throughout the lifecycle to assure the base PFD is not exceeded.

3 Alarm IPL

Next to basic process control, operator response to alarms is probably one of the most frequently used IPLs in the process industry. Poor alarm management is a leading cause of unplanned downtime and contributes to over $20 billion in lost production every year [1]. Assuring the effectiveness of the alarm IPL is critical to SIS health because operator response to an alarm may be used to establish both the target PFD for a SIF and reduce the demand rate on the SIF. ANSI/ISA 18.2 [2] is considered RAGAGEP for alarm management for process industries. The alarm management lifecycle includes some key steps to monitor and improve the performance of an alarm system. An ineffective alarm management system is characterized by a high number of alarms, stale alarms and nuisance alarms. For a successful alarm response the operator must do three things: detect the alarm, diagnose the situation and respond appropriately. The alarm IPL fails because [3]:

Detection	Diagnose	Respond
• Undetected alarms • Operator overload	• Alarm ignored • Insufficient training • Incorrect diagnosis • Poor descriptor	• Incorrect action • No action • Untimely action

Implementation of ISA 18.2 begins with two important steps. The first step is deployment of an alarm management software on the BPCS to capture alarm history and perform metric analysis. Data should be collected for a minimum of 30 days, then a baseline analysis is to be completed to benchmark alarm performance. Figure 3 (next page) illustrates some common alarm system diagnostics that identify three types of nuisance alarms (repeating, fleeting and stale) and overall annunciate rate. The data is representative of a continuous process prior to implementation of an alarm management program for a 29 day period.

Parallel to benchmarking, organizations should develop an alarm management philosophy. The philosophy defines objectives for alarm management for each step of the lifecycle and establishes the foundation for procedures, practices and metrics for alarm management. Alarm classifications are defined as part of the philosophy. Data collection, reporting and record retention is established for each classification. A LOPA Alarm classification should be established for alarms that are considered IPLs. During the rationalization process, IPL alarms should be assigned the appropriate classification. Monitoring reports for IPL alarms should be routinely prepared and reviewed for alarms with this classification.

The philosophy also establishes guidelines for monitoring alarm system performance targets and action limits. Some suggested metrics for LOPA IPL alarms include:

- IPL alarm annunciation frequency
- IPL bypass/out of service time
- IPL standing alarm time

Metric analysis for the LOPA IPL alarms should consider both the frequency of the alarm annunciation, and the cause for each event. Frequency should be recorded by cause, compared to the frequency assigned in the LOPA, and LOPA updated if actual frequency exceeds assumed frequency. If initiating events are identified that are not already in the LOPA, the LOPA should be updated to reflect the previously unidentified initiators and SIL target revised for any associated SIFs.

For example, the alarm 470-FT114 in Figure 3 annunciated 4,533 times in the 29-day sampling period. This calculates out to a frequency in excess of 59,000 times per year. This frequency far exceeds the frequencies of 0.1 to 1E-3 reported in literature [4]. If this is a true process upset, then the LOPA underestimates the demand on the SIF. If it is a nuisance alarm, then operators become desensitized and will likely fail to recognize a true upset and the IPL is ineffective.

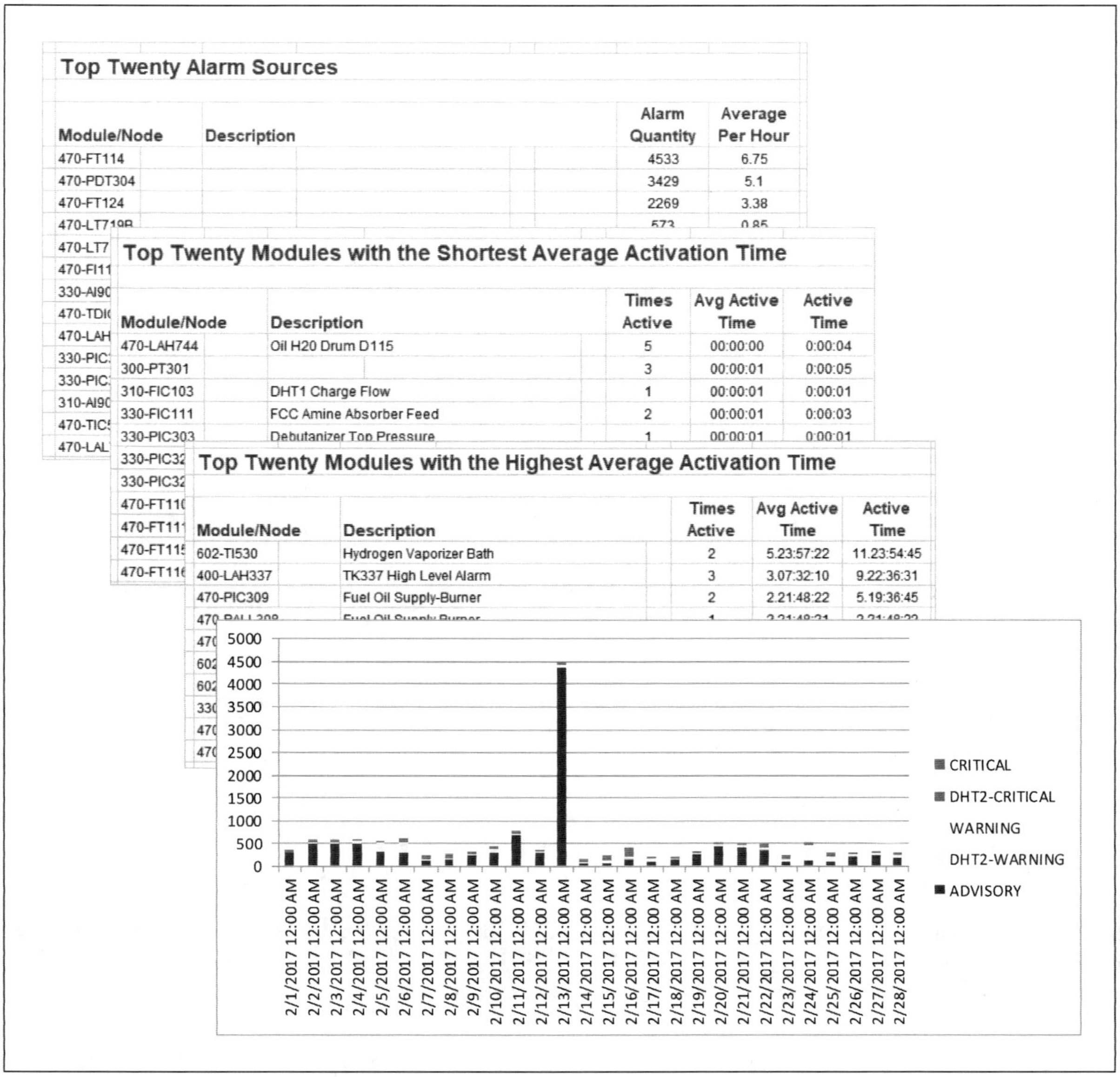

Figure 3. Sample Alarm Diagnostics

.

Analysis should also consider alarm failure. When an alarm is bypassed, or it is annunciated for long periods of time (stale), the alarm is effectively failed in place. Each occurrence should be recorded, out of service time tallied, and compared to PFD listed in the LOPA. If assumed PFD is exceeded, corrective action should be taken to reduce the out of service time or LOPA updated to reflect actual conditions.

The two alarms in the sample LOPA were assigned a PFD of 0.1 which is typical for the PFD cited in literature for human IPLs. The alarm 602-TI530 in Figure 3 annunciated twice for a total activation time of 11 hours and 24 minutes (rounded) in the 29-day sampling period. Considering only the annunciated out of service data point, contribution to overall PFD in 0.017 or nearly 20% of the assumed failure rate for an alarm IPL.

4 Safety Instrumented Function

The Safety Instrumented System (SIS) is a complex protection layer designed to prevent an incident and/or mitigate the consequences of an incident. Each Safety Instrumented Function (SIF) within the SIS is designed to detect a specific hazard condition and act to bring the process to a safe state. In the chemical process industry, the majority of these SIFs operate in low demand mode where the demand is no more than twice the proof test interval. Safety Integrity Level (SIL) of a low demand mode SIF is set to provide a target risk reduction (Table 1) in accordance with the requirements set forth in IEC 61511 [5].

Table 1. Safety Integrity Levels and Associated Parameters (Low Demand Mode)

Safety Integrity Level (SIL)	PFD_{AVG}	Required Risk Reduction
4	$\geq 10^{-5}$ to $< 10^{-4}$	> 10,000 to ≤ 100,000
3	$\geq 10^{-4}$ to $< 10^{-3}$	> 1,000 to ≤ 10,000
2	$\geq 10^{-3}$ to $< 10^{-2}$	> 100 to ≤ 1,000
1	$\geq 10^{-2}$ to $< 10^{-1}$	> 10 to ≤ 100

Alarm IPLs are usually assigned a PFD of 0.1, or RRF of10 (RRF= 1/PFD). This is less than the minimum PFD that would be assigned to a SIF. The SIF in Figure 2 has a PFD of 8.33E-3 or RRF of 120. This is 12 times the risk reduction of an alarm IPL.

In many systems, SIF trip points are defined as alarm points on the sensor elements and are usually not annunciated. In this case, the data collection and analysis tools used for alarm management can be utilized for SIFs as well. The alarm philosophy document should define a SIF Alarm classification during the rationalization process, and SIF alarms need to be assigned the appropriate classification even if they do not annunciate to the operator panel (log only). Routine monitoring and analysis is then conducted for the SIF alarm as described above for IPL alarms. Some suggested metrics for alarms and SIF IPLs include:

- SIF trip frequency
- SIF bypass time
- SIF standing alarm time
- SIF demand rate
- SIF response time

5 SIF Demand Rate

A SIF is considered to be in low demand mode when the demand frequency is twice the proof test interval and no more than once per year. SIFs with higher demand rate are considered high demand mode or continuous mode and have different design criteria. If a SIF is designed for low demand mode, and it is actually operating in high demand mode, the risk reduction delivered by that SIF may be significantly lower than required.

As noted above, an alarm IPL is considered in the calculation of target RRF for the SIF and can be used to reduce the demand on the SIF if the alarm IPL is shown to be effective. For example, the LOPA in Figure 2 has a SIF demand frequency of 0.12 when alarm is not considered to reduce demand, and 0.012 when the alarm is considered [6]. The sample LOPA identified two alarms that would serve as the Alarm IPL layer against initiating events; therefore, the performance data for both alarms must be considered together to fully assess the alarm effectiveness. In addition, SIF trip history should be maintained to confirm that the actual demand on the SIF is less than the LOPA determined demand. A near miss analysis should be completed on every SIF trip to determine cause of the trip and reasons why demand reducing IPLs were ineffective.

6 SIF Response Time

Process Safety Time is specified in the SIF Safety Requirements Specification (SRS). The SIF response time must be less than half the process safety time[1]. SIF trip time must be verified at proof test to confirm that it is acceptable. In addition, it is recommended that data capture provisions are in place so that SIF trip time can be determined in the event of a trip.

7 SIF Proof Testing

In low demand mode, the SIF components operate as static devices, therefore, failures may remain hidden unless diagnostic or routine proof testing is completed. The manufacturer safety manual will suggest one or more proof test procedures and provide the proof test coverage associated with each test. The tests must be completed according to the proof test schedule identified in the SIF SRS. "Proof test intervals that extend beyond 15% of the period mandated will start to impact the integrity of the SIF; so, track proof test intervals as a leading indicator." [7] Once failures are identified, repairs must be completed in a timely manner to restore the SIF and return to service.

Reliability for a SIF in low demand mode is expressed as average probability of failure on demand (PFD_{avg}). This is established via probability calculations based on the SIF design, proof

[1] Based on the Nyquist theory that in order to adequately reproduce a signal, it should be sampled at 2x the highest frequency that is to be recorded.

test plan, test frequency, and anticipated useful life. Data for the calculations is provided from industry databases, FMEDA analysis, or SIL certification documents. The failure rate data that is the basis for PFD_{avg} calculations are identified in terms of safe or dangerous, and detectable or undetectable failures by diagnostics or proof testing. It is important to record proof test results and periodically analyze the data so that the actual component failure rates can be compared to the values used in the SIL verification.

Proof test effectiveness is highly dependent on the discipline of the maintenance program. Are the technicians well trained? Are the tools they use in good condition and properly calibrated? Are there procedures in place to assure replacements parts are 'in kind' (i.e. identical model) with the design? Are utility services (power, air, hydraulic fluid) kept clean and well maintained? Are repairs correctly performed? Maintenance discipline can be assessed with the Site Safety Index (SSI) model (https://www.exida.com/SSI). Routine audits of performance factors and periodic re-assessment is recommended.

8 Useful Life

The calculations used to verify SIF probability of failure based on the assumption that the underlying failure rate data is representative of the useful life of components. We refer to the period of useful life as the SIF mission time. The useful life of a component is determined based on the portion of life where the random failure rate is predictable and before wear out begins. Therefore, each component in a SIF must be refurbished to like new condition or replaced at the end of its useful life. The date of installation and dates of refurbish/replacement must be included in the SIF history along with serial number.

9 Summary

This paper has discussed a variety of metrics that are necessary to confirm the effectiveness of the SIS layer of protections. Table 2 provides a summary of the metrics and the potential failure they indicate.

Table 2. Metric and Indication Summary

Category	Metric	Indication
IPL Alarm Performance	Alarm annunciation frequency and event cause analysis	Initiating event not considered in LOPA. Initiating frequency higher than considered in LOPA
	Alarm bypass/out of service time	Compromised alarm IPL. Potential alarm PFD is higher than design; SIL target understated. Potential increased demand on SIF.

Category	Metric	Indication
	Alarm standing alarm time	Compromised alarm IPL. Potential alarm PFD is higher than design; SIL target understated. Potential increased demand on SIF.
SIF Performance	SIF trip frequency (demand rate)	Potential SIF demand higher than design basis. Possible that SIF operating in high rather than low demand mode; SIF may not provide desired risk reduction.
	SIF bypass time	Compromised SIF IPL; SIF is not providing desired risk reduction.
	SIF standing alarm time	Compromised SIF IPL; SIF is not providing desired risk reduction.
	SIF response time	SIF responding slower than intended and may not perform within the Process Safety time.
SIF Health	Adherence to proof test schedule	Potentially compromised SIF IPL; SIF is not providing desired risk reduction.
	Adherence to useful life scheduled replacement/refurbish	Failure rate not in predictable range; potential increased failure rate.
	Comparison of failure history to verification assumptions	Potential failure rate is higher than considered in design; SIF not meeting SIL target and is not providing desired risk reduction.
Maintenance Performance	Adherence to training/competency program	Proof test not being performed correctly. Proof test results recorded incorrectly or not recorded at all.
	Tools maintenance and calibration logs	Proof test executed inaccurately.
	Field condition audits	Potential that supporting utilities (air, power, hydraulic) will not be available when required. External neglect suggests internal neglect.
	Post repair proof test results	Repair may not have restored SIF to service properly.

Routine metric monitoring should be established along with target and action limits for each metric.

10 Conclusion

Assuring the SIS layer of protections remain robust during the operation and maintenance phase of the lifecycle is a difficult task that must include many indicators to monitor effectiveness. Assuring the performance of alarm IPLs reduces the demand on SIFs and assures that the SIL target is supported by operational discipline. Failure to perform proof testing regularly, monitor proof test results, and replace components before wear-out can result in SIF failure. A well trained and well-equipped operations and maintenance staff along with disciplined record keeping and regular performance analysis can assure the reliability of SIS protection layer.

11 Notation

ALARP	As Low As Reasonable Practicable
BPCS	Basic Process Control System
CCPS	Center for Chemical Process Safety
FMEDA	Failure Modes Effects and Diagnostic Analysis
IPL	Independent Protection Layers
LOPA	Layer of Protection Analysis
PFD	Probability of Failure on Demand
PFD_{avg}	Average Probability of Failure on Demand
RAGAGEP	Recognized And Generally Accepted Good Engineering Practices
RRF	Risk Reduction Factor
SIF	Safety Instrumented Function
SIL	Safety Integrity Level
SIS	Safety Instrumented System
SRS	Safety Requirements Specification

12 References

[1] T. Stauffer, N. P. Sands, and D.G. Dunn, "Alarm Management and ISA-18-A Journey, Not a Destination," https://www.exida.com/Resources/Whitepapers (accessed February 21, 2019).

[2] "ANSI/ISA-84.2-2016 Part 1 (IEC61511-1 Mod) Management of Alarm Systems for the Process Industries," Instrument Society of America (ISA), Research Triangle Park, NC, 2016.

[3] T. Stauffer, N. P. Sands, and D. Strobhar, " 'Closing the Holes in the Swiss Cheese Model'- Maximizing the reliability of Operator response to Alarms," 13th Global Congress on Process Safety, March 2017.

[4] CCPS. *Layers of Protection Analysis Simplified Process Risk Assessment*, Table 6.5 Center for Chemical Process Safety, American Institute of Chemical Engineers, New York, NY 2001.

[5] IEC 61511-1 Ed 2.0, "Functional Safety: Safety instrumented systems for the process industry sector – Part 1: Framework, definitions, system, hardware and application programming requirements," IEC, Table 4, Geneva, Switzerland 2016.

[6] D. Chastain-Knight and L. Stewart, Considering Multiple Initiating events in a LOPA, Mary Kay O'Conner Process Safety International Symposium, October 2018.

[7] D. Chastain-Knight and J. Jenkins. "Proof Test Prudently. Understand how to effectively evaluate low demand safety instrumented functions," Chemical Processing, September 2018, https://www.chemicalprocessing.com/articles/2018/proof-test-prudently/ (accessed February 21, 2019).

Crediting Check Valves as IPLs?
Testing Protocol to Better Understand Check Valve Reliability

Jody E. Olsen P.E.
JE Olsen Consulting LLC
Anchorage, AK
jodyolsen.jeoc@gmail.com

Prepared for Presentation at
American Institute of Chemical Engineers
2019 Spring Meeting and 15th Global Congress on Process Safety
New Orleans, LA
March 31 – April 3, 2019

Crediting Check Valves as IPLs?
Testing Protocol to Better Understand Check Valve Reliability

Jody E. Olsen P.E.
JE Olsen Consulting LLC
Anchorage, AK
jodyolsen.jeoc@gmail.com

Keywords: check valve, safeguards, independent protection layer, IPL, process hazard analysis, PHA, layer of protection analysis, LOPA

Abstract

Conventional process safety wisdom assumes that check valves are not reliable safeguards. Experience indicates that check valves are prone to failure and that they may fail undetected. Therefore, the conservative assumption is that check valves may be listed in process hazard analyses as safeguards, but they are rarely considered to meet the standards required of an independent protection layer (IPL). Independent protection layers must be effective, independent, and auditable. Although independence is readily achievable by check valves, confirming and routinely auditing effectiveness is rarely pursued. And maintenance practices for check valves are often insufficient. Little data is available from operating companies regarding failure and leakage rates for various check valve types in various service applications or at various stages of service life. This paper examines a testing protocol that was put in place in 2014 for the purpose of testing check valves in order to apply layer of protection analysis (LOPA) credit to these valves for reverse flow scenarios. In order to understand check valve performance expectations, leakage allowances for new check valves are reviewed. Industry guidance and standards regarding consideration of check valves as safeguards or IPLs are also discussed. The analysis of new valve standards and the assessment of process safety requirements are the basis for establishing the pass/fail thresholds for the tests. The goal of sharing this information is that the discussion will stimulate others to consider the opportunity and the need to set-up similar testing and to begin gathering and sharing a larger body of data on check valve performance in various applications. Accumulation of check valve performance data and sharing of that data should lead to better understanding of check valve performance by type, size, age, and service. Better performance may be achieved where maintenance is improved and design and selection learnings are applied. In instances where requirements are met and credit is due, check valves may be credited in PHA and LOPA.

1 Introduction

Check valves are a curious equipment component in process plants and operations. Check valves are by their very nature a safeguard; they prevent flow in the wrong direction. However, standard practice during process hazard analysis is to assume that the check valve will fail and give no credit to the safeguard. This long-standing practice seems prudent given experience with serious incidents involving check valve failure.

However, does the assumption of failure of the check valve become a self-fulfilling prophecy? Does the assumption of failure become an enabler of failure? Does failure to recognize actual failure modes and frequencies prevent improvement in check valve performance and reliability?

In most applications, check valves are not causes of hazard scenarios, they are preventative safeguards.

> As per Baldas/Carithers: Check valves are almost exclusively used as safeguards, rather than as active design features. . . . It is not typical that the failure of a check valve is the cause of a [Layers of Protection Analysis,] LOPA scenario; the cause of the reverse flow comes from some other failure [or initiating event]. A check valve is typically a failure to protect, not a cause [1].

CCPS Guidelines for Initiating Events and Independent Protection Layers in Layer of Protection Analysis (CCPS IE/IPL), 2015 [2], generally supports that characterization. In Section 4.3.3 and Appendix D, a distinction is made between check valves operating in low demand mode versus high demand mode [2]. Check valves operating in high demand mode, such as internal to equipment, tend to be design features and may be initiating events. Whereas check valves operating in low demand mode are generally operating as a safeguard.

> *Note: API 521 lists check valves as "causes" for over-pressure scenarios; however, the discussion regarding reverse flow scenarios is much the same [3].*

As such, we list check valves as safeguards in process hazard reviews, but check valves rarely meet the standards required for considering the check valve or check valves in series as an independent protection layer (IPL). Check valves are recognized for having the potential to fail, to fail undetected (ie., latent failure), and to fail suddenly or catastrophically. Check valves are also recognized to operate in severe service because moving parts are present in the flow stream. However, lack of testing and inspection data remain a primary obstacle to being able to better assess and quantify the actual performance and failure modes and frequencies for check valves.

Although inspection, maintenance, and testing (IMT) are considered essential to assure the performance and reliability of any piece of mechanical equipment, check valves generally do not receive the same attention as other equipment and components [4]. The lack of routine inspection and/or maintenance may in part explain why check valves are considered unreliable.

This paper advocates changing the paradigm. Where check valves are required to satisfy corporate risk criteria, they need to not only be inspected, but also tested for leakage rates in order to assure their reliability. Where defects are found, check valves should be repaired and returned to acceptable condition. That means that the installation configuration must allow for future inspection, maintenance, and testing, this work must be planned and scheduled, and then performed as scheduled. History on the performance of check valves in various applications, for various types and sizes, and at various stages of their service life will allow better engineering design of check valves and better selection of check valves for the particular application.

Improving selection and assuring proper function of check valves via an appropriate mechanical integrity program allows for more robust process safety design by sharing the burden of protection across more disparate safeguarding devices and independent protective layers. Understanding the actual performance of check valves allows for attributing appropriate credit, where it is due, and also allows resources to focus attention on hazards that have a larger actual gap between target risk levels and current mitigated risk.

This paper discusses an example testing protocol that was written and implemented for several new check valves that were installed in an oil and gas processing plant in 2014. In preparation for that testing, standards that define maximum allowable leakage rates for new valves were reviewed. Guidance from industry publications on considering check valves as safeguards or IPLs was also reviewed. The analysis of valve standards and the assessment of process safety requirements were the basis for establishing the pass/fail thresholds. Key take-aways and experiences from the preparation and testing are summarized.

While testing check valves is necessary to allow for potential crediting of check valves as independent protection layers (IPLs), inspection programs must also be in place to identify insipient failures. Inspections reveal conditions that are precursors to failures such as mechanical damage, fouling, debris, and looseness that are not yet manifested as leakage. The particular application, hazards, and consequences must be thoroughly understood in order to consider PHA and/or LOPA credit for individual or redundant check valve installations.

2 History

Early publications of process safety hazard analysis methodology including CCPS LOPA (2001) [5] and CCPS Guidelines for Hazard Evaluation Procedures, Table 7.4, (2008, 1992) [6] cautioned practitioners that check valves should not be credited as safeguards or IPLs. With the publication of later references, such as CCPS IE/IPL in 2015, guidance has changed to allow for the usage of check valves as IPLs [2]. From CCPS IE/IPL:

> Section 1.3 . . . At the time of CCPS LOPA (2001), check valves were not generally considered to be valid IPLs due to a lack of data supporting their reliability. Since that time, understanding of check valve reliability has improved, assisted by more data that substantiates their reliability. (Refer to Appendix D for Example Reliability Data Conversion for Check Valves for more information.) Based on this data, check valves have been included as IPLs in this book [2].
>
> Section 1.7. . . Appendix D . . . Check valve design has improved over the years, and data exist that indicate that check valves can be effective IPLs when properly specified and maintained [2].
>
> Section 5.2.2.3 . . . check valves can be considered to be IPLs when properly specified, designed, installed, and maintained [2].
>
> *Note: The data referenced in Appendix D [2] is based on 1987 data. However, the PFD analysis and incorporation into LOPA guidance came later.*

Little has changed since the initial publication of PHA and LOPA guidance with respect to opinions, assumptions, and maintenance practices regarding check valve reliability. As such, practices and assumptions in the process safety arena have remained essentially the same as well. Many check valves remain untagged equipment items indicating that the check valve will not be included in inspection or maintenance programs and no history will be tracked as to the condition of the valve. All check valves will not require inspection and testing, but failure and repair history should be tracked, at a minimum, for all safeguards.

New guidance in API 570 [7] May 2017 addendum requires that critical check valves "shall" be included in inspection or testing programs [4]. While some operators have migrated to include critical check valves in basic inspection and maintenance programs, very few operating companies test their check valves. Therefore, very few operating companies credit their check valves as IPLs. While many standards exist for quantifying leakage rates for new check valves provided by manufacturers, little to no quantitative data is available on check valve leakage with respect to type, size, age, or service after check valves are placed in service. Lack of performance data rather than specific data indicating unacceptable leakage rates is a primary reason for lack of movement towards potential crediting of check valves as IPLs. This situation applies to not only older check valves that may not be well configured for testing, but also to newer check valve installations.

3 PHA or LOPA Credit – When to Credit and How Much

When considering giving a check valve or check valves qualitative (order of magnitude) credit as a PHA safeguard or semi-quantitative credit as an independent protective layer for a hazard scenario, there are several options for crediting that depend on the equipment and process configuration.

Check valves that meet IPL criteria may be considered for crediting as IPLs where the check valve or valves are:

a) Single check valve - credited as stand-alone IPL
b) Dual check valves in series - credited as single IPL
c) Combination IPL comprised of upstream PSV or PSV setting and downstream check valve(s) (where check valve leakage rates are assumed and/or tested)

Inspection and testing protocol including threshold leakage rates must be defined depending on which of the above configurations is applicable. When establishing threshold leakage acceptance rates, it is important to consider the amount of leakage that is deemed tolerable within the particular process system in question. Process equipment rating and capacity, reactivity hazards, relief device capacity (where applicable), upstream rotating equipment impacts, and other factors for a given scenario must be considered. As with all IPLs, check valve IPLs must prevent the consequence of concern.

It is important to note that option c) references a single IPL that is comprised of a specific PSV or PSV setting and a specific check valve or check valves in series. Inspection, maintenance, and testing of all components are required to credit the single IPL. These components may not be used separately as independently credited IPLs. Likewise, in option b) two check valves may be credited as a single IPL and both check valves require independent inspection, maintenance, and testing.

3.1 Crediting Check Valves, Single or Dual Check Valves, as IPL

Several reference documents address potentially crediting check valves as safeguards and/or IPLs. CCPS IE/IPL [2] Section 5.2.2.3 and Table 5.3.2, addresses crediting a single maintained check valve as an IPL. CCPS IE/IPL distinguishes check valves operating in low demand mode as IPLs whereas check valves operating in high demand mode are considered initiating events. However, the calculated probability of failure is the same.

> *Note: In CCPS IE/IPL [2], the probability of failure evaluation for a single check valve in either mode, high demand as initiating event or low demand as IPL, is based on the same data set. The probability of failure on demand (PFD) calculated for a single check valve IPL is the same as the initiating event frequency (IEF) for a high demand check valve, 0.1. CCPS IE/IPL also calculates IEF for dual check valves in the high demand mode of 0.01. The data set used for the probability of failure calculation is based on Reliability Data Book for Components in Swedish Nuclear Power Plants, RKS/SKI 85-25, p 79 (Bento et al. 1987). Calculations are shown in Appendix D of CCPS IE/IPL.*

CCPS IE/IPL [2] highlights the fact that some check valve leakage may be expected. Therefore, the user must define tolerable leakage rates and assure through inspection and leakage testing that the check valve can be expected to perform as anticipated and protect against the scenario. Additionally, for consideration as an IPL, the check valve should be operating in clean, non-fouling, non-corrosive service [2].

An IPL with calculated PFD for dual check valves is not explicitly discussed in CCPS IE/IPL [2] Section 5.2.2.3. CCPS IE/IPL cautions the users to consider design, operability, and function of two check valves in series to assure that both respond appropriately and independently. CCPS IE/IPL [2] also reminds the user to consider common cause failures.

Although, CCPS IE/IPL [2] does not show a separate dual check valve IPL category, it is assumed that the PFD for a dual check valve IPL will be similar (0.01) to the initiating event frequency (IEF) for dual check valves in high demand mode. Crediting dual check valves with this PFD assumes that the check valves are individually inspected and are unlikely to experience common cause failures.

Crediting of check valves that meet IPL criteria is also discussed by Baldas and Carithers [1]. They calculated check valve IPL credit for multiple valves in series based on a single check valve PFD = 0.1, or Risk Reduction Factor (RRF) = 10. Using this assumption and an assumption for common cause failure rates of 1 in 10, Baldas and Carithers [1] calculated a PFD for dual check valves in series of 0.0181 (RRF = 55) which rounds to approximately two orders of magnitude. They further showed via the same calculation that more check valves in series would not yield a substantially lower combined probability of failure on demand; therefore, no more than two orders of magnitude credit should be attributed to check valves in series, even if more than two check valves, that are maintained, are present in series [1]. This conclusion is also supported in API 521 (2014) [3] Section 4.4.9.3.3. "a) Because of potential common mode failures the user is cautioned against taking a larger credit for more than two check valves in series that are inspected and maintained."

It is notable that CCPS IE/IPL [2] and API 521(2014) [3] differ on guidance regarding single check valve performance and probability of failure on demand (PFD). Although API 521 [3] only addresses check valve(s) working together with PSV(s) to comprise the IPL and does not explicitly address mathematical PFD for check valves, inherent unreliability is implied in API 521 (2014) [3] because this standard assumes that a single check valve, regardless of service, history, or testing, is not sufficient to be credited for even partially restricting flow to the PSV. In other words, this guidance assumes a single check valve should receive no credit for being in place, therefore, PFD = 1 (RRF = 1). API 521 (2014) [3] dictates guidance that all single check valves should be assumed to completely fail and not reduce reverse flow to the PSV. This conclusion is fundamentally different from the position supported by data and PFD calculations in CCPS IE/IPL [2] Appendix D. Additional data should be gathered that is specific to the check valve type, size, age, and service in order to better define the PFD for a given check valve.

Although CCPS IE/IPL [2] provides a basis for crediting either single or, by extension of IE discussion, dual check valves when IPL requirements and other special considerations are satisfied, few companies have moved toward applying credit to check valves as IPLs. In "Check Valves as Safeguards: Friend or Foe in Process Hazard Analysis", Modi [16], Table 9.1 provides a sampling of credit given to check valves by operating companies of various size and sector. Based on this overview, very few companies attribute any credit to a single check valve as a safeguard in PHA analyses. A number of companies attribute a single order of magnitude credit to dual check valves safeguards in PHA assessments. But LOPA credit is rarely attributed even to dual check valves.

As with any IPL, check valves considered for crediting as an IPL must meet all required criteria, including assuring 1) effectiveness, 2) independence, and 3) auditability. The decision to credit either a single check valve or dual check valves in series as a safeguard or as an IPL is based on individual company protocol. This protocol should be based on historical experience, understanding of check valve failure modes, reliability and performance, understanding of hazard scenarios involved, and company risk tolerance. The lack of credit taken for check valves represents a conservative approach. However, this conservative approach is also attributable to lack of progress in understanding check valve performance and failure modes and frequencies in various process services.

Refer to Table 1 for a summary of check valve probability of failure guidance based on several references discussed above. This table summarizes potential PHA or LOPA credit that may be given when check valves are maintained, inspected, and tested and considered with regard to special considerations stipulated in the guidance documents.

Table 1. Potential PHA and LOPA Credit

OM = Order of Magnitude		PFD = Probability of Failure on Demand IEF = Initiating Event Frequency RRF = Risk Reduction Factor				*Scenario Likelihood Reduction
		Independent Protection Layer		Initiating Event		
Device	**Safeguard**	**PFD**	**RRF**	**IEF**	**RRF***	**Reference**
Single Check Valve (Low Demand Mode)	1 OM	0.1	10			Guidelines for Initiating Events and Independent Protective Layers in Layer of Protection Analysis, Chap 5, 5.2.2.3, Table 5.32 based on Appendix D [2]
Dual Check Valves (Low Demand Mode)	Not Addressed	Not Addressed	Not Addressed			CCPS IE/IPL [2], Not addressed
Single Check Valve (Low Demand Mode)	1 OM	0.1	10			Baldas/Carithers [1], Assumed Value
Dual Check Valves (Low Demand Mode)	2 OM	0.01	100			Baldas/Carithers [1], Calculated, Assumption: No common cause failures
Dual Check Valves (Low Demand Mode)	~ 2 OM	0.0181	55			Baldas/Carithers [1], Calculated, Assumption: Common cause failures
Three Check Valves (Low Demand Mode)	2 OM	0.0107	93			Baldas/Carithers [1], Calculated, Assumption: Common cause failures
Single Check Valve (Low Demand Mode)	0 OM	1	1			API 521, 2014, 6th Ed, [3] Implied based on assumption that single check valve will always fail
Single Check Valve (High Demand Mode)	NA			0.1	10	CCPS IE/IPL [2], Chap 4, 4.3.3, Table 4.11 based on Appendix D [2]
Double Check Valve (High Demand Mode)	NA			0.01	100	CCPS IE/IPL [2], Chap 4, 4.3.3, Table 4.12 based on Appendix D [2]
All values assume that appropriate inspection and testing is conducted and that special considerations have be reviewed for the scenario and for the application of the generic probability of failure value given.						

3.2 Crediting Check Valve(s) and upstream PSV(s) as IPL

CCPS IE/IPL [2] not only references the possibility of crediting check valves as IPLs, it also addresses the possibility of crediting a PSV in conjunction with a check valve in Sections 3.2.1 and 5.2.2.2.2 as per configuration c) above. API 521 [3] also references configuration c). However, CCPS IE/IPL [2] and API 521 [3] differ regarding guidance as to whether the combination PSV/check valve IPL may be credited with only a single check valve in place that limits reverse flow to the PSV. API 521 (2014) [3] does not allow for crediting a single check valve with reducing flow to the PSV. API 521 (2014) [3] directs the user to assume "complete" check valve failure when a single check valve is in place, even if maintained and inspected. CCPS IE/IPL [2] references a single check valve in scenarios involving an IPL comprised of a PSV and a check valve.

> *Note: Regardless of these discrepancies in guidance, it is also important to recognize that the guidance provided in API 521 [3][17] only applies to configuration c) above. Configurations a) and b) are not related to relief scenarios. Configurations a) and b) consider the check valve or check valves as an IPL on its own merits.*

As per CCPS IE/IPL [2], when a PSV, or PSV setting, is used in combination with a check valve or check valves as an IPL, the probability of failure on demand (PFD) attributed to the combination PSV(s) and check valve(s) IPL should be equal to the PFD of the component having the greatest probability of failure. CCPS IE/IPL [3] references the single check valve PFD of 0.1, provided in Table 5.32, as the overall PFD for the single check valve/PSV combination IPL as an example.

While check valve PFD is discussed directly or indirectly in CCPS IE/IPL [2] and API 521[3], frequency of inspection and/or testing of check valves is not defined. Frequency of inspection and testing is left to the user to define. Shut-down requirements for testing of check valves must also be considered and planned in conjunction with defining inspection and test frequency.

4 New Check Valve Seat Leakage Requirements – Industry Standards

In order to understand the potential performance capability for check valves in operational service, we must first understand the performance expectations for new check valves. Standards have been established that define valve performance requirements including seat leakage rate allowances. In the United States, block valves and check valves are generally tested by manufacturers to one of the following standards or to an internal manufacturer standard that meets or exceeds these standards: API 598 [8], ISO 5208 [9], MSS SP-61 [10], or API 6D [11][12]. EN-12266-1 [13] is an equivalent European standard which superseded BS 6755-1 [14]. The standard chosen depends primarily on the valve materials of construction, location jurisdiction, industrial sector, or customer requirements.

Metal-to-metal seated valves of all types, including check valves, are expected to have some leakage. Resilient seated valves, also called soft seat valves, have more stringent leakage standards, usually indicated as zero leakage or no visual leakage. Most of these industry standards allow for higher seat leakage rates through metal seated check valves as compared to other types of metal seated block valves. However, some manufacturers design check valves to standards comparable to other valve types. Other valve manufacturers may conform to the standards but exceed the minimum requirements.

API 598 [8] addresses gate, globe, plug, ball, butterfly, and check valves. API 598 (2016) Table 5 defines maximum allowable seat leakage rate by valve size and type. ISO 5208 [9]: Table 4 (see Appendix A) references a range of leakage rates referred to as Rate A (zero leakage rate) through Rate G, including Rate A, AA, B, C, CC, D, E, EE, F, and G. ISO 5208: Table 4 Note 4 refers users to Rate G for check valves tested using liquid and to Rate EE for check valves tested using gas. Rate G essentially mirrors the maximum allowable leakage rates from API 598 [8] for check valves using liquid test, and likewise regarding Rate EE for gas test. MSS SP-61[10]: Table 4 (Appendix B) refers to leakage rates for all "full open" and "full closed" valves types included within the scope of this standard, including gate, ball, plug, globe, or butterfly valves. As per 5.7.2, MSS SP-61 allows for 4 times the leakage rate shown in MSS SP-61: Table 4 for check valves. API 6D [12] states that leakage rates for metal seated valves shall not exceed ISO 5208 Rate D. ISO 5208 Rate D allowable leakage rates are slightly more stringent that MSS SP-61 for block valves. API 6D Ed 24, 2014 [11] added an allowance for check valves to meet ISO Rate G. Previous API 6D [12] editions did not allow for higher leakage rates from check valves. Recent updates to major standards have resulted in convergence on leakage limits. Tables below show maximum allowable valve leakage rates for various valve sizes as defined by API 598 [8], MSS SP-61 [10], and ISO 5208 [9] using consistent units of measure. Table 2 and Table 3 reference tests conducted using liquid and gas, respectively. Manufacturer performance standards for two example manufacturers are also shown.

Table 2. Valve Seat Leakage Allowance by Standard [8][9][10][11], Liquid Test

								All Mn (RS)	(Req) MnY						Mnfr Y
								Mn X (RS)					Mnfr X		
								SP-61 (RS)			MSS SP-61 Blk		MSS SP-61 Chk		
								API 6D (RS)			API 6D Blk				API 6D Chk
		All RS	Non-Chk	Check		Non-Chk	Check	API 598 (RS)		API598 Blk					API 598 Chk
		API 598	API 598	API 598	MSS SP-61	MSS SP-61	MSS SP-61	ISO 5208	ISO 5208	ISO 5208	ISO 5208	ISO 5208	ISO 5208	ISO 5208	ISO 5208
								Rate A	Rate B	Rate CC	Rate D	Rate E	Rate EE	Rate F	Rate G
			16.00	3	0.00	0.167	0.668	0.000	0.01	0.08	0.10	0.30	0.39	1	2
Units above =>			drops/ml	cc/ min/ nom diam in	cu cm/ nom diam inch	cu cm/ min/ nom diam inch	cu cm/ min/ nom diam inch	No visually detectable leakage	cu mm /sec / DN	cu mm /sec / DN	cu mm /sec / DN	cu mm /sec / DN	cu mm /sec / DN	cu mm /sec / DN	cu mm /sec / DN
Nom Pipe Diam		Soft Seat			Soft Seat			Soft Seat							
DN	inches	ml/min	ml/min	ml/min	ml/min	ml/min	ml/min	ml/min	ml/min	ml/min	ml/min	ml/min	ml/min	ml/min	ml/min
	1	0	0.00	3	0.0	0.167	0.668	0.00	0.0150	0.120	0.150	0.450	0.59	1.50	3.0
50	2	0	0.00	6	0.0	0.334	1.34	0	0.0300	0.240	0.30	0.90	1.17	3.00	6.0
80	3	0	0.38	9	0.0	0.50	2.00	0	0.05	0.38	0.48	1.44	1.87	4.8	9.6
100	4	0	0.50	12	0.0	0.67	2.67	0	0.06	0.48	0.60	1.80	2.34	6.0	12
150	6	0	0.75	18	0.0	1.00	4.01	0	0.09	0.72	0.90	2.70	3.51	9	18
200	8	0	1.00	24	0.0	1.34	5.34	0	0.12	0.96	1.20	3.60	4.68	12	24
250	10	0	1.25	30	0.0	1.67	6.68	0	0.15	1.20	1.50	4.50	5.85	15	30
300	12	0	1.50	36	0.0	2.00	8.02	0	0.18	1.44	1.80	5.40	7.02	18	36
400	16	0	2.00	48	0.0	2.67	10.69	0	0.24	1.92	2.40	7.20	9.36	24	48
500	20	0	2.50	60	0.0	3.34	13.36	0	0.30	2.40	3.00	9.00	11.70	30	60
600	24	0	3.00	72	0.0	4.01	16.03	0	0.36	2.88	3.60	10.80	14.04	36	72
900	36	0	4.50	108	0.0	6.01	24.05	0	0.54	4.32	5.40	16.20	21.06	54	108
1200	48	0	6.00	144	0.0	8.02	32.06	0	0.72	5.76	7.20	21.60	28.08	72	144

Table 3. Valve Seat Leakage Allowance by Standard [8][9][10][11], Gas Test

						Mnfr Y (RS)			Mnfr Y			
						Mnfr Y (RS)			Mnfr X			
						MSS SP-61 (RS)			MSS SP-61 Chk			
						MSS SP-61 (RS)			SP-61 Blk			
			Check	Non-Check	Check Valves	API 598 (RS)	API 598 Blk				API 598 Chk	
		API 598	API 598	MSS SP-61	MSS SP-61	ISO 5208	ISO 5208	ISO 5208	ISO 5208	ISO 5208	ISO 5208	ISO 5208
			Given			Rate A	Rate AA	Rate C	Rate D	Rate E	Rate EE	Rate G
			1.50	0.1	0.40	0.000	0.18	3	30	300	470	6000
Units for above =>			SCF / hr/ nom pipe diam inch	SCFH per nom pipe diam inch	SCFH per nom pipe diam inch	No visually detectable leakage during test	cu mm/ sec/ DN	cu mm/ sec / DN	cu mm/ sec / DN	cu mm/ sec / DN	cu mm/ sec / DN	cu mm/ sec / DN
Nominal Pipe Diameter		Soft Seat		Soft Seat		Soft Seat						
DN	inches	SCFH	SCFH	SCFH	SCFH	SCFH	SCFH	SCFH	SCFH	SCFH	SCFH	SCFH
	1	0	1.50	0.10	0.40	0.000	0.0006	0.01	0.10	0.95	1.49	19.05
50	2	0	3.00	0.20	0.80	0.000	0.0011	0.02	0.19	1.91	2.98	38.10
80	3	0	4.50	0.30	1.20	0.000	0.0018	0.03	0.30	3.05	4.78	60.96
100	4	0	6.00	0.40	1.60	0.000	0.0023	0.04	0.38	3.81	5.97	76.20
150	6	0	9.00	0.60	2.40	0.000	0.0034	0.06	0.57	5.72	8.95	114.30
200	8	0	12.00	0.80	3.20	0.000	0.0046	0.08	0.76	7.62	11.94	152.40
250	10	0	15.00	1.00	4.00	0.000	0.0057	0.10	0.95	9.53	14.92	190.50
300	12	0	18.00	1.20	4.80	0.000	0.0069	0.11	1.14	11.43	17.91	228.60
400	16	0	24.00	1.60	6.40	0.000	0.0091	0.15	1.52	15.24	23.88	304.80
500	20	0	30.00	2.00	8.00	0.000	0.0114	0.19	1.91	19.05	29.85	381.00
600	24	0	36.00	2.40	9.60	0.000	0.0137	0.23	2.29	22.86	35.81	457.20
900	36	0	54.00	3.60	14.40	0.000	0.0206	0.34	3.43	34.29	53.72	685.80
1200	48	0	72.00	4.80	19.20	0.000	0.0274	0.46	4.57	45.72	71.63	914.40

Values shown in Table 2 and Table 3 are approximate as calculated by author. Refer to Appendix C and D for expanded tables that show leakage rates for all ISO 5208 [9] performance levels. Refer to Appendix E and F for simplified overview graphics of allowable leakage rates.

Leakage rates for various types of block valves, including gate, ball, plug, globe, and butterfly, as well as check valves, are included in the standards referenced above. A separate standard, ANSI/FCI 70-2, Control Valve Seat Leakage [15], specifically addresses control valves. This standard is a useful point of comparison as control valves operated as basic process control systems (BPCS) may also be considered for credit as IPLs, as per CCPS IE/IPL [2] 5.2.2.1, Safety Control Loop (Normal Operating Control) Table 5.12, if they meet IPL criteria.

> *Note: CCPS IE/IPL [2] Section 5.2.2.3, Check Valves as IPLs, references API 598 (2009b), MSS SP-61(2009), and ANSI/FCI 70-2 as applicable standards that address check valve leakage rates. Specific leakage rate classes from ANSI/FCI 70-2, Class I through Class VI are replicated in Check Valves as IPLs section of CCPS IE/IPL 5.2.2.3. This reference is misplaced with respect to check valves, as ANSI/FCI 70-2 is specific to control valve leakage. Manufacturers of check valves surveyed confirmed that they refer to API 598, ISO 5208, API 6D, or appropriate MSS standard for requirements on allowable leakage rates and testing protocol. Some manufacturers institute more stringent corporate leakage standards.*

The standards referenced stipulate not only maximum allowable leakage rate criteria, but also testing requirements and procedures, including test pressure and duration by valve type.

As noted, all mechanical valves are prone to leak, regardless of valve type or style. Zero leakage rates are only achieved where resilient or soft seats are included in the valve design. Valves designed for zero leakage rates using soft seats include various categories of valves, including check valves. However, temperature and materials compatibility issues as well as wear and maintenance requirements must be considered when soft seating materials are specified.

5 In-Service Check Valve Seat Leakage Assumptions

API 521 [3] is considered the definitive reference regarding sizing and capacity of relief equipment. API 521 [3] provides guidance on assumptions for check valve leakage rates that dictate PSV relief capacity requirements for reverse flow scenarios. Many users reference this guidance given the lack of leakage rate data that is available in industry from other sources. Section 4.4.9.3.1 of API 521 (2014) [3] discusses several scenarios for reverse flow conditions, including complete check valve failure, severe leakage, and normal leakage.

5.1 Complete Check Valve Failure

As per API 521 (2014) [3], complete check valve failure is considered to be "potential for gross reverse flow because the check valve does not function at all (e.g. it is stuck wide open or the internals are gone)." "Complete check valve failure is assumed for all check valves in series that are not inspected or maintained and for a single check valve regardless if it is inspected and maintained." [3]

The assumption of complete failure means that the normal operation of the check valve has no limiting effect on the reverse flow rate. API 521 (2014) provides three options for estimating reverse flow through a completely failed check valve [3].

a) Assume no flow resistance in reverse direction. [3]
b) Assume the same flow resistance in reverse through the check valve as in the forward flow direction. [3]
c) Assume the failed check valve has an orifice equal to the check valve flow area without internals. [3]

CCPS IE/IPL 5.2.2.2.2 Case 2 [2] speaks to this scenario. However, since the check valve is uninvolved in a protective capacity in this scenario, the IPL is comprised of only the PSV, not a combination of the check valve and PSV.

> *Note: If assumption b) or c), above, is used for sizing the PSV, in the event of removal or modification of the check valve, the PSV capacity would have to be revalidated through the MoC process.*

5.2 Severe Check Valve Leakage

As per API 521 (2014) [3], severe check valve failure is considered to be "potential for significant reverse flow because of check valve seat damage or obstruction." API 521 (2014 [3], 2008 [17]) advises that it is the responsibility of the user to determine the appropriate method for estimating reverse flow, but offers the following guidance. "Where no specific experience or company guidelines exist," API 521 (2014) [3] directs the user to assume complete failure of the smallest check valve and to assume severe failure of all other check valves in series.

Two alternatives are provided in API 521 (2014) [3] for assuming reverse flow rate for check valves having severe leakage. The following assumed reverse flow rate may be used for sizing relief devices upstream of inspected and maintained check valves [3].

1) Assume that the check valve orifice diameter is 10% of the check valve nominal diameter (1% of nominal flow area) and calculate reverse flow [3]. (See API 521 (2014) [3] for properties and parameters assumptions. Note: API 521 (2008) [17] references the same flowrate calculation assumption without parameter details provided.)
2) Treat check valve as an orifice sized to pass 10% of normal forward flow. Back calculate orifice size. Then calculate reverse flow [3]. (See API 521 (2014) [3] for properties and parameter assumptions.)

These leakage rate assumptions for severe check valve failure provided in API 521 (2014) [3] are conservative and form the basis of PSV design calculations where leakage rates are not known or quantitatively tested.

CCPS IE/IPL [2] Section 5.2.2.2.2 Case 1, PSV with check valve IPL, references similar language that is taken from API 521 (2008) [17]. (Note: a single check valve is referenced in Case 1 as described in CCPS IE/IPL [2].)

> From CCPS IE/IPL [2] . . . one may estimate the reverse flow through series check valves as flow through a single orifice with a diameter equal to one-tenth of the largest check

> valve's nominal diameter. A lower value may be used if a condition-monitoring system for the check valves . . . is installed to monitor the condition to ensure that leakage rate is below the capacity of the low-pressure side of the relief device. [2]

This reference to API 521 (2008) [17] guidance alludes to "severe" leakage initially and potentially "normal" leakage as the "lower value." In practice, most users assume the severe leakage rate as described above and in API 521 (2008) [17], or as described in API 521 (2014) [3], assumption 1) or 2), because leakage rates are rarely, if ever, tested directly.

Note: in CCPS IE/IPL [2] Section 5.2.2.2.2 Case 1, the following statement is made:

> If the ANSI/API 521 (2008) [or 2014] guidance is followed, the PRV sizing basis would be reduced to handle just the flow from a slight reverse leak through the check valve.[2]

The word slight in this sentence is misplaced as this assumption for leakage represents the "severe" failure case having "significant" leakage rates. Leakage rates discussed later in this paper in conjunction with testing protocol also support the position that this reverse flowrate assumption represents "severe" rather than "slight" or "normal" leakage of the check valve.

In order to utilize this guidance and credit the entire IPL, both the relief valve(s) and the check valves, must be maintained via company preventative maintenance (PM), inspection, and repair programs. It is important to note that this guidance is provided in lieu of known quantitatively tested check valve leakages rates, which explains the conservative assumptions that are used in API 521 (2014 [3], 2008 [17]) for assumed leakage rates.

5.3 Normal Check Valve Leakage

As per API 521 (2014) [3], normal check valve failure is considered to be "potential for minor reverse flow due to normal check valve wear." API 521 (2014) [3] advises that normal check valve leakage may be assumed when two of more check valves are in series that are inspected and maintained *and* demonstrate reliability. API 521 (2014) [3] does not allow for assumption of normal leakage rates through a single valve, even if inspected and maintained. (Note: Guidance changed in API 521 (2014) [3], as API 521 (2008) [17] *did* allow consideration of a single inspected and maintained check valve to limit reverse flow.)

API 521 (2014) [3] does not provide the user with assumptions for normal leakage rates. For normal leakage, the user is advised that if a pressure relief device is involved in the IPL, the relief device loads and sizing including the check valve reverse flow leakage rates must be determined by the user. API 521 (2008) [17] alluded to methods for establishing actual leakage rates such as through condition-monitoring using pressure indicators but did not reference specific assumptions or "normal" leakage values.

Having reliable, repeatable check valve leakage data along with historical failure modes and failure frequency data associated with the check valve type, size, age, and service is essential for users planning to assume normal leakage rates. Establishing normal leakage rate data is essential for application of either combination PSV/check valve IPLs or standalone check valve IPLs.

Normal leakage should be evaluated and considered relative to new check valve leakage standards.

5.4 Alternative Characterization of Seat Leakage

An alternative characterization to the three check valve leakage conditions described in API 521 [3] would be complete failure; degraded condition; and like-new condition. Like-new condition implies that maximum allowable seat leakage rates given by industry standards, such as API 598 [8], ISO 5208 [9], or MSS SP-61 [10], are met. Degraded condition encompasses leakage rates ranging from like-new to complete failure. Leakage rates characterized by API 521[3] as normal or severe would fall within the lower and upper ranges of degraded performance, respectively.

6 Testing Design Consideration (General)

In order to implement a quantitative leak testing program for check valves, the process and mechanical configuration should be considered in the design phase of new projects and modifications to existing installations.

Key considerations:

- Isolation requirements both downstream and upstream of the check valve
- Tap locations, for connecting pressure indication devices/transmitters
- Tap locations, for connecting pressure source, return line, and bleed line

Each check valve assembly involved in the testing program, whether a single check valve or dual check valves in series, must be configured with sufficient upstream and downstream taps or connections to allow for testing each check valve individually. Isolation of the check valve for testing must also be considered. Many legacy installations lack adequate spacing for required connections or isolation locations.

Taps are required in order to monitor pressure both upstream and downstream of all individual check valves that will be tested. For testing purposes, these locations should be configured with pressure transmitters.

Where an external pressure source is required to deliver the test pressure, taps are also required to provide that pressure source to the downstream side of the check valve. Taps must be provided on the normally upstream side of the valve, as well, in order to allow for return flow during the test if a flow through test is performed. Connections must be provided to bleed the pressure from both sides of the check valve following the test.

Additionally, isolation requirements must be considered. In order to assure that the leakage measured is going in the direction assumed, such as in reverse across the check valve, the system on both sides of the check valve must be tight and leak tested in advance of performing the leak test on the check valve. Obtaining and proving a tight system may be achieved by using double block and bleed valving on process piping upstream and downstream of the check valve(s) or by providing isolation points for installing blinds both upstream and downstream. As always, the costs and challenges of getting an effective double block and bleed leak test must be weighed against safety risk and operational requirements for opening and blinding. Any laterals that tie into the check valve piping involved in the testing have the same isolation requirements. Therefore, these laterals should be minimized or eliminated where possible unless the lateral provides the source pressure.

The entire testing configuration and procedure steps must receive a full process hazard review before any external connections are made and before any procedural steps are implemented. Likewise, if testing pressure is provided via a process or utility pressure source through an alternate (abnormal) valve alignment, the valve alignment and procedural steps must be evaluated via process hazard review prior to implementation.

7 This Test – Purpose and Description of Configuration

7.1 Purpose

The check valve configuration and the testing procedure described here was put in place in 2014. The reason for testing the check valves was to apply layer of protection analysis (LOPA) credit to these valves for reverse flow scenarios. An added benefit of testing the check valves was to better understand their performance and reliability and to quantitively assess their performance over time through leakage testing.

7.2 Description of Configuration

Five check valves, two different types, were installed and tested in two different process locations. One location involved testing a single axial flow check valve in a higher pressure service downstream of a gas compressor. (See Figure 1.) This check valve was in series with an existing wafer check valve on the discharge of the compressor setting. Procedures were envisioned to test the existing check valve using process pressure. Testing of both the new and existing check valves would allow for crediting as dual check valves rather than a single check valve. Testing procedures for the existing check valve are not included in the scope of this discussion. The other location involved two sets of dual check valves, an axial flow check valve in series with a wafer check, on two parallel process trains in lower pressure gas service. (See Figure 2.)

The check valves being tested were purchased and installed in the same year. The testing procedure was implemented at the time of installation and commissioning to serve as baseline performance testing. Follow-up testing was planned at intervals consistent with the company LOPA protocol based on the credit that was intended for these check valves. Testing would occur during planned and scheduled shut-down opportunities, either full or partial plant outages.

The testing procedure involved connecting an external pressure source to the system, nitrogen bottles, and supplying pressure at prescribed levels to the downstream side of the check valves. The nitrogen bottles were fitted with a regulator to supply the appropriate pressure. The nitrogen skid and connected equipment was protected by a PSV skid designed to protect the equipment involved in this procedure. The check valve piping assembly was fitted with ports to allow for pressuring the downstream side of the valve and relieving the upstream side of the valve through the return line. Pressure transmitters were available on both the downstream and upstream side of the check valve(s). The pressure transmitters were reconfigured during the testing period to allow for gathering of data at very short time intervals. The leakage could be measured one of two ways. a) Using a rotometer in the outlet flow path from the upstream side of the check valve to atmospheric vent location or b) Closing the outlet path and monitoring the pressure transmitter on the upstream side of the check valve for pressure build-up of closed upstream chamber and using gas laws to calculate the leakage rate associated with the pressure build-up. Plan a) would be used initially. And Plan b) would be used in the event that the rotometer method was unsuccessful or not definitive. The check valve configuration was in an enclosed module, but the nitrogen bottle source and the return line discharge from the rotometer were both outdoors. Refer to Figure 1 for a simplified depiction of the testing configuration for Test 1: Single Axial Check Valve, Higher Pressure Gas Service downstream of Compressor.

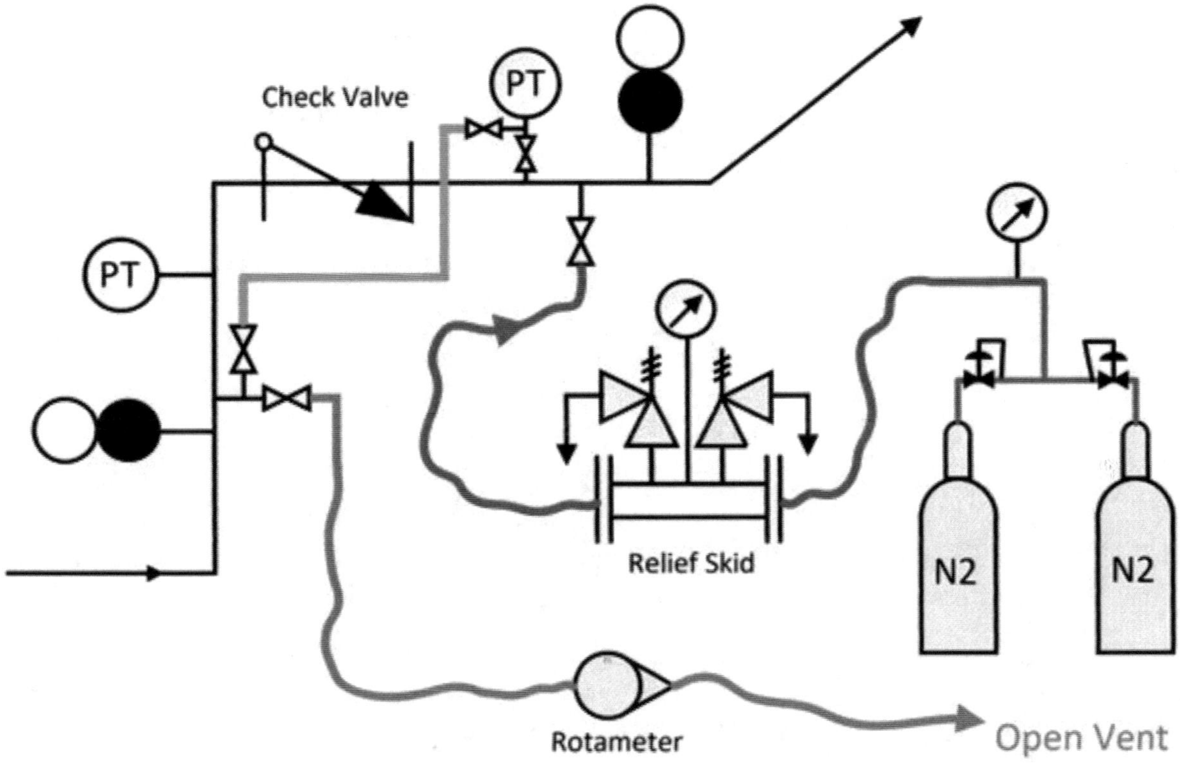

Figure 1. Simplified sketch of Test #1 configuration. Test single check valve using nitrogen.

The testing protocol for dual valves vs the single valve was similar. The primary difference was that having two valves added complexity and steps to the configuration and procedure. Additionally, in this particular configuration, isolation of the location was attempted via double block and bleed rather than using blinds.

Because of the number of testing points involved and a desire to minimize cost of additional pressure transmitters (PTs), a tubing manifold and valving was designed, built, and installed that would allow for usage on certain pressure transmitters in more than one testing configuration. However, achieving a successful leak test of this manifold proved difficult. Successful leak tests were also required of double block and bleed valving to the process. Refer to Figure 2 for a simplified depiction of the testing configuration for Test 2: Dual Check Valves (Axial and Wafer Checks), Lower Pressure Gas Service.

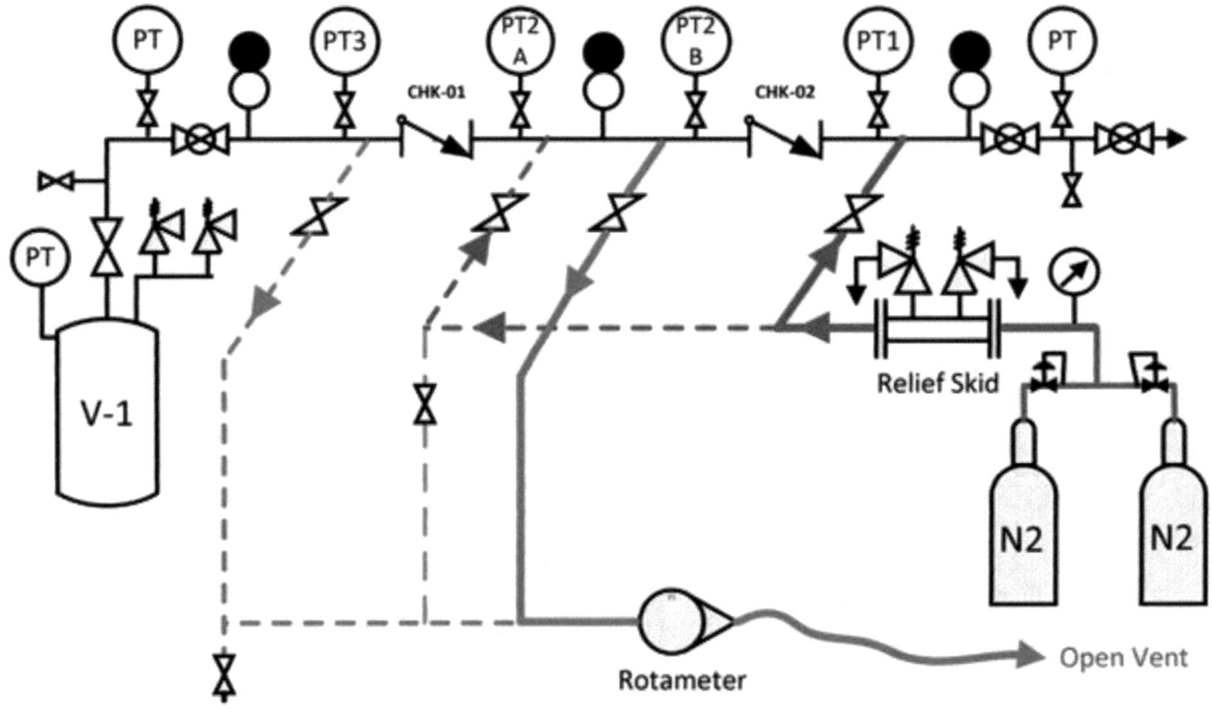

Figure 2. Simplified sketch of Test #2 configuration. Test dual check valves, axial and wafer. Nitrogen bottles used to pressure downstream side of check valves, one at a time.

8 Defining Maximum Allowable Leakage Thresholds

So what level of leakage is the threshold for "acceptable?" In order to answer this question, the hazard scenario including consequence severity must be understood. The engineer should recognize the concept of process safety time with respect to progression towards the hazardous event. In selecting the threshold for allowable leakage also called maximum allowable leakage rate (MALR), the engineer or other personnel who set up the testing procedure should look at several different points of reference for expected and/or acceptable leakage.

In the sample test procedure described here, Test 1, the following thresholds were identified in the procedure as points of reference. Rates shown in Table 4 are based on standards for new check valve allowable seat leakage, rotometer size selected, physical constraints of the equipment included in the test configuration, API 521 [3] guidance for "severe" check valve failure, and normal as well as expected maximum forward flowrates. A pass/fail limit was selected based on review of these rates.

Table 4. Test 1 Flowrates and Leakage Rates Range

Description of Flow Rates and Leakage Thresholds Considered: Test 1	Rate	Units
Maximum Process Flow Rate (forward flow), ~ 18 MMSCFD	750,000	SCFH
Normal Process Flow Rate (forward flow), 8.5 MMSCFD	354,165	SCFH
Calculated Reverse Flow Rate based on API 521 (2014), Severe Leakage Assumption, Method 1) 10% orifice	185,120	SCFH
Calculated Reverse Flow Rate based on API 521 (2014), Severe Leakage Assumption, Method 2	Not Calculated	SCFH
Arbitrary limit considered, 10% of normal forward flow rate	35,416	SCFH
Testing Limitation: Choked flow condition during test	4,050	SCFH
Arbitrary limit considered, 1% of API 521 Method 1	1,851	SCFH
Rotometer selected for Leakage Test: Maximum measurable flow rate	57	SCFH
Maximum Allowable Leakage Rate, New Check Valve, as per API 598 and ISO 5208	9.0	SCFH
Rotometer selected for Leakage Test: Minimum measurable flow rate	5.9	SCFH
Maximum Allowable Leakage Rate, New Check Valve, as per MSS SP-61	2.4	SCFH

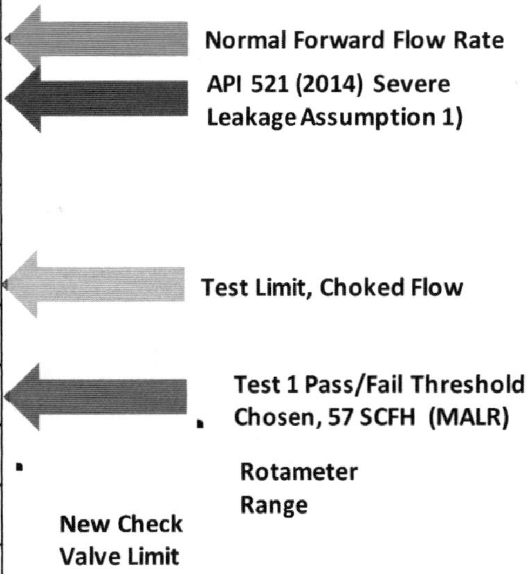

Review of this range of leakage and normal flow rates highlights a key point. The leakage rate(s) associated with API 521 [3] "severe leakage" guidance which are also referenced in CCPS IE/IPL [3] Section 5.2.2.2.2 Case 1 are four to five orders of magnitude larger than the acceptable leakage rates for new check valves prescribed by numerous standards for valve leakage. The check valve has failed from a reliability standpoint long before the API 521 [3] "severe" failure leakage rates are approached. This situation is implied given that the failure mode is described as "severe" with leakage defined as "significant" in API 521 [3]. The reference to "slight" leakage in CCPS IE/IPL [3] Section 5.2.2.2.2 Case 1 is in error. Leakage rates within the vicinity of these "severe" failure values may be unlikely in many check valve configurations, as the leakage may be either far below these values (ie., "normal" leakage potentially represented by values much closer to new valve leakage) or above representing "complete" valve failure. In order to experience "severe" leakage, failure modes involving significant debris or foreign object preventing closure or possible freeze up from Joule-Thompson flashing would likely be involved. These failures modes could result in fixing the check valve element at a location off but somewhat near the seat.

The specific technical origin of these API 521 [3] leakage rate assumptions for "severe" failure are not clear. Current API 521 committee members are only aware that the values were adopted from a member company participating on the committee. It is possible that due to lack of leakage rate data, very high leakage rate assumptions were chosen using an abundance of caution rather than based on data supporting that check valves commonly fail in a manner that would lead to this range of leakage rates.

Since little to no data is available for typical "normal" check valve leakage rates, establishing an appropriate initial limit is challenging. However, as was made clear by defining the various thresholds, a limit well below the rate referenced in API 521 [3] and near or somewhat above new valve allowable leakage rates is an appropriate limit to assure that the check valve is performing its function as intended. Where process safety considerations, such as relief valve capacity or other upstream limitations, do not drive the maximum allowable leakage rate limit, other issues should be considered. Impacts of reverse flow on upstream rotating equipment should be considered and may define the MALR.

The MALR for Test 2 was defined based on similar considerations as with Test 1 configuration.

9 Test Method and Procedure

Preparation for and conducting of the testing procedure involved several steps. Below is a description of the primary steps involved in the test procedure for testing dual check valves.

1) Process Hazard Analysis (PHA for testing configuration)
    a. Safeguards and IPLs were considered that remained in place for the entire system as well as safeguards and IPLs specific to the testing equipment.
        i. A PSV skid was designed and installed for use as an IPL for the equipment installed for the testing.
    b. Alarm setpoints were adjusted on transmitters (PTs) including on the upstream chamber to indicate build-up of pressure through leakage.
    c. Board Operators were engaged in monitoring potential pressure build-up.
2) Energy Isolation and Safe-Out: Performed as per standard energy control procedures.
3) Equipment installed for the test:
    a. Nitrogen bottle skid
    b. Regulators, from Nitrogen bottle skid
    c. Pressure gauges at Nitrogen Bottle Skid (scaled for each testing procedure)
    d. Hose, 5000 psig, inspected and tested. Installed with appropriate safety devices and signage.
        i. Connecting supply side:
            1. N2 skid and PSV skid
            2. PSV skid and check valve downstream tap
        ii. Connecting return (vent) side
            1. Check valve upstream tap to rotometer (or alternatively, use SS tubing)
            2. Rotometer to atmospheric vent point
    e. PSV skid, sized and with setpoint specific to each test configuration
    f. PTs/PITs on existing process piping (PITs preferred for onsite pressure display.)
        i. Reset appropriate alarm setpoints for the testing procedure
        ii. Broadcast of data reset from once every 5 minutes to once per second for this testing procedure
    g. Meter, rotometer sized for at or above anticipated leakage rate and MALR
4) Configuration approval and sign-off before each use. Assure engineer sign-off of current job package.
5) Verify initial positions: valve positions, blind positions, as required by procedure

6) Pre-Start-up Safety Review: Verify as installed equipment and initial valve/blind positions. Visual inspection and checklist review.
7) Perform leak testing on installed testing equipment (peripheral equipment).
8) Perform leak testing of downstream double block and bleed valves. If leakage found, safe-out equipment and roll downstream blind.
9) Perform leak testing of upstream block valve (by connecting pressure source between upstream check valve (CHK-01) and upstream block valve. If leakage found, safe-out equipment and roll upstream blind between vessel and first check valve.
10) Prepare for leak test of downstream check valve (CHK-02). Operator resets Alerts and Alarms on PTs. Flowmeter is connected to check valve upstream connection and vents to atmosphere. Pressure source is connected to check valve downstream connection.
11) Assure PTs at zero pressure reading. Note time for start of test.
12) Perform leak testing as per procedure on check valve (CHK-02). Set pressure regulator at prescribed test pressure and pressure up to test pressure on the downstream side of downstream check valve.
13) Note and record pressure reading. (PT data available for further analysis at a later date.) Also note and record pressure readings downstream and/or upstream of block valves to assure no leakage.
14) Record flowmeter flow rate data.
15) Review acceptance criteria. If below acceptable leakage rate (including below measurable rate), conclude test. If above, consider either modified rotometer configuration or determine rate using pressure build-up method.
16) Wait and take another data set after specified time interval.
17) Following conclusion of test, de-pressure the downstream and upstream side of the check valve through tap and connection to rotometer vent line. (Note: that the rotometer should register instantaneous flow even if no flow was detected during the check valve leak test.)
18) Disconnect testing equipment.
19) [If dual check valves being tested, repeat procedure for upstream check valve (CHK-01).]
20) Return process equipment to original setpoints and data broadcast frequencies.
21) Reverse energy isolation procedures.
22) Restart process as per normal start-up procedures.

10 Alternative Testing and Assessment Options

The procedure described represents a direct testing method using an external pressure source. In addition to this testing method, other alternatives should also be considered as appropriate. In some instances, use of existing process pressure may be a viable manner for applying a pressure source for testing certain check valves particularly on parallel equipment. In this case, quantification of the leakage rate could be accomplished by one of several methods. Pressure build-up analysis using SCADA systems may be most convenient if upstream isolation is available. Acoustic monitoring equipment should also be considered as a simple alternative which allows for assessing seat leakage rates and check valve performance. Given the nature of check valve operation, some type of shut-down, whether full plant turnaround, unit or train outage, or parallel equipment shut-down will be required to assess check valve performance.

Although not quantitative, diagnostic, or predictive, when check valves are engaged, there may be opportunities to confirm gross closure of the check valve by operators or mechanics. Operations or maintenance personnel may have some indication of closure of check valves based audible sound and on response of equipment that cannot tolerate back spinning. Equipment that cannot tolerate reverse flow, including certain rotating equipment, may be immediately or catastrophically damaged by incidences of check valve failure and loss of containment may occur. In some instances, check valve failure on spared pumps may be detected following start-up when flow rate and/or discharge pressure are lower than expected.

11 Key Take-Aways and Findings from Check Valve Leakage Test

Because the testing involved only a single instance of testing five separate new check valves, the data from the testing is not presumed to represent a wider population of in-service check valves. Because of this fact and because much of the specific data from the testing is not available for sharing*, the results of the check valve testing will be discussed in general terms.

1) The testing procedure worked relatively well in both instances. The testing of the single valve (higher pressure) was more easily accomplished due to the simplicity of the testing procedure. Only a single valve was being tested and isolation was accomplished with blinds rather than double block and bleed valving.
2) In the case of the single axial check valve test (higher pressure application) isolated by spectacle blinds:
    a. The rotometer was used for measuring the leakage rate through the check valve.
    b. Leakage rates were found to be below the scale of the rotometer, therefore less than 5.9 SCFH.
        i. This rate of leakage was well within the acceptable range for new check valve allowable leakage criteria, so no further investigation or calculation of actual leakage rates was investigated.
        ii. During relief of the trapped pressure from the downstream system through the rotometer following the test, the functionality of the rotometer was confirmed.
3) In the case of the dual check valves, one axial check valve and one wafer check valve on each train (lower pressure application) isolated by block valves:
    a. Initially achieving a good leak test was challenging due primarily to the tubing manifold planned for use. Usage of the tubing manifold was abandoned in preference to direct connections (and re-positioning of PTs).
    b. Using valving to isolate the test section from the process also required additional steps in the leak test procedure. Leak testing of block valves was successfully accomplished. No blinds were rolled.
    c. Check valve leakage rates were determined using PTs and pressure build-up calculations rather than with the rotometer.
    d. Leakage rates were calculated for both the axial check valve(s) and the wafer check valve(s). Both valves passed the threshold stipulated in the procedure protocol of less than 38.0 SCFH.
    e. Wafer check leakage was higher than axial check valve leakage. Both were within acceptable levels of leakage established for this test.

4) As anticipated, selecting rotometers that meet the actual range of leakage was challenging. However, the rotometers selected were within the "ballpark" of leakage ranges. With a little more time and test data, rotometers could be effectively "dialed in" that would be in the range of the actual leakage.
5) Pressure build-up testing proved to be a viable and simple alternative method to determine leakage rates.
6) System and block valve leak tests must be conducted in advance of check valve leak testing to assure that isolation from other potential leakage points have been successfully achieved.
7) The testing requires either a full or partial outage to accomplish.
8) As with any new procedure, appropriate process hazard review for the connected equipment and procedure as well as pre-start-up safety review must be included in the procedure. A personnel job safety analysis (JSA) must be conducted as well.

* Descriptions of results and findings from the testing are the general recollections of the author. Specific results from the testing were not made available for sharing in this paper.

12 Key Take-Aways from Review of Reference Material

As noted within this paper, several discrepancies were found in reference material. The following inconsistencies should be resolved within reference documentation in order to prevent confusion for users and practitioners.

1) Crediting a Single Check Valve as IPL and using an appropriate PFD:
    a. CCPS IE/IPL [2] lists a single check valve as a candidate IPL in Chapter 5 and provides a generic PFD value of 0.1 (RRF = 10) based on data and calculations provided in Appendix D. Special considerations are listed that must be satisfied. Initiating event frequencies (IEFs) are also given for check valves in high demand modes of 0.1 and 0.01 for single and dual check valves, respectively, in Chapter 4 based on the same data in Appendix D.
    b. API 521 (2014) [3] does not allow for crediting a single check valve with any reduction in flow rate to the upstream PSV, even if the check valve is inspected and maintained. This assumption equates to a PFD for a single check valve of 1, or RRF = 1. No data is given to support this assessment of PFD = 1. Note: API 521 (2008) [17] did allow for potential credit of single check valves to limit reverse flow to the PSV when the check valve was appropriately inspected and maintained.
    c. CCPS IE/IPL [2] references API 521 (2008) [17] language in 5.2.2.2.2 Case 1 which refers to a "series check valves," but also references a single check valve in the Case 1 description.

The discrepancies in CCPS IE/IPL [2] and API 521 (2014) [3] regarding check valve failure likelihood should be reconciled. Check valve probability of failure on demand (PFD) should be based on testing and operational data gathered for failure mode and frequency by valve type, size, and service.

Check valve leakage rates should be determined based on testing by valve type, failure mode, size, and service. Check valve leakage rates for "severe" and "normal" failure should be better defined and consistently applied.

2) Leakage rate assumption(s) cited by API 521 (2008) [17] and replicated in CCPS IE/IPL [2] Section 5.2.2.2.2 Case 1 represent "severe", significant, or high rates of leakage. The check valve has failed from a reliability standpoint at a much lower rate of leakage (4 to 5 orders of magnitude less leakage). The reference to "slight" leakage in CCPS IE/IPL [2] 5.2.2.2.2 Case 1 should be corrected.

Little to no data is currently available regarding check valve typical or "normal" leakage rates for check valves that have been in operation. Likewise, little data is available regarding check valve failure modes and frequencies. This need was the driver for designing and performing these field tests and for writing this paper.

3) Numerical estimates for "normal" leakage rates are not offered by API 521 [3][17] (or by CCPS IE/IPL[2]). API 521 [3][17] indicates it is the responsibility of the user to determine normal leakage rates.
4) Existing industry equipment reliability databases such as CCPS process equipment reliability database (PERD) have very little data available on check valve performance including failure modes and frequencies. PERD offers guidance for data collection in a manner that supports reliability analysis. Usage of PERD and other industry resources is recommended.
5) Testing of check valves that have been in service is a prerequisite to not only understanding and improving check valve performance, but also to crediting check valves as IPLs or safeguards.

13 Conclusions

Check valves perform an important and often critical function in process equipment. Therefore, check valve performance and reliability should be a foremost concern for operating companies. Check valve performance and reliability can and should be managed through maintenance programs that include inspection, testing, and repair of check valves. Check valves have historically not received proper attention in inspection and maintenance programs. And testing programs to quantify leakage have rarely been instituted.

More and better data on check valve failure modes, failure likelihood, and performance (aka, leakage) can lead to better design by manufacturers, better selection for the given application by operators, better piping configuration design, and better performance of the check valves.

Currently, discrepancies exist in some core industry reference material regarding characterization of check valve probability of failure on demand. Those discrepancies should be reconciled. (See prior Section 12 for description of these discrepancies.)

Little data is available regarding check valve performance and failure modes by type and in given service applications. And little progress has been made on gathering more modern data on check valve performance. The author encourages operating companies to consider testing and inspection of check valves in critical service as part of their mechanical integrity program, particularly when designing and installing new installations where appropriate isolation and testing ports may be included in the piping arrangement. The author further encourages the sharing of check valve inspection and testing data more broadly, such as through the CCPS Process Equipment Reliability Database (PERD), in order to achieve industry-wide progress towards improved check valve reliability. Better practices, better data, and better understanding of performance can lead to changing expectations regarding check valve performance.

When check valves are thoroughly and routinely inspected, tested, and maintained, PHA and/or LOPA credit for the check valve(s) as a safeguard or IPL, respectively, may be achieved. As always, operators must consider applying this credit with a thorough understanding of the check valve failure modes and their ability to prevent the hazard consequence from being realized.

14 Acknowledgements

The author wishes to thank Robert Coover, Praxair, Inc.; Fred Henselwood, NOVA Chemicals Corporation; Hal Thomas, exida; Jan Hayes, RMIT University; Barry Bennett, Stepan Company; Angela Summers, SIS-Tech Solutions; Forrest Carithers, Wyatt Field Service Company; Loren Stewart, exida; and Jennifer Mize, Eastman for their review, comments, and contributions on this paper and presentation.

The author also wishes to thank American National Standards Institute (ANSI) for providing permission on behalf of International Organization for Standardization (ISO) to reproduce Table 4 from page 11 of ISO 5208:2015 and Manufacturers Standardization Society of the Valve and Fittings Industry (MSS) for providing permission to reproduce Table 4 from page 4 of MSS SP-61-2013.

15 References

15.1 Cited References

[1] A. Baldas and F. Carithers, "Check Valves and Risk Assessment: Should You Take Credit?" Available at https://bluefieldsafety.com/2017/09/check-valves-and-risk-assessment-should-you-take-credit/, Accessed on December 23, 2018.

[2] CCPS. *Guidelines for Initiating Events and independent Protection Layers in Layer of Protection Analysis.* Center for Chemical Process Safety, American Institute of Chemical Engineers, New York, NY, 2015.

[3] American Petroleum Institute (API). *API Standard 521: Pressure-relieving and Depressuring Systems.* 6th Edition. API Publishing. Washington, D.C. 2014.

[4] C. Patschkle, "Solving the Mysteries of Critical Check Valves," Available at: https://inspectioneering.com/journal/2018-06-28/7745/solving-the-mysteries-of-critical-check-valves, Accessed January 10, 2019.

[5] CCPS. *Layer of Protection Analysis: Simplified Process Risk Assessment.* Center for Chemical Process Safety, American Institute of Chemical Engineers, New York, NY, 2001.

[6] CCPS. *Guidelines for Hazard Evaluation Procedures.* 3rd Edition. Center for Chemical Process Safety, American Institute of Chemical Engineers, John Wiley and Sons Inc., New York, NY, 2008.

[7] American Petroleum Institute (API). API Standard 570: Piping Inspection Code: In-service Inspection, Rating, Repair, and Alteration of Piping Systems, 4th Edition. API Publishing. Washington, D.C. 2016.

[8] American Petroleum Institute (API). ANSI/API Standard 598: Valve Inspection and Testing. 10th Edition. API Publishing. Washington, D.C. 2016.

[9] International Standard Organization (ISO). ISO 5208: Industrial valves - Pressure testing of metallic valves. 4th Edition. ISO Publishing. Switzerland. 2015.

[10] Manufacturers Standardization Society of the Valve and Fittings Industry, Inc. (MSS). Standard MSS SP-61-2009: Pressure Testing of Valves. MSS Publishing. Vienna, Virginia, 2009.

[11] American Petroleum Institute (API). *ANSI/API Standard 6D: Specification for Pipeline and Piping Valves.* 24rd Edition. API Publishing. Washington, D.C. 2014.

[12] American Petroleum Institute (API). *ANSI/API Standard 6D: Specification for Pipeline Valves.* 23rd Edition. API Publishing. Washington, D.C. 2008.

[13] European Standard. EN-12266-1:2012: Industrial valves – Testing of metallic valves – Part 1: Pressure tests, test procedures and acceptance criteria – Mandatory requirements. European Committee for Standardization. Brussels. 2012.

[14] British Standards Institute (BSI). BS 6755-1: 1986. Testing of valves - Part 1: Specification for production pressure testing requirements. 1986.

[15] American National Standard Institute (ANSI). ANSI/FCI 70-2-2006 Control Valve Seat Leakage. Fluids Controls Institute. Cleveland, OH. 2006.

[16] K. Modi, Check Valves as Safeguards: Friend or Foe in Process Hazard Analysis," Available at http://jclrisk.com/wp-content/uploads/sites/3/2017/05/Check-Valves-as-Safeguards-JCL-Risk-Author-Kay-Modi-April-25-2017-at-SynGas-Conf.pdf, Accessed on December 23, 2018.

[17] American Petroleum Institute (API). *API Standard 521: Pressure-relieving and Depressuring Systems.* 5th Edition. API Publishing. Washington, D.C. 2008.

15.2 Other References

[18] ValMatic, "Design and Selection Criteria of Check Valves," Available at: http://www.valmatic.com/pdfs/DesignSelectCriteriaCV_9-1-17.pdf, Accessed on Dec 27, 2018.

[19] B. Ellis, "The Basics of Check Valves, Is sizing for the application worth the cost?" Pumps and Systems, July 24, 2018, Available at: https://www.pumpsandsystems.com/basics-check-valves, Accessed January 10, 2019.

[20] M. Shorts, "Understanding Check Valves: Sizing for the Application, Not the Line Size," Waterworld, May 1, 2007, Available at: https://www.waterworld.com/articles/print/volume-23/issue-5/editorial-feature/understanding-check-valves-sizing-for-the-application-not-the-line-size.html, Accessed January 15, 2019.

[21] M. Johnson, "The Misunderstood Check Valve," Available at https://www.dft-valves.com/pdf/literature/general-information/CheckValveBacktoBasicsArticle.pdf , Accessed on December 23, 2018.

[22] VB Valves & Automation, "Leak testing of valves," Available at: http://www.vbvalvesautomation.com/wp-content/themes/vb/pdf/Testing%20Standard.pdf, Accessed December 23, 2018.

[23] Global Supply Line, "Leakage Acceptance Rates Comparison Metal & Soft Seated Valves ISO 5208/API 598/API 6D/MSS SP-61/FCI 70-2," Available at http://globalsupplyline.com.au/wp-content/uploads/2014/10/Valve_Leakage_Rates_Test_Std.pdf, Accessed on December 27, 2018.

[24] Milwaukee Valve Company Engineering Department, "Technical Bulletin, Valve Seat Leakage Standards and Related MVTB-06-002," Available at:https://www.milwaukeevalve.com/data/files/ValveLeakage.pdf, Accessed on Dec 23, 2018.

[25] ValvTechnologies, "Valve Testing Standards," Available at: http://presentation.valv.com/pdfs/v1-v3/v1-1/features/testing-standards/valve-testing-standards.pdf, Accessed December 27, 2018.

[26] K.L. Mc Elhaney, "Failure Modes and Causes for Swing and Lift Type Check Valves," Oak Ridge National Laboratory, 1997. Available at https://www.osti.gov/biblio/621557, Accessed on December 27, 2018.

[27] D.A. Casada, M.D. Todd, "A Characterization of Check Valve Degradation and Failure Experience in the Nuclear Power Industry," NUREG/CR-5944, ORNL-6734, Oak Ridge National Laboratory, Office of Nuclear Regulatory Research, Washington D.C., 1993. Available at https://www.nrc.gov/docs/ML0403/ML040340445.pdf, Accessed Dec 27, 2018.

[28] K.L. Mc Elhaney and R.H. Staunton, "Reliability Estimation for Check Valves and Other Components," Lockheed Martin Energy Research Corporation, Oak Ridge National Laboratory, 1996. Available at:

https://inis.iaea.org/collection/NCLCollectionStore/_Public/27/066/27066030.pdf?r=1&r=1, Accessed on December 27, 2018.

[29] U.S. Nuclear Regulatory Commission, "Component Reliability Datasheets 2015 Update," Available at: https://nrcoe.inl.gov/resultsdb/publicdocs/AvgPerf/ComponentReliabilityDataSheets2015.pdf, Accessed December 27, 2018.

[30] CCPS. *Guidelines for Process Equipment Reliability Data, with Data Tables.* Center for Chemical Process Safety, American Institute of Chemical Engineers, New York, NY, 1989.

[31] Fisher, Emerson Process Management, "Using Acoustic Emission to Determine In-Service Valve Seat Leakage, White Paper, D352212X012, May 2013," Available at: https://www.emerson.com/documents/automation/white-paper-using-acoustic-emissions-to-determine-in-service-valve-seat-leakage-fisher-en-136046.pdf, Accessed on January 15, 2019.

[32] P. Boughton, "Quantification of through valve leakage using acoustic emission technology," EngineerLive, October 24, 2016, Available at: https://www.engineerlive.com/content/quantification-through-valve-leakage-using-acoustic-emission-technology, Accessed on January 15, 2019.

[33] American Petroleum Institute (API). ANSI/API Standard 598: Valve Inspection and Testing. 8th Edition. API Publishing. Washington, D.C. 2004.

[34] Milwaukee Valve Engineering Department, "Technical Bulletin, Gate Valve Installations and Leakage MVTB-05-001 Rev A," Available at: https://www.milwaukeevalve.com/data/files/GateInstallationLeakage.pdf, Accessed on December 23, 2018.

[35] Manufacturers Standardization Society of the Valve and Fittings Industry, Inc. (MSS). Standard MSS SP-42: Corrosion-Resistant Gate, Globe, Angle, and Check Valves with Flanged and Butt Weld Ends (Classes 150, 300, & 600). MSS Publishing. Vienna, Virginia, 2013.

[36] Manufacturers Standardization Society of the Valve and Fittings Industry, Inc. (MSS). Standard MSS SP-71: Gray Iron Swing Check Valves, Flanged and Threaded Ends. MSS Publishing. Vienna, Virginia, 2005.

[37] Manufacturers Standardization Society of the Valve and Fittings Industry, Inc. (MSS). Standard MSS SP-80: Bronze Gate, Globe, Angle, and Check Valves. MSS Publishing. Vienna, Virginia, 2013.

[38] Manufacturers Standardization Society of the Valve and Fittings Industry, Inc. (MSS). Standard MSS SP-125: Check Valves: Gray Iron and Ductile Iron, In-Line, Spring-Loaded, Center-Guided. MSS Publishing. Vienna, Virginia, 2018.

[38] American Petroleum Institute (API). API Standard 570: Piping Inspection Code: In-service Inspection, Rating, Repair, and Alteration of Piping Systems, 3rd Edition. API Publishing. Washington, D.C. 2009.

16 Appendices

16.1 Appendix A – ISO 5208:2015, Table 4 – Maximum allowable closure test leakage rates

Table 4 — Maximum allowable closure test leakage rate

Test fluid	Unit leakage rates	Rate A	Rate AA	Rate B	Rate C	Rate CC	Rate D	Rate E	Rate EE	Rate F	Rate G
Liquid	mm³/s	No visually detectable leakage for the duration of the test	0,006 × DN	0,01 × DN	0,03 × DN	0,08 × DN	0,1 × DN	0,3 × DN	0,39 × DN	1 × DN	2 × DN
	drops/s		0,000 1 × DN	0,000 16 × DN	0,000 5 × DN	0,001 3 × DN	0,001 6 × DN	0,004 8 × DN	0,006 2 × DN	0,016 × DN	0,032 × DN
Gas	mm³/s		0,18 × DN	0,3 × DN	3 × DN	22,3 × DN	30 × DN	300 × DN	470 × DN	3 000 × DN	6 000 × DN
	bubbles/s		0,003 × DN	0,004 6 × DN	0,045 8 × DN	0,340 7 × DN	0,458 4 × DN	4,583 7 × DN	7,129 3 × DN	45,837 × DN	91,673 × DN

NOTE 1 The leakage rates only apply when discharging test fluid to the atmosphere.

NOTE 2 The closure leakage rate that applies is either that identified in a valve product standard or a leakage rate identified in a purchaser's valve procurement purchase order that is more stringent than that specified in the product standard.

NOTE 3 For the definition of visually detectable leakage see 2.12.

NOTE 4 There is a correspondence between the leakage rate acceptance values of API 598 and the leakage values Rate A as applied to DN ≤ 50, Rate AA-Gas and Rate CC-Liquid for other than metal seated check valves and for check valves Rate EE-Gas and Rate G-Liquid. Rates A, B, C, D E, F and G correspond to values in EN 12266-1.

16.2 Appendix B – MSS SP-61-2013, Table 4 – Units of Leakage per NPS/DN

TABLE 4
Units of Leakage per NPS/DN

Liquid		Gas	
Per NPS	**Per DN**	**Per NPS**	**Per DN**
10 cc/hr	0.4 cc/hr	0.1 SCFH 2.88 SCIM	120 cc/hr
0.167 cc/min.	6.6×10^{-3} cc/min.	47.2 cc/min.	2cc/min.
2.66[(a)] drops/min.	0.11[(a)] drops/min.	1180[(b)] bubble/min.	50[(b)] bubble/min.

TABLE 4 NOTES:

(a) For information only. Based on 16 drops per cc, which is the equivalent of a spherical shaped drop having an approximate diameter of 3/16 inch (0.5 cm).

(b) For information only. Based on 25 bubbles per cc, which is the equivalent of a spherical shaped bubble having an approximate diameter of 5/32 inch (0.42 cm). See Section 2.4, Note (1).

GENERAL NOTES:

1. 1 ml = 1 cc (For information: Small liquid leak rates are typically measured in "ml" and "cc" within the U.S.)
2. SCFH = Standard cubic feet per hour
   (14.7 psia @ 60 °F) (1.01 bar @ 16 °C)
3. SCIM = Standard cubic inch per minute
   (14.7 psia @ 60 °F) (1.01 bar @ 16 °C)

16.3 Appendix C – Valve Seat Leakage Allowance by Standard [8][9][10][11], Liquid Test

(RS) Resilient Seat = Soft Seat

(Std) = Manufacturer Standard Specification (no special request)

Nom Pipe Diam		All RS	Non-Check Valves	Non-Check Valves	Check Valves	Check Valves
		API 598	API 598	API 598	API 598	API 598
		0.000	16.00	2.00	3	16.00
Units above =>			drops/ml	drops/min/nom pipe diam	cu cm/min/nom pipe dia inch	drops/ml
		Soft Seat				
DN	inches	ml/min	ml/min	drops/min	ml/min	drops/min
	1	0	0.00	0	3	48
less	1.5	0	0.00	0	4.5	72
50	2	0	0.00	0	6	96
65	2.5	0	0.31	5	7.5	120
80	3	0	0.38	6	9	144
100	4	0	0.50	8	12	192
150	6	0	0.75	12	18	288
200	8	0	1.00	16	24	384
250	10	0	1.25	20	30	480
300	12	0	1.50	24	36	576
350	14	0	1.75	28	42	672
400	16	0	2.00	32	48	768
450	18	0	2.25	36	54	864
500	20	0	2.50	40	60	960
600	24	0	3.00	48	72	1152
700	28	0	3.50	56	84	1344
750	30	0	3.75	60	90	1440
800	32	0	4.00	64	96	1536
900	36	0	4.50	72	108	1728
1000	40	0	5.00	80	120	1920
1050	42	0	5.25	84	126	2016
1200	48	0	6.00	96	144	2304

Nom Pipe Diam			Non-Check Valves	Non-Check Valves	Check Valves	Check Valves
		MSS SP-61	MSS SP-61	MSS SP-61	MSS SP-61	MSS SP-61
		0.00	0.167	16.00	0.668	16.00
Units above =>		cu in/min/nom pipe dia in	cu cm/min/nom pipe dia inch	(calced from ml/min)	cu cm/min/nom pipe dia inch	(calced from ml/min)
		Soft Seat				
DN	inches	ml/min	ml/min	drops/min	ml/min	drops/min
	1	0.0	0.167	2.67	0.668	10.69
less	1.5	0.0	0.251	4.01	1.00	16.03
50	2	0.0	0.334	5.34	1.34	21.38
65	2.5	0.0	0.42	6.68	1.67	26.72
80	3	0.0	0.50	8.02	2.00	32.06
100	4	0.0	0.67	10.69	2.67	42.75
150	6	0.0	1.00	16.03	4.01	64.13
200	8	0.0	1.34	21.38	5.34	85.50
250	10	0.0	1.67	26.72	6.68	106.88
300	12	0.0	2.00	32.06	8.02	128.26
350	14	0.0	2.34	37.41	9.35	149.63
400	16	0.0	2.67	42.75	10.69	171.01
450	18	0.0	3.01	48.10	12.02	192.38
500	20	0.0	3.34	53.44	13.36	213.76
600	24	0.0	4.01	64.13	16.03	256.51
700	28	0.0	4.68	74.82	18.70	299.26
750	30	0.0	5.01	80.16	20.04	320.64
800	32	0.0	5.34	85.50	21.38	342.02
900	36	0.0	6.01	96.19	24.05	384.77
1000	40	0.0	6.68	106.88	26.72	427.52
1050	42	0.0	7.01	112.22	28.06	448.90
1200	48	0.0	8.02	128.26	32.06	513.02

Nom Pipe Diam		(Std, RS) Mnfr Y (Std, RS) Mnfr X (RS) MSS SP-61 (RS) API 6D (RS) API 598		(If Request) Mnfr Y						~API598 Non-Check	
		ISO 5208	ISO 5208	ISO 5208	ISO 5208	ISO 5208	ISO 5208	ISO 5208	ISO 5208	ISO 5208	ISO 5208
		Rate A	Rate A	Rate AA	Rate AA	Rate B	Rate B	Rate C	Rate C	Rate CC	Rate CC
		0.000		0.006	16	0.01	16	0.03	16	0.08	16
Units above =>		No visually detectable leakage during test		cu mm /sec / DN	drops/ml	cu mm /sec / DN	drops/ml	cu mm /sec / DN	drops/ml	cu mm /sec / DN	drops/ml
		Soft Seat									
DN	inches	ml/min	drops/min	ml/min	drops/min	ml/min	drops/min	ml/min	drops/min	ml/min	drops/min
	1	0.00	0	0.009	0.14	0.0150	0.24	0.0450	0.72	0.120	1.92
less	1.5	0.00	0	0.014	0.22	0.0225	0.36	0.0675	1.08	0.180	2.88
50	2	0	0	0.018	0.29	0.0300	0.48	0.0900	1.44	0.240	3.84
65	2.5	0	0	0.023	0.37	0.0390	0.62	0.1170	1.87	0.312	4.99
80	3	0	0	0.03	0.46	0.05	0.77	0.14	2.30	0.38	6.14
100	4	0	0	0.04	0.58	0.06	0.96	0.18	2.88	0.48	7.68
150	6	0	0	0.05	0.86	0.09	1.44	0.27	4.32	0.72	11.52
200	8	0	0	0.07	1.15	0.12	1.92	0.36	5.76	0.96	15.36
250	10	0	0	0.09	1.44	0.15	2.40	0.45	7.20	1.20	19.20
300	12	0	0	0.11	1.73	0.18	2.88	0.54	8.64	1.44	23.04
350	14	0	0	0.13	2.02	0.21	3.36	0.63	10.08	1.68	26.88
400	16	0	0	0.14	2.30	0.24	3.84	0.72	11.52	1.92	30.72
450	18	0	0	0.16	2.59	0.27	4.32	0.81	12.96	2.16	34.56
500	20	0	0	0.18	2.88	0.30	4.80	0.90	14.40	2.40	38.40
600	24	0	0	0.22	3.46	0.36	5.76	1.08	17.28	2.88	46.08
700	28	0	0	0.25	4.03	0.42	6.72	1.26	20.16	3.36	53.76
750	30	0	0	0.27	4.32	0.45	7.20	1.35	21.60	3.60	57.60
800	32	0	0	0.29	4.61	0.48	7.68	1.44	23.04	3.84	61.44
900	36	0	0	0.32	5.18	0.54	8.64	1.62	25.92	4.32	69.12
1000	40	0	0	0.36	5.76	0.60	9.60	1.80	28.80	4.80	76.80
1050	42	0	0	0.38	6.05	0.63	10.08	1.89	30.24	5.04	80.64
1200	48	0	0	0.43	6.91	0.72	11.52	2.16	34.56	5.76	92.16

Nom Pipe Diam		MSS SP-61 >ISO D API 6D Non-Check		MSS SP-61<<ISO E		Mnfr X > ISO E MSS SP-61 Chk >E		Mnfr X << ISO F MSS SP-61 Chk<<F		(Std) Mnfr Y API 6D, 2014 Check ~API 598 Check	
		ISO 5208	ISO 5208	ISO 5208	ISO 5208	ISO 5208	ISO 5208	ISO 5208	ISO 5208	ISO 5208	ISO 5208
		Rate D	Rate D	Rate E	Rate E	Rate EE	Rate EE	Rate F	Rate F	Rate G	Rate G
		0.10	16	0.30	16	0.39	16	1	16	2	16
Units above =>		cu mm /sec / DN	drops/ml	cu mm /sec / DN	drops/ml	cu mm /sec / DN	(calced)	cu mm /sec / DN	drops/ml	cu mm /sec / DN	drops/ml
DN	inches	ml/min	drops/min	ml/min	drops/min	ml/min	drops/min	ml/min	drops/min	ml/min	drops/min
	1	0.150	2.40	0.450	7.20	0.59	9.36	1.50	24	3.0	48
less	1.5	0.225	3.60	0.675	10.80	0.88	14.04	2.25	36	4.5	72
50	2	0.30	4.80	0.90	14.40	1.17	18.72	3.00	48	6.0	96
65	2.5	0.39	6.24	1.17	18.72	1.52	24.34	3.9	62	7.8	125
80	3	0.48	7.68	1.44	23.04	1.87	29.95	4.8	77	9.6	154
100	4	0.60	9.60	1.80	28.80	2.34	37.44	6.0	96	12	192
150	6	0.90	14.40	2.70	43.20	3.51	56.16	9	144	18	288
200	8	1.20	19.20	3.60	57.60	4.68	74.88	12	192	24	384
250	10	1.50	24.00	4.50	72.00	5.85	93.60	15	240	30	480
300	12	1.80	28.80	5.40	86.40	7.02	112.32	18	288	36	576
350	14	2.10	33.60	6.30	100.80	8.19	131.04	21	336	42	672
400	16	2.40	38.40	7.20	115.20	9.36	149.76	24	384	48	768
450	18	2.70	43.20	8.10	129.60	10.53	168.48	27	432	54	864
500	20	3.00	48.00	9.00	144.00	11.70	187.20	30	480	60	960
600	24	3.60	57.60	10.80	172.80	14.04	224.64	36	576	72	1152
700	28	4.20	67.20	12.60	201.60	16.38	262.08	42	672	84	1344
750	30	4.50	72.00	13.50	216.00	17.55	280.80	45	720	90	1440
800	32	4.80	76.80	14.40	230.40	18.72	299.52	48	768	96	1536
900	36	5.40	86.40	16.20	259.20	21.06	336.96	54	864	108	1728
1000	40	6.00	96.00	18.00	288.00	23.40	374.40	60	960	120	1920
1050	42	6.30	100.80	18.90	302.40	24.57	393.12	63	1008	126	2016
1200	48	7.20	115.20	21.60	345.60	28.08	449.28	72	1152	144	2304

Values shown in Appendix C are approximate as calculated by the author.

16.4 Appendix D – Valve Seat Leakage Allowance by Standard [8][9][10][11], Gas Test

							Mnfr Y (RS) Mnfr Y (RS) MSS SP-61 (RS) MSS SP-61 (RS)						Mnfr Y SP-61 Blk	Mnfr X MSS SP-61 Chk			
			Non-Check	Check	Non-Check	Check Valves	API 598 (RS)	API 598 Blk	API 598 Blk						API 598 Chk		
		API 598	API 598	API 598	MSS SP-61	MSS SP-61	ISO 5208	ISO 5208	ISO 5208	ISO 5208	ISO 5208	ISO 5208	ISO 5208	ISO 5208	ISO 5208	ISO 5208	ISO 5208
			Given	Given			Rate A	Rate AA	Rate AA	Rate B	Rate C	Rate CC	Rate D	Rate E	Rate EE	Rate F	Rate G
			4.00	1.50	0.1	0.40	0.000	0.18	3.00	0.30	3	22.3	30	300	470	3000	6000
Units for above =>			bubbles/ sec / pipe diam inch	SCF / hr/ nom pipe diam inch	SCFH per nom pipe diam inch	SCFH per nom pipe diam inch	No visually detectable leakage during test	cu mm/ sec/ DN	bubbles/ sec / DN	cu mm/ sec / DN	cu mm/ sec / DN	cu mm/ sec / DN	cu mm/ sec / DN	cu mm/ sec / DN	cu mm/ sec / DN	cu mm/ sec / DN	cu mm/ sec / DN
Nominal Pipe Diameter		Soft Seat			Soft Seat		Soft Seat										
DN	inches	SCFH	bubbles/ min	SCFH	SCFH	SCFH	SCFH	SCFH	bubbles/ min	SCFH	SCFH	SCFH	SCFH	SCFH	SCFH	SCFH	SCFH
	1	0	0	1.50	0.10	0.40	0.000	0.0006		0.0010	0.01	0.07	0.10	0.95	1.49	9.53	19.05
less	1.5	0	0	2.25	0.15	0.60	0.000	0.0009		0.0014	0.01	0.11	0.14	1.43	2.24	14.29	28.58
50	2	0	0	3.00	0.20	0.80	0.000	0.0011	9	0.0019	0.02	0.14	0.19	1.91	2.98	19.05	38.10
80	3	0	12	4.50	0.30	1.20	0.000	0.0018	14	0.0030	0.03	0.23	0.30	3.05	4.78	30.48	60.96
100	4	0	16	6.00	0.40	1.60	0.000	0.0023	18	0.0038	0.04	0.28	0.38	3.81	5.97	38.10	76.20
150	6	0	24	9.00	0.60	2.40	0.000	0.0034	27	0.0057	0.06	0.42	0.57	5.72	8.95	57.15	114.30
200	8	0	32	12.00	0.80	3.20	0.000	0.0046	36	0.0076	0.08	0.57	0.76	7.62	11.94	76.20	152.40
250	10	0	40	15.00	1.00	4.00	0.000	0.0057	45	0.0095	0.10	0.71	0.95	9.53	14.92	95.25	190.50
300	12	0	48	18.00	1.20	4.80	0.000	0.0069	54	0.0114	0.11	0.85	1.14	11.43	17.91	114.30	228.60
350	14	0	56	21.00	1.40	5.60	0.000	0.0080	63	0.0133	0.13	0.99	1.33	13.34	20.89	133.35	266.70
400	16	0	64	24.00	1.60	6.40	0.000	0.0091	72	0.0152	0.15	1.13	1.52	15.24	23.88	152.40	304.80
450	18	0	72	27.00	1.80	7.20	0.000	0.0103	81	0.0171	0.17	1.27	1.71	17.15	26.86	171.45	342.90
500	20	0	80	30.00	2.00	8.00	0.000	0.0114	90	0.0191	0.19	1.42	1.91	19.05	29.85	190.50	381.00
600	24	0	96	36.00	2.40	9.60	0.000	0.0137	108	0.0229	0.23	1.70	2.29	22.86	35.81	228.60	457.20
750	30	0	120	45.00	3.00	12.00	0.000	0.0171	135	0.0286	0.29	2.12	2.86	28.58	44.77	285.75	571.50
900	36	0	144	54.00	3.60	14.40	0.000	0.0206	162	0.0343	0.34	2.55	3.43	34.29	53.72	342.90	685.80
1200	48	0	192	72.00	4.80	19.20	0.000	0.0274	216	0.0457	0.46	3.40	4.57	45.72	71.63	457.20	914.40

Values shown in Appendix D are approximate as calculated by the author.

16.5 Appendix E – Valve Seat Leakage Allowance Overview [8][9][10][11], Liquid Test

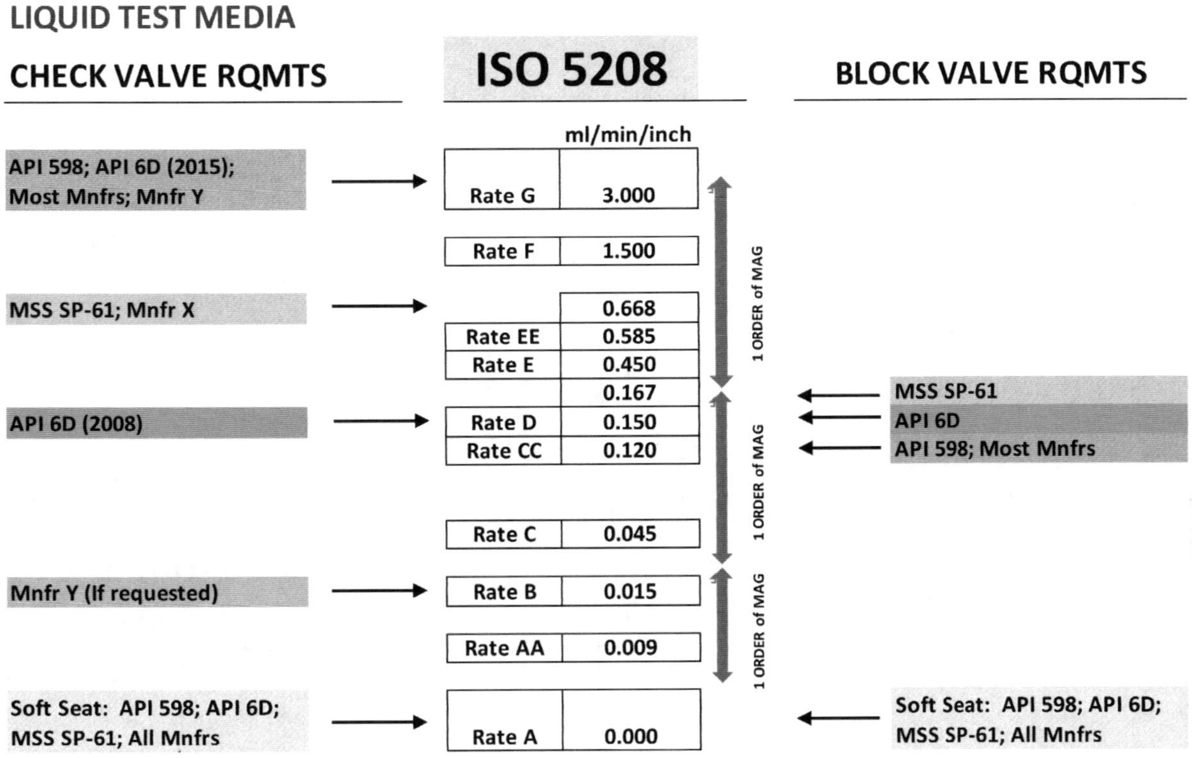

16.6 Appendix F – Valve Seat Leakage Allowance Overview [8][9][10][11], Gas Test

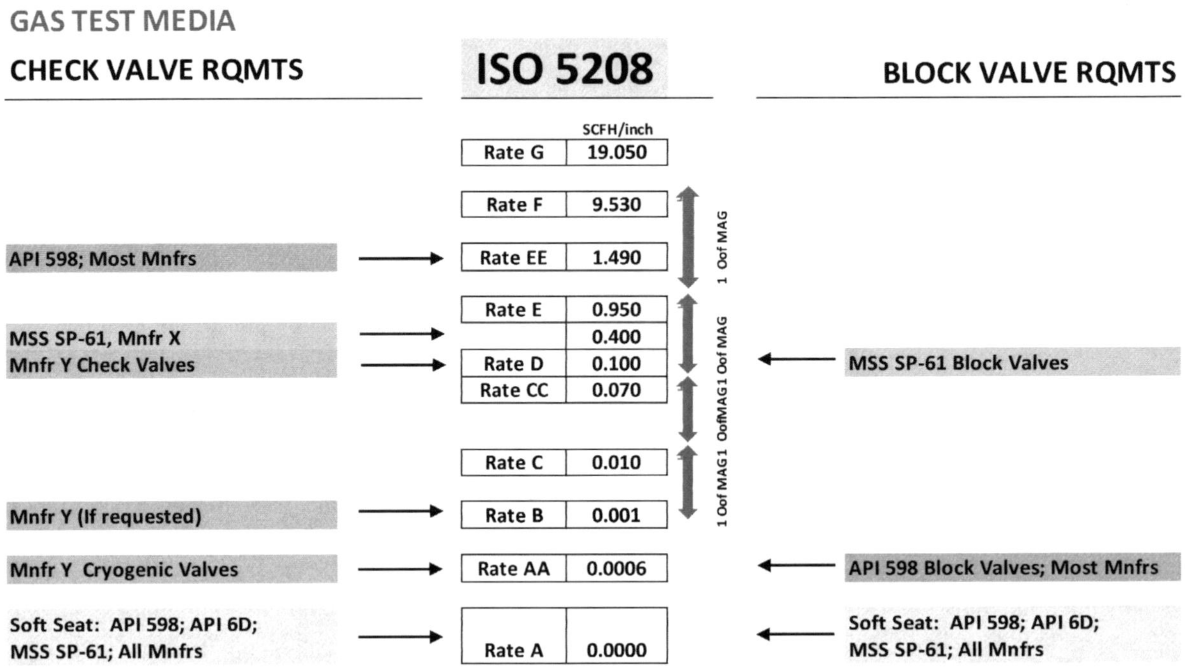

551253 Safeguard Meters within Chevron

Chevron has developed a framework for implementing the concept of "safeguard meter" within our operating assets. The meter(s) will display leading indicators related to key safeguards that prevent or mitigate incidents at Chevron facilities. These indicators may provide signals that a safeguard's performance is deviating from accepted norms and require attention to remain reliable. Examples from several operating areas and the approach for deployment across the company will be shared and discussed.

The framework includes identification of the most important safeguards to be measured and addresses how assurance against established performance standards is measured and displayed.

The discussion will also address how leadership response and culture should be addressed in any deployment that provides visibility of real time leading performance indicators within an operating asset.

Trust is Our Currency

Paul Gathright
Ascend Performance Materials
Cantonment, FL
jgathr@ascendmaterials.com

Prepared for Presentation at
American Institute of Chemical Engineers
2019 Spring Meeting and 15th Global Congress on Process Safety
New Orleans, LA
March 31 – April 3, 2019

Trust is Our Currency

Paul Gathright
Ascend Performance Materials
Cantonment, FL
jgathr@ascendmaterials.com

Keywords: trust, value, tiger, constraint, customer, fast

Abstract

"Liberty not only means that the individual has both the opportunity and the burden of choice; it also means that he must bear the consequences of his actions." Friedrich Hayek

In process safety, our goal is to keep process energy contained, to *keep the tiger in the cage*. To do so effectively, a team effort is required, including operations, maintenance, engineering, management, contractors, and other relevant organizations. We, as process safety leaders and program owners, are selling a product, driving a transaction that includes ideas, investments, and resource allocation.

We could, as many do, use the regulatory requirements to facilitate the transaction. This 'push' method of sales is less effective than a 'pull' approach where we represent the value of our product to the buyers: our management teams, operations, maintenance, engineering, communities, and shareholders. The currency of this more effective exchange is trust.

In economics, the Theory of Constraints has the underlying premise that the constraints of an organization are either throughput, operational expense, or inventory. In process safety, these three constraints equate to risk reduction, action items associated with process safety programs, and resources used to develop and drive these actions. Minimizing these constraints requires trust within the organization, unleashing the ability of the team to innovate and act in a way that maximizes throughput (risk reduction).

Developing company or site programs that are too restrictive, limiting good engineering judgement in favor of always or never requirements, erodes trust. For example, an operator may perceive a requirement that the PHA not allow safeguard of an operator response to an alarm as a sign of distrust in the operator's ability to perform his assigned duties. The opposite, such as always allowing an operator response to alarm to be used a safeguard, would be reckless, likely indicating a general distrust in process safety as a value-added program.

This paper will explore the approach of process safety as a product, where trust is the currency of the exchange, thus leading to greater throughput, as measured by risk reduction. When trust is present, team members feel more empowered, and hold themselves more accountable. They are better able to demonstrate that the tiger is in the cage.

1 A Journey of Understanding

"Real knowledge is to know the extent of one's ignorance." Confucius

To start with, a truth, my journey toward understanding process safety has been rewarding, though filled with examples of tripping over my own assumptions. In most cases the assumption was that others understood what I was asking or trying to explain. I would be so convinced in my ability to articulate the question or convey the information and so compelled by my own words that I would fail to seek challenging feedback; thus leading to an outcome different from my intent, but amazingly aligned with the perception of my customers. Who was right? Well that depends. The message I intended to convey was generally correct. The interpretation was in good faith. Who is responsible for the miscommunication? That also depends. The receiver of the message could have asked clarifying questions. Ultimately, I was to blame. As the 'owner' of the message, I had a responsibility to seek challenging feedback to drive understanding.

My career started prior to the existence of OSHA PSM. As I moved through my various roles: product and application development, technical service, process engineer, operations, turnaround management, and process technology; I had multiple opportunities to engage in process safety initiatives. In most cases, these opportunities were uneventful. The learnings came from the memorable engagements, the changes that did not go as intended, the PHAs that exposed my ignorance of the process, and the realization that a safeguard was not as effective as I had thought.

An epiphany occurred when I was leading an effort to implement a process control project. This project involved some complex voting logic. For simplicity, the core team had three roles: the process engineer (me), the operator, and the process control engineer. Initially, this team would align on an idea and leave the table with each agreeing on a single plan. The process control engineer would complete the assigned task and set up on a simulator. We would come together and realize that each of us had a different perception of what had been 'agreed' on. The controls engineer had programmed his version of the truth. There were no wrong versions of the truth, only different perceptions of a single message. I decided to slow down, take the time to develop very detailed logic narratives, detailed description in paragraph form and detailed logic diagrams. We would use this forum to develop our perceptions. Immediately we started achieving our objectives on each control loop on the initial pass. By slowing down, we went faster, much faster.

We used these same detailed documents in the MOCs and for the project teams to develop implementation plans. The project came in under budget, ahead of schedule and with a quality-assurance checkout rate of 100%. Again, we were going faster. From this experience, I started to realize that other MOCs were similarly constrained. In most cases, I realized the need to slow down, focus on scoping, engage the correct subject matter experts, and ensure that field implementation exactly aligned with the approved scope. Again, dramatic improvement in results. Finally, I realized that this was the intent of the process safety programs all along. However, unnecessary administrative burden had obscured this intent.

What does this have to do with Trust? My ignorance of the intent of process safety and the administrative 'noise' inside the programs had led me to distrust the programs to add value. The process safety professionals responsible for developing and maintaining the process safety programs had added much of the administrative 'noise' to address prior gaps, sometimes referred to as *bolt-ons*. Their intentions were good, and they were working very hard to add value, but the product was evolving to be more confusing and perceived to be of low value.

I have learned from my experience. My epiphany was so profound that I changed the direction of my career to process safety, working to help others see the value in the product that is process safety.

2 Process Safety is a Product and a Service, with Sellers and Buyers

Per OSHA, "the major objective of process safety management (PSM) of highly hazardous chemicals is to prevent unwanted releases of hazardous chemicals especially into locations that could expose employees and others to serious hazards."[1] The premise of this paper is that the objective of process safety is to keep process energy contained, to *keep the tiger in the cage*. This is, per the citation above, in alignment with the objective of the regulation.

Effectiveness in process safety requires buyers and sellers, treating process safety as a product (or service), where the buyer and seller willfully engage in the transaction. The buyers are the stewards of the processes (e.g. operating units) containing the hazardous chemicals, the persons who own the responsibility keep the process energy contained. The sellers are those responsible for setting process safety policy and educating the buyers in those policies. Management is responsible for ensuring that the buyers and sellers are empowered and accountable.

There is another, often used, approach, where the seller forces the buyer to take the product, citing regulation or policy as the driver. This approach can also be effective in the short term and may even work to convince some buyers of the value of the product. However, it may also result in a disdain for the product if the value of or delivery of the product is obscured by inefficiencies or waste, lacks empathy, or destroys value (either perceived or real). This can occur when programs drive action without effectively communicating value.

Some organizations use the language of Risk Based Process Safety to convey this premise, and I have done so many times. Simply stating or writing that a program is 'risk based' fails to recognize the buyer's point of view. A feedback process is required that ensures the buyer is a willing participant in the exchange. This requires patience and persistence, as well as an openness to challenge. Most of all, it requires TRUST. The buyer must trust that the program is in good faith, with the intent of reducing risk and improving the likelihood of *keeping the tiger in the cage*. The seller must trust that the buyer has a genuine desire to *keep the tiger in the cage* and seek feedback from the buyer on the value of the product, and its delivery, such that future transactions can further improve the likelihood of *keeping the tiger in the cage.*

We do not have to compromise compliance. Period! TRUST is the currency of the exchange. There will almost certainly be instances where the buyer does not see the value and yet action is required due to regulations or policy. In these instances, the seller has a responsibility to state the requirement to drive immediate action, but also has the responsibility for committing to ongoing efforts to demonstrate value. Trust is earned. This occurs when the seller actively shows commitment to demonstrating value to the buyer, even if after implementation. Trust is earned when the buyer effectively implements the action, even though value is not yet understood, but with a commitment from the seller to drive understanding.

Each day, consumers willfully engage in purchases of goods and services. In most cases, the buyers have choices in the products and services they buy. When making a first-time purchasing decision, such as deciding to try a new restaurant in the area, the buyer may be influenced by advertising, word of mouth, or availability (just driving by). Trust may not appear to be a key factor. The seller may be even be able to finagle the buyer into an initial purchase. Repeat business does rely on an element of trust, where prior value is still deemed worthy of the price. A compliance or policy driven approach to selling a process safety product will often result in a transaction, albeit the buyer may not view the transaction positively.

"The most important single central fact about a free market is that no exchange takes place unless both parties benefit." Milton Friedman

All stakeholders benefit when we successfully *keep the tiger in the cage*.

3 The Value of having Customers 'Pull' for Process Safety Initiatives

"The people when rightly and fully trusted will return the trust." Abraham Lincoln

Customers buy products they like. Sellers who satisfy the needs of the customer have repeat customers. When the price is correctly set, both parties profit. Few people smile when paying taxes and even fewer seek out opportunities to pay taxes. Most people seek out products they value. Satisfied customers (the buyers) will seek more of the product when value is realized; they will 'Pull.'

I adamantly believe that process safety programs add value to the buyer. I spent most of my career as a buyer of process safety. Early on, when the primary message was compliance, I perceived process safety as a burden, but was willing to comply. Acting because of compliance is an example of a 'Push' transaction. Once I was able to see through the noise and realize the discipline that process safety provides, I was a willing buyer, 'Pulling' for more support provided by these programs.

As a process safety leader, I try to avoid using compliance as a driver of action. In instances where I do have to fall back on compliance, I internalize the experience. It means I have an opportunity and obligation to improve my messaging. I know there was value in the required action, and my failure to demonstrate the value is an opportunity for me to improve.

Now there are instances where the buyer does not want to listen, so the argument could be made that in these instances I am putting undue pressure on the seller. My assessment is that the seller should take ownership, as should the buyer. When we internalize failure to assess how "I" could have affected a better outcome, we improve. This has been my experience! Blaming others, as I have tried many times, rarely leads to improvement. Take ownership, involve others, ask for help, but do not settle.

Sometimes the 'Push' and 'Pull' approaches work in parallel. For example, usage of seat belts by drivers increased because of both approaches. In my opinion, education to inform drivers of the advantages of wearing seatbelts has had the most impact, though it was slower than the impact of changing laws. Thus, compliance will always remain a driver of change. The value of process safety, primarily via a reduction in residual risk, should be the focus of the transaction. This will create a "Pull" which correlates to more process safety champions in the organization.

The amount of 'currency' (trust) in circulation grows as champions are developed. Note, champions are developed, not appointed. Let us say we have an operator on Shift A develop into a champion because of his work in a PHA, on various MOCs, and by leading an incident investigation team. That operator will more quickly gain the trust of other operators than a typical process safety professional. As the 'currency' continues to expand, the value continues to grow. As the value grows, employees internalize the value by virtue of feeling their safety is valued by others, thus further expanding the 'currency.' Trust is the currency!

4 Understanding the Constraints

"Conditions impact the behavior of people." Etiyahu Goldratt

Since Dr. Eliyahu Goldratt introduced the Theory of Constraints (TOC) in his novel The Goal, in 1984, the term has become commonplace in manufacturing industries. TOC has evolved since this 1984 work, but throughput, operational expense and inventory remain focus areas for constraints. 'Bottlenecks' and 'critical path' are examples of terms used to identify constraints. When a turnaround manager is trying to reduce the duration of the outage, the focus should be on the critical path, the sequence of steps that are setting the overall duration. When output rate is constrained by the availability of raw materials, then increasing throughput must relieve that constraint; otherwise, the effort is a waste.

Where are the constraints in your process safety program?

Regarding throughput, I propose that risk reduction is our key measure of throughput. As we administer PHA, MOC and other PHA programs, we should be able to demonstrate a measurable reduction in residual risk. We should design our programs to have the following as their primary output:

- Identifying and understanding the risks in our processes (the scenarios that can cause harm);
- Identifying the controls to prevent or mitigate those risks; controls that are either in place or ones that are needed;
- Ensuring effectiveness measures are in place for each control.

As we identify gaps, actions items will result to drive closure of these gaps. Each of these action items demand resources to drive closure. These are usually shared resources, with responsibilities for many other action items. These competing action items are often outside the purview of process safety. Effective organizations prioritize well, with priority assigned based on value. When the action item owner does not understand the value of the action item, then the item will be prioritized accordingly.

Waste and clutter have a way of revealing themselves within the TOC. Value must be real, and perceived as value by the buyer, otherwise the associated tasks will stall. Even when the buyer realizes value, other high or higher value tasks compete for resources. In these instances, the seller should demonstrate patience and empathy if attempting to negotiate on priority. The seller erodes trust when acting selfishly to drive priority.

There will be circumstances where hard deadlines, compliance obligations, or other drivers may set priorities. In such instances, both buyers and sellers should seek common ground to build trust.

Minimizing these constraints requires trust within the organization, unleashing the ability of the team to innovate and act in a way that maximizes throughput (risk reduction). Process safety leaders need to be open to feedback regarding constraints within programs, realizing that not all mistakes can be "programmed out."

"Innovation has nothing to do with how many R & D dollars you have. When Apple came up with the Mac, IBM was spending at least 100 times more on R & D. It is not about money. It's about the people you have, how you're led, and how much you get it." Steve Jobs

5 Good Engineering Judgement

"If you want people to take action, you must refrain from giving them the answers." Etiyahu Goldratt

Daily, our operating sites rely on good engineering judgement to identify, assess, and manage risk. Our process safety programs should embrace this and provide criteria such that good engineering judgement leads to lower residual risk. Constraining good engineering judgement through too many prescriptive requirements stymies innovation and presumes that we know the best answers. Good engineering judgement is a team effort, often including engineers, operators, maintenance, contractors, management, and possibly others.

For example, many check valves are unreliable as protective layers due to variables such as fouling service. Because of this, some organizations forbid their use as protective layers. However, to forbid their use as protective layers constrains good engineering judgement. I have been directly involved in process safety events where well designed, well maintained check valves acted properly and prevented the escalation of the event. These check valves were evaluated against alternative means of protection and deemed most practical for the service. They were Fit for Purpose. I implore the reader not to focus on the example to engage in a discussion on check valves. Rather, it serves as a personal example of where constraints can be self-inflicted. In the check valve example, the scenario was well understood, thoroughly evaluated and was not constrained by cost of alternative solutions. Good engineering judgement prospered, yielding a good result.

When an organization feels the need to constrain processes due to lack of trust, then they should focus on efforts to regain trust as a currency. Adding constraints as a replacement for trust is not a viable way to create long-term value. Some constraints can help the team navigate more effectively, yielding good engineering judgement.

6 Leveraging Good Practices

"An expert is not someone who gives you the answer; it is someone who asks you the right question." Etiyahu Goldratt

When seeking solutions to process safety challenges, the quality of the solution is directly proportional to the problem statement. A college professor of mine would frequently, and emphatically, lecture our class on the importance of defining the problem. He would say something to the effect of 'the challenge is not finding the solution; the challenge is that you did not understand the problem.' We would laugh at his enthusiasm for this message, but life and experience has re-enforced his message.

To continue with the message from the previous section, good engineering judgement, the team or individual evaluating solutions should clearly define the problem, even seeking guidance and feedback on the proposed problem statement. In my experience, once the problem statement is clear, solutions arise. Sometimes through existing good practices, either internal or external, such as RAGAGEP (Recognized and Generally Accepted Good Engineering Practices). We are fortunate in our industry to have a solid foundation of RAGAGEP. Each passing year, the body of RAGAGEP grows. Use good engineering judgement to select RAGAGEP that is fit for purpose.

Sometimes it is difficult to determine which RAGAGEP, if any, is most applicable to our needs. Much of the discovery leading to the existing RAGAGEP came at a steep price, as many are 'written in blood.' This is unfortunate, but even more reason why we should heed the lessons from these practices. The benefits of RAGAGEP are numerous.

The key to success is aligning the problem with a viable solution, thus the need to understand the problem first. Then using good engineering judgement to evaluate possible solutions, including applicable RAGAGEP.

7 Trust – Perception vs. Intent

"We don't see things as they are. We see them as we are." Anais Nin

To the person perceiving, perception really is reality. Have you heard someone say, in a frustrated or angry voice, "that is not what I intended?" Have you said that yourself? I have. It can be frustrating to have someone perceive your message in way that is hurtful to him or her, when that was not your intent. I have actually said things, intended as a compliment, only to have the message perceived as an insult. I have also been on the other side of that exchange.

What is a proper way to respond when you realize your message is perceived incorrectly? Well, from experience, emotional replies such as "That is NOT what I intended!" only exacerbate the concern. This can lead to a spiraling effect that has both parties focusing on their anger and losing track of the message.

I have found that politely and sincerely taking ownership of the difference works in essentially all cases:

- If I am the seller - "I see. I can understand how you perceived my message as such. I did not intend to lead you to that conclusion. I would like to try to state it a different way. Thank you for helping me see that perspective."
- If I am the buyer – "I now realize that was not your intent. Let's try this again; can you help me understand your intent?"

By taking ownership and diffusing the emotion, you are taking a step toward a trusting relationship. It is the burden of the seller to compel the buyer to engage in a fair exchange.

8 Where Value was Created – The Author's Story

The following are examples of approaches I have employed in the past to achieve improved results in process safety. Each example supports the premise of treating process safety as a product or service, with buyers and sellers, as well as the value of trust as the currency of exchange. Each example also resulted from previous failed attempts where compliance or policy were drivers, yielding *vicious compliance* at best.

a. Approach: PHA Recommendations presented to management using an Event Tree; illustrating the scenario from initiating event to all resulting consequences, clearly showing identified gaps. Result: Management had clearer understanding of the recommendation, was more engaged in the dialogue, took ownership of the recommendation, assigned appropriate resources to address, and actively engaged with PHA team to ensure understanding in evaluating potential solutions. In short, trust improved.
b. Approach: Gatekeeper on MOC to evaluate scope prior to progressing for review and approval. Focus was on ensuring the scope painted a clear picture of the proposed change, with redlined PSI (drawings, procedures, etc.) and that reviewers assigned were appropriate to the scope. When modifications were needed, MOC was sent back to owner / originator for modifications. Result: In the short term, this resulted in some delays. Within a few months, the MOC process was functioning much more effectively and constraints were being relieved. This also provided the gatekeeper (me) an opportunity to coach / mentor the MOC owner. Value creation from the MOC process increased, subject matter experts felt more engaged, employees were more inclined to engage in the MOC process, seeing the value first hand. Process safety and MOC champions began to emerge, taking initiative to help others based on their experience. Champions are developed, not appointed. Overall, the organization realized a noticeable increase in trust.
c. Approach: For procedures used as protective layers, the shift leader would present a scenario, written on a note card, to the operator. The operator would then have to articulate the appropriate response back to the shift leader in an acceptable amount of time. Result: These 'drills' helped keep process safety critical procedures fresh in the mind of operators by also helped expose effectiveness gaps in these protective layers; which were then addressed. There was some initial resistance because these were viewed as "got-you" drills. The operators started to 'pull' for the opportunity to demonstrate their competency once the operators realized the purpose of these drills.
d. Approach: In response to process safety events, verify the PHA identified the event and compare the actual event to the hypothesized event. This allows for updating the PHA real-time and for the organization to calibrate their understanding of the process. Other PHA's may also need updating. Result: Improved overall PHA: better quality scenarios, more precise and

accurate identification of controls (safeguards), and better implementation of effectiveness programs for controls. Within a year, PSE rates of each severity began to decline. Employees and managers began to have increased trust of the PHA (thus the PHA team).

e. Approach: Removing the 'noise' from operating procedures, focusing on specific steps in the procedure and consequences of deviation, significantly reducing the number of pages in a procedure. Result: More ownership of procedures by operators as measured through proposed changes that came directly from the operators. Operators, engineers, and managers agreed that procedure quality and adherence to procedures improved.
f. Approach: Started holding shift lunch sessions with operators and maintenance to have them ask questions regarding things they do not understand. This took some prompting and seeding with examples, but they began to bring forth questions about MOCs or investigations they were concerned about, safeguards they did not understand, procedures they thought were wrong or inefficient. Result: Operators and maintenance personnel began to feel more empowered to raise concerns and bring forth ideas. Their questions and challenges helped improve effectiveness of some safeguard, identified where higher reliability safeguards were available, and even some scenarios previously not evaluated in a PHA. Their ownership of their procedures improved. Operators began to request assignment to a PHA.

9 Summary

Ronald Reagan said "Trust, but verify." This is consistent with the theme of this paper, where trust is the currency of an **effective** process safety program. In our operating facilities, we do have turnover, but most transactions are occurring amongst people and organizations with a significant level of familiarity. If trust is demanded (e.g. "because I said so"), but not earned (via the 'verify' or 'effectiveness' step), then trust will be eroded and be very difficult to maintain. Long-term trust relies on integrity. In an organization where integrity is scarce, trust will also be scarce.

Process safety is a product, which is the key premise of this paper. The exchange of a product requires sellers and buyers. In order for both the seller and buyer to benefit from the exchange, both parties must understand the value and the constraints. When the exchange is mutually beneficial, then we improve the probability of *keeping the tiger in the cage.*

10 References (examples given to show format).

[1] Process Safety Management Guidelines for Compliance, Purpose: Available at https://www.osha.gov/Publications/osha3133.html Accessed on December 29, 2018

Employee Training – Beyond Compliance

Brian R. Desrosiers, CSP, OHST
Environmental Resources Management (ERM), Inc.
99 East River Drive, 3rd Floor, East Hartford, CT, 06108
brian.desrosiers@erm.com

Prepared for Presentation at
American Institute of Chemical Engineers
2019 Spring Meeting and 15th Global Congress on Process Safety
New Orleans, LA
March 31 – April 3, 2019

Employee Training – Beyond Compliance

Brian R. Desrosiers, CSP, OHST
Environmental Resources Management (ERM), Inc.
99 East River Drive, 3rd Floor, East Hartford, CT, 06108
brian.desrosiers@erm.com

Keywords: Training, Compliance, Management, Accelerative Learning

Abstract

For facilities covered under the Occupational Safety and Health Administration's Process Safety Management (PSM) Standard or the United States Environmental Protection Agency's (USEPA's) Risk Management (RMP) Rule, training is often paramount for communicating the hazards of the process to the employees. This also includes the "how to comprehend" operating procedures, in order to ensure consistent operation. Under the PSM and RMP rules, it's only the employees who interact with the process that receive training. Subsequently, the corresponding elements of the training focus on the hazards of the process, the safe work practices, and an overview of standard operating procedures.

However, since culture and workplace involvement are key pillars in the Risk Based Process Safety approach, we must consider communicating more than just the regulatory minimum. If the training for personnel (on the elements of process safety) also include their pragmatic application methods and interconnectivity with other elements, we can increase employee buy-in to the process safety culture.

This paper provides an overview of the authors' experience on how to conduct these trainings to keep the personnel engaged and provide focus on key areas to ensure a meaningful retention and implementation. He will also share their experience with position appropriate scope and responsibilities. Their recommendation is that "Even positions such as Plant Managers, who may not have direct interaction with the program, should be trained on the expectation of the Process Safety program and how it potentially affects the organization, independent of any regulatory regime."

As an example, this paper includes an overview of the proposed changes to the USEPA's RMP rule on training requirements for management personnel and how we can meet these changes. Through robust training in the elements throughout all levels of an organization, PSM culture can continue to evolve as employees continue to better understand the elements and the management systems process.

1 Introduction

Training is identified as one of the most effective communication tools for controlling risk. In process safety, training should (?) be used to communicate hazards, ensure consistent operation of the system, and drive positive culture by showing the importance of Process Safety Management programs., When a facility meets the threshold quantities that trigger the Occupational Safety and Health Administration's (OSHA's) Process Safety Management [1] or the United States Environmental Protection Agency's (USEPA's) Risk Management Plan [2] rules (aside from the revisions we will discuss later), training is a required element for compliance with these programs. When we look at the requirements of these standards, they only require that training be provided for "Each employee presently involved in an operating process, and each employee before being involved in operating a newly assigned process." Since this only applies to employees who directly operate the process, how do we educate and inform other employees up to management on the importance of these processes and how they could be affected by them? This paper will provide guidelines on who should be trained, what they should be trained on, the frequency of the training, and how to provide effective training that is sustainable.

2 Hazards of Improper Training

In January1966 at a liquefied petroleum gas (LPG) tank located on a farm adjacent to a refinery in Lyon, France, an operator was draining water from a pressurized propane tank when some of the LPG was spilled leading to formation of a vapor cloud. The vapors flashed over causing the storage tank to be engulfed in flames. The resulting fire led to the tank rupturing and causing a Boiling Liquid Expanding Vapor Explosion (BLEVE). The event caused 18 fatalities and 81 injuries. [3]

The incident investigation identified that a lack of training led to improper interaction with the process due to a lack of knowledge regarding procedures and the hazards. If we take the lessons learned from this event, we realize the importance of providing effective training to personnel. Additionally, if we also communicate the hazards of the process to other staff throughout various levels of the organization, we can drive a better understanding of the process, the risks, and why PSM is critical for managing risk.

3 Beyond Compliance Training Model

Process Safety Management is a system that requires buy-in and involvement from every level of an organization that operates a covered process. From operators, maintenance, engineering, and all the way up to senior management, we need to ensure that the hazards and expectations related to safe process operation are communicated appropriately. Once we identify the need to communicate and train all these individuals, two questions arise, "what content should we include in the training for each level?" and "how often should they be trained?" To assist with this, we can use the common Plan-Do-Check-Act model with assistance from industry standards.

3.1 Training Content

To properly identify and document specific training needs, the use of a Training Needs Matrix can identify training elements based upon employee job title. To start, identify all the job titles in the organization who directly interact, oversee, or have responsibility for the covered process. This includes senior management since they will typically approve budgets and have overall responsibility for the safety performance of the facility. However, since they are not directly operating the process, we can tailor the training content to be specific to their job tasks. Overburdened training with content not relevant to the attendee is a real risk with EHS trainings, and avoiding this where possible will create a more effective training program. Guidance for effective training can be found in the American National Standard Institute (ANSI) / American Society of Safety Engineers (ASSE) Z490.1 standard for the *Criteria for Accepted Practices in Safety, Health and Environmental Training.* [4] Table 1 includes an example of a Training Needs Matrix that provides insight into topics that could be covered at each level of the organization.

Table 1 – Sample Training Needs Matrix

Job Classification	Training Topics	Method of Delivery
General Employees	Hazards of covered process (i.e., chemical hazards, process hazards.)	Video / Classroom / PowerPoint
Operators	Elements outlined in PSM Standard / Elements of PSM	On-the-job-training / Classroom
Maintenance	Hazards of the Process / Maintenance Activities / Safe Work Practices / Design Code Review	On-the-job-training / Classroom
Engineering / Environmental, Health and Safety (EHS)	Elements of PSM / Design Codes Review	Video / Classroom / PowerPoint
Senior Leadership / Plant Manager	Elements of PSM / Hazards of the Process / Overview of PSM Enforcement	Video / Classroom / PowerPoint
Corporate Leadership	Overview of Covered Process Types / Overview of Importance of PSM	Short Recorded Video / Webinar

In the next section, we will cover each of these rows in detail to provide an overview of the topics and delivery methods.

3.1.1 General Employees

The general employees group consists of any personnel who do not interact with the covered process. This includes other manufacturing personnel on the floor, administrative staff, or any other personnel who could be affected by improper safety precautions, or a release of highly hazardous chemicals from the process. These employees need be made aware of the hazards of the process, how to detect any potential releases, alarm systems / emergency procedures, and any special or unique hazards such as electrical classification. Commonly processes with flammable liquids require electrically classified areas; however, general employees are not typically trained on what this entails. Often these employees violate the rule by bringing cell phones and other non-intrinsically safe equipment into the area, presenting a potential ignition source. However, effective communication of special safety precautions makes safety around the covered process everyone's responsibility. Guidelines for this can be outlined in OSHA's Hazard Communication Standard codified under 29 CFR 1910.1200. [5]

Due to the need for consistency in the information provided in this training, pre-recorded sessions can be provided to personnel when they first arrive at the facility that covers these elements. A classroom-based training also can be effective as it allows for personnel to ask questions, which can be critical to the learning process. PowerPoints can provide visual aids during these classroom trainings with diagrams and pictures. If the system has unique alarm systems, recorded sounds of the alarms can help employees become more familiar with the multiple types in the event of an emergency.

3.1.2 Operators

Since operators are the most involved in the process, their training must be the most robust and effective. OSHA's PSM and the USEPA's RMP standards provide required elements that operators must be trained on. These include:

1. Overview of the process and in the operating procedures;
2. Specific safety and health hazards;
3. Emergency operations (including shutdown); and
4. Safe work practices applicable to employee's job tasks. [6]

If we want to inform our operators of more than the bare minimum, we would also want to consider providing them an overview of the elements of Process Safety Management, including safety systems. As the first line of operations, it is critical for operators to see what the elements are, how they interact, and why each is required for the continued safe operation of the process. Specific emphasis should be directed towards the elements that operators will have the most responsibility for. This includes, but is not limited to, Management of Change, Process Safety Information, and Operating Procedures.

Operator training will often be delivered in various ways to be effective. A combination of on-the-job training and classroom-based learning should be considered to ensure open communication between new and experienced operators. If results show that the current training delivery model is not effective, the facility should consider getting insight from operators on how they feel the training would be the most effective.

3.1.3 Maintenance

Maintenance personnel must be trained on maintenance procedures, safe work practices, and any standard operating procedures if they are going to interact with the system. However, similar to operators, we will want to consider delivering training on the elements of the PSM program that could impact their work activities the most. This includes an overview of the available Process Safety Information, as well as an overview of the Management of Change process, the Contractor Management program, and other areas the site identifies as impactful.

One other aspect to consider communicating to maintenance, is an overview of the Recognized and Generally Accepted Good Engineering Practices (RAGAGEP) that have been implemented on the process. Since maintenance personnel are going to be doing some of these inspections and preventative maintenance activities, giving them an overview of the basis for these can help provide insight into why it is critical.

The training delivery methods should mirror those of the operators. Since maintenance personnel will have direct interaction with the process, we want to ensure that their training is robust and effective to verify that they have the knowledge and skills to maintain the system safely.

3.1.4 Engineering / Environmental, Health and Safety (EHS)

In facilities with poor process safety culture, we often find that since PSM/RMP is an OSHA/EPA regulation, the responsibility lies with just the EHS group. However, this could not be further from the truth. PSM requires specialties from multiple departments to be successful, and one of the most significant supporters of the program along with EHS should be the engineering department. Engineering should be providing assistance with elements including, but not limited to, design code reviews on MOCs, documentation of Process Safety Information, and establishing preventative maintenance schedules.

The EHS group will help ensure that overall program implementation is being adhered to in accordance with OSHA/EPA regulations. However, at larger facilities with a covered process (or multiple processes), the system compliance is often left to a PSM Coordinator, or similar role. Often, the rest of the EHS team is not as familiar with the requirements of PSM; however, they should be trained in recognizing compliance issues during their walkthroughs.

Trainings for these two groups can be conducted together to help save company time. However, the focus for these sessions should be on the elements of process safety as well as the Recognized and Generally Accepted Good Engineering Practices (RAGAGEP) requirements identified for the process. The elements of process safety should include a focus on how they will directly impact the group and what they need to know to drive compliance. The RAGAGEP review will assist the team in ensuring that field observations meet design requirements and help them become more familiar with the process. For example, in ammonia refrigeration systems, the continuous ventilation system should be in operation at all times. If, the EHS/Engineering groups are trained on this then if they observe the fan is off due to a failure, it can be reported and repaired appropriately. Competency is a critical aspect of PSM, and ensuring that these two groups largely responsible for compliance are competent in the expectations of the program drives success.

3.1.5 Site Senior Leadership / Plant Manager

Since Senior Site Leadership and the Plant Manager might be required to sign PSM documents, and have overall responsibility for the safety, well-being, and enforcement issues at a facility, we must consider them in this training matrix. However, as we begin to climb the chain of command at a company, finding the time to deliver trainings for leadership can be difficult, especially when it is outside their technical field, or new to a PSM-covered facility. Time is a precious commodity in business, so when we identify what we want to communicate to senior leadership, we must be as effective as possible while still delivering meaningful content. Culture should be driven from the top down, to set the proper expectations; managers should be trained on the program.

In the Training Needs Matrix, we identified three focus areas for this groups training:

1. Elements of PSM;
2. Hazards of the Process; and
3. PSM Enforcement.

Though this may seem like a significant amount to cover, we can compress each element to cover the key points that we need to communicate to management. Do we need to give them an overview of all Process Safety Information? No, we do not. But, we should let them know that it needs to be maintained, who is responsible, and where it is located. The same process should be applied to each element, we only give a brief summary.

Hazards of the process should be a bit more detailed. An overview of the risks of the process as well as any special or unique hazards (e.g. electrical classification) should be communicated to leadership. They need to be aware of the consequences of not adhering to these safety precautions if they are going to drive enforcement. An employee walking into an alcohol processing area with a cell phone might not sound bad to an untrained person, but explain to them how it could create a flash fire. This becomes a new level of awareness and competency at the leadership level. We should also consider showing them the worst-case and alternative-case scenarios developed, if required by the EPA RMP rule. This will alert them to the potential negative impacts improper operation of the process could create for the surrounding community.

Lastly, we should consider communicating to leadership potential enforcement citations and fines that could impact the business if the program is not adhered to. A common issue in EHS and PSM is getting funding for projects from management since there is no direct return on profitability. However, if we can show an example of how a similar process was cited for thousands of dollars for not following the PSM Standard, it drives the cost-benefit of approving funds. At times, senior leadership does not know how serious the monetary impact of non-compliance can be to a site until it is too late.

For the delivery method to this group, consideration should be given to the impact that a PowerPoint session can provide during a portion of a site leadership meeting. Since a majority of those sitting in the training will not have any experience with PSM, we should allow for the ability to ask questions as they arise. Not only does this promote learning, but it also might initiate brainstorming sessions and strong discussions on ways leadership can assist in influencing these elements that EHS may not have considered. This training ideally should be no longer than 30 minutes – 1 hour, however, always include 10-15 minutes for questions. We do not want to overload the leadership team with too much information at once.

3.1.6 Corporate Leadership

The final group we want to reach is corporate leadership. Though this group is the farthest removed from the day-to-day operations within the chemical facility, their leadership in the organization is essential to drive a stronger culture towards process safety and an understanding of the risks associated with these processes. This is critical since this group will drive the appropriate risk criteria for the company.

If a company has the same types of covered processes (e.g. ammonia refrigeration, flammable liquids processing), we can give a brief summary of the hazards associated with these processes to demonstrate the importance of these systems that have been or need to be implemented. Another way to demonstrate the importance of PSM is to include examples of catastrophic releases in a related industry or similar process to demonstrate the potential negative impact on business continuity.

Since delivery of this training in a classroom will most likely not be possible, we consider hosting a webinar to deliver this training to corporate leadership. We will want to make it concise, but meaningful with the session being no longer than 30 minutes – 1 hour. Webinars still allow for the questions to be asked at points during the presentation. If this option is not feasible, consider a recorded video that can be sent to the corporate team so that they can watch it when it is convenient for them. At times, coordinating schedules can be difficult, so this method (though less effective) can still begin the process of communicating PSM and beginning to build the culture.

3.2 Training Frequency

The frequency for these trainings will vary depending on the maturity level of the process safety culture at the corporate and at the site level. When developing the training needs assessment consider how comfortable you, as a process safety lead, are with the level of knowledge and competency related to your covered processes and begin to make a determination from there. Also, consider that the level of maturity will vary at different levels of the organization. For example, corporate leadership might be very knowledgeable and have a strong culture for PSM whereas a site might not, and vice versa. Table 2 contains examples for how often these trainings should be provided based on culture levels.

Table 2 – Frequency of Training by Culture Level

Job Classification	Strong Culture	Moderate Culture	Poor Culture
General Employees	Triennially	Annually - Biennially	Annually
Operators	Triennially (By Regulation)	Annually – Semi-Annual	Semi-Annual - Annually
Maintenance	Triennially (By Regulation)	Annually - Biennially	Semi-Annual - Annually
Engineering / EHS	Triennially	Annually - Biennially	Semi-Annual - Annually
Senior Site Leadership / Plant Manager	Triennially	Annually - Biennially	Quarterly - Biannually
Corporate Leadership	Triennially	Biennially	Semi-Annual

Strong Culture – Leadership, operators, and general employees have a robust knowledge of the process, the hazards associated with the process, and PSM elements.

Moderate Culture – Leadership, operators, and general employees have a fair amount of knowledge around the process, the hazards, and PSM elements.

Poor Culture – Leadership, operators, and general employees have little to now experience with PSM-related processes.

3.3 Continued Effectiveness

Once the training has been delivered, establish a follow-up procedure to verify the effectiveness of the training program and content. Consider having informal interviews with corporate leadership to determine if the topics in the training are still fresh in their mind. With operators, try to obtain feedback on the program to see if they have any thoughts on how to modify the training program to be more effective. Feedback from training attendees can be one of the most effective ways for growing an effective PSM training matrix.

4 Instructional Shift to gain Success

Working with corporations where the focus has been "Beyond Compliance", it has been observed, that this mindset brings about a drastic change in the safety culture of the organization which takes care of compliance related needs.

Going back to basics, when we say "training", the human mind instantly assumes this to be a one-way street, where the teacher is instructing the students. This is rather a hierarchical approach, that is now evolving to a different level, even in schools. While schools are dealing with educating children and young adults, the industry intends to provide the same "knowledge sharing in concept" to adults.

However, when the desired end result is education, it has been noted that the approach to educate children and adults is different. An instructional shift or approach called "Accelerative Learning" [7] has shown some outstanding results. What does this mean?

Training adults is different from teaching children. While a conventional one-way instruction medium may work for children (similar to the conventional training approach in the industry), the approach has its gaps when the receptive audience is adults. For adults, the biggest impact has been when the instruction medium has been interactive. The interaction is a combination of –

- A two-way dialogue between the instructor and the audience
- Input from the attendees in the "training class", that are able to share their experience
- Audio, visual and kinesthetic adjustments to the educative environment.

What this approach provides is an assurance to the audience (the attendees of the training) –

- Their intelligence has not been challenged
- Respects their thoughts, and encourages them to provide input
- Brings about experience and knowledge sharing.

While this approach is not novel; on the other hand it is been widely practiced in various different areas and has shown success. Why? The focus is the human mind, the focus is the adult age group; the focus is the human mindset.

This technique is not any ordinary class room session or folks reviewing a Computer Based Training (CBT). It rather focuses on the human senses related to audio, visual and touch. The technique is known to stimulate the enterprising and ever-thinking human brain; and takes into account human attention span; emotional mind and is a two-way interactive learning session.

While the audience absorbs the content, and connects it with real-life situations, literally embracing the concept of safety, the instructor absorbs more knowledge of their areas of concern, and is able to alter the course path appropriately.

The eventual outcome has been success in the staff embracing a robust safety culture.

These are not any ordinary training sessions, rather they are learning sessions, where every attendee not only absorbs the subject matter, but also interacts with the other attendees to enhance the learning process. For a regular classroom session, this evolves into an interactive journey of ideas being shared and collaboration in understanding the concepts. At the same time, the instructor understands the moods and mannerisms of those around him and has to be able to transmit the messaging in several different ways. To enhance this learning journey, the instructor makes use of –

- Various pieces of music, where the tempo controls the desired outcome,
- Colors and other visual aids, that are pleasing to the human eye and are organized in a color scheme, and
- Kinesthetic tools to stimulate the learning process.

5 Proposed Rule Changes

The USEPA proposed changes to any covered process that exceeds the threshold quantities outlined in 40 CFR 68 In the May 30, 2018 proposed changes, it notes that there will be modifications to the Training element of the RMP rule.[8] The proposed change would require training for employees "operating a process" to "involved in operating a process."[3]

This revision would require that supervisors responsible for directing process operations and with operational responsibilities to be trained on the RMP program accordingly. With the implementation of the training guidelines established above, we can help drive compliance towards these changes.

While in the PSM/RMP world, the conventional training techniques have generated success; a bigger successful outcome has been observed using the Accelerative Learning approach. This approach has shown higher knowledge retention rates.

6 Positive Impacts on Process Safety Commitment

Under the Risk Based Process Safety approach, the first section of the model contains five (5) critical elements of a successful program. Often these pillars are known as the cornerstone of process safety excellence. [9]

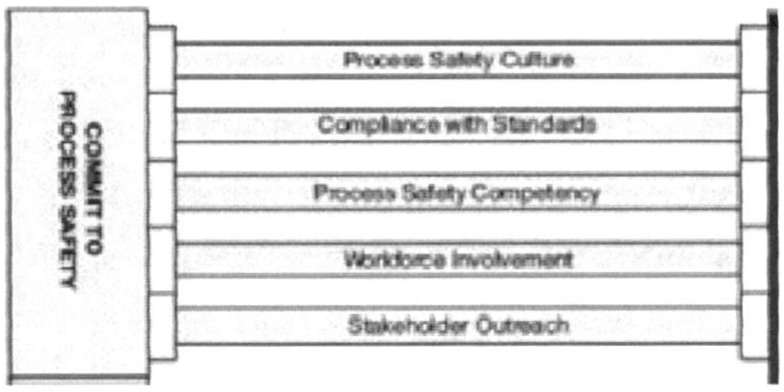

Guidelines for Risk Based Process Safety – Center for Chemical Process Safety, Page 8.

The reason why these pillars are first in the program is that without these foundations in place, the program will likely start to see failures with root causes ultimately being in these pillars. However, by using the Training element beyond compliance to communicate and expand the knowledge of process safety to all aspects of the organization can help improve these first five pillars. Some examples include:

- Process Safety Culture – Culture can improve with a robust training model by increasing knowledge and sharing of responsibility around the program. When corporate and senior management are aware of the requirements of the program, they can drive a positive influence into the rest of the organization by stressing the importance of safety of the process being everyone's responsibility.
- Compliance with Standards – With the communication of the elements to certain groups, the Compliance with Standards pillar will have a wider group of knowledgeable individuals trained in PSM and RAGAGEP requirements, which can drive better compliance through inspections.
- Process Safety Competency – One of the most important impacts this training model can provide is a stronger sense of competency related to the covered process, PSM, and RAGAGEP throughout all levels of the organization. With a strong focus on senior management, this can drive the importance of competency down to general employees.
- Workforce Involvement – When all levels of an organization are trained in the process, there is more involvement and a feel that process safety is everyone's responsibility.
- Stakeholder Outreach – Communication of worst-case scenarios to senior management can help begin the discussions for a stronger Stakeholder Outreach to the neighboring community and environment.

To ensure continued effectiveness of the program and continuous improvement, we must periodically evaluate our Training element as a whole. Management should conduct routine assessments and audits of the program to ensure that the training needs assessment does not need to be modified. This includes ensuring that content, frequency, and identified parties are adequate to meet the expectations of the goals you have set forth in your program.

7 Conclusion

If a company strictly uses the regulatory requirements to design and implement their process safety management system, they will find that their culture and competency among the organization will not grow. If we put the training strictly on the operators and not the other key stakeholders in the program, a company is not taking full responsibility for managing the risks associated with a process that contains highly hazardous chemicals. Poor training and process competency among the organization will ultimately lead risks in communication and process management.

If we start to model our training program to reach out to all stakeholders in the program, we can promote a stronger team-based approach to managing the process. As personnel become more familiar with PSM and the process, they will have more stake in ensuring the safe operation with promising benefits to culture, workforce involvement, and stakeholder outreach. Take the time to develop a robust training program appropriate to the level of culture at the company, and the program will not just be the responsibility of the "PSM Coordinator," but everyone.

8 References

[1] Occupational Safety and Health Administration, Process Safety Management of Highly Hazardous Chemicals Standard, 29 CFR 1910.119
Available at web link
https://www.osha.gov/lawsregs/regulations/standardnumber/1910/1910.119
Accessed on January 8, 2019

[2] United States Environmental Protection Agency, Risk Management Plan Rule, 40 CFR 68, Available at web link https://www.ecfr.gov/cgi-bin/text-idx?c=ecfr&SID=b843807afdc641b203ffec44aa671d36&rgn=div5&view=text&node=40:16.0.1.1.5&idno=40#se40.16.68_110
Accessed on January 8, 2019

[3] Health and Safety Executive, Refinery Fire at Feyzin. 4th January 1966
Available at web link http://www.hse.gov.uk/comah/sragtech/casefeyzin66.htm
Accessed on January 4, 2019

[4] American National Standards Insstitute (ANSI) / American Society of Safety Engineers (ASSE) Z490.1 – 2016 Edition, *Criteria for Accepted Practices in Safety, Health and Environmental Training*

[5] Occupational Safety and Health Administration, Hazard Communication Standard, 29 CFR 1910.1200
Available at web link
https://www.osha.gov/lawsregs/regulations/standardnumber/1910/1910.1200
Accessed on January 8, 2019

[6] Occupational Safety and Health Administration, Process Safety Management of Highly Hazardous Chemicals Standard, 29 CFR 1910.119
Available at web link
https://www.osha.gov/lawsregs/regulations/standardnumber/1910/1910.119
Accessed on January 8, 2019

[7] Israni, Chris & Watson, Colleen, "Applying the Concept of Accelerative Learning for Design and Delivery of Process and Personnel Safety Leadership Programs in Oil & Gas Assets" , SPE Health, Safety, Security, Environment, & Social Responsibility Conference - North America, 18-20 April, New Orleans, Louisiana, USA

[8] United States Environmental Protection Agency. Proposed Rule "Federal Register Volume 83, No. 104," 40 CFR Part 68, pp. 9, May 30, 2018.

[9] Center for Chemical Process Safety, *Guidelines for Risk Based Process Safety, Section 3, Figure 2. 2015 Edition*

AFPM/API Advancing Process Safety: Reevaluating the Permit to Work Process through the Human Reliability Lens

Morgan Walker
Phillips 66
San Francisco Refinery
1380 San Pablo Ave.
Rodeo, CA 94572
morgan.t.walker@p66.com

Ben Manns
ExxonMobil
22777 Springwoods Village Pkwy
Spring, TX 77389
ben.l.manns@exxonmobil.com

Jeffrey Hazle
Marathon Petroleum Corporation
19100 Ridgewood Pkwy
San Antonio, Texas 78259
jlhazle@marathonpetroleum.com

Lauren Coughlin and Andrew Broadbent
API
1220 L Street NW
Washington, D.C. 20005
coughlinl@api.org and broadbenta@api.org

Prepared for Presentation at
American Institute of Chemical Engineers
2019 Spring Meeting and 15th Global Congress on Process Safety
New Orleans, LA
March 31 – April 3, 2019

AFPM/API Advancing Process Safety: Reevaluating the Permit to Work Process through the Human Reliability Lens

Morgan Walker
Phillips 66

Ben Manns
ExxonMobil

Jeffrey Hazle
Marathon Petroleum

Lauren Coughlin and Andrew Broadbent
API

Keywords: Human Reliability; Human Factors; Human Organizational Performance; HOP; Permit To Work; PTW; Safe Work Permit; SWP; Process Safety; Industry Programs; Process Safety Performance; Process Safety Metrics; Lessons Learned; Practice Sharing

Abstract

This paper represents the key learnings from the efforts of the Human Reliability Subgroup of the AFPM/API Advancing Process Safety (APS) program to review the Permit To Work (PTW) process through a human reliability lens. The HRSG reviewed the PTW process as a case study in applying human performance principles to an established process safety system.. The research and data provided in this paper demonstrate the importance of effective permitting practices as these tasks relate to process safety, as well as the potential for human error during the permitting process that may lead to process safety events. The HRSG surveyed refining and petrochemical companies through the APS network on PTW practices, using the material provided by these companies to develop AFPM Practice Sharing documents. The HRSG made four observations in reviewing how companies apply human performance considerations to their PTW systems: (1) sites can optimize the use of permitting resources by ensuring a fit-for-risk (fit-for-purpose) PTW program; (2) while a permit system may be robust, it must be executed as designed to be effective; (3) an effective auditing process verifies conformance with the PTW process and provides for continuous improvement; and (4) PTW systems must clearly define and communicate roles and responsibilities and provide a means to verify the competency of individuals involved in the PTW process.

1 Introduction

From 2016 to 2018, the Human Reliability Subgroup of the Advancing Process Safety network, consisting of experienced safety professionals, reviewed the application of human performance principles to the formal Permit To Work program. The group's goal was to learn how sites address the possibility for human failure in a process safety system that is ubiquitous across the refining industry and required both by regulations and industry codes and standards.[1] The group surveyed US refineries[2] and collected representative practices that apply human performance considerations to the Permit To Work (PTW) system. The collected practices were published as "Practice Sharing" documents and made available to members of the refining and petrochemical industries. The practices demonstrate methods to ensure competency in the PTW process, facilitate more valuable audits, and define the scope and application of the system to reduce time spent on ineffective or redundant administrative procedures.

The HRSG is a component of the Advancing Process Safety (APS) joint initiative between the American Petroleum Institute (API) and the American Fuel and Petrochemical Manufacturers (AFPM). The APS initiative has evolved and changed based on industry needs. Today, APS administers six programs: Industry Learning and Outreach, Hazard Identification & Practices Sharing, Process Safety Regional Networks, Human Reliability, Mechanical Integrity, and Process Safety Site Assessment. Oversight of APS is provided by the Process Safety Workgroup and the Process Safety Advisory Group. The Process Safety Workgroup of APS created the Human Reliability Task Force in 2015 and tasked the group with addressing the incidence of human performance-related process safety challenges across industry. In late 2015, the Task Force was promoted to the level of Subgroup within the APS network, and it began work on an effort to produce effective tools to help refineries and petrochemical manufacturers apply human performance understanding to critical operational tasks.

The HRSG consists of corporate and site occupational and process safety leaders, training coordinators, and human performance specialists from the US refining and petrochemical industries. The group's composition is heavily favored toward refining company representatives, but several petrochemical company representatives also participate, and the group's petrochemical membership is growing. The key findings and practices from the 2016 to 2018 work effort originated from US oil refinery sites and corporate offices of companies with US oil refining operations.

This paper represents the key learnings from the HRSG PTW effort. It is presented as a case study in applying a human performance lens to an established process safety system. A similar study in human performance could be applied to other process

[1] All US refineries complete permits to work or "safe work permits" per OSHA Regulations (Standards – 29 CFR). The OSHA Voluntary Protection Program also includes some permit-related requirements.
[2] 15 refining and petrochemical companies contributed to the development of the "Human Performance considerations related to Permit to Work" AFPM Practice Sharing series.

safety activities. The HRSG surveyed API and AFPM member companies through the APS network on PTW practices, publishing select practices as AFPM Practice Sharing documents. The surveyed sites apply human performance considerations in various ways. The HRSG did not seek to create a comprehensive PTW standard; rather, it gathered and published a suite of resources that could be adopted by site and corporate leaders. Readers are encouraged to review the Practice Sharing documents with their own sites, resources, and safety cultures in mind. It would not be practical or effective to implement every procedure, checklist, and training document collected, but site leaders are encouraged to select the most appropriate practices and lessons learned for their own situations.

2 The Link Between Human Performance and Process Safety

Human Reliability and Human Organizational Performance aims to pivot the attempts to eliminate human error toward acknowledgement and appreciation of the principle that human error is inevitable. Through improving process systems, the effect of error and the associated risks can be reduced. According to the Department of Energy (DOE), "About 80 percent of all events are attributed to human error. Roughly 20 percent of events involve equipment failures. When the 80 percent human error is broken down further, it reveals that the majority of errors associated with events stem from latent organizational weaknesses (perpetrated by humans in the past that lie dormant in the system), whereas about 30 percent are caused by the individual worker touching the equipment and systems in the facility, as shown in Figure 1. Clearly, focusing efforts on reducing human error will reduce the likelihood of events," (DOE-HDBK-1028-2009, 1-10). Further, Todd Conklin, Human Organizational Performance expert, expands on this belief, stating that "Contemporary wisdom says that the average skilled worker, who works with their hands, makes 5 to 7 errors per hour" (9).

Figure 1: Event Causes from DOE *Human Performance Improvement Handbook*

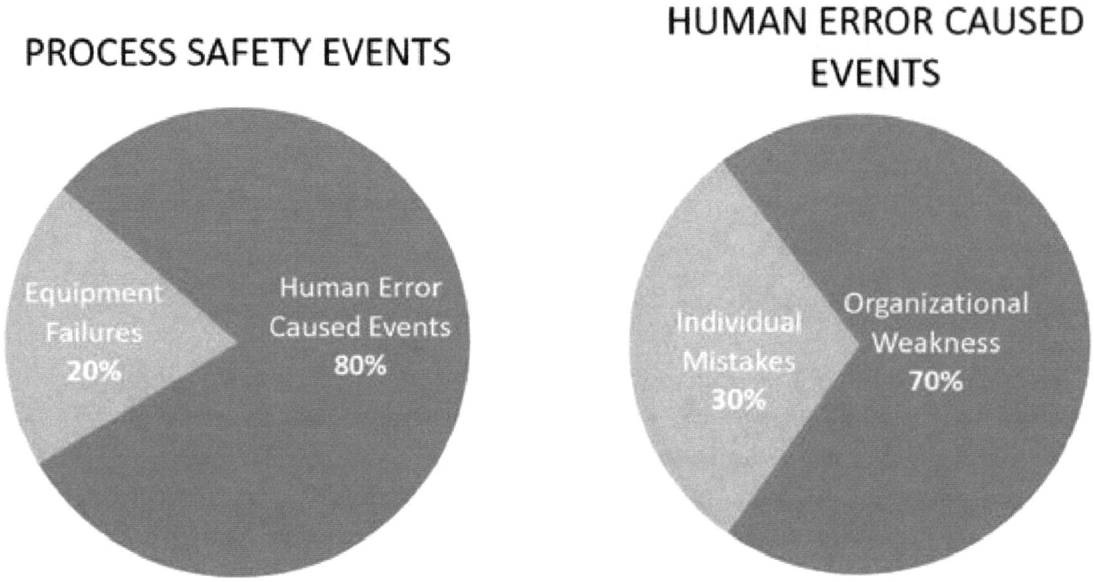

Since APS is dedicated to reducing process safety events, the HRSG reviewed relevant process safety data. Based on data from API RP 754 data collection, APS recognizes the link between human reliability errors and process safety events. In 2016, human error was attributed as a causal factor in 62% of the process safety events that occurred under "normal operating mode" at refineries and petrochemical plants, according to data submissions from API 754 Process Safety Event Collection. Of these events, 70% were attributed to "Organizational Weakness," and 30% were attributed to "Individual Mistakes." Multiple types of events during normal operating mode can include human error as a causal factor in the event, and percentages of events citing human error as at least one cause of the event is shown in Figure 2.

Figure 2: API RP 754 Events Citing Human Error as a Cause of the Event

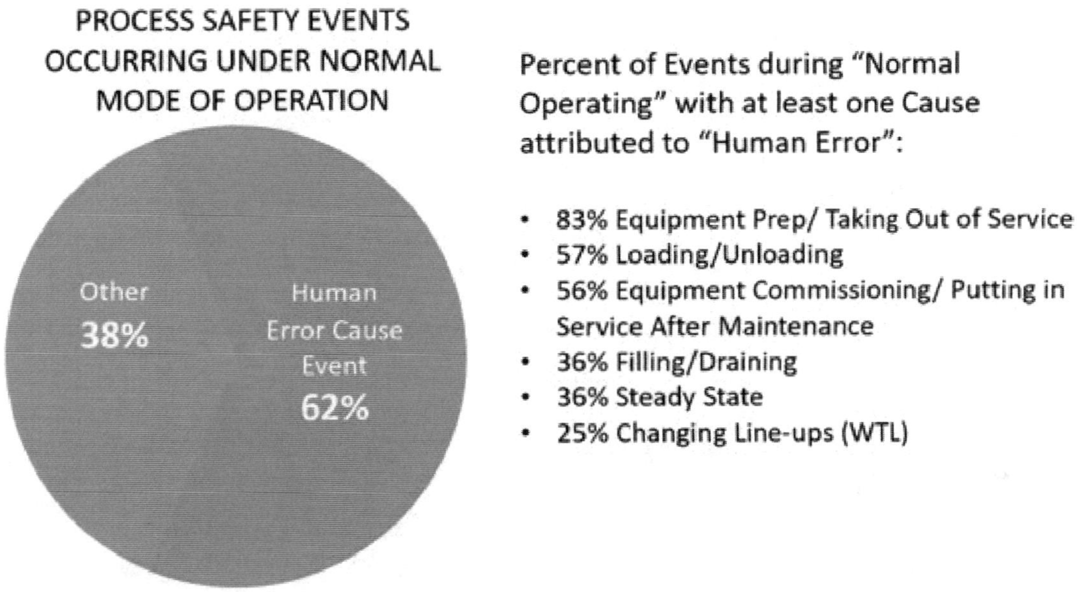

Beyond the refining and petrochemical industries, De Felice, et al. (2012) found through statistical modelling that over 90% of accidents in nuclear industries, over 80% of accidents in chemical industries, over 75% of maritime accidents, and over 70% of aviation accidents can be attributed to human error.

These statistics illustrate the importance of putting systems in place that are designed to reduce the potential for the occurrence of human errors, and to mitigate the consequences when (inevitable) human errors do occur. To best set up the worker and sites for safety success, it is important to reevaluate the practices and procedures in place through a human reliability lens and improve, as necessary.

3 The Link Between Human Performance and Permit to Work

The HRSG recognized numerous company and region-specific definitions of a PTW, but the group adhered to the consensus of the chemical engineering community when defining a permit to work.

> ***Permit to Work.*** *A formal written and documented procedure that authorizes certain people to carry out specific work within a specified time frame. It is used to control certain types of work that are potentially hazardous and forms an essential part of safe systems of work for many maintenance activities and allows work to begin only after safe procedures have been defined. It provides a clear record that all foreseeable hazards have been considered. Where the proposed work is assessed as having a high risk, strict controls are required detailing the*

work and how it will be done. It may require a declaration from those involved in a shift handover of the procedures or extensions to the work that is required to be done. Before process equipment or machinery is returned to service, it may also require a declaration from the permit originator that it is ready for normal operational use.

—Oxford Dictionary of Chemical Engineering

PTW or "safe work permitting" is a component of multiple regulatory requirements[3] affecting industrial workplaces. A work permitting process is included in regulatory and industry standards applying to the following types of hazardous work:

- Hot Work
- Confined Space Entry
- Control of Hazardous Energy (Lockout/Tagout - LOTO)
- Live Electrical Energy

As a critical component of various elements of process safety, poor execution of the PTW system can contribute to process safety failures. The Chemical Safety Board (CSB) cited a failed PTW as a contributing factor in the 2005 Valero Refinery Delaware City, DE accident and included PTWs as a part of its guidance on hot work (*CSB*, 2006 and 2014). In the case of the Delaware City Accident, the CSB reported that an employee participated in a high-risk activity, presumably to avoid filling out a permit to work for a confined space entry.

While there have been multiple case studies that evaluate human reliability applied to PTWs outside of the United States,[4] the HRSG did not discover any similar work from a US refinery in its literature review.

The HRSG reviewed the final assessment entries of the API Process Safety Site Assessment Program (PSSAP) to evaluate the site performance related to PTW systems. PSSAP is a third-party assessment program for refineries and chemical facilities created in 2010. PSSAP assessments use guidelines called Protocols to review a facility's management of process safety.[5]

Two PSSAP Protocols include specific questions about site PTW practices: Safe Work Procedures and Management of Change. PSSAP assessment data demonstrate trends in the application of PTW systems.[6] Assessment questions and scoring methodology pertaining to PTW are listed below and are available in Appendix A. The

[3] 29 CFR 1910.146, 29 CFR 1901.1206, 29 CFR 1910.252(a)(2)(iv), 29 CFR 1910.147, NFPA 70E

[4] See M. Jahangiri et al. (2015) and A. Ghasemi, F. Atabi, and F. Golbabaei (2015).

[5] PSSAP maintains Protocols related to Leadership, Operating Practices, Mechanical Integrity Fixed Equipment, Safe Work Procedures, Management of Change, Process Hazard Analysis, and Facility Siting. An additional Protocol assesses hydrofluoric alkylation processes in accordance with API 751. Forthcoming Protocols will assess facilities based on Incident Learning and Product Storage and Transfer.

[6] PSSAP data is blinded and aggregated to ensure anonymity of the site for which the individual assessments were written.

PSSAP data for these PTW questions shows that there is opportunity for improvement. Observed trends related to PTW are discussed in the following section.

4 Observations of the HRSG and Practice Sharing Resources

The HRSG identified examples of practices from sites that address observed PTW trends. The HRSG reviewed practices that address both systemic and individually driven human error. The observations of the HRSG were anecdotal, but trends from PSSAP data support the Subgroup's observations. Between their own site experiences and information from PSSAP data, the members of the HRSG identified six opportunities for improvement that may exist within the PTW system as executed by refining and petrochemical sites and compiled resources to address those opportunities.

The HRSG identified a total of 30 individual company practices that represent human reliability considerations on PTW systems. The practices collected were in use at one or more refining or petrochemical sites. The HRSG found that the selected practices offer potential enhancements to key aspects of the PTW system as utilized in the refining and petrochemical industries. The practices were categorized as follows:

- Establishing scope of activities managed outside of a PTW document – practices to manage some lower risk work through a potentially more effective fit-for-purpose process without compromising safety;
- Energy isolation (Lockout-Tagout) – practices to help ensure that sources of energy have been effectively isolated to avoid exposing workers to the hazards of energized / pressurized systems;
- Joint job walks – practices to help ensure more effective joint job walks in which key information is transferred from the permit issuer to the permit receiver, and from the permit receiver to the team executing the work;
- Safeguard verification and stop work triggers – practices to help ensure key safeguards are in place and in effect before initiating higher risk tasks, as well as identifying for discussion with the work team those foreseeable conditions/situations for which the work team is expected to "stop work" and seek further guidance from their supervisor;
- Permit-related auditing process – practices to verify that key aspects of the permitted work are effectively communicated to the work team, the permit process is executed as expected with verified safeguards, and the documentation associated with the permit is properly developed and closed out; and
- Permit issuer/receiver roles (e.g., competence, training, and execution) – practices to help ensure personnel performing these roles have a clear understanding of their responsibilities, and have the key competencies and experiences expected of the roles to help them be successful in carrying out their duties.

These practices were organized into Practice Sharing documents as shown in Figure 3.

Figure 3: Organization of HRSG Permit to Work Practices Sharing Documents

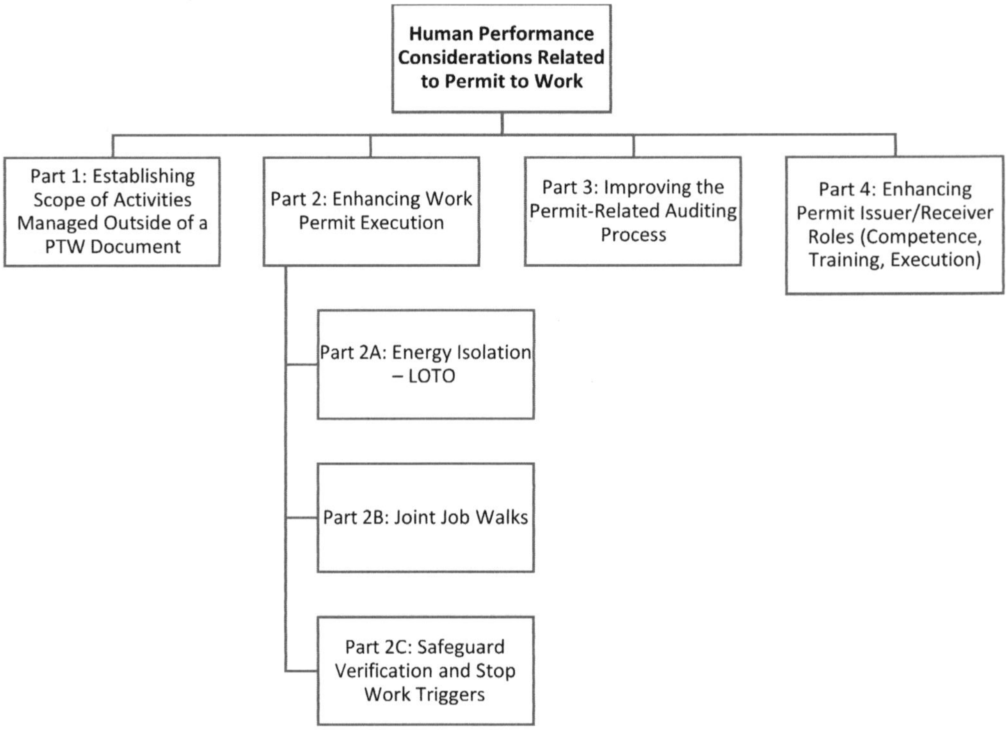

The Practice Sharing documents collected by the HRSG are intended for use by management, supervisors, and workers in both the process and mechanical disciplines that oversee the development and execution of the refinery or petrochemical site's PTW system.

4.1 Sites can optimize the use of permitting resources by ensuring a fit-for-risk (fit-for-purpose) PTW program.

Early in its review of the PTW system, the HRSG observed that there is potential value in narrowing the scope of the formal PTW process based on complexity and risks of performing site tasks. Members observed that the value of a PTW system may be reduced if permits are required for low-risk tasks because the permitting exercise could draw resources away from higher-risk tasks. Regulations require permits for some defined tasks, discussed above, but other PTW requirements are determined by site and company.

The HRSG determined that a way to improve the execution quality and human reliability of the PTW process is to consider the best way to manage PTW resource demands. Part 1 of the Practice Sharing series focuses on consideration of work activities that could be managed outside of a site's PTW document. Sometimes, PTW requirements can be time-consuming such that the effort to complete the PTW process is disproportionate to the risk involved in the work. Permit issuers, from their perspective, may sometimes perceive the need to rush to get the permit issued for work to begin. With often multiple permits to issue on any given day, it is possible to overburden available resources with formal permitting of low-risk tasks. Simply stated, by excluding well-understood-low-risk tasks from the formal PTW process, it is possible to devote more attention to the permitting of complex or higher risks activities.

Some companies have established practices that control a defined set of lower risk work activities through a managed process that is fit-for-risk, outside of the formal permit process. In doing so, they have the potential to free-up time to allow the permit issuer to place more emphasis on tasks with higher risk. A presumption in each of the examples provided is that a PTW is required for any task not specifically identified for hazard management via the alternate work process. Establishing the scope of activities managed outside of a permit to work document in an optimal manner has the power to transform the way work is performed to best manage risks.

Part 1 of the Practice Sharing series provides five site-specific company practices, summarized in the table provided in Appendix B. The practices contained in the document provide some explanation or system-wide trends in determining the scope of PTWs, but the key takeaway is that there is no universally accepted definition of which tasks truly require the completions of a PTW, and sites should consider the costs and benefits of alternate control methods when developing their PTW procedures.

4.2 While a permit system may be robust, it must be executed as designed to be effective.

PTW is a mature system in the refining and petrochemical industries. All companies that participated in the HRSG and all sites observed by the PSSAP assessment team have PTW procedures in place. Despite the maturity of the PTW system, permit-related errors still occur. These errors can largely be attributed to mistakes in the execution of the permit, rather than flaws in the related procedures. Failures in execution often provide the reasoning behind labelling poor PTW practices as a "pencil-whipping exercise."[7] Execution failures include failing to identify known hazards, mislabeling job requirements, or improperly logging a completed and closed permit. Another issue observed by HRSG members is that aspects of the PTW may not be fully communicated to the workers responsible for the work. While it is not uncommon for a foreman to be responsible for the PTW, there can be challenges in delivering the information on hazards, safeguards, and task requirements to workers.

Execution challenges can be identified by observant site assessors or review of a site's past PTW audits. Part 2 of the Practice Sharing series, divided into three subsections, directly addresses issues in PTW execution, providing tools that sites use to prevent or mitigate the potential for human error. Part 2 is divided into three

[7] This is a phrase that the HRSG and API PSSAP assessors hear and use frequently.

subcategories, each organized as their own Practice Sharing document: Energy Isolation – LOTO, Joint Job Walks, and Safeguard Verification and Stop Work Triggers.

Part 2A provides tools to enhance execution of energy isolation/LOTO PTWs. The document includes five site-specific practices. Some of the sites surveyed by the HRSG go above and beyond the LOTO PTWs required by regulation. The sites have created more thorough work permits for electrical distribution that require the participants to completely understand all segments of the PTW. To ensure that this is the case, sites verify knowledge of specific permitting processes through certifications and examples. Some sites have found that the best way to communicate the expectations of a correct LOTO PTW is through visual aids.

The HRSG identified the Joint Job Walk (JJW) as an opportunity to improve the execution of PTWs. A JJW is required by PTWs for several tasks, such as LOTO or other forms of energy isolation. Like the larger challenge of correct "execution," JJWs present an opportunity for human error if they are considered a mundane or unnecessary task. Part 2B of the Practice Sharing series provides resources specific to enhancing the JJW portion of PTWs. The Practice Sharing document includes three site-specific company practices. These practices are tools intended to make the JJW portion of the PTW process more robust. Some of the practices are widely accepted practices, such as the inclusion of an Energy Isolation Log with completed PTWs. In another practice, a set of questions help guide the JJW during a PTW. The final shared practice applies the requirement to perform a Job Loss Analysis (JLA) for JJWs. In reviewing these shared practices, sites may identify helpful tools to apply in their own JJW procedures, such as using a color-coding system to differentiate hazards listed on a JLA.

Failure to recognize changing conditions can result in a failure to stop work when necessary. Even in a PTW process that is otherwise executed correctly, a changing condition can create significant challenges. Changing conditions can include a wide variety of situations, such as equipment malfunctions, adjacent work, or a leak in an unrelated area. The workers completing work need to recognize these changes, secure the work, and contact the permit issuer to reconsider the hazards before work progresses. Part 2C of the Practice Sharing series directly addresses safeguard verification and stop work though many other elements in the series can help a site build a PTW system that responds to change as well. The Practice Sharing document includes five site-specific company practices. While PSSAP assessors have observed that all sites use checklists of various sorts, many sites do not apply the same format to checklists across various PTW operations, which can lead to confusion and human error.[8] The Practice Sharing document includes a sample safeguard verification checklist for higher consequence tasks. The checklist is a helpful tool for workers to verify the health of safeguards at the beginning and during work. Some sites prefer to expand their safeguard verification steps beyond paperwork reviews. Rather than simply acknowledging them on a permit, some

[8] While not a significant reason for sites to score below average on site assessments, PSSAP assessors cited this challenge frequently in their observations on Safe Work permit-related questions.

sites chose to have a face-to-face conversation about potential Stop Work Triggers to ensure that workers fully understand their opportunities and obligation to stop work if necessary. In another practice, a site deploys a separate review team when a stop work (or time-out) is triggered, this team provides a "fresh set of eyes" to help the work team review the process and determine a path forward. Part 2C clearly shows that there are many different methods to confront human error in executing PTWs that include safeguard verification and stop work triggers.

4.3 An effective auditing process verifies conformance with the PTW process and provides for continuous improvement.

Both the personal experiences of the HRSG members and review of PSSAP data suggest that sites have a great opportunity for improvement in their PTW audit systems. Of the PSSAP questions related to PTW, the questions related to audits and the development of metrics to measure correct completion of PTWs represent the two lowest scoring categories among sites assessed by the program. PSSAP assessors review the sustainability of audit programs as well as the timing and frequency of audits. Sites that score poorly on audit-related questions may have incomplete audit programs; requirements that only a limited subset of site employees may contribute to audits; or inadequate review of final audit metrics. Sites that use PTW metrics, including audit results, to produce lagging and/or leading indicators of program health tend to score well on the audit and metrics-related questions. Sites that meet or exceed PSSAP questions have robust audit programs that create statistically significant metrics that are then reviewed and used to produce continuous improvement.

Like the PTW process, auditing is a mature process in the refining and petrochemical industries. The errors that arise in executing PTW audits are frequently attributed to human error, rather than inadequate procedures. However, some sites have implemented audit procedures and tools that address the potential for human error when conducting, reviewing, and presenting PTW audit results. Part 3 of the Practice Sharing series provides tools to enhance the PTW audit process.

As the above discussion of limitations of some PTW audit processes indicates, audits can be performed too infrequently, or the audit process can be incomplete or otherwise insufficient. While the HRSG did not develop guidance on audit sample size or frequency, it thoroughly addressed how PTW audits can be structured, potentially reducing opportunity for systemic human error. Part 3 of the Practice Sharing series is focused on improving the audit process around the PTW process. The Practice Sharing document contains six site-specific company practices. The tools provided in the practice can make a PTW audit more valuable by asking the right questions and providing the most helpful tools to auditors. The document begins with an example of a PTW audit program. Another practice is a reference in the form of a pocket card for use in the field during an audit. Finally, as when completing a PTW, it is important to identify the roles and responsibilities in an audit process. The HRSG collected an example of a roles and responsibilities list for a PTW audit. To address the challenge of continuous improvement, the HRSG collected a sample audit protocol that can serve as a benchmark for other sites conducting audits.

4.4 PTW systems must clearly define and communicate roles and responsibilities and provide a means to verify the competency of individuals involved in the PTW process.

Finally, the HRSG observed the importance of defining and communicating the roles and responsibilities in the PTW process. PSSAP Protocol questions specifically address this issue. Sites that score well on the questions have defined, communicated, and sustainable role and responsibility information contained in the PTW systems.

Part 4 of the published Practice Sharing series offers sample practices used to help sites define roles and responsibilities in a PTW system by addressing competence, training, and execution. The Practice Sharing document contains eight site-specific comprehensive company practices, broken into subparts to make for 32 individual practice elements (including checklists, forms, and comprehensive procedures). The practices represent components of formal site or company-wide PTW processes. The processes set standards and expectations for workers who issue and receive permits, verified through certification processes. The surveyed sites provided additional layers of verification to ensure quality in execution of PTWs. The HRSG determined that such personnel competency can improve the PTW process beyond that which a simple review of the layout and contents of PTW forms can provide. The practices cover experience and certification requirements for those authorizing work ranging from simple tasks that do not require a formal PTW document up to those tasks involving Highly Hazardous Entry (HHE) and/or special hot work permits.

In the sampled practices, roles and responsibilities are clearly defined. This is a quality of PTWs that is fundamental to their success. Some sites surveyed by the HRSG found that including hands-on training can increase competency, so such training programs are provided as sample practices in the document. Another observation made by the HRSG was that there can be different levels of certifications for different responsibilities within the PTW system. Accordingly, a sample practice that includes these levels of certification is provided in Part 4 of the published Practice Sharing series.

5 Accessing the Permit to Work Practice Sharing Documents on the AFPM Safety Portal

Practice Sharing documents are meant to share information on process safety practices in order to help improve process safety performance and awareness throughout industry. The goal is to capture and share knowledge that could be used by companies or sites when developing new process safety practices or improving existing ones. The practice being shared has been used by an industry member, but this does not mean it must be used or that it will produce similar results at any other site. Rather, it is an option to consider when implementing or adjusting programs and practices at a site.

These tools are only available to AFPM member companies and associate members via the Safety Portal.[9]

6 References

[1] *DOE-HDBK-1028-2009, Human Performance Improvement Handbook, Volume 1.* US Department of Energy. (2009). Available at https://www.standards.doe.gov/standards-documents/1000/1028-BHdbk-2009-v1/@@images/file. Accessed January 27, 2019.

[2] Conklin, Todd. *Pre-Accident Investigations: An Introduction to Organizational Safety*. 1st Edition. CRC Press. (2012).

[3] API RP 754 Data, Tier 1 and 2, Refining and Petrochemical. (2016).

[4] F. De Felice, A. Petrillo, A. Carlomusto, and A. Ramondo. Human reliability analysis: a review of the state of the art. IRACST – Int J Res Manage Technol (IJRMT), 2 (2012), 35-41.

[5] *A Dictionary of Chemical Engineering*. Oxford University Press. (2014).

[6] "CSB Issues Case Study in 2005 Valero Refinery Delaware City, DE, Accident; Report Notes Inadequate Nitrogen Asphyxiation Hazard Awareness Training and Improper Confined Space Rescue Actions." *CSB.* (2006). Available at https://www.csb.gov/csb-issues-case-study-in-2005-valero-refinery-delaware-city-de-accident-report-notes-inadequate-nitrogen-asphyxiation-hazard-awareness-training-and-improper-confined-space-rescue-actions/. Accessed January 27, 2019.

[7] "CSB Emphasizes Existing Resources Available on Hot Work Safety." *CSB.* (2014). Available at https://www.csb.gov/csb-emphasizes-existing-resources-available-on-hot-work-safety/. Accessed January 27, 2019.

[8] M. Jahangiri, N. Hoboubi, A. Rostamabadi, S. Kshavarzi, and A. Hosseini. "Human Error Analysis in a Permit to Work System: A Case Study in a Chemical Plant." *Safety and Health at Work.* 7, 1, (2016), 6-11.

[9] Ghasemi, F. Atabi, and F. Golbabaei. "Human Error Classification for the Permit to Work System by SHERPA in a Petrochemical Industry." *Journal of Occupational Hygiene Engineering*. 2, 3 (2015) 56-73. (Farsi)

[10] AFPM Safety Portal. Available at https://safetyportal.afpm.org/. Accessed on January 27, 2019.

[9] Contact the American Fuel & Petrochemical Manufacturers if you have trouble accessing the Safety Portal.

7 Appendix

7.1 Appendix A: Process Safety Site Assessment Questions Related to Permit to Work

Sites are scored from 0 to 4, with a score of 3 indicating that a site is meeting the question, and a score of 3.5 or 4 indicating that a site is going above and beyond.

PSSAP Protocol 4: Safe Work Procedures

Protocol Questions
04.01.01.01 Work permit procedures are in place for: o Hot work. o Hot taps, stopples, welding on in-service process equipment, etc. o Energy isolation/LOTO o Confined space Inert entry. o Confined space entry. o Excavation procedures (e.g., to prevent cutting into underground lines or piping). o Crane operations & lifting (e.g. to prevent damage to process equipment). o Live flare work. o Vacuum Trucks o Use of freeze plugs or equivalent as a method of isolation o Line break and vessel opening o Other unique "higher hazard" work permit procedures
04.01.01.02 Work permit procedures are audited: o Paper audits are routinely conducted. o Field audits are routinely conducted. o Results of audits are used to update procedures. o Results of audits are communicated to the work force. o Permitted jobs are executed in compliance with all permit requirements o Direct feedback is provided to inform individual workers and contractors of any deficiencies noted during audits of their areas.
04.01.01.03 Metrics are developed to evaluate field execution of safe work permit procedures and permitted jobs to effectively measure correct completion and execution of safe work permits such as: o % of safe work permits completed correctly o % compliance with safe work permits based on field audits of permitted jobs o % of safe work permits where JSA/TRA hazards and mitigation measures are included in the work permit requirements o Number of incidents/near misses related to failure of work permit requirements o Number incident investigation where inadequate work permitting was determined to be a contributing factor/root cause o Permit issued without proper approval o No gas test (initial or subsequent) conducted for hot work or CSE o Number of permits closed out without signatures
04.01.01.04 Work permit procedures include: o Preparation of detailed job packs where appropriate.

o Gas testing where required. o Joint site visits by operations and maintenance where appropriate. o Risk assessments, JSAs/TRAs, or other methods of evaluating the risks associated with the permitted job. o Cancelation of permits under emergency situations. o Work continuance across multiple shifts if the site allows permits to endure across more than one shift. o Discussion of the toxic or flammable properties of involved process streams.
04.01.01.05 Work permit procedures clearly define (including signature requirements and authorities). o Roles and responsibilities for personnel preparing permits. o Roles and responsibilities for personnel issuing permits. o Roles and responsibilities for personnel conducting gas testing. o Roles and responsibilities for personnel receiving permits. o Roles and responsibilities for operations personnel whose area the permit applies to. o Other workers doing the permitted job.
04.01.01.06 Special considerations are given to: o Multiple jobs at the same time in adjoining areas. o Limiting the number of higher risk jobs in a single time frame based on resources. o Limiting number of non-essential people in plant areas when higher risk work is conducted.
04.01.01.07 Systems are in place to correct/revise or develop safe work procedures, safe work permit/field audit checklists and/or training based upon o Developed metrics for the safe work permit procedure process o Incidents or Near Misses o Incident investigation root causes and recommendations and/or PHA recommendations o Changes in regulatory requirements

PSSAP Protocol 5: Management of Change

Protocol Question
05.01.02.05.04 Are Safe work practice and work permitting procedures addressed in the MOC process?

7.2 Appendix B: Permit to Work Classification Matrix

Complete company practices can be found in Part 1 of the "Human Performance considerations related to Permit to Work" Practice Sharing series, available on the AFPM Safety Portal.

Tasks marked (X) require a formal PTW, and unmarked tasks require an alternative control method such as a verbal authorization and/or joint job walk.

Task Category	Task	Co. 1	Co. 2	Co. 3	Co. 4	Co. 5
Cold Work	Cold Work on new equipment without connection to process	X				
	Cold work on proven low risk isolated equipment and on new equipment without connection to process		X	X		
	Cold work on proven Low Risk Isolation Option isolated equipment	X				
	Cold Work, not in process area or on process equipment		X			
Hot Work	Hot Work in an open area classified as non-hazardous (e.g. fabrication work in designated fabrication area)		X	X		
	Maintenance activities (inc. hot work) taking place in a permanent shop building and documented in a maintenance procedure		X	X		
	Set-up work for Supplied Breathing Air/Hot Work				X	
Mechanical	Equipment health monitoring and analyzer duties consisting of walk-throughs, inspections, viewing/recording readings, verifying validation results, flow adjustments, calibration and minor repairs which would not affect process control.	X	X	X	X	
	HVAC service rounds: filter replace, visual inspection, thermostat check, inlet/outlet temp split check, cleaning coils	X		X		
	Machinery lube oil samples	X				
	Pump or motor preventive maintenance		X	X		
Electrical	Area lighting replacement, light bulbs change (not repair work) when reenergized; PM on emergency lights or egress/exit lights	X	X	X		
	Electrical and instrument troubleshooting, UPS battery checks, resetting breakers	X	X	X	X	
	Electrical/Power Distribution - Substation visual walkthrough inspection			X		
	Field instrument work, including breaking containment (if meets Low Risk Isolation		X	X		

	Option requirements; Safety Instrumented System Override Risk Assessment required)					
	Running conduit and pulling wire (non-process connected equipment where no LOTO is required and not working at heights)	X	X			
	Turnaround (TAR) or Ring-fenced Projects - Electrical work, once isolated and verified de-energized, cold work, setting up or removing temporary screwed piping, not tied to process		X			
	Use of battery powered, "intrinsically safe" tools, or equipment “rated/electrically classified” for the area being used, tools and/or equipment.		X	X	X	X
Measurement	Asbestos - Sampling of Asbestos Containing Material (ACM) by a qualified person			X		
	Fugitive Emissions 1st attempt at leak repair	X	X			
	Gauging / Sampling tanks	X				
	Hydro testing, outside process areas		X			
	Integrity testing, repair or alteration			X		
	LDAR (fugitive emissions) - VOC monitoring from grade or fixed, permanent platform or deck, ladder or approved scaffold.	X		X		
	Leak testing, hydro testing, outside process areas		X	X		
	Steam trap survey and steam/tracing work, not tied to process		X	X		
	Vibration testing on rotating equipment			X		
	Visual or ultrasonic testing on piping, vessels and fixed equipment, using intrinsic tools and equipment.	X		X		
Safety	Chemical deliveries outside of classified/process areas, including fuel deliveries			X		
	Industrial Hygiene exposure monitoring and sampling	X		X	X	
	Insulation or Fireproofing work - no known leak	X	X			
	Manual transport of high-pressure cylinders by hand or dolly (when Vehicle Entry is not required)	X				X
	Overhead crane inspections and PM	X				
	Safety Reviews, field observations, job site visits, audits, inspections, validating drawings / prints / procedures, taking measurements, estimating costs	X		X	X	
General	Animal control activities		X	X		
	Applying Herbicides to Vegetation using hose reel if truck is not driven inside firewall or inside a unit					X

	Chemical service suppliers performing routine duties and monitoring of cooling towers	X		X		
	Commercial building type work (such as drywall, flooring or carpet laying, moving furniture etc.)		X	X		
	Connecting or disconnecting vessels at the docks				X	X
	Digging with a hand shovel (non-excavation digging), pouring concrete, digging, erection of structural steel, hand threading bolts, use of steel claw hammer				X	X
	Drain cleaning with hose (not rodding)			X		
	Environmental recovery well work outside of process areas, not entering Waste Management Unit			X		
	General housekeeping activities	X	X	X	X	X
	Hand cleaning equipment using wire brush – Not for potentially lead coated surfaces	X	X	X		
	Hand threading bolts				X	
	Janitorial / Delivery				X	X
	Job preparation: Set-up work – Moving / positioning tools, materials and/or equipment, erection of scaffolding (only if hazardous energy is not involved)	X				X
	Marking by aerosol can, piping layout, drafting/sketching/hang tags				X	
	Painting and hand cleaning equipment at grade with brush or roller, except on live process or on hot surfaces					
	Phone system and computer installation and repair	X		X		
	PM / Repair / Replace access control room door hardware, barrier arms, cameras, badge readers and components	X				
	Set up at grade or fixed platforms without lifting tools and equipment overhead			X		
	Vacuum storm water and sanitary sewers (including servicing of sanitary holding tanks and portable toilets)	X				

Conducting Quality Process Safety Management Audits - Globally

Tekin Kunt, Carolina Del Din, and John Michael (Mike) Munsil[1], PSRG Inc., Houston, TX

Abstract

Process safety regulations are present in many countries and, although all of them focus on maintaining the safety of workers, the public and the environment their implementation is achieved in different ways. Further, there are a host of NGO-based requirements that are incorporated by reference. Complicating this is the fact that most global companies require the use of a 'corporate language' that may not be well understood at the local level. Added to this is the problem of a local safety culture that does not support a performance-based standard and the lack of adequate support at the local level in implementation and development. However, there are practices that can be applied at the Corporate and Local levels and by the auditor, that can minimize the difficulties and support the implementation and continued development of performance-based process safety management.

Situation

Process safety regulations vary from country to country. Although all these regulations focus on maintaining the safety of workers, the public, and the environment, their implementation differs and the quality of their enforcement varies greatly between countries.

The United States of America

In the United States Process Safety Management (PSM) is a regulation administered by the U.S. Department of Labor's Occupational Safety and Health Administration (OSHA). Its major object is to prevent unwanted releases of hazardous chemicals especially into locations that could expose employees and others to serious hazards. In addition to the OSHA PSM Standard, the USEPA Risk Management Plan (RMP) Rule implements Section 112(r) of the 1990 Clean Air Act amendments. RMP requires facilities that use extremely hazardous substances to develop a Risk Management Plan. Just as for OSHA PSM, some states have also been granted delegation of EPA's authority to implement and enforce the risk management program rule. While PSM is an occupational health program intended to protect workers, RMP is intended to protect the environment and the community.

---

[1] Presenting Author

Europe

In Europe, the Seveso Directive covers establishments where dangerous substances may be present (e.g. during processing or storage) in quantities exceeding a certain threshold; its object aims to protect both people and the environment. Since this is a directive, it requires member states to achieve a particular result without dictating the means of achieving that result; this means the directive lays down certain results that must be achieved but each Member State is free to decide how to implement it. Further, in Europe compliance is required with the ATEX directive, consisting of two EU directives describing what equipment and workspace are allowed in an environment with an explosive atmosphere. The UK has unique process safety regulations/agencies including the Control of Major Accidents Hazards (COMAH) regulation and the Health & Safety Executive.

These are only a few examples of how process safety regulation is differently implemented all over the world.

Non-governmental Organizations and RAGAGEP

In addition to the regulations implemented by the countries, states and, local agencies, there are guidelines that have practical application within industry. These guidelines are developed by non-governmental organizations (NGOs) committed to supporting process safety in industry.

CCPS (the Center for Chemical Process Safety) is a not-for-profit, corporate membership organization within AIChE (American Institute of Chemical Engineers) that identifies and addresses process safety needs for a variety of facilities involved with handling, storing, using or processing, and transporting hazardous materials. CCPS brings together manufacturers, government agencies, consultants, academia and insurers to lead the way in improving industrial process safety.

API is another organization that supports process safety. The mission of API (American Petroleum Institute) is to promote safety across industry. Although its focus is primarily in the United States, its work has expanded to include a growing international dimension.

These organizations develop consensus codes that are widely used by certain PSM-covered industries.

Existing industry codes and standards constitute the large majority of information comprising RAGAGEP (recognized and generally accepted good engineering practices). Therefore, when auditing global companies, the audit team must keep in mind that the scope of the process safety management audit may be considerably wide, needing to verify the conformance to local regulation, to RAGAGEP and to the corporate standards.

Likewise, the need to verify compliance with multiple regulations and standards often results in a requirement to prepare multiple reports. This is due to the fact that the reports may be submitted to enforcing agencies that require a fixed table of contents or differing sets of information.

When the safety management system is completely supported, in particular by the corporation, with clarification of concepts and, above all, promoting safety by funding investments, both the management and the employees are invested in the safety management system and in its compliance. However, without support, the facility rarely embraces the management system and instead it tends to only comply with local regulation.

Complications

Complying with process safety requirements can be challenging even without the added complications that are inherent in any application of a standard (or multiple standards) on an international scale. The authors have observed the following additional complications.

Multiple Languages

It is not uncommon for a multinational corporation to standardize on one language (often, but not always English) for official documentation and sharing of information. Although this may be necessary it often causes complications at the local level. As you would expect we have found that requirements not expressed in the local language are not as well complied with, or understood, as they might otherwise be. As a result, it is not uncommon for audit findings to be issued that arise from simple confusion or translation issues related to "corporate-speak" in the official language. The base complication here is that corporate requirements are issued as the transmission of information rather than shared communication.

Cultural Issues

The authors find that the local Safety Culture is not generally well-developed. Local staff tend to be more comfortable with prescriptive regulation than performance-based regulations. As a result, in the absence of a specific checklist, actions are either not performed or do not meet the intent of the performance standard.

As a further complication, we find that the local staff may reject an international standard out of hand as not applying to them. Although our experience indicates that the vast majority of local staff are well-intentioned and desire to comply, performance-based standards simply confuse them. When you add in the simple fact that many international standards are in part based upon a support infrastructure that may not exist locally, there is a tendency to passively resist implementation rather than aggressively seek compliance.

Somewhat in line with the tendency for local staff to be more exposed to prescriptive regulation rather than performance-based regulation, we find that training verification is generally not competency-based. That is, training competency is supported by attendance sheets rather than by tests or other competence verification.

Inadequate Local Support

It is our common finding that original implementation of process safety at the local level either is sporadic or patchy, or has been accomplished by third-party experts. As a result, the implementation has been emplaced without due consideration for the local support and infrastructure necessary to assist the process.

Complications that arise as a result include that:

- There is often little to no training for the local "competent person".
- There are no internal (or local) Subject Matter Experts available to support the implementation and continuing evolution of the system.

As a result, we commonly find that management of the process safety system is not hands-on. Instead, it either does not exist in a functional manner or is a token implementation that exists solely in response to external stimuli, such as audits.

Our base conclusion here is that process safety at the local level is commonly not well-supported by the global corporation and that accountability is not directed to the source of the complication, the corporate body responsible for global implementation and support.

Resolution - Best Practices in International Auditing

Introduction

Based on complications arising from the current status of PSM audit for global companies, we have observed the following best practices. These best practices either alleviate complications mentioned in the previous section provide the necessary steps to not add any further complications. We have divided the proposed best practices in three area based on the ownership: corporate, local site, and external auditors.

Corporate

In conducting Process Safety Management (PSM) audits for a global corporation, there are a few best practices that global organization can implement to make these external audits a success.

Refer to corporate standard rather than country-specific (OSHA/SEVESO)

One area that a corporate PSM organization need to do before conducting PSM audits is to set up the protocol i.e. the standards that will be verified against the current practices at the local sites. The protocol should aim for a performance-based auditing instead of compliance-specific. Instead of using a simple checklist, it should describe how the performance will be improved in the system and set up the audit protocol based on this guidance.

Since the local sites might be subject to different country-specific standards as well as overseen by local governance, it is imperative that the global organization sets up its corporate standards and the other rules that will be followed when checking against performance of a local site. These standards need to be published ahead of time and already translated/explained in the local language that a site is subject to. There are different ways of coming up harmonized standards for a corporation although they might not be easily adopted by a local site. The best way to address this shortcoming is to involve PSM leaders from each country that a corporation operates in.

Regional PSM Manager

Since the Regional PSM Manager is responsible for multiple countries in a geographic region, he/she is critical in conveying the concerns and needs of the region to the corporate PSM organization as well as transmitting corporate standards to the region effectively.

Internal Stakeholder

Region PSM Manager is both an internal stakeholder representing the local site as well as an external enabler for the local site to understand and meet global standards. As such, we believe that it is best practice to have the Region PSM Manager to be involved in all phases of the external audit process including the on-site visit at the local plant. Although keeping the balance between internal needs and external demands is difficult, it is an important part of the learning process for both the local site as well as the global PSM organization in a company. As external auditors, it is critical to identify and work closely with the Regional PSM Manager. If this role has not been established effectively or not supported properly, this fact should be taken into consideration as a possible finding in an external audit process.

Emphasis on visual/graphical communication rather than text-based communication

Another area of conducting effective global audits is communication. Communication in different geographic areas should rely more on visual aids than verbal/written communication assuming that the local site is in a country where English is not the de-facto communication language. Even when the language has not been perceived as an impediment, it is well accepted that "a picture worth thousand words".

Operating procedures with pictures

For instance, more and more we observe local sites documenting their important operational procedures in pictures so that anyone can follow them with ease. This practice also cuts the time needed for training both operators and engineers.

Video-based training

Another important tool available for global PSM organization is to prepare and disseminate global standards through video-based training. Video-based training, once it is done with a global scope in mind, cuts down the time needed to provide timely support to the local sites. Again, if these video-based training contents are prepared using more visual cues/information instead of text then it removes the need to translate the materials to multiple languages.

Proper training support should be provided by the global organization so that the local site can receive ongoing distance-learning/training so that audits as well as other PSM activities can be delivered remotely and effectively.

Global Tools

One area of deficiency that we have observed in global audits is the lack of continuing hazard recognition and sustained risk analysis processes. The corporate PSM team can help local sites by providing global tools that are supported locally.

These global tools are especially important for proper action tracking and resolution of not only audit findings but also all PSM related activities discovered during risk analysis, MoC, PSSR, and incident investigation. Without a global tool with proper local support, tracking action items become a bottleneck leading to repeat findings in subsequent audits. It is important that a unified action tracking system is provided to the local site so that all PSM-related actions exist in one location.

- Maintenance scheduling and tracking
- Asset tracking and integrity system
- Distance learning/training
- Action tracking and resolution
- Performance-based auditing
- One tracking tool for PSM-related actions

Local Site

Local site should treat an external audit as a learning opportunity as well as a broad benchmarking tool against similar sites in the corporation and as well as the industry in general. There are few best practices observed over different sites in a corporation and multiple corporations.

Use of RACI matrix

One area of difficulty for a local site as cited before in the paper is that the perception of audit taking away valuable resources in the day-to-day operations. Moreover, it has been observed that the same perception exists over other PSM elements. In the light of these challenges, usually one person in the local-site has been appointed as the PSM responsible person. This person usually does not have all the tools, training, background, and/or resources to address all PSM elements effectively. One way to address this challenge is to set up a RACI (Responsible/Accountable/Consulted/Informed) matrix where there is a different person (department) responsible for one of each PSM element. For instance, Operations can be responsible for the Workforce Involvement element in PSM. This clear division of responsibilities can make PSM supervisor's job more manageable.

Skill matrix-based training and Competence Verification

Another best practice that a local site should employ is development and implementation of skill matrix-based training and competence verification. When the local site does not have a clear map of competence building, it is very difficult to address findings from audits. These competency maps should be accompanied with a well-defined competence verification step so that the site can assess the gaps in knowledge and address them based on the training map.

Third-party audits by global auditing company with local sensitivity

Since the local site most often pays for the external audit expenses, it is very important for the site to work with an external audit services provider with global resources but local capabilities. For instance, having external auditors with a firm understanding of local rules/regulations as well as local customs and possibly language. In cases where multiple external auditors are engaged, it is a good practice to have at least one of them be proficient with the local culture.

The Auditor

External auditors should allow for adequate preparation time before visiting the site for a PSM audit. These preparations should include reviewing site's known hazards and prior audit reports if applicable. Having a pre-audit call with the site is also a good practice. These preparations before conducting the site visit allows auditors to prioritize and allocate their time to most pressing issues. It also allows them to review as much documents as possible before being on-site so that the time while visiting the site can be spent on interviewing everyone involved in PSM elements as well as facility observations.

Once on-site, auditors should never stop asking questions. All answers should be trusted but also verified by observations as well. During the on-site visit, auditors should never stop communicating with site personnel even when they are reviewing documents. It is a best practice to go over the documents with the relevant personnel so that both the intent as well as execution of PSM elements can be evaluated.

As part of these ongoing communications with on-site personnel, one best practice that our team found very useful is to have daily briefings at the end of the day. In most cases, the site manager is interested in these meetings so that he/she can find out where things are in terms of the audit as well as provide important input that can speed up discovery of related issues. The daily briefings conducted at the end of each day are also useful for auditors to verify their findings. In some cases, the findings might not be correct due to misunderstandings, especially where different cultures/languages are involved. These daily briefings also make the Exit Meeting where all the findings are discussed much easier since most of the findings were already shared with the on-site team.

19 AIChE Spring Meeting
15TH GLOBAL CONGRESS ON PROCESS SAFETY
NEW ORLEANS • 2019

High performance in Process Safety – Balanced Approach

Coen van Driel
TÜV SÜD Process Safety
Mattenstrasse 24
CH-4002 Basel - Switzerland
Coen.vandriel@tuev-sued.ch

Prepared for Presentation at
American Institute of Chemical Engineers
2019 Spring Meeting and 15th Global Congress on Process Safety
New Orleans, LA
March 31 – April 3, 2019

High performance in Process Safety – Balanced Approach

Coen van Driel
TÜV SÜD Process Safety
Mattenstrasse 24
CH-4002 Basel - Switzerland
coen.vandriel@tuev-sued.ch

Keywords: high performance, Process Safety Competency, Leadership, balance, excellence in performance organization, culture, risk management, human factor

Abstract

A high-performance process safety culture in which people are actively aware of where they are related to best practices goes beyond compliance. It is a culture where there is a continuous challenge of the current procedures and behaviors by trying to identify blind spots. A culture in which competency development is not seen as a cost but as core to make the experience driven journey to high-performance culture a success. A culture in which balance leadership creates an environment in which proactively the risks are managed. Leadership keeping the balance, focus, rhythm, and discipline. This committed culture has risk based thinking at its core.

1 Introduction

The industry can realize technically advanced plants and major innovation steps, but realizing a continuous high performance in process safety remains a challenge for all stakeholders & organizations.

The silo mentality, approaching safety from an occupational safety perspective, or an asset integrity/reliability perspective, or a process safety (process safety and asset integrity do deal not only with safety but cover health, safety, environment and security issues with the primary goal to avoid loss of containment) perspective, is one of the reason for this challenge. For example, the incident "Deep water Horizon" was caused by this silo mentality in which the stakeholders had the perception that the operations were safe but unfortunately it resulted in a major incident. The operations were safe from an occupational statistics perspective but there were major gapes in the process safety and/or asset integrity processes.

Another aspect for causing incidents in the industry is the continuous drive to improve the financial performance of the organizations. The drive for continuously improving performance should be at the core of the culture of the organization. But due to the financial markets to focus is on financial performance which can result is less investment in for example maintenance, safety, and integrity of the plants.

And with the advanced plants, the chemical production becomes more complex, the need to drive continuous improvement in process safety — not just for industry, but for all stakeholders — becomes increasingly urgent. Continuous improvement cannot only be realized with better rules and regulations resulting in procedures (management systems). The success of continuous improvement lies in the hands of the behavior and competences of the people in the organization. Despite rigorous rules and regulations, severe incidents continuous to happen. Is the focus on compliance out of balance to improve?

An example that perfectly shows that the focus on compliance is out of balance is the challenge of climbing a summit like the Mont Blanc which rises 4,808 m (15,774 ft.) above sea level.

Nowadays the summit is ascended by an average of 20,000 mountaineer-tourists each year. It could be considered a technically easy, yet arduous, ascent for someone who is well-trained and acclimatized to the altitude. From l'Aiguille du Midi (where the cable car stops), Mont Blanc seems quite close, being 1,000 m (3,300 ft.) higher. But while the peak seems deceptively close, La Voie des 3 Monts route (known to be more technical and challenging than other more commonly used routes) requires much ascent over two other 4,000 m mountains, Mont Blanc du Tacul and Mont Maudit, before the final section of the climb is reached and the last 1000 m push to the summit is undertaken.

Each year climbing deaths occur on Mont Blanc, and on the busiest weekends, normally around August, the local rescue service performs an average of 12 missions, mostly directed to aid people in trouble on one of the normal routes of the mountain. Some routes require knowledge of high-altitude mountaineering, a guide (or at least an experienced mountaineer), and all require proper equipment. All routes are long and arduous, involving delicate passages and the hazard of rock-fall or avalanche. Climbers may also suffer altitude sickness, occasionally life threatening, particularly if they do not acclimatize to it. [1]

The mayor of one of the towns near the Month Blanc is responsible for the wellbeing of all the people on the mountain and the safety performance. Due to the many incidents, he is requesting more stringent rules and regulations for the safety of the climber's. The guides that lead groups of climbers to the summit must follow the rules and regulation, as well as the people that do the attempt on their own. As in the industry, if something happens with one of the climbers the guide must defend his responsibility for taking the action in court. Despite that there are rigorous procedures in place and compliance is enforced many, incidents and casualties are counted for.

In the example is spoken about compliance but what is meant by compliance. The definition of compliance is: the fact of obeying a particular law or rule, or of acting according to an agreement [2]. And according to the CCPS definition of "compliance with standards" is: *the processes for maintaining adherence to applicable standards, codes, regulations, and laws (standards), the attributes of a standards system, and the steps an organization might take to implement the standards element. Standards is a system to identify, develop, acquire, evaluate, disseminate, and provide access to applicable standards, codes, regulations, and laws that affect process safety.* In conclusion knowledge of and conformance to standards can help a company (1) operate and maintain a safe facility, (2) consistently implement process safety practices, and (3) minimize legal liability. How can it happen that compliant companies still have incidents and not achieving high performance?

The pitfall of focusing on compliance is that the risk on incidents can increase because people do not think anymore but execute tasks. Another pitfall of too must focus on being compliant is that the system becomes a goal instead of a means for improvement. Both pitfalls show that there is more to become successful in process safety than being compliant.

In the example of the Mont Blanc the same mayor show that the culture of climbing the Mont Blanc is changed and more and more driven by money greed of the guides. Also, the perception of the climbers has changed from seeing the Mont Blanc as a challenging mountain to climb to an easy mountain to walk up, which is influenced by declining competencies of climbers and guides. To improve the safety performance on the mountain also has a cultural or human aspect to it. So, the safety performance cannot improve by only focusing just on the system and compliance but also requires attention to change the mindset of the climbers (human factor). There is a need in changing the trend around the Mont Blanc as well in the chemical industry to get on a journey to high-performance in these settings.

An interesting inside is the vision of Viktor Roggeveen [3] in which he expresses that safety does not exist but risk's does. Risk is something tangible that can be measured to continuously improve. The core of high performance in process safety is managing risks to prevent incidents from happening. In a high-performance process safety culture, there a balance between process safety management system and the human/culture dimension which is enabled by leadership. In such a culture the organization makes a step beyond compliance. Focus on compliance gives a culture in which people in the organization forget to think because of blindly following the processes and procedures. This increases the risk on incidents happening. Leadership at all levels in the organization can achieve success by keeping the balance between management system and human factor creating a high-performance culture.

An essential element in the journey to a high-performance process safety culture is competency management. In other words, having and continuously working on the right and balanced mix of competencies at all levels of the organization. Realizing that right and balanced mix of competencies is an experience driven transformation journey resulting in a high-performance process safety culture.

2 Process Safety Excellence in performance

Understanding what is meant with high performance is a pre-requisite for success and this is further explained in this chapter. It's also critical to monitor the leading indicators to operational excellence because they provide a window into the pace of improvement within your organization and whether that pace is sufficient to either build or sustain a high performance in process safety. The third aspect in is finding the right rhythm and discipline to continuously improve.

2.1 High-Performance defined

In the business world and in the chemical sector, there is continuous strive for improving performance. Like great athletes, even high-performing organizations must continuously strive to improve and "work on their game." Without the passion for improvement, they are unlikely to remain high performers for long. After all, there's no shortage of ambitious business leaders who are taking notes on their fellow high performers and working hard to ensure that their own companies eventually reach the top echelons of organizational excellence [4].

In the view of Katzenbach and Smith [5] the essence of a high-performance organization is the power of teamwork. *" Wisdom lies in recognizing a team's unique potential to deliver results and in understanding its many benefits--development of individual members, team accomplishments, and stronger companywide performance".* As Basketball legend Michael Jordan slam dunks the same point: "Talent wins games, but teamwork and intelligence win championships."

Although teamwork is an essential element of high-performance, the picture is not complete to make high-performance successful. Dr. André de Waal MBA (HPO Center) [6] has formulated the five success factors for high-performance organization:

- Quality of Management/Leadership
- Open and action driven culture
- Long term focus for the organization
- Continuous improvement as the base for the organization
- Quality of the employee's

Many views of high-performance organization, as the ones above, are focusing on the human aspect of high-performance. This is not yet the complete picture. People in society and organizations also need boundary conditions, for example knowing what their responsibility is, what the structure is, and the rules and regulations. In a high-performance process safety culture the balance between the system and the human factor is embedded in the organization and continuously facilitated by leadership (see figure 1) [7]. On the one hand, there is focus needed on the technical soundness of the installation and a management system to manage the barriers. On the other hand, having a culture in which there is focus on continuously improving with good set of values and beliefs. This supported by craftsmanship (good balanced set of competencies) at all levels of the organization. All will not work if the client and stakeholders do not appreciate the effort of the organization.

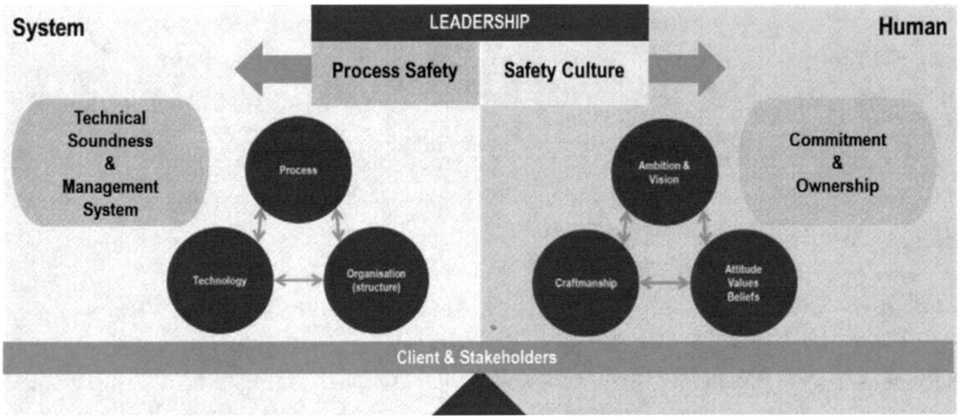

Figure 1: System and human related elements. Both are mutually dependent.

In addition to make high performance work in the chemical industry risk thinking should be at the core of the system and culture (human behavior) (figure 2). It links nicely with the view of Mr. Roggeveen who talks about risk instead of safety. In the integrative system, there is a strong base of a risk-based safety culture supporting the different processes of design integrity, technical integrity, operational integrity and emergency response. Those processes handle the issues and threats for safety, process safety, and asset integrity using the risk approach. It enables one focus in the organization on continuously managing the issues and threats on the likelihood and consequences, not only the ones that are easily in the view, but also the unlikely ones with high consequence. The performance management system, discussed in more detail in section 2.3, based on the Plan – Do – Check – Act (PDCA) circle and using measurement & metrics (KPIs) enables a high performance for all three aspects. Viewing them as part of the same system allows for greater risk mitigation and added operating value.

Figure 2: Risks management at the core of the performance of the organization.

All elements, system and human facilitated by leadership, to create an environment of high-performance are there to realize a value creation for the internal and external stakeholders. The value creation is not only to realize benefits for the shareholders (the financial results) but more importantly the non-financial benefits like sustainability/continuous development, innovation, satisfied external stakeholders (nearby residents, local authorities, etc.), brand name, process safety, and competencies/talent of the organization.

2.2 Driver for Operational Excellence

In the industry many companies have been investing in operational excellence program, particulary to improve their cost base. The market is beginning to see that there are many insights into driving towards operational excellence.

Another development in the market is that there is a growing corporate risk awareness and environmental citizenship concerns driving businesses to invest in more sustainable production. In order to select the right approach towards operational excellence addressing all these different goals, it is vital for organizations to have a coherent overview of the risk and value opportunities and their chance of success. To have an all-encompassing approach that is including safety, process safety, operational, and financial improving is often struggling for companies to identify the correct initiatives and fail to make a serious impact on the bottom line.

In developing an all-encompassing approach the first step is understanding the current situation and defining the picture of the future. How can you accomplish that?

This is the textbook example of a 5 cent question with a 5 dollar answer. At one level, the answer is simple. If the organization is outperforming its peers on key operational metrics and winning or retaining business as a result, then it can probably claim to have achieved operational excellence. There are firms that provide detailed objective benchmark reports that can be used to identify where your organization falls on the ruler.

But external benchmarks only tell part of the story. They're simply providing a snapshot of performance for a sampling of companies against a set of lagging indicators, and the performance bar is always shifting because no organization is standing still.

For operational excellence in process safety that snapshot is not enough. To realize high-performance in process safety "Knowing the Unknown" is a critical successfactor but difficult to work with. In the theory of Taleb the popular notion of the black swan is introduced: there are unpredictable deviations in every complex system, such as a production process or a planning process. In other words saying that it is difficult to find the unknown because there will always be unpredictable deviations which leads to incidents. It should be the challenge of every organisation and individual in that organisation to continuously seek for the unknown. Two aspects are key in this "unkown" and continuousy look for it.

Many organisation have built auditing program to find the unkown. Beside the fact that companies drowning in the number of audits executed in companies, audit have only one view of reality. The definition as given by the IIA's: auditing is an independent, objective assurance and consulting activity designed to add value and improve an organization's operations. It helps an organization accomplish its objectives by bringing a systematic, disciplined approach to evaluate and improve the effectiveness of risk management, control and governance processes. Despite this definition in many of the companies the focus is not on adding value but on compliance of the processes and risk management. Another pitfall of auditing is that many improvement actions are defined but none are executed. In that sense auditing gives you the snapshot that you are compliant but misses the step beyond compliance.

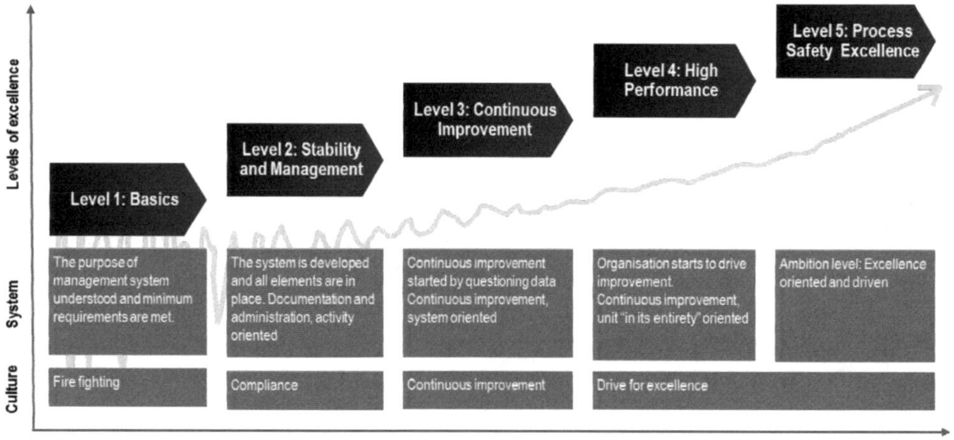

Figure 3: Levels of excellence in process safety performance

Moving beyond compliance is to have a journey to operational excellence in place. Each journey starts with a good travel plan knowing the stages of your travel. In figure 5 the journey towards operational excellence or high performance in process safety is divided in 5 levels from basics, to high performance and with the ultimate goal to be excellent. These levels or stages can be different per company because each journey is different. Ownership in the board of the organisation is more important than if it is 3 or 5 or more levels. In this case the 5 levels have a system component and a culture component with criteria for each of the topics of process safety.

A company at the first level needs to understand the purpose of a performance management system and its activity is oriented to meet the minimum requirements of process safety. At the second level a company's thinking is process oriented and starts with the stabilization of the developed management system using the documentation and the management of safety procedures and processes. To reach the third level, a company should be closing the loop of the performance improvement based on sets of data driven by the people in the organization. At level four, the culture and system are in perfect balance to realize high performance driven by the leadership at all levels of the organization. At the fifth level, which is more an ambition, the company is excellence driven and perfects the application of process safety management to realize the vision of process safety excellence and reach the main target of the introduced process safety management system.

The second essential element of the journey towards operational excellence is a good preparation which means knowing what is expected, knowing your competences, knowing the way you want to travel, knowing your travel plan, etc. In the context of operational excellence in process safety is done by a holistic in-depth assessment of the organization on the cultural and system elements. The toolbox for this kind of assessments have a range of the lean six sigma tools in it like data analysis, safety gemba's (understanding the way of working in the field), process mapping or value stream mapping, cultural assessment, case-studies, multi-disciplinary process challenge workshops, performance system assessment, technical evaluation, visualization of the results and so on. Using those tools in such a way that the stakeholders like management in the organization own the outcome of the assessment is the ultimate goal. This gives then the base to design the journey and realize continuously improvement during the journey.

Finally, the distinguishing aspect of the journey towards operational excellence is that employees start to "live" process safety in their day-to-day operations and the technical processes reach the desired safety level. In this way, the idea of PSE will be sustainably part of the DNA of an organization.

2.3 Measuring success of process safety excellence in performance

The essence of why the chemical industry has survived is that the products are produced with good quality, with optimal margin and with minimum risk of an incident happening. People should be able to understand that and convert that into an integrative risk approach using all aspects of process safety, occupational safety and asset integrity (reliability). Striving for excellence in operational performance (high-performance) using a risk management approach is crucial for further success.

Business leaders crave knowledge about the key elements of organizational performance. After all, in the corporate world, these are the elements that make or break companies every day. How can excellence in operational performance be achieved using a risk management approach? The methodology, seen in Figure 3, integrates the existing knowledge of the hazard analysis into the work processes and links it in to performance management. The advantage is that everybody in the organization gets to know their most relevant risk and have that integrated into their daily business. Secondly, the performance management is then built upon the risk meaning the indicators are related to the risk in the plant.

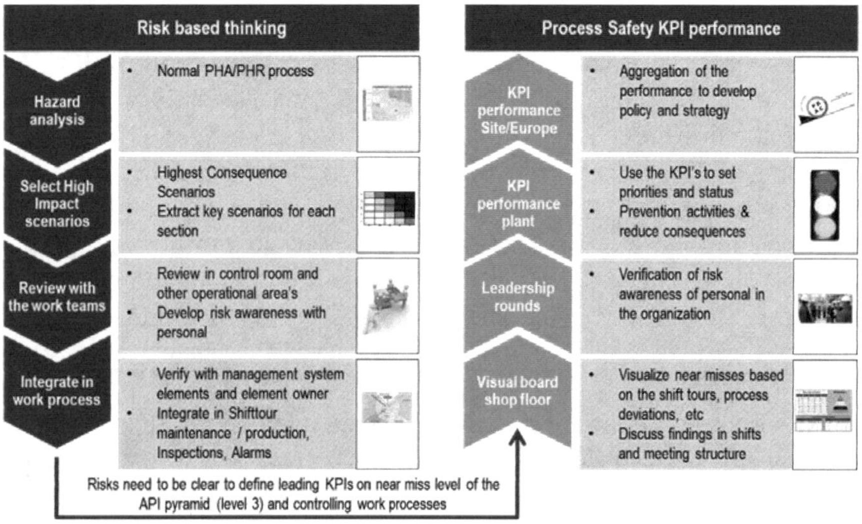

Figure 4: risk based thinking as foundation for Safety performance

The core of performance management is closing the loop of the continuous improvement cycle Plan-Do-Check-Act. As mentioned in the previous paragraph, one of the fundamentals of high-performance organization is continuous improvement in combination with leadership taking the ownership to put energy in the cycle to excel. This energy of leadership balances between using the performance management system and the human aspects as having a vision, the facilitating and process safety competencies and values and beliefs. The elements of the performance management system are defined process, strategy and planning, short interval control, leading and lagging Key Performance Indicators (KPI's), reporting, and formulating actions to improve performance. Those elements can become successful by the commitment of all people in the organization from all levels.

As is known in the industry, risk comes from multiple points and all should be proactively managed as major incidents rarely result from a single cause or failure. Rather, multiple failures coincide and collectively result in an exceptional event with severe, and potentially fatal, consequences. The model, which is based on the Swiss cheese model, in figure 5, visualizes that when different barriers fail in an organization a major incident can happen. Unfortunately, there are many incidents in the industry where this has happened with catastrophic consequences.

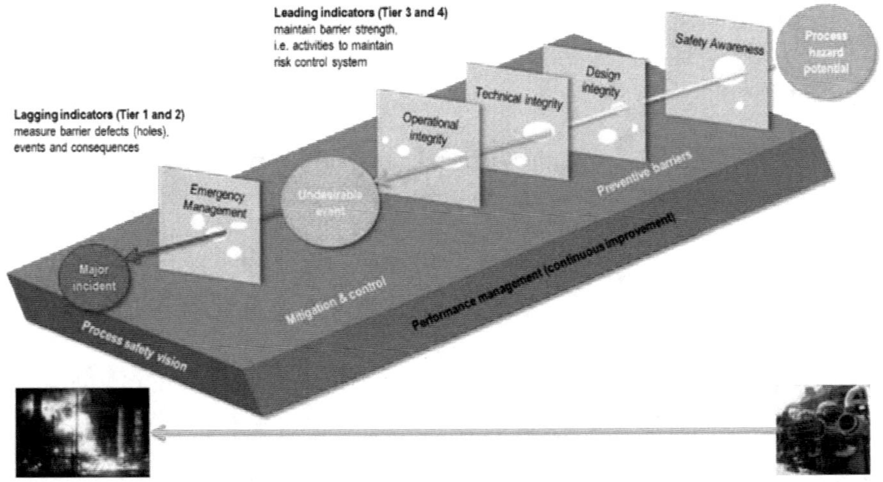

Figure 5: process safety control system

As a human being, we prefer no risks, while we continuously take risks in our daily lives: by crossing the street, driving to work, going on ski holiday and as mentioned in the introduction, climbing the Mont Blanc. Whatever we invest in safety, a zero risk is unachievable goal. Yet it is not only coincidence, when things are going well. With balanced leadership, coincidence can be organized.

The success of an organization depends on the financial performance and even so important, the success in the field of occupational safety and process safety. Even though the frequency of the incidents in the industry is low, the impact of such an incident is very high. After the severe incidents, the governments insist on more and stricter control mechanisms. On the other hand, we all know that the risk of zero is only achievable in theory.

The biggest risk of risk management is blindly trusting the rules and regulations. Stricter control and tighter procedures often lead to a counterproductive effect. From research is seen that with extra control and supervision the risk on incidents increases. Debit is that nobody in the end is taking responsibility because everybody thinks somebody else is responsible for the overall picture. Beside this, in the execution behavior start to develop that can best be described as: ‘as long I am following the rules I am doing well’.

The challenge for organizations is to find the balance between using the risk management systems, safety technical solutions and the human factor. The human factor having the proper competencies, driven by values and beliefs and facilitated by leadership at all levels in the organization. That challenge can be won by using the balanced approach which is further explained in the next chapter.

3 Balanced approach to realize excellence in performance

3.1 Balanced approach in process safety

As in "Yin Yang" the concept of the balanced approach with system and human is about duality forming a whole. In the balanced approach, nothing is completely system or completely human. It should always contain aspects of the system site and the human site to realize high performance or excellence in process safety. To give an example, the idea of "Last Minute Risk Review - LMRA" can only work when the system and human aspects are in balance. The system of LMRA cannot work without a certain belief, craftsmanship or leadership of the human site. Defining and installing the system elements of process safety in the operation management system is only one half or the concept. The other half is that the people in the organization embrace the way of working defined in the operation management system. Ultimately, it is about creating an environment in which it is possible for the employees to develop their skills, the employees to take ownership of the tools and procedures (management system) and a natural drive for excellence in process safety develops. Leadership is the link that makes the balance in process safety excellence in performance possible. In the next paragraph this role is further explained.

Another aspect of the balance approach is that is not a one stop exercise but a way of living or as earlier set a journey. A famous person in history said the following:

"Life is like riding a bicycle. To keep your balance, you must keep moving" (Albert Einstein).

Riding a bike is not about the intensity of energy you put into the bike but of balance, focus, and rhythm. In that way, you realize the most effective and efficient way the different stages of your biking journey. In the context of process safety excellence in performance this is also all true. As leadership at all levels of the organisation keeping the balance needs movement. The movement in continuously questioning the status quo and designing the next stage of improvement of the current situation. In the paragraph 2.2 drive for operational excellence the 5 stages are explained to reflect against and it is about moving on that ladder to operational excellence. In the next paragraph's the role of leadership is explained, the design of a transformation journey and the importance of developing competencies.

3.2 Leadership to keep the balance for excellence

The role of leadership is continuously developing, as the organization is developing part of its environment. The way value is created by leadership is changing. From a top down approach to a facilitating approach. The main role of leadership is to create an environment in which others can become successful, resulting in a high-performance organization. Also, the appreciation of scarcity in the world makes that we should value more what the world can offer us. Make optimal use of the talents and competencies of the individual and organization.

The conceptual model of balanced leadership as visualized in figure 4 can create an environment in which people will become successful. Role of leadership at all levels in this process is to create an environment in which the experience driven transformation journey to high performance process safety culture is possible by keeping the balance between the left side of the model "system/technical soundness" and the right side of the model "human commitment and ownership".

The management system is the organization in which boundaries conditions, in which people are daily working at all levels, are formally described. It contains the description of the core process and secondary processes with roles and responsibilities. The organization has the opportunity here to jointly formulate what is expected from the organization and the individuals in the organization to become successful. The formal communication chain of planning and reporting based on the formulated KPI's is the performance management system that enables an organization to visualize the pathway to success and the realization of that success. The risk based performance management system, as described in paragraph 2.3, starts with the risk scenarios and convert that in working processes with KPI's to continuously improving the performance. The CCPS looks at management systems in the vision 2020 as "Vibrant Management Systems" which are engrained throughout the organization. Vibrant systems readily adapt to the organization's varying operations and risks. The core of the system side of the organization is having and maintaining a technical sound asset base. The pots and the pans for a chemical organization or in other words the technically advanced plants that produce the products with optimal quality for the customers in a safe way.

In a certain way, you can say that the left side of the model is about the structure and the right side about culture. The right side of the model is going a step further. It is about the mix of competencies (craftsmanship), the total behavior of people, value and beliefs, the results or value creation and ambitions of the stakeholders in relation to the environment. The human side of the model shifts the focus to the "why" (vision) of the organization. The development of the vision and getting the vision alive is creating a base of meaning. The elements of the model at the left side bring live to an organization. The employees and the leaders at all levels use the system elements to make the organization successful without losing the "why", the what and how of the organization. In the values and beliefs of the organization a sound risk appetite is there to operate the organization is a safe way for all stakeholders.

According to the vision 2020 of the CCPS leadership has a major impact on culture: "in a Committed Culture, executives involve themselves personally, managers and supervisors drive excellent execution every day, and all employees maintain a sense of vigilance and vulnerability". The value that is created by a committed culture is in both safe and reliable operations by doing "the right thing, the right way" every time. People have the feeling and motivation that they can improve and they are in the starting position to take the steps for improvement. Therefore, developing the process safety competencies of the people in the organization is an essential role of leadership.

The base of the model are the different stakeholders in the organization like customers, suppliers, employees, owners, and external stakeholders with their formal interests and goals. In the end, it is about creating value for the environment in which the organization is acting. The value created in delighting the customer, profit for the organization, developing the talent of the individual's and producing in a safe way.

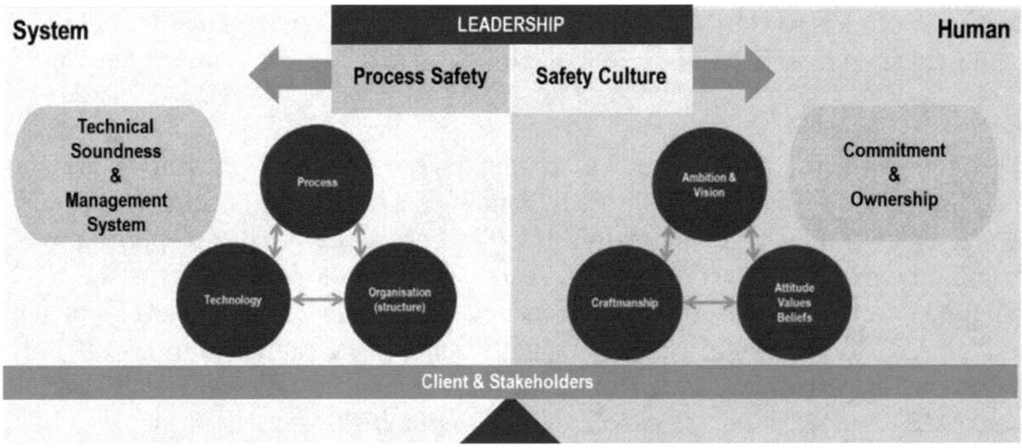

Figure 6. System and human related elements. Both are mutually dependent. If one fails, the other might not be able to compensate. Leadership must ensure the right balance.

The dimensions of balanced leadership that makes an organization successful[7]:

- Creating a shared ambition level
  - Long term relationship with all stakeholder (employees, external, contractors)
  - Precondition is to create clear objectives which are shared by the internal and external stakeholders
- Establishing clear but not too narrow boundary conditions
  - Boundary management with a shared management system and technical conditions of the assets
  - Effective working performance management, continuous improvement based on risks in the DNA
- Clear expectation management and discipline in operational execution
  - Open and action driven culture – integrity
  - Wanting to close the loop in performance (action), communication, dialog, environment where people can become successful
- Craftsmanship – adequate mix of (technical) competencies
  - Competency complementary, taking responsibility, flexible, sharp
  - Exchange of knowledge to create a high level of craftsmanship by widely supported awareness that all technical conditions must be there, but if this is met the human factor the quality and level determines
- Sustainable value creation
  - Creating a positive footprint for the environment in reducing risk, financial benefits and development of the stakeholders
  - Knowing the internal and external stakeholders and their needs – the ambition of the company is directly related to the needs of the different stakeholders

The role for leadership is to understand that the road to success is to keep the balance between system driven and human driven performance improvement to realize the results. Leadership adapts to facilitative leadership competencies, like results focus, managing the boundaries, motivational learning, challenging, etc., that enables them to realize the results by continuously facilitating the transformation to a high-performance organization with the adequate mix of competencies.

3.3 Journey: facilitating continuous closing the loop – managing risks

Stagnation is decline for organizations. An organization falls asleep having mindset that they are the best and having the perception that you have the most advance technical installation and excellent organization. Then competition is lurking to take over your position as the leader in the market. Severe incidents happen because leadership and people in the organization are not anymore sharp to question current situation. There is no sense of urgency to innovate and to work as a team to continuously improve. Based on Kotlers [11] managing change we see the pitfalls for success of change as: no sense of urgency, not having a power coalition, lack of a vision, lost the empowerment of others to act on the vision, no plan and short-term wins, declaring victory too soon, and lack of embedding new way of working.

The key to success of a high-performance organization is to set on an experience driven transformation journey using competency development as the enabler. Establishing a platform for innovation with competency development by continuously creating new knowledge and transforming the organization. The good news for good leaders is that using the balanced leadership approach will facilitate the transformation journey to a high-performance organization. Leadership in transformation in the words of the American president Dwight Eisenhower: "the art to getting someone else to do something you want done because he wants to do it."

The base for transformation is the essence of life and in the words of Darwin: "It is not the strongest or the most intelligent that will survive but the ones most adaptive to change" (Darwin 1808-1882).

The first step in an experience driven transformation journey is that people are aware of their improvement potential in a way that they admit themselves that they can improve. Something like an "ah" moment and becoming aware of the blind spots. I understand now where we can improve in system, culture, technical status and competencies and believe in it that we can improve.

How to get to that moment? Together with the people in the organization visualizing fact based the improvement potential by looking at the situation from different perspectives like the technical soundness, management system and the behavior/culture point of view. Different tools, from brown paper workshops about the processes to case studies support the process in realizing the visualization of mutually understanding the gaps in the system and the individual usage of the system by those people involved indicate the process safety culture within a company.

To understand the culture, it is necessary to get an understanding of the attitudes, values and beliefs of the people in the organization from leadership to the shop floor. This information can be gathered by standard methods such as questionnaires but also with advanced tools like Katalyst, anonymous online workshops in which the approach unveils the attitudes, values and beliefs behind the façade people tend to show.

All the perspectives at the end give a realistic picture of the current safety culture by visualizing the blind spots and creating an environment in which it is safe to become conscious incompetent (the "AH" moment) as an organization and person.

The second step is in cooperation with leadership in the organization developing a transformation roadmap to realize the vision or in other words the picture of the future. It is a roadmap which is owned by the people and leadership in the organization because they have developed it themselves only facilitated by the consultants. Starting journey with the organization and realizing high performance is a stage by stage process. It is like climbing a mountain to go first to base camp and that camp by camp to reach the summit. Now, you are at the summit the decent is the tricky part of the journey and evenly important to get back to base camp – the goal/objective of the organization.

Getting back to base camp is the moment to take a step back and rethink your vision and objectives. Making the plan for the next stage of the journey by assessing the situation

3.4 Process Safety Competency definition

In process safety, human factors have contributed to the causes of several incidents in the past, in a variety of safety critical industries. For example: On March 23rd, 2005, the BP Texas City refinery experienced a catastrophic process accident. It was one of the most serious U.S. workplace disasters of the past two decades, resulting in 15 deaths and more than 170 injuries. Both the BP and the U.S. Chemical Safety and Hazard Investigation Board reports [8] identified numerous technical and human failings at the refinery and within corporate BP.

As Stephen Pryor mentioned in his quote in the Vision 2020 the human factor is about the knowledge and using the knowledge to understand the risk and act to it.

"Process safety is not something for the leader...and the safety experts to know; everybody in the workforce has to become more and more knowledgeable about understanding what the risks are, helping us to identify risks, and making sure that they understand and execute all of our procedures properly and consistently to avoid those risks." — Stephen Pryor, ExxonMobil Chemical (quote from Vision 2020).

Not only those incidents create a sense of urgency for developing process safety competency in the organizations and industry. It is, also, getting attention for many other reasons like the brain drain through aging of the workforce, the loss of knowledge in acquisitions, and too much focus on compliance that people forget to think. In addition, regulators and industry associations like the CCPS are more and more putting process safety competence on the agenda. Therefor companies begin to realize that investing in process safety competence is an asset for the company and correlates with the process safety performance.

The following definition of competency gives the boundaries of what is meant by a competency:

The definition of the European Qualifications Framework (EQF): 'Competence' means the proven ability to use knowledge, skills and personal, social and/or methodological abilities, in

work or study situations and in professional and personal development. In the context of the EQF, competence is described in terms of responsibility and autonomy (EC, 2008).

In other words, competency has two sides, the measurable technical/ management system competencies and the non-measurable competencies as social, human and personal competencies. In this context, the balance between system/ technical related competencies and human related competencies makes a distinctive difference in the success. This is also expressed in the Vision 2020 of the CCPS:

Intentional Competency Development to ensure that all employees who impact process safety are fully capable of meeting the technical and cultural requirements for their jobs. The bottom line: no matter how good the culture or management system is, or how well the company adheres to standards, it takes highly competent employees to implement those systems or standards. And that requires intentional competency development.

Another perspective on competencies and the framework that has been developed by the UN is looking at three elements: core values, core competencies, and managerial competencies. In this framework, the technical competencies or craftsmanship are missing. But the different angle is the core values which people are evaluated on. This fits nicely with the model of balanced leadership.

In the definition of Prof P. Dehnbostel [9] there are three components that form the base for occupational development of the mix of competencies of people.

- Technical Competence constitutes a person's ability and readiness to process tasks and problems in an autonomous, professionally appropriate and methodical manner and to evaluate the result.
- Social Competence describes a person's ability and readiness to work together with others in a target oriented manner, understand the interests and social situations of others, deal with and communicate with others in a rational und responsible way.
- Personal Competence describes a person's ability and readiness to develop further and to shape his or her own life in an autonomous and responsible manner within the respective social, cultural or occupational context.

Measuring competencies can be done across a 4-tier scale, based on a combination of the tiers. This allows a more granular determination of the competency required across a varied workforce. These tiers are defined as follows:

<u>Awareness</u>

Has knowledge of theory and displays conceptual understanding. Actively participates in discussions regarding the skill. Performs routine tasks with significant supervision. Learns how to do things.

<u>Basic application</u>

Performs fundamental and routine tasks. Requires occasional supervision. Increases functional expertise and ability. Works with others.

Skilled application or proficiency

Independent contributor. Integrates work with other disciplines. Frequently mentors or coaches' others. Assesses and compares alternative options.

Mastery or expert

Advanced experience in the skill. Applies creative solutions to complex problems. Defines and drives critical business opportunities and needs. Represents the organization internally and externally on critical issues. Sets standards within the organization. Recognized as subject matter expert.

The process safety competencies framework should consist of a mix of Craftsmanship competencies, ranging from technical/management system competencies to personal and social competencies, and core values and beliefs. This defined for all levels of the organization from the workforce on the shop floor to the board of directors. In the IchemE framework [10] the roles are defined and the technical/ management system competencies. This is a starting point for developing the complete process safety competencies framework including the personal and social competencies and core values/ beliefs. The TÜV SÜD Process Safety competency model starts from the same technical/ management systems competencies but in the transformation process based on competency development the personal/ social competencies and core values/ beliefs will be jointly developed with the client. The current DNA and the vision of the company forms the base for the development.

3.5 Contribution of Process safety competency to high-performance

As craftsmanship and attitude/behaviors are essential aspects to make a high-performance process safety culture successful, competency development is making a difference in the success of a high-performance culture. Naturally in the process industry there is more attention to develop the technical skills of the people in the organization. In this context, the balance of competencies – technical/ management system competencies, social/ personal competencies and core values/ beliefs makes the difference.

There are some actuals threats impacting the process safety performance. For example, as in recent crises years, there has not been enough attention to competency development and/or a lot of senior people are leaving the companies. Another threat is that people do not know the assets anymore because the assets are too reliable.

The opportunity is there to invest in the competency development in an experience driven transformation journey facilitated by leadership. Such an experience driven transformation journey for competency development has two components:

A) Individual development: current competencies – experience driven development plan – execute journey – evaluate and update picture
B) Organizational development: current culture, knowledge, and structure (roles and succession planning) – organizational journey – execute journey – evaluate and adjust

Integral embedded in the first component (A) is the learning process supported by the different pedagogic tools like coaching, experience learning settings, training on the job, classroom training, e-learning, and learning by doing. The learning process depends on: prior knowledge, motivation, memory, understanding, intellectual engagement. Leadership can support the learning process by continuously challenging to prevent normalization from happening. The Organizational development (B) is supported by the transformation approach and leadership.

3.6 Practical suggestions to get started

Keep in mind that many transformation improvement journey's fail during the process mainly because of ten reasons:

- No sense of urgency
- Failure to get right/enough people that want to lead the change
- Permit obstacles to block the new image
- Failure to generate short term wins
- No clear image of the results of the stages in the journey towards excellence
- Declare victory to soon
- Under communicating the image of success by a factor 100
- Not anchoring the changes in the DNA of the organizations culture

Taking those aspects of change in to account, as a leader, will give you continuously success in realizing the journey towards process safety high performance or excellence.

The following are elements to incorporate in the journey:

- An operational management system is an integral system with HSE, Asset Integrity, Process safety and Performance improvement as the main pillars
- Management ensures participation of the entire organization in understanding the drivers of risks
- Risk based thinking on the same criteria is essential for the success to create a proactive risk based culture – everyone understands how activity and risks impact the overall operational reliability of the organization
- Leadership keeps the balance between structure and culture to create a high-performance culture
- Performance improvement is driven by a good mix of key indicators and a culture of understanding the world behind the indicators
- Operational excellence is established by knowledge capture and competency development

The last practical suggestions to have a continuous self-reflection as a leader in the organization to have a clear picture of yourself as a balanced leader. What can I do differently and what am I doing good to keep the balance in the organization?

4 Conclusion

A high-performance process safety culture in which people are actively aware of where they are related to best practices goes beyond compliance. It is a culture where there is a continuous challenge of the current procedures and behaviors by trying to identify blind spots. A culture in which competency development is not seen as a cost but as core to make the experience driven journey to high-performance culture a success. A culture in which balance leadership creates an environment in which proactively the risks are managed. This committed culture has risk based thinking at its core.

In summary, excellence in process safety performance is not a matter of intensity but of balance, focus, rhythm and discipline.

5 References.

[1] https://en.wikipedia.org/wiki/Mont_Blanc

[2] Cambridge dictionary

[3] Roggeveen, Victor, "Veiligheid bestaat niet, risico's wel", Veilig gedrag? – Daar hebben we het nog wel over, 10/2016.

[4] http://www.amanet.org/training/articles/the-essentials-of-high-performance-organizations.aspx

[5] J.R Katzenbach, D.K. Smith, "The Wisdom of teams", October 2015.

[6] Dr Andre de Waal, "High performance Organisation", May 2016

[7] C. van Driel, K. Hoeksema, G. Swarte (redactie), "Talentleidershap – Koersen op duurzame waardeontwikkeling, May 2017.

[8] U.S. Chemical Safety and Hazard Investigation Board. Investigation Report - Refinery Fire and Explosion and Fire. BP Texas City March 23, 2005

[9] P. Dehnbostel, "Competence – conceptual introduction from a pedagogic and scientific point of view", Booklet on the ECCE 8's special session on process and plant safety "Process Safety Competence – European Strength degrading to Weakness?", pp. 24-28, 2012,

[10] IChemE, "Process Safety Competency – a Model", IChemE Safety Centre Guidance, 2015

[11] John Kotler, "Managing change: The Power of Leadership", Balanced Scorecard – Insight, experience & ideas for strategy-focused oragnizations, Harvard Business School Publishing, February 2001.

19 AIChE Spring Meeting
15TH GLOBAL CONGRESS ON PROCESS SAFETY
NEW ORLEANS • 2019

Implementing Process Safety Critical Roles to Reduce Risk

Andrew Wilson
DuPont Sustainable Solutions
Level 4, 225 St Georges Tce. Perth. WA. 6000. Australia
andrew.j.wilson@dupont.com

David Harwood
Woodside Energy Limited
Mia Yellagonga, 11 Mount Street, Perth WA 6000, Australia
dave.harwood@woodside.com.au

Prepared for Presentation at
American Institute of Chemical Engineers
2019 Spring Meeting and 15th Global Congress on Process Safety
New Orleans, LA
March 31 – April 3, 2019

Implementing Process Safety Critical Roles to Reduce Risk

Andrew Wilson
DuPont Sustainable Solutions
Level 4, 225 St Georges Tce. Perth. WA. 6000. Australia
andrew.j.wilson@dupont.com

David Harwood
Woodside Energy Ltd

Keywords: Energy Institute Framework, Operational Discipline, Process Safety Critical Role, Process Safety Management

Abstract

In late 2014 Woodside commenced the implementation of the Energy Institute Process Safety Management Framework to transform how process safety is managed in the organization. The framework consists of four focus areas and twenty elements which formed a structure around which a cultural change to a "line led, risk based" approach was made. The transformation was focused on four key improvement opportunities; (i) consistent multi-disciplined risk assessments, (ii) efficient management of safety critical devices, (iii) better recognition of excursions beyond integrity limits, and (iv), lift and formalize the capability of people to lead and deliver superior process safety performance. The first three focused on tools and methodologies, the PSM 'system'. The capability piece considered the people aspect of process safety, so that people understood and were able to execute their role in managing process safety.

This paper discusses the capability workstream in Woodside's Process Safety Management Initiative, the 'how' of establishing and implementing process safety critical roles. It then considers the impact of the effort, in terms of a reduction in process safety risk, using both leading and lagging indicators.

1 Introduction

In late 2014 Woodside commenced the implementation of the Energy Institute Process Safety Management Framework to transform how process safety is managed in the organization. This initiative included a cultural change to a "line led, risk based" approach. This paper discusses the 'how' of establishing and implementing process safety critical roles. It then considers the impacts of the effort, in terms of a reduction in process safety risk, using both leading and lagging indicators.

2 Implementing Process Safety Critical Roles

The Energy Institute (EI) Framework consists of four focus areas, the first being 'Process Safety Leadership'. Process Safety Leadership consists of five elements, including Element 3 – 'Employee selection, placement and competency, and health assurance'. The framework states that "control of operations depends on having competent people in position. Management must ensure that existing and new personnel have the required competencies and are fit for work." Woodside's approach to PSM had traditionally been technical in nature, with a strong emphasis on physical barriers to prevent process safety events. The EI framework was chosen by Woodside because of its strong emphasis on management responsibility, and the extension of Process Safety Management (PSM) beyond technical considerations to the role of people in achieving process safety excellence. This extension necessitated a cultural transformation that was undertaken with the support of DuPont Sustainable Solutions.

2.1 Process Safety Critical Role Procedure

The requirement that "management must ensure that existing and new personnel have the required competencies" seems reasonable and relatively simple, however the implications for an organization are significant. What are the required competencies? Who needs to have them? How does management ensure that new and existing personnel have them?

To answer these questions in a transparent and robust manner, Woodside produced a 'Process Safety Critical Role Procedure'. This procedure;

- Defines more than twenty roles that are process safety critical (e.g. operator, technician, engineering team leader, maintenance superintendent, drilling supervisor, project manager, asset manager, VP operating unit etc.)
- Defines the required competencies for each role; for each of the four focus areas (process safety leadership, risk identification, risk management, review and improvement) a role can have one of four competency levels – awareness, knowledge, skilled, mastery
- Defines the method for demonstrating required competencies
- Extends the application of Management of Change requirements to people movements that involve process safety critical roles, or if there is an organizational restructure that impacts on process safety critical roles or the management of process safety risks

2.2 Defining Process Safety Critical Roles

Having decided to implement Process Safety Critical Roles (PSCRs), a fundamental question then arose for Woodside – which roles are process safety critical? Some roles were obviously critical. Legacy roles such as 'Engineering Authority' and 'Technical Authority' were traditional guardians of Technical Integrity, and therefore had well defined competencies that transferred easily to process safety. Similarly, the Asset (or Plant) Manager is clearly the line manager accountable for process safety performance of a facility, and is a PSCR.

Other roles were defined as PSCR based on a philosophy that the decisions people in those roles made were clear barriers to process safety events. Hence operations and maintenance technicians are PSCRs, as are Operations and Maintenance Superintendents, and Operations and Maintenance Managers.

Some functional roles were also defined as process safety critical – e.g. process safety advisor. Other functional roles were not; although an Operations Excellence Manager may influence process safety outcomes through improved operational discipline for example; their decisions are not a barrier to process safety events.

Having defined what the PSCRs were, it was then necessary to define the required process safety competencies for each role, and develop a sustainable approach for ensuring that everybody in those PSCRs has the required competency;

- **Awareness** – understands that there are PSM requirements that need to be met, and the importance of those requirements to prevent low probability, high consequence events from occurring.
- **Knowledge** – able to implement PSM requirements by following required processes and procedures. Able to use the Woodside PSM framework to identify and manage process safety risks.
- **Skilled** – able to lead PSM activities, ensure that practitioners conform with Woodside requirements and improve team performance.
- **Mastery** – expert that takes a company wide view to improve performance, reduce risks, and able to deal with new, challenging and ambiguous process safety issues.

2.3 Learning and Development Program

Based on the requirements of the Process Safety Critical Role (PSCR) Procedure, a Learning and Development program was developed (refer to Table 1).

Table 1. PSM Learning and Development Program

Competency Level	Learning Approach	Competency Assessment
Awareness	Attend a half day course that introduces Woodside's approach to PSM, later replaced by an E-Learning module	Course attendance
Knowledge	Attend a half or full day interactive workshop for each of the four focus areas that describes in detail how Woodside managers that aspect of PSM	Quiz at end of workshop
Skilled	Attend a full day interactive workshop, with follow up coaching to enable demonstration of required competencies	Formal competency assessment requiring demonstration of competencies
Mastery	Coaching to enable demonstration of required competencies	Formal competency assessment requiring demonstration of competencies

In 2016 250 PSM courses were delivered to 4000 people, and in 2017 50 courses were delivered to 800 people. 200 leaders in Skilled and Mastery roles have received PSM coaching. In 2017 92% of people in Process Safety Critical Roles had been assessed as competent. All courses are now delivered by Woodside leaders and practitioners with no support required from DuPont.

2.4 Line led, risk based

Key documents were developed such as the 'Process Safety Management Procedure' and 'Process Safety Critical Role Procedure' that are now embedded in the Woodside Management System. These documents translate the EI Framework into Woodside specifics and language, and are supported by relevant tools and practices. The cultural aspect of the transformation was turning this paper based system into reality, so that process safety at Woodside is "line led, risk based".

"Line led" is requiring line managers (assets, projects and functions) to work with the mindsets and behaviors that visibly demonstrate a commitment to process safety, and integrating PSM with other business priorities such as production, cost and people. Process safety is more difficult to manage than workplace safety because the probability of events is very low, and signals that reveal deficiencies can be weak. Keeping process safety requirements front of mind can be problematic when faced with numerous short term issues. Line managers are better able to implement process safety systems because they have the authority and access to resources to drive operational discipline and focus; 'whatever interests my boss fascinates me'.

"Risk based" is about making better decisions, and allocating resources and focus based on risk. Improving risk assessments enables better risk based decision making, and Woodside has achieved more consistent and better quality risk assessments through updated tools and processes, and developing competency through the PSCR learning and development program. After benchmarking with peers the Woodside Risk Matrix was updated and fully socialized to enable more consistent understanding and productive decision making based on risk.

3 Discussion/Results

3.1 Tier 1 and Tier 2 Events

Like its peers, Woodside tracks and benchmarks PSM performance by counting Tier 1 and 2 events as defined by the API 754 standard. These lagging indicators show a steady improvement, with 0 events in 2017 (refer to Figure 1). By their very nature these events are low probability, and hence a lack of events is no guarantee that future events won't occur. Unfortunately, in 2018 there was a Tier 1 and a Tier 2 event. Neither of the events caused any injuries, but are a demonstration that despite strengthening PSM systems, eliminating process safety events completely is an ongoing challenge .

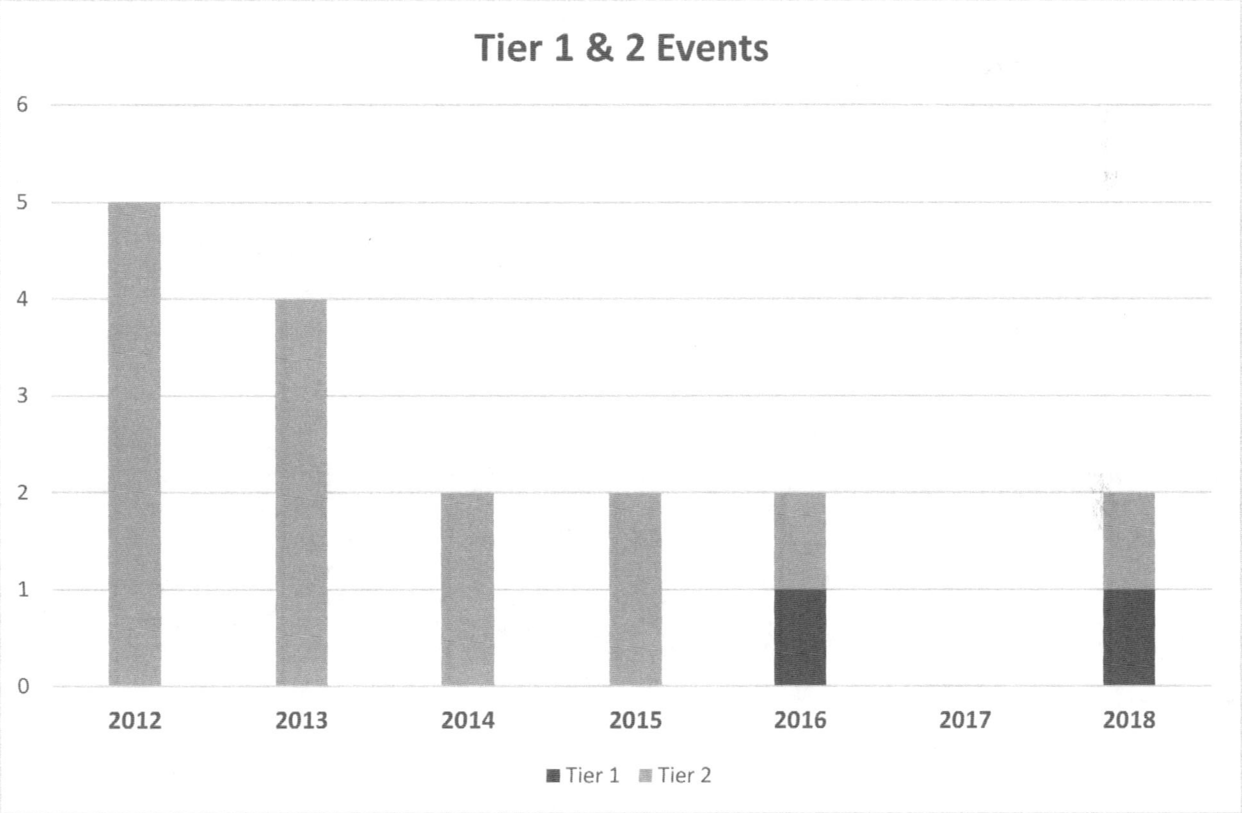

Figure 1. Woodside Tier 1 and 2 Process Safety Events.

3.2 Risk Reduction

Because of their low probability, the best indicator of future PSM performance is not historical events, but the strength and robustness of PSM systems that effectively reduce risk. Operational

Discipline is required to drive improved implementation of risk reduction processes, which are a leading indicator of future performance.

A suite of process safety metrics was developed to measure the implementation of the PSM Framework. Two examples below demonstrate risk reduction through improved operational discipline, as well as the challenges of improving further and sustaining the change.

3.2.1 Priority Current Risk Deviations

A key leading indicator of risk is the number of 'Priority Current Risk Deviations'. These are deviations from standards or original design that are assessed as currently (before future mitigations are put in place) 'High Risk'. They are captured and managed in the MOC system and may for example be identified because of equipment failure, a recognition of inadequate design through a planned HAZOP[1] or LOPA[2] study, or a learning from an incident. A deviation may take time resolve due to technical issues, shutdown opportunities or waiting for the delivery of spare parts.

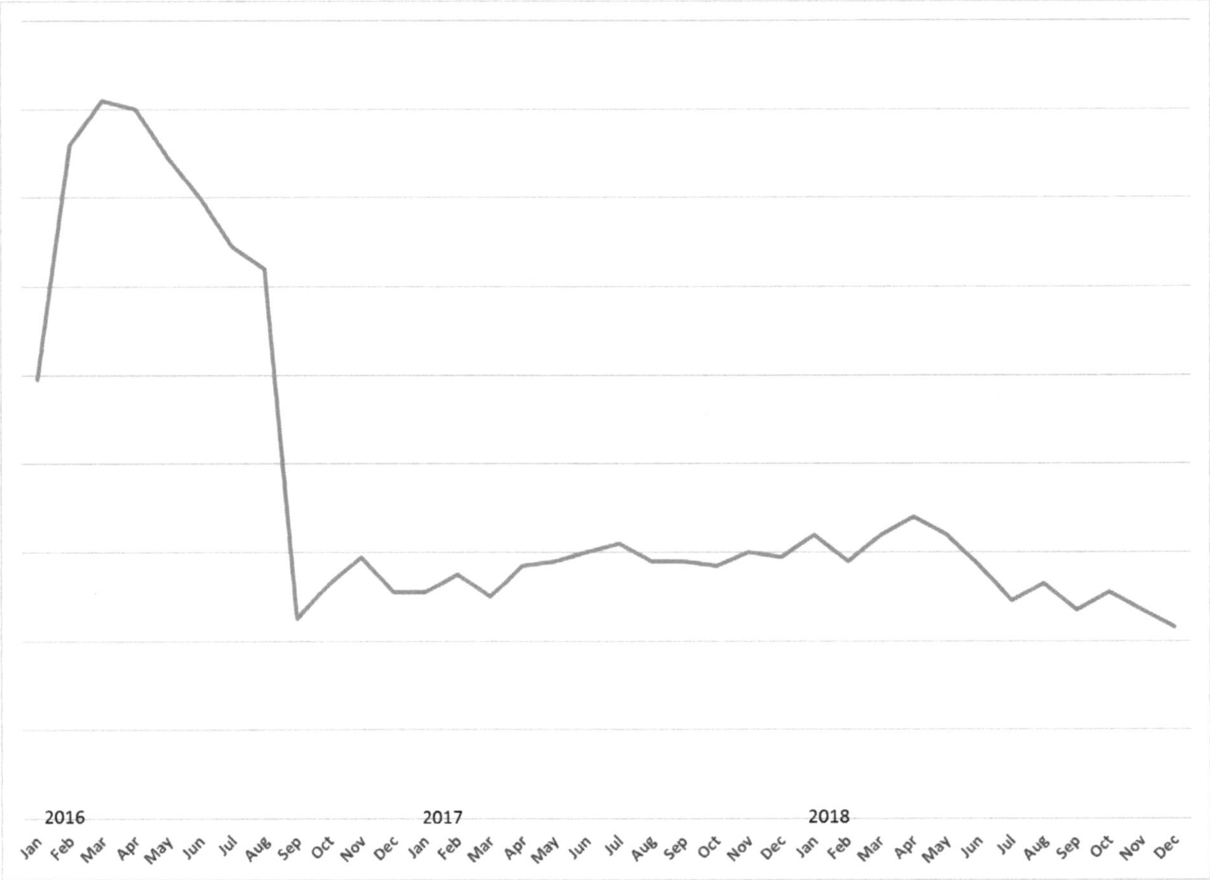

Figure 2. Woodside Priority Current Risk Deviations.

[1] HAZOP - Hazard and Operability Study
[2] LOPA - Layers of Protection Analysis

A major effort in 2016 by the assets to lift the bar by focusing on resolving Priority Current Risk Deviations (line led, risk based) reduced the number of deviations by more than 50% (refer to Figure 2). This work continued in 2017 in parallel with the introduction of LOPA studies and many HAZOP reviews. These powerful LOPA studies revealed risks that weren't previously recognized. Hence new deviations were raised that offset the close out of backlog items. The focus on Priority Current Risk Deviations remains with the goal of achieving zero and sustaining a culture of Operational Discipline that is risk based, rigorous and does not tolerate deviance.

4 Summary and Conclusion

Woodside undertook a major transformation of its PSM approach based on the EI Framework that included increasing the capability of its people. A 'line led, risk based' mindset was introduced through a learning and development program based on adult learning principles that included E-Learning, instructor led training and follow up coaching.

Demonstrable improvements were realized based on lagging indicators (Tier 1 and 2 events) and a suite of leading indicators including Priority Current Risk Deviations

When Emergency Response Goes Fatally Wrong: Lessons Learned from Investigating Two Major Incidents

Scott Davis
Gexcon US
4833 Rugby Avenue Suite 100
Bethesda, MD
sgdavis@gexcon.com

Prepared for Presentation at
American Institute of Chemical Engineers
2019 Spring Meeting and 15th Global Congress on Process Safety
New Orleans, LA
March 31 – April 3, 2019

When Emergency Response Goes Fatally Wrong: Lessons Learned from Investigating Two Major Incidents

Scott Davis
Gexcon US
4833 Rugby Avenue Suite 100
Bethesda, MD
sgdavis@gexcon.com

Keywords: emergency response planning, lessons learned, case studies, major explosion incidents

Abstract

Emergency response plans are an essential, yet oftentimes overlooked, layer of protection for facilities where all other safeguards failed to prevent an incident. While catastrophic accidents such as large releases of chemicals, fires or explosions are devastating for the process industry, experience investigating numerous incidents has shown that a lack of an effective emergency response plan can lead to an unnecessary and tragic escalation of the initial incident. More specifically, investigation of two recent incidents: (1) 2013 West Texas Explosion and (2) 2015 explosion on the FPSO Cidade São Mateus; will demonstrate how fatal consequences resulted from the lack of emergency planning in the first case study and from deficiencies in the emergency response plan or implementation of the plan in the second case study. More specifically, the FPSO explosion resulted in 9 fatalities, eight of which were responding to a leak of a light-end hydrocarbon, and the West Texas explosion that resulted 15 fatalities, most of whom were responding to the fire that preceding the devastating explosion. Lessons will show that the fatal consequences could have been avoided with the proper emergency planning. In addition, while emergency response plans consider "maximum credible" scenarios, past events have shown that low probability high consequences should also be at least considered.

1 Introduction

Emergency response plans are an essential, yet oftentimes overlooked, layer of protection for facilities where all other layers of protection failed to prevent an incident. An emergency plan specifies procedures for handling sudden or unexpected situations, where the ultimate objectives are to be prepared to: (1) prevent fatalities and injuries; (2) reduce damage to buildings, stock, and equipment; (3) protect the environment and the community; and (4) accelerate the resumption of normal operations [1]. Developing the emergency response plan consists of five major steps (see Figure 1). The first step is to assess the vulnerability of the site and determine what hazards pose a threat to the installation. Such an exercise will help determine not only what "credible" events can occur at a given facility, but also what catastrophic or major events can occur. It should be noted that major emergencies are generally rare events.

Next, it is necessary to develop your emergency response plan and determine what means are currently available to execute the plan and what is still needed. At the planning stage, it is important that several groups be asked to participate, such as: subject matter experts and process engineers; the health and safety group; and even appropriate municipal officials should also be consulted since control may be exercised by the local government in major emergencies and additional resources may be required. Once the emergency response plan is developed, the necessary facilities and equipment need to not only be provided but tested and maintained. Communication, training and periodic drills will ensure adequate performance if the plan must be carried out. Lastly, there should be a periodic review of the emergency response plan.

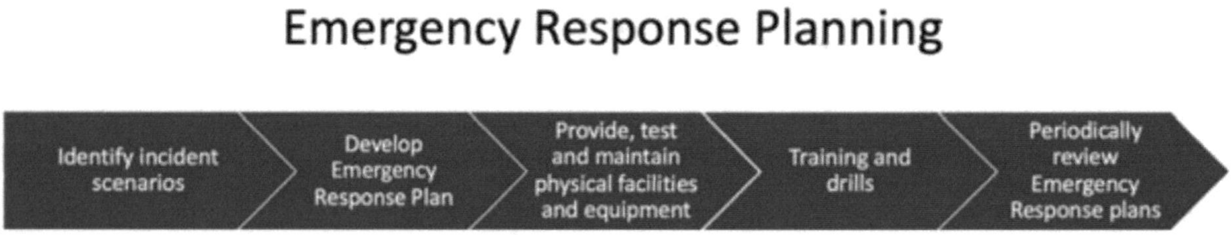

Figure 1: Steps of Emergency Response Planning

While the major benefit is to provide guidance during an emergency, developing the plan has other advantages. For example, previously unrecognized hazardous conditions, which would aggravate an emergency situation, could be discovered during the emergency planning and means put in place to mitigate or eliminate these hazards. The planning process may also identify deficiencies, such as the lack of resources (equipment, trained personnel, supplies), which can be easily corrected before an emergency occurs.

While catastrophic accidents such as large releases of chemicals, fires or explosions are devastating for the process industry, experience investigating numerous incidents has shown that a lack of an effective emergency response plan can lead to an unnecessary and tragic escalation of the incident. Chaos can quickly occur during an emergency if there are lack of resources and trained personnel coupled with an urgent need for rapid decisions and a very short period of time. The stress of the situation can lead to poor judgment resulting in severe losses, and normal channels of authority and communication cannot be relied upon to function routinely. Therefore, it is imperative to eliminate the "it can't happen here" mentality, as emergencies can occur and preplanning is necessary.

The importance of emergency planning will be discussed within the context of two recent incidents: (1) 2013 West Texas Explosion and (2) 2015 explosion on the FPSO Cidade São Mateus. These two case studies will demonstrate how the lack of emergency planning resulted in devastating consequences in the first case study and from deficiencies in the emergency response plan or implementation of the plan in the second case study.

2 Background of Incidents

2013 West Texas Explosion

On April 17th, 2013, an explosion occurred at the West Fertilizer Company storage and distribution facility in West, Texas. The explosion at West Fertilizer resulted from an intense fire in the seed and fertilizer area of the facility that led to the detonation of approximately 30 tons of ammonium nitrate stored inside a wooden receiving bin. The explosion occurred while emergency services personnel were responding to a fire at the facility. Fifteen people were killed, more than 250 were injured, and numerous buildings were damaged or destroyed.

The seed and fertilizer building (Figure 2) was built in 1958 and stored fertilizer and seed, which were distributed to farmers. Figure 3 shows the layout of the seed and fertilizer building, which included an ammonium nitrate (AN) storage bin neighboring the seed room, as well as an empty bin used for ammonium nitrate storage, and bins containing ammonium nitrate, K-Mag (fertilizer mix of potassium, magnesium and sulfur), Potash, ammonium sulfate, DAP (diammonium phosphate), and a bin for potassium chloride along the west wall. This building was capable of receiving rail shipments directly from the rail car by means of a rail spur junction and an underground conveyor belt and bucket elevator; the bucket elevator dumped directly into the northern AN storage bin, or onto a conveyor belt feeding the bins along the west wall.

Figure 2: West Fertilizer Facility – viewed from the west (upper) and east (lower)

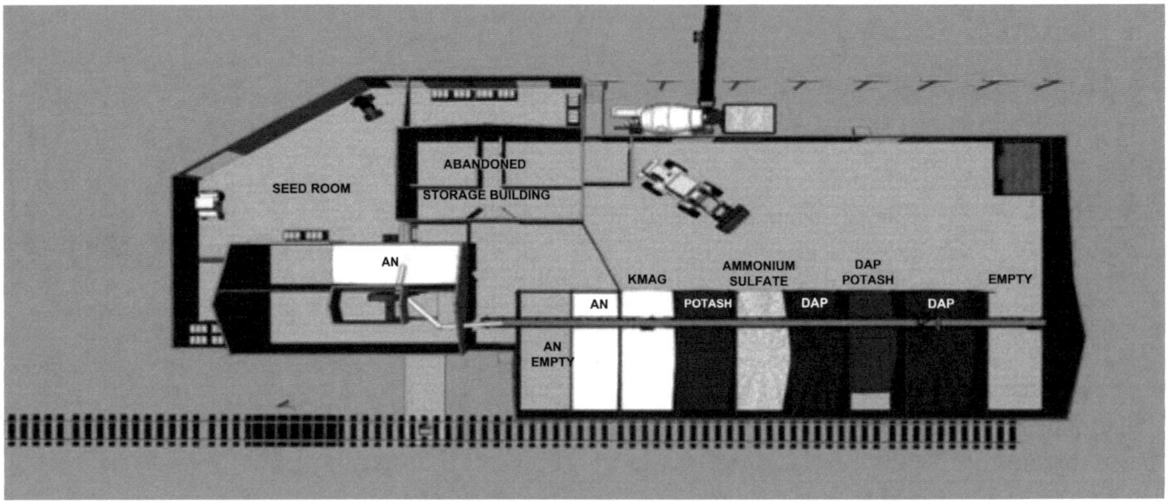

Figure 3: Layout of the Seed and Fertilizer Building.

On April 17th, 2013, at approximately 7:29 pm, a fire was reported, and fire fighters responded to the fertilizer plant [1]. All operations were already stopped for the day and employees had left the plant at about 5:30-6:00 pm. GexCon obtained several videos that recorded the progression of the initial fire [3]. Figure 4 shows the initial fire as viewed from the east and west; note that the highest point on the building represents the location of the bucket elevator by the northern AN storage bin. The image shows that the fire is isolated to the northern section of the building, which includes the seed room and portions of the northern AN storage bin containing about 30 tons of ammonium nitrate.

At approximately 7:50 pm, a massive explosion occurred at the fertilizer plant. Figure 5 shows images from a video taken from the high school parking lot to the southeast and Figure 6 the ensuing blast damage and crater from the detonation. Based on the videos from the day of the incident, there were actually two explosions: one smaller blast followed by a larger blast roughly 300 msec later.

Figure 4: Close up of the Seed and Fertilizer Building during fire as viewed from east (left image) and viewed from the west (right image)

Figure 5 – View of the explosion (viewed from southeast)

Fifteen people were killed as a result of the explosion, which included voluntary firefighters, and over 250 people were injured [3]. Damage was extensive throughout the neighboring community, which included homes, schools, an apartment building and a nursing home.

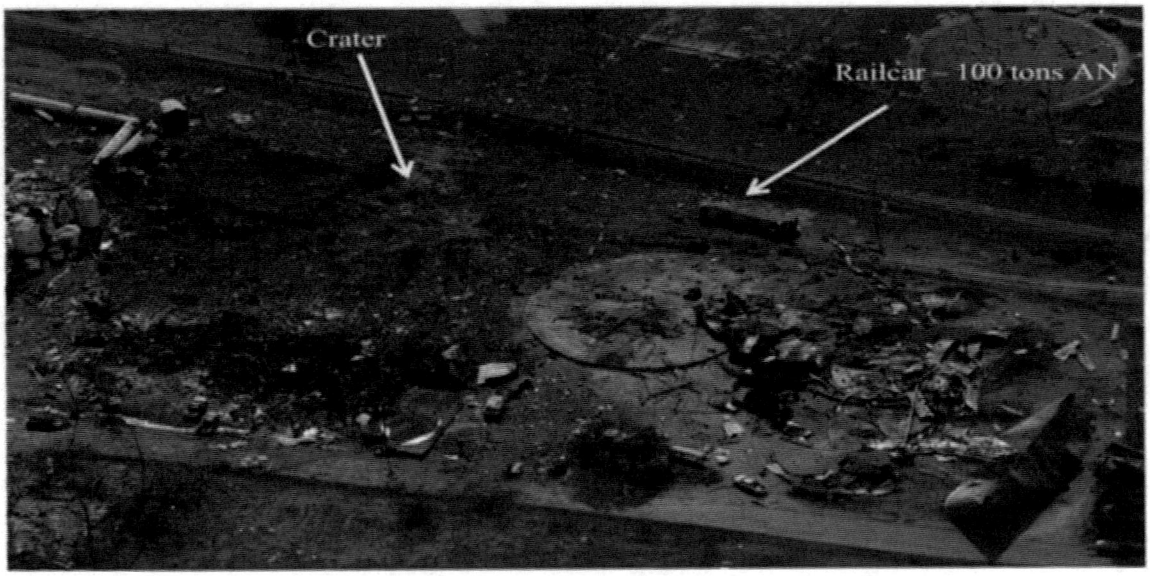

Figure 6: Blast damage and resulting crater

2015 explosion on the FPSO Cidade São Mateus

Due to the confidentiality of matters in this case, all references are made to two publicly available reports regarding the incident [5], [6]. The Cidade São Mateus was a Floating, Production, Storage and Offloading of Oil and Gas Export (FPSO), with a capacity to produce 4,000m³/day of oil and 10,000,000m³/day of gas as shown in Figure 7.

Figure 7: 2015 FPSO Cidade São Mateus.

On February 11, 2015 there was a leak of condensate (light-end hydrocarbon liquid) discovered in the pump room at a spade blind flange connection. Figure 8 shows that the pump room is situated just forward of the engine room and aft of the storage tanks. The layout and levels of the pump room is given in the left image of Figure 9, while the leaking spade blind flange is in the right image.

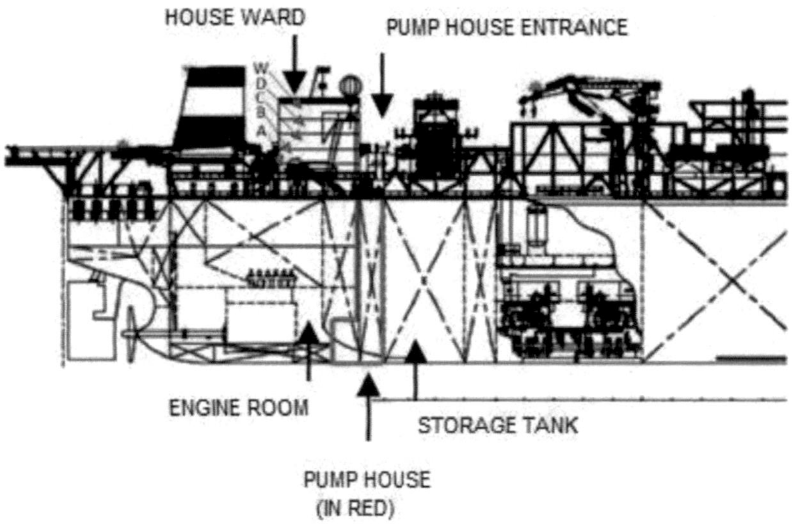

Figure 8: Cutaway side view of the pump room on the FPSO Cidade São Mateus.

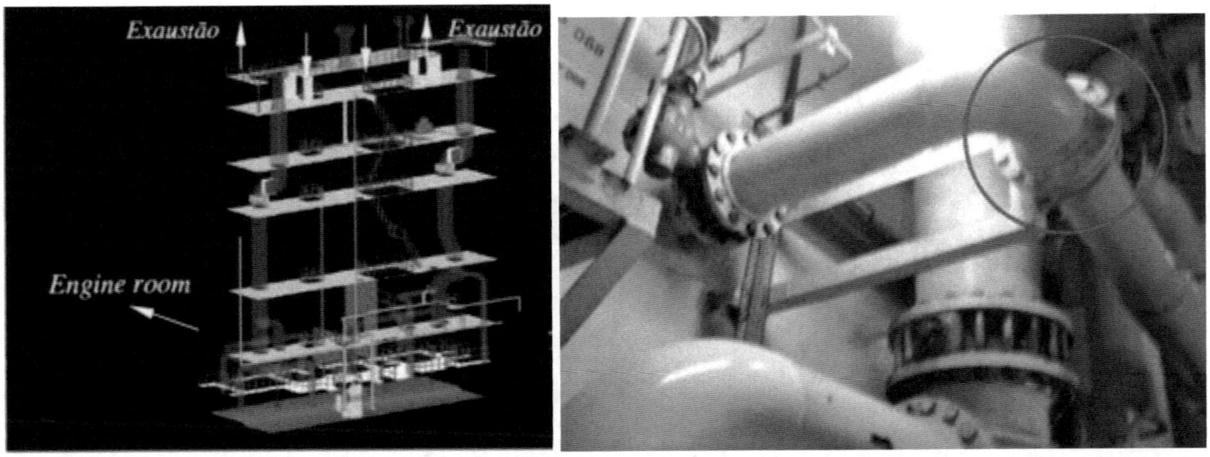

Figure 9: Model of the pump room on the FPSO Cidade São Mateus (left) and spade blind flange (right).

Just prior to the leak, the operations crew was attempting to drain condensate center cargo tank 6 (6C) into an alternate tank and they used an alternate pump alignment (stripping pump). However, the valve on discharge side of stripping pump was closed and they were pumping against a spade blind that was inadequate for such pressures, which ultimately failed the blind and seals. The gas detectors in the pump room began to alarm, and ultimately all three were alarming.

The ventilation was interlocked to shutdown per "confirmed gas" in the pump room. Emergency response teams entered the pump room to attempt clean up and repair the flange. There were three separate entries, and on during the third entry the explosion occurred. Figure 10 through Figure 12 show the level of blast damage to the pump room. Nine people were killed, of which eight were first responders.

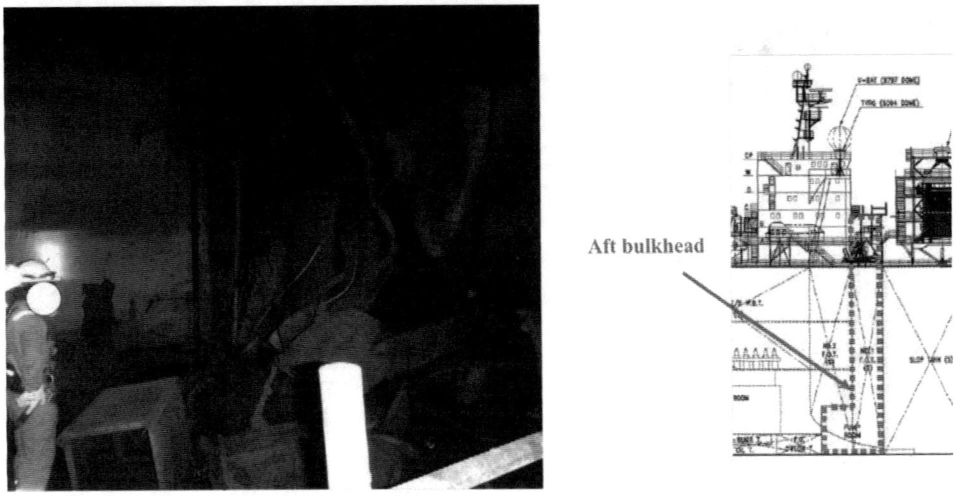

Figure 10: Damage to level above the pumps – complete bulkhead failure.

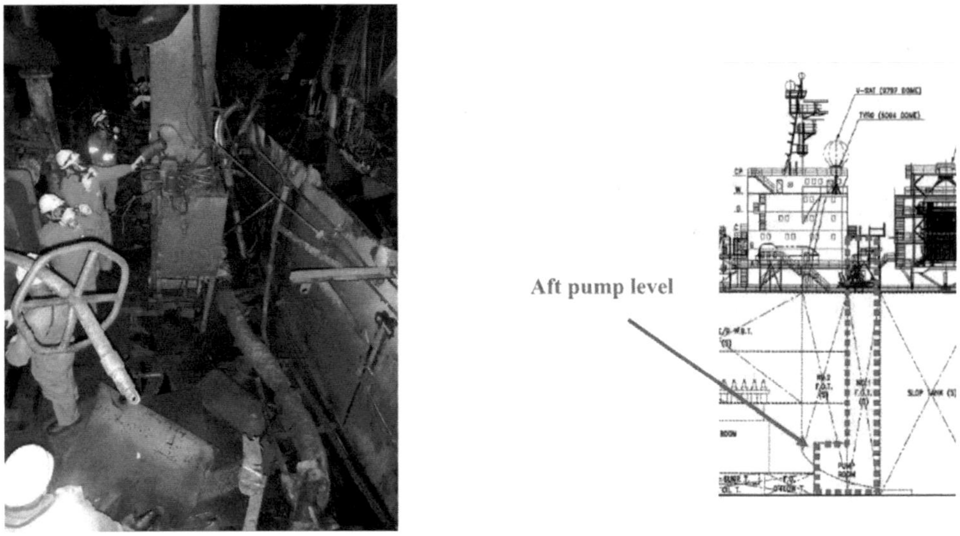

Figure 11: Damage to the engine room.

Figure 12: Damage to upper level of pump room.

3 Emergency Response

Response during the 2013 West Texas Explosion

As the fire occurred 20 minutes prior to the explosion, fire personnel were deployed to scene. Of the 15 people killed, 12 were first responders (8 volunteer firefighters and emergency services), and the other individuals were good Samaritans who supported the emergency response at the fertilizer plant. One of the volunteer fire fighters, who died in the blast, was employee of West Fertilizer. In fact, he was the plant foreman and knew what was stored in the facility. The emergency response is outlined in Figure 13.

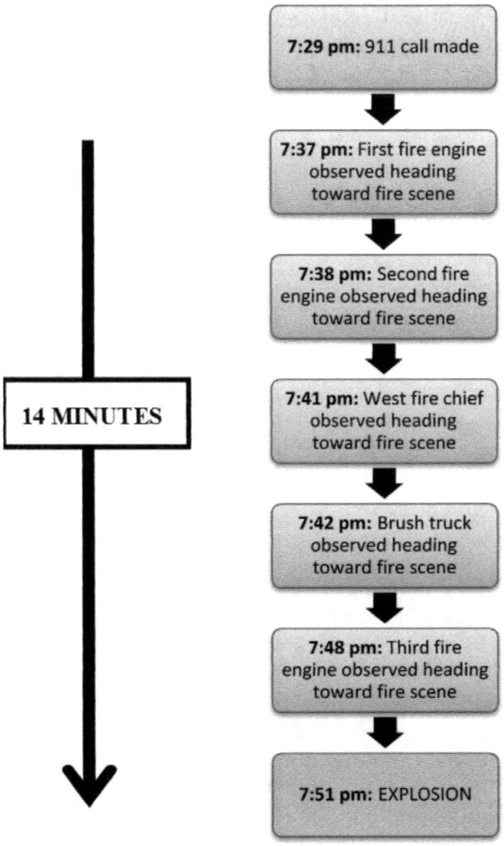

Figure 13: Emergency Response Plan at West Texas

Plant personnel and management did not understand the unique explosive nature of ammonium nitrate. More specifically, while the conditions necessary for detonation are not precisely known, the data suggest that detonations are promoted by severe shock or by heating with confinement, such as the massive fire at the plant. In fact, pools of molten ammonium nitrate may form in the event of fire, and if the molten mass becomes confined (e.g., in the elevator pit) it could explode. These hazards were not known nor understood by plant personnel and emergency responders. Therefore, responders were unaware of the potential of ammonium nitrate to detonate, had no knowledge of how long the fire had been burning, no robust incident pre-planning process in place, and no previous ammonium nitrate-related fire emergency training or drills. Hence, the strategy and tactics utilized by the West Volunteer Fire Department were not appropriate for the rapidly developing situation and they were exposed to unnecessary and extreme risk that proved fatal.

Response during 2015 explosion on the FPSO Cidade São Mateus

The first detector alarmed at 11:31, and by 11:38 all three detectors reached the over limit levels. The 1st team entered the pump room between 11:49-11:54 to assess the gas alarms, and they reported a leak at the flange and a 2-meter diameter pool under the flange. A 2nd team entered the pump room around 12:07 to assess the damage and whether repairs could be made. Despite the confirmed gas on all three detectors, the 2nd team still entered, and a personal gas detector alarmed at 100% LEL. The 2nd team exited around 12:15 and the ship was partially demobilized around 12:20 for lunch. The 3rd team enters around 12:27 to clean and repair, with ladders, wrenches, fire hose and absorbent blankets. The explosion occurred around 12:38 and killed four members in the pump room and four members at the top of the pump room on watch. Due to the failure of the wall into the engine room, the utilities operator was fatally injured in the engine control room.

There was a detailed Emergency Response Plan on the FPSO, along with the muster points for each person involved in the emergency response as shown Figure 14.

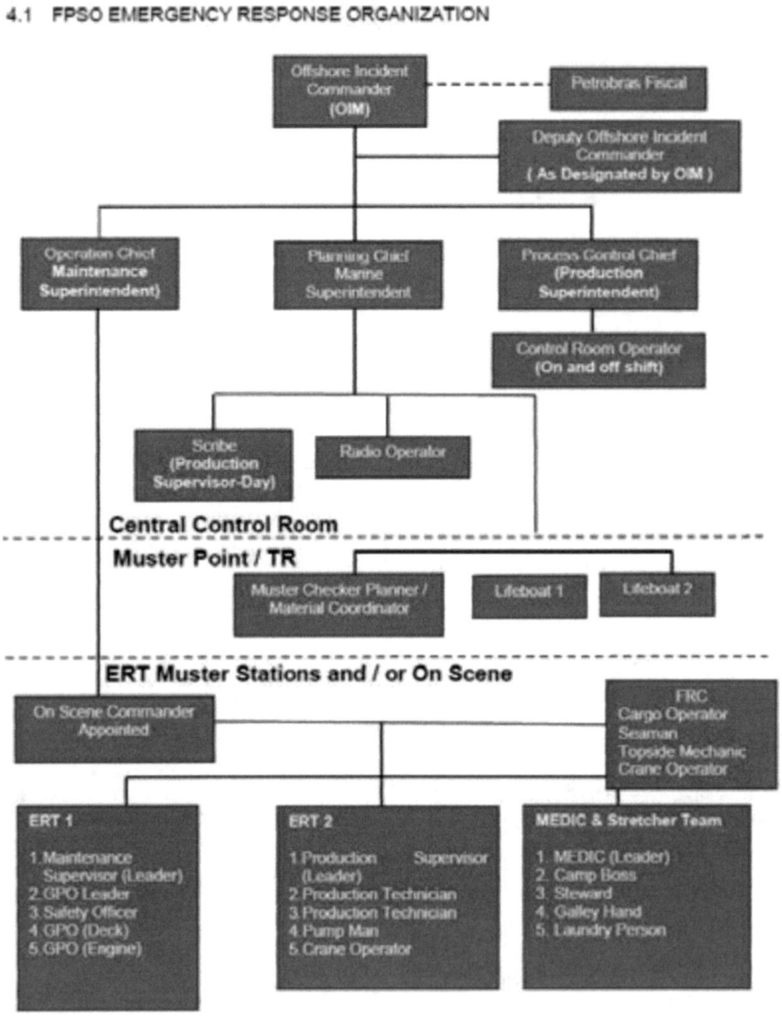

Figure 14: Organization Chart and muster points of the emergency response team

The Offshore Incident Commander is the Offshore Installation Manager (OIM) and would have the roles of implementing a command structure that would satisfy the organizational needs to protect life, environment and the installation, as well as deploying quick and efficiency actions to control incidents, as well as managing response actions to ensure that the escalating potential, risks of injury to personnel and damages to the installation are minimized. In the event of a leak in potentially explosive atmospheres, the emergency response must consider the possibility of the event escalating to a fire or explosion scenario. Thus, systems and procedures should be provided for the crew to control the situation and mitigate the risks of escalation from explosive atmospheres.

Investigation showed that the vessel was not designed for condensate, but for crude and gas. Yet despite the fact that there was confirmed gas detection within the pump room and on personal gas detectors, the crews entered three times. On the third entry, the explosion occurred. It was also reported that an explosive atmosphere in the pump or engine room was not included as an accident scenario in the emergency response plan, and there were no clear instructions of this particular scenario. Despite the knowledge that there was confirmed gas detection on the fixed sensors installed at the bottom of the pump room and also on personal detectors, Emergency Response Teams were sent into the pump room to evaluate the scenario in an explosive atmosphere.

The actions of Command essentially played down the risk of the condensate leak and allowed personnel to enter the pump room on multiple occasions to clean the spill without first performing a risk analysis, and without additional monitoring the room's atmospheric conditions. This noncompliance with the applicable emergency procedure was not only observed during entry operations, but also through the partial or total demobilization of the people irrespective of their muster points with an ongoing confirmed gas detection in the pump room. This demobilization resulted in two members being injured as they were found in a damaged elevator shaft after the blast.

4 Analysis

While deficiencies in the emergency response plans resulted in unnecessary and tragic consequences, the reasons each were lacking are very different and distinct.

For the West Texas Explosion, the Texas State Fire Marshal incident investigation [2] concluded that neither West Fertilizer nor West Fire Department had pre-fire and pre-incident plan at the facility. -Therefore, the recommendations provided were to ensure that local fire departments implement an Emergency Response Plan, with a particular focus on high-hazard occupancies. It was identified that fire departments visit and pre-incident plan responses to commercial structures in their jurisdiction. Further, they concluded that it is imperative that high-risk occupancies containing hazardous materials are identified and appropriate emergency response strategies and tactics are developed through pre-plan documents and SOP directives. Despite the extremely low likelihood of ammonia nitrate detonations, the recommendations state that high-hazard and high-risk occupancies containing heavy fire loads and hazardous materials should be identified by Fire Departments and all members made aware of their potential dangers. This will better prepare them for not only credible hazards, but also those high consequence hazards which may be found in the event of a fire and also provide them with insight as to when it is prudent to attack a fire or evacuate the area. More specific recommendations regarding the emergency response planning included:

- Develop a detailed Emergency Response Plan and Evacuation Plan for the area within 1 mile of existing ammonia nitrate buildings
- Develop standard operating procedures/guidelines to train and educate members on firefighting and emergency scene strategies and tactics for high-hazard occupancies
- Develop pre-incident plans by visiting the commercial and high-risk properties within the response jurisdiction.
- Utilize risk management systems on all emergency responses to determine when to employ alternative tactics.

- Incorporate incident management systems in every emergency incident, training exercise and basic operational process of the fire department to provide appropriate command and control for conducting operations.
- Implement a strategic incident action plan (IAP) for all emergency operations through a coordinated application of appropriate emergency scene tactics for the presented dangers
- Ensure that an incident safety officer (ISO), a person who is independent from the incident commander, is appointed and effectively utilized at every incident that would require an ISO.
- Establish a hazardous materials program and actively train members on the tactics to use when an incident involving hazardous materials occurs

In contrast, for the 2015 FPSO event, there was a detailed Emergency Response Plan for the vessel. Despite this plan, investigations concluded that there were significant deficiencies in the plan that needed to be remedied, and largely included reference to personnel exposure, failure to identify the accident scenarios for the emergency plan, training and mobilization. The execution of the plan failed to properly identify the potential high consequence and catastrophic result of the actions of the emergency responders. More specifically:

- Exposure to personnel
    - In the event of a leak in potentially explosive atmospheres indoors, the plan must consider the possibility of event escalation to a fire and/or explosion. Thus, systems and procedures should be provided once there is the formation of an explosive atmosphere. To ensure the success of mitigation actions, emergency procedures and resources for emergency response should be pre-planned, drawing on the use of all available resources, including the time of scenario assessment.
    - To determine that all functions involved in emergency response, including land-based support, have periodic theoretical training in emergency response procedures.
    - Set in the emergency response plan the available material and human resources to the response of each accidental
- Failure to identify accident scenarios or not included from the risk scenarios
    - Establish criteria and procedures for the migration of risk scenarios identified in the risk analysis and safety studies for emergency response plans. In addition to the selected scenarios, the procedures should contain scenarios that represent challenges to response actions, such as fatalities, and multiple injuries, among others.
    - Establish periodically simulated emergency covering all identified accident scenarios, including scenarios with multiple injuries and fatalities. Such simulations should include the activation of external resources to inform platform that response structure is effectively tested and that corrective measures arising from simulated ensure continuous improvement of the response actions of all entities involved.
- Demobilization of meeting points
    - Periodically coordinate all staff regarding the hazards involved in the assumptions made in the response procedure to the emergency unit.

  - Prohibit the release of people to normal activities before the full control of the emergency.
- Ignition of the explosive environment
  - As the ignition was introduced by the action of the emergency responders within explosive atmospheres, review the procedures related to emergency response and safe work practices considering the possible ignition sources identified in existing rules

5 Conclusion

The lessons from two specific incidents show that critical deficiencies in the emergency response plan or implementation of the plan can result in fatal consequences that could have been avoided. The 2013 West Texas explosion shows how important it is to not only recognize high risk events, but also have the appropriate plans in place to avoid unnecessary escalation of an already tragic incident, where 12 of the 15 people fatally injured were first responders. By preparing first responders for not only credible hazards, but also those high consequence hazards, can provide them with crucial insight as to when it is prudent to attack a fire or evacuate the area. In addition, not properly following or playing down the risk of a high consequence event was observed in the 2015 FPSO event, where eight of the nine crew members killed by the blast were first responders. The failure to recognize the potentially high consequence and catastrophic event, led to an emergency response that unnecessarily exposed responders and ultimately led to an exacerbated event. This study confirms that emergency response plans are an essential layer of protection for facilities where all other safeguards failed to prevent an incident, which could have prevented unnecessary and tragic escalation of the incident.

6 References

[1] Preparedness for Emergency Response - FEMA Training - FEMA.gov

[2] Texas State Fire Marshal, "Firefighter Fatality Investigation FFF FY 13-06," West, TX, 2013

[3] Davis, SG, DeBold, TF, & Marsegan, C (2017). Investigation Findings and Lessons Learned in the West Fertilizer Explosion. Journal of Fire Science. Volume 35, Issue 5, September 5, 2017: pp. 379-395.

[4] "A Public Health Report on Injuries Related to the West (Texas) Fertilizer Plant Explosion," June 24, 2014.

[5] Brazilian Navy Directorate of Ports and Coasts, Maritime Accident Safety Investigation Report, FPSO Cidade São Mateus - Explosion followed by flooding and casualties, February 11th, 2015.

[6] Investigation report of the explosion incident occurred on 11/02/2015 in the FPSO Cidade São Mateus, Brazilian National Agency of Petroleum, Natural Gas and Biofuels, Superintendence of Operational Safety and the Environment, August 2015.

19 AIChE Spring Meeting
15TH GLOBAL CONGRESS ON PROCESS SAFETY
NEW ORLEANS • 2019

Major Explosion at a Chemical Company in China: the Causes and Lessons Learned

Wei Wang
China University of Petroleum (East China)
No. 66, West Changjiang Road, Huangdao District, Qingdao, China
15621097585@163.com

Yi Liu
China University of Petroleum (East China)
No. 66, West Changjiang Road, Huangdao District, Qingdao, China
liuyi@upc.edu.cn

Dongfeng Zhao
China University of Petroleum (East China)
No. 66, West Changjiang Road, Huangdao District, Qingdao, China
zhaodf@vip.sina.com

Yanchun Wang
China University of Petroleum (East China)
No. 66, West Changjiang Road, Huangdao District, Qingdao, China
Wychun25@163.com

Jian Chen
China University of Petroleum (East China)
No. 66, West Changjiang Road, Huangdao District, Qingdao, China
836771528@qq.com

Hanhai Dong
China University of Petroleum (East China)
No. 66, West Changjiang Road, Huangdao District, Qingdao, China
donghanhaiupc@126.com

Prepared for Presentation at
American Institute of Chemical Engineers
2019 Spring Meeting and 15th Global Congress on Process Safety
New Orleans, LA
March 31 – April 3, 2019

Major Explosion at a Chemical Company in China: the Causes and Lessons Learned

Wei Wang
China University of Petroleum (East China)
No. 66, West Changjiang Road, Huangdao District, Qingdao, China
15621097585@163.com

Yi Liu
China University of Petroleum (East China)

Dongfeng Zhao
China University of Petroleum (East China)

Jian Chen
China University of Petroleum (East China)

Yanchun Wang
China University of Petroleum (East China)

Hanhai Dong
China University of Petroleum (East China)

Keywords: Fishbone Diagram; Analytic Hierarchy Process; Process Safety Management; Incident Investigation

Abstract

On July 12, 2018, a devastating explosion incident occurred in Yibin City Hengda Technology Co., Ltd. in Sichuan Province, resulting in 19 deaths and 12 injuries, and impacting an area of about 1,800 square meters. This was one of the most disastrous explosions that occurred in a chemical plant in China in the past five years. The incident reflects the general deficiencies of process safety management in small and medium-scale companies. Therefore, it is significant to analyze the causes of the incident and put forward countermeasures to avoid such kind of incidents.

According to the characteristics of complexity, diversity, and uncertainty about the influencing factors of the explosion, this paper presented a systematic analysis from qualitative and quantitative perspectives. The fishbone diagram model was used to identify influence factors from six aspects of people, design, management, environment, safety prevention, and materials. Furthermore, an assessment index including 26 factors was determined. The analytic hierarchy process method was used to analyze the importance degree of each factor quantitatively, and the key factors leading to the incident were obtained. The results showed that the sum of weights of three main aspects, including people, design, and safety prevention is nearly80%, so they were the dominant causes leading to this incident. Lack of chemical industry experience, non-standard process design, and incomplete automatic control system were the focus of prevention among 26 factors.

Process safety management is the application of management systems to the identification, understanding, and control of process hazards to prevent incidents from occurring. This analysis considered process hazard analysis, pre-startup safety review, operation procedures, training, and emergency plan and response, which were the main elements in process safety management framework. Finally, we proposed safety precautions for the identification, storage, production, and use of hazardous chemicals, as well as recommendations for process equipment management.

1 Introduction to the event

At 18:42:33 on July 12, 2018, an explosion occurred in Yibin Hengda Technology Co., Ltd. in the industrial park of Jiang'an County, Sichuan Province, causing 19 fatalities and injuring 12 people. The explosion caused the second and third workshops to catch fire, and the combustion area is $1800m^2$ [1]. The first workshop, warehouse, drying room, analysis room, office building, and other buildings had been damaged to some extent. The fire department dispatched 42 fire engines and more than 200 firefights. It took more than two hours to extinguish the fire on the spot as shown in Figure 1. It was estimated that the total energy of the explosion in this incident was equivalent to 235 kg of TNT. The direct economic loss was 41.42 million yuan.

At the time of the incident, the company was producing imazamox and triazole. Triazole is a pesticide, and imazamox is an herbicide. The process of the incident was the dehydration of butyramide in the process of synthesis of imazamox. At about 11:30 on July 12, the company collected 2 tons of chemicals labeled as raw materials (actually the explosive chemical sodium chlorate, not the raw material required for the production of butyramide). At around 14:00, the deputy director of the company's second workshop received raw materials from the warehouse. At around 15:30, the workers in the second workshop raised the raw materials for production through the lift to the third floor of the second workshop. At around 16:00, the 2R301 kettle for dehydration of butyramide was completed, and the feed was in an empty state. Before 17:20, the 2R301 kettle was completed. At about 6 pm, the dehydration completed. At 18:42:33, it was due to the on-site handover time. A chemical explosion occurred in the 2R301 kettle on the third floor of the second workshop. The direct cause of the incident was that the operator replaced the unlabeled sodium chlorate with butyramide and put it into the 2R301 kettle for dehydration, which caused the explosion and fire. Sodium chlorate is a kind of explosive chemical, which is easy to form an explosive mixture when mixed with butyramide. The explosion led to the disintegration of 2R301 kettle. The high-temperature toluene vapor rushed out rapidly formed an explosive mixture with the external air and produced a secondary explosion, which also caused the explosion and combustion of the materials such as sodium chlorate, toluene, and methanol stored at the workshop site.

Figure 1. Extinguishing at the scene.

According to the investigation, in this incident, the second workshop was the core area of the explosion, and the overall structure was severely damaged as shown in Figure 2. The platform on the third floor collapsed. The reactor with a height of nearly two meters and a diameter of about one meter fell from the upper level.

Figure 2. Severely damage to the second workshop.

2 Incident evaluation index system based on the fishbone diagram

Fishbone Diagram is also named as cause and branches figure analysis. Kaoru Ishikawa, a Japanese management master, invented it for finding out the root cause of incident [2]. It indicates the relationship between a problem and its underlying causes. It is a qualitative analysis tool to help identify incident causes [3]. With step by step in-depth study to analyze its influencing factors,

Fishbone Diagrams can systematize complicated incident causes. When starting the specific report, we first analyze the factors which affect the safety most and then followed by departure from the overall reason to find the basic reason and to detect and determine the main reason [4,5,6].

After the incident, the government set up a team of experts to identify risk factors from six aspects: human, design, management, environment, safety prevention, and material and establish an incident evaluation index system with 26 elements. The fishbone diagram that induced the explosion is shown in Figure 3.

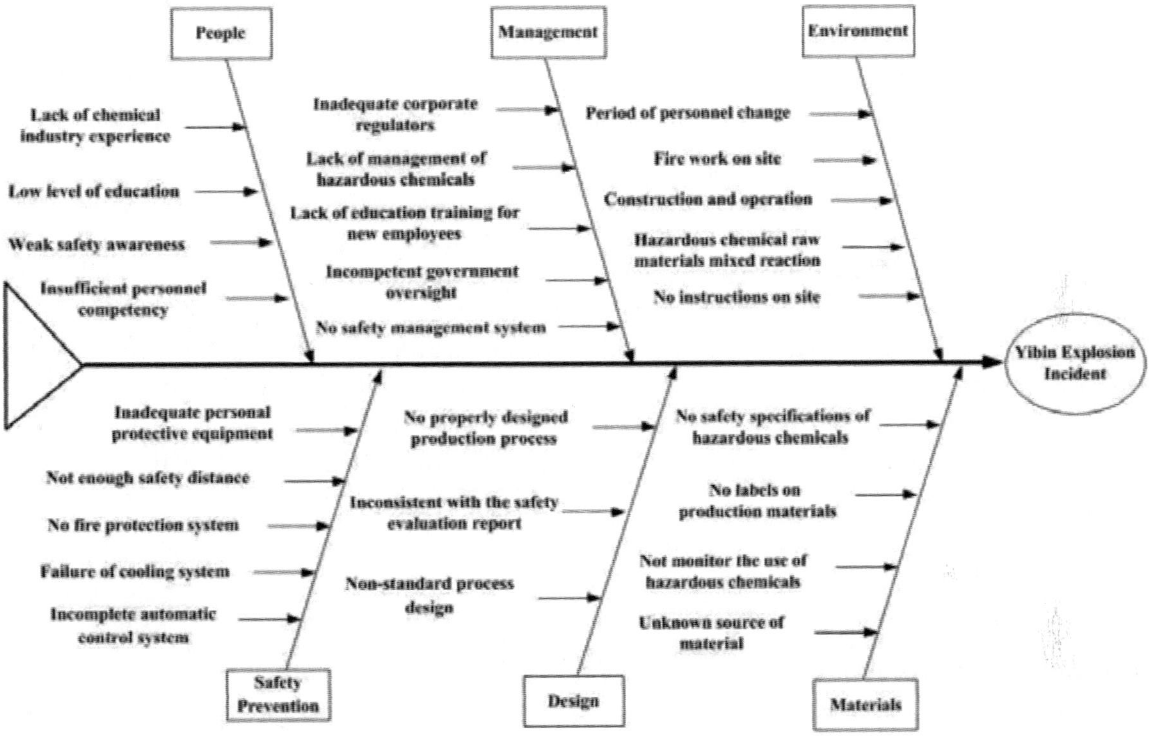

Figure 3. Fishbone diagram of Yibin explosion incident [7].

2.1 People

Neither the safety managers nor the operators demonstrated proficiency in the requirements of process safety competency. There were two main aspects below:

- Corporate management - The general manager of the enterprise had responsibility for the mechanical processing and production contract. However, he did not have appropriate experience in the chemical industry. The technical director did not have experience with management of process safety and production systems. The corporate legal person responsible for finance and logistics also had no experience in chemical engineering.

- Operators - Among the 19 people who died in the incident, 16 were operators, and most of these operators were local farmers. They lacked the basic knowledge of chemical process safety production, and they did not understand the process safety risks in the production process.

2.2 Design

The second workshop where the incident occurred was not constructed and installed strictly by the design drawings, and it was originally planned to have nine sets of 5000L tanks and one set of 3000L tanks and produce 5-nitroisophthalic acid. As built, the workshop had three sets of 5000L tanks, seven sets of 3000L tanks, and eight sets of 2000L tanks were set up. The product was also different from the original, but imazamox and triazole. A similar situation existed in the third workshop. The production technology of imazamox and triazole used by the company had no formal technical source, and the enterprise also did not entrust professional institutions for process calculation and construction drawing design. The technical director of the enterprise only planned the required production equipment and operation process according to his experience, but he knew nothing about the process safety risks of related technologies. It was during the production of imazamox that the distillation kettle used for the dehydration of butyramide by toluene first exploded. According to the *Notice on Further Strengthening the Safety Design Management of Hazardous Chemicals Construction Project*, Hengda company needed to start construction in April 2018, but the company had already started construction in July 2017. Also, the fire separation design of each building did not meet the requirements of the *Petrochemical Enterprise Design Fire Code (GB50160).*

2.3 Management

The lack of process safety culture is the main cause of management failure. Safety culture is developed through the interactions of factors such as safety policies and safety goals, the discipline of the organization, the behavior and attitude of the company staff, and the competence of workers. Improper operating discipline and procedures can lead to a negative process safety culture and an increased likelihood of process safety incidents. The enterprise had not formulated the post safety responsibility system, safety management rules, and regulations. Furthermore, the enterprise manager and the person in charge of each department only had the oral authorization, and there was no formal appointment document to clarify the post responsibility. The safety management organization was not complete. The company's safety and environmental protection department had only one safety officer who did not have safety management capabilities. There were also no full-time or part-time safety officers in each workshop. There was a lack of management of hazardous chemicals in the process safety management. The technology provider concealed some chemical name marks for confidentiality, and there was no safety specification. Required knowledge and competence connected with safety for workers were the pivotal content of safety training. New employees did not have security education directly work or internship. The government had not strictly regulated the safety departments of local key hazardous chemicals companies.

2.4 Environment

The factory built the workshop while producing, and used the polypropylene pipe which was easy to produce static electricity to transport flammable and explosive liquid materials. The building materials, hardware, and chemicals in the warehouse were mixed and piled with hidden dangers. The incident occurred during shift time, resulting in a large number of casualties. In the on-site operation workshop, there was neither operation instructions nor warning signs for hazardous chemicals. At the time of the incident, the second workshop produced two products at the same time. This practice of producing multiple produced at the same time in the same workshop has excellent security risks.

2.5 Safety prevention

Compared with the 20 elements of CCPS, China drew on the process safety management experience of the Occupational Safety and Health Administration (OSHA) and further optimized the formation of 12 elements for the production of Chinese chemical companies. After that, a series of process safety technologies were developed and promoted for the key elements of risk management and equipment integrity. In 2013, the SAWS issued *Guidance for enhancing chemical process safety management.* The Ministry of Emergency Management, which was integrated by the State Administration of Work Safety(SAWS) in 2018, attaches great importance to the supervision of hazardous chemicals, hazardous chemical processes, and major hazardous sources, referred to as "Two Priorities and One Major." If an enterprise violates The relevant provisions, it shall be punished according to the upper limit of economic punishment according to law.

In the production process of imazamox and triazole, the nitrification, oxidation, diazotization, hydrazine hydrate, hydrogen peroxide, methanol, and other dangerous chemicals and processes involved were all "Two Priorities and One Major" projects, which needed strict supervision. However, the company's automatic control system, flammable and toxic gas alarm system and fire water system and other safety facilities were not installed. Once there was a danger, the enterprise itself could not deal with an emergency. The lack of an automated control system led to more than ten operators performing manual operations around the reactor, which was an important cause of major casualties.

2.6 Materials

According to the investigation, the safety evaluation report made by the enterprise before the incident did not fully consider the process safety information, such as the lack of description of dangerous chemicals and the risk identification of the process, so the enterprise had great potential risks. Workers didn't know the source of raw materials, and many chemical production raw materials were not marked. Workers knew nothing about the risks of these dangerous chemicals, and company officials failed to communicate information about the hazards of the process to employees. The workers compiled codes to replace the names of products and raw materials, and these numbers were not abbreviated as specifications. The raw materials used by Hengda Company, such as hydrazine hydrate, nitric acid, toluene, hydrochloric acid were dangerous chemicals or precursor chemicals, but the company still purchased them illegally without relevant licenses. In the production process, regulators did not supervise the management of hazardous chemicals.

3 Determination of the weights of the incident incentive index based on the analytic hierarchy process

The Analytical Hierarchy Process (AHP) is a decision-making method proposed and used by American operations researcher T.L.Saaty in the mid-1970s [8]. AHP is a process that breaks down a problem in a hierarchical matrix with increasing level of detail to determine the most significant causes of an incident. According to the obtained fishbone diagram of the Yibin explosion incident, an incremental layer structure for analytic hierarchy analysis was constructed, as shown in Fig. 4, where the A, B, and C layers illustrated the primary effect and hierarchy of causes.

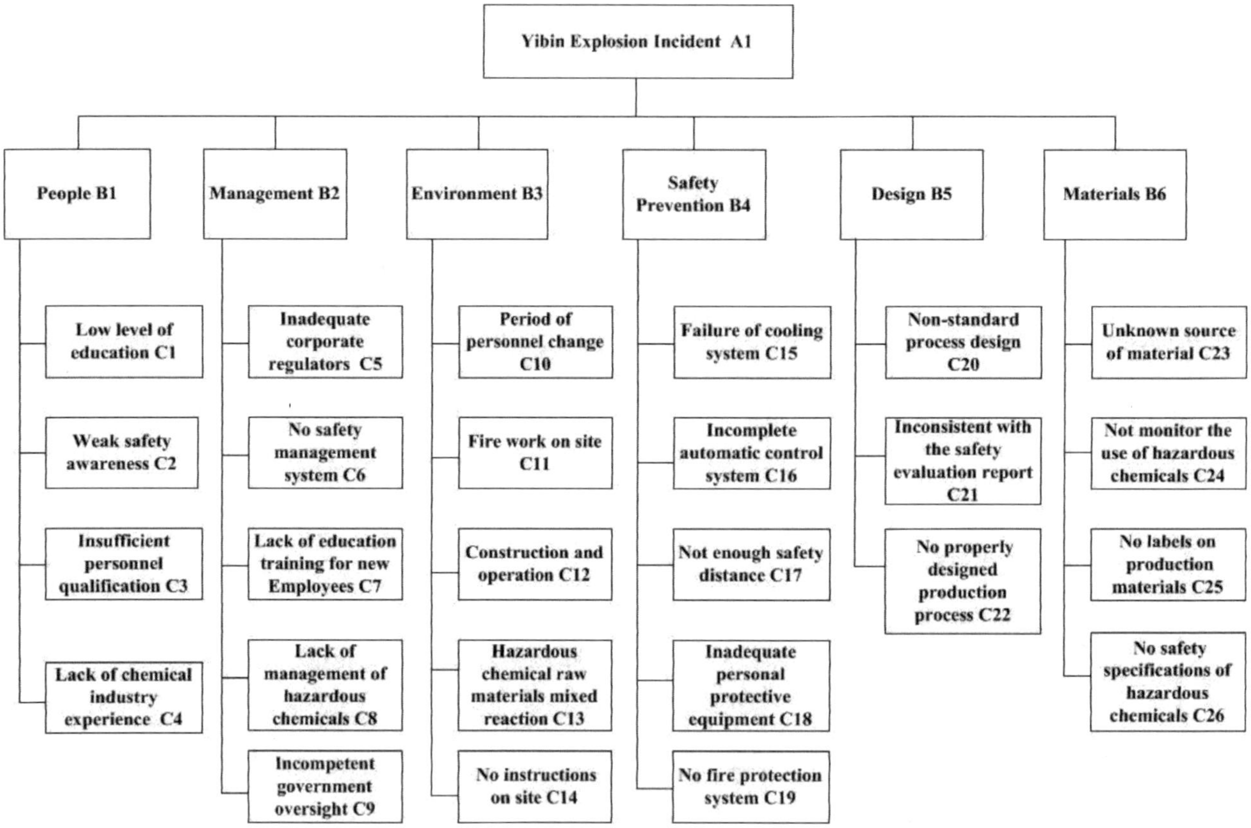

Figure 4. Hierarchical structures of safety evaluation system for Yibin explosion incident.

3.1 The single-layer index weights of the influencing factors

According to the actual data collected and expert opinions, we used 1-9 scale to construct the judgment matrix (A) of the first-level evaluation index, as shown in table 1, and obtained the influence of various factors in layer B on layer A. λ_{max} is the maximum eigenvalue, CI is the consistency index, RI is the average random consistency index, and CR is the random consistency ratio; When CR<0.1, it passed the consistency test [9]. According to table 1, the weights of design is the largest, followed by people and safety prevention, with the sum of the weights of the three as high as 0.8. It can be seen that the leading cause of this incident is due to the low level of process design, operators' insufficient qualification of chemical safety production and non-compliant protection measures. Therefore, enterprises should pay attention to intrinsic safety design, improve the safety awareness of staff by popularizing safety knowledge and operational guidelines, and finally achieve the goal of safe production.

Table 1. Judgment matrix of primary evaluation indicators and weights results

A	B1	B2	B3	B4	B5	B6	Weights	Consistency	Ranking
B1	1	2	3	1/2	1/3	7	0.1645	λ_{max}=6.4698 CI=0.094 RI=1.26 CR=0.0746<0.10 Satisfy consistency	3
B2	1/2	1	2	1/3	1/4	6	0.1060		4
B3	1/3	1/2	1	1/4	1/5	5	0.0700		5
B4	2	3	4	1	1/2	8	0.2545		2
B5	3	4	5	2	1	9	0.3810		1
B6	1/7	1/6	1/5	1/8	1/9	1	0.0239		6

3.2 The comprehensive weights of the influencing factors

We built each judgment matrix of layer C to layer B and calculated the single layer weights of layer C, and finally, we obtained the comprehensive weights of incidents caused by bottom influencing factors. We considered the product of these two layers as the comprehensive weights of the sub-factor. For example, the comprehensive weights of C1 is equal to the single-layer weights of C1 to B1 multiplied by the single-layer weights of B1 to A. The comprehensive weights result and ranking of each sub-factor were shown in table 2.

Table 2. Single and synthetic weights of C layer indexes

		Single weights	Consistency	Synthetic	Ranking
B1	C1	0.0843	λ_{max}=4.1375 CI=0.0458 RI=0.9 CR=0.0515<0.1 Satisfy consistency	0.0139	16
	C2	0.2120		0.0349	8
	C3	0.1544		0.0254	12
	C4	0.5492		0.0904	3
B2	C5	0.1556	λ_{max}=5.2550 CI=0.0638 RI=1.12 CR=0.0569<0.1	0.0165	15
	C6	0.0566		0.0060	21

	C7	0.4735	Satisfy consistency	0.0502	5
	C8	0.2578		0.0273	11
	C9	0.0566		0.0060	22
B3	C10	0.6253	λmax=5.3598 CI=0.08995 RI=1.12 CR=0.0803<0.10 Satisfy consistency	0.0438	7
	C11	0.0901		0.0063	19
	C12	0.0435		0.0030	23
	C13	0.1976		0.0138	17
	C14	0.0435		0.0030	24
B4	C15	0.0781	λ_{max}=5.1717 CI=0.042925 RI=1.12 CR=0.0383<0.1 Satisfy consistency	0.0199	13
	C16	0.5895		0.1501	2
	C17	0.1271		0.0324	9
	C18	0.0781		0.0199	14
	C19	0.1271		0.0324	10
B5	C20	0.6833	λ_{max}=3.0246 CI=0.0123 RI=0.58 CR=0.0236<0.1 Satisfy consistency	0.2604	1
	C21	0.1168		0.0445	6
	C22	0.1998		0.0761	4
B6	C23	0.1242	λ_{max}=4.1891 CI=0.0630 RI=0.9 CR=0.0708<0.1 Satisfy consistency	0.0030	25
	C24	0.2626		0.0063	20
	C25	0.5465		0.0131	18
	C26	0.0667		0.0016	26

As can be seen from table 2, the full weights of C16 (Incomplete automatic control system) and C20 (Non-standard process design) was greater than 0.1. The full weights of C4 (Lack of chemical industry experience) and C22 (No properly designed production process) were very close to 0.1. The combined weights of these four factors reached 0.577, which was more than half of the total. Therefore, these four influencing factors were the primary causes of the explosion incident in Yibin, Sichuan, as strict control and prevention were given priority.

Among them, both C20 and C22 belong to the design aspect, and enterprises should entrust qualified formal companies to design the process and production plan. In the method of safety design, the operation and design should be analyzed in real time to check the status of dangerous materials, effectively reduce the operating temperature and pressure, and control the influence of hazards factors in the process on chemical equipment. The equipment used in production shall meet the strength and anti-corrosion performance required by the process, and shall be designed and manufactured by qualified factories.C4 is a human problem. Therefore, it is necessary to conduct chemical safety education and operation training for operators before entering the plant. Many of Hengda's workers were the relatives and friends of management personnel. They were generally poorly educated and skilled in operation, and had not received any training in chemical safety. The workers did not go through systematic learning, but the workshop director directly taught them operating skills. C16 is a problem of safety prevention. The automatic control system can not only improve the operation efficiency of production equipment but more importantly, it can protect the safety of chemical personnel and reduce the loss caused by incidents. Enterprises should follow the safety production regulations, install automatic control systems, toxic and harmful gas alarm systems, and fire protection systems to effectively prevent incidents.

The weights of C7 (Lack of education training for new employees), C10 (Period of personnel change) and C21 (Inconsistent with the safety evaluation report) were all around 0.05. The weights of C2 (Weak safety awareness), C19 (No fire protection system) and C17 (Not enough safety distance) also exceeded 0.03. These five factors were important causes of incidents and should be effectively controlled.

C7 is a new employee training management problem. Enterprises should pay attention to the educational training of employees before entering the plant and implement the three-level safety training education of the plant, workshop, and team. C10 involves change management. At the time of the incident, it was during the shift time of employees, the change of personnel and the change of materials and equipment, which had great potential risks. Enterprises should establish a change management system, conduct comprehensive training for the staff who play a vital role in the change management process and clarify the position and responsibility of each person in the change system.

4 Lessons learned from the causes of the incident

The Yibin explosion was a pharmaceutical chemical incident. Through the investigation and analysis of historical disasters in the chemical industry in China, it was found that the proportion of pharmaceutical chemical incidents in the chemical industry was huge [10]. It was characterized by a random change of chemical process, small scale, low level of process control, involving dangerous materials and complex chemical reaction. Especially in small and medium-sized chemical enterprises, the lack of process safety management is an essential cause of such incidents. From the perspective of process safety management, this paper summarized and put forward suggestions for preventing such incidents.

- Process Hazard Analysis - In the design stage, the risk identification and assessment tool system should be applied to identify the risks of materials in the process of technology. In the review stage, HAZOP, SIL are used to review the process, layout and material flow. Companies should use scientific models to calculate the explosion-proof grade and safety distance of buildings. By adding the explosion-proof wall and the explosion-proof film, the building close to the installation area can reach the required safety factor. After the device is put into operation, it needs to be compared with the hazard analysis in the design stage, and the hazard that is inconsistent with the design should be analyzed again. Hengda's actual operating equipment, raw materials, and products were inconsistent with the original plan, and the company had not discovered and controlled the real hazards [11].

- Pre-startup Safety Review - The company shall organize a safety inspection team, whose task is to confirm whether the improvement measures and safety guarantee measures in the process hazard analysis report have been implemented as required; Whether adequate training of personnel in operation, maintenance, and management of process equipment has been completed; Whether necessary safety information of process equipment is completed and updated; Whether necessary equipment testing and checking have been completed and confirmed; Whether the process equipment is constructed and installed according to the design requirements [12].

- Operation Procedures - The incident occurred during the change management phase, and there was no review or other risk assessment during the shift. Staff should review the operating procedures as often as possible to ensure that the operating procedures reflect current operating conditions, including changes in chemicals, process equipment, and facilities. The enterprise shall confirm the adaptability and effectiveness of the operating procedures and ensure that the operators understand and master the operating systems [13].

- Training - A complete process safety management is inseparable from the participation of operators, and the human factor is the most unstable factor in process safety management. Therefore, Enterprises should establish strict production operation disciplines and require operators to master process safety information, equipment integrity management points, and emergency response policies and finally improve workers' ability to control equipment safety operations [14].

- Emergency Plan and Response - Enterprises need to establish a set of overall emergency plans, which usually stipulate in written form how the factory should deal with abnormal or urgent situations. Also, enterprises should establish emergency response teams, traditionally composed of enterprise personnel, but also include external staff; The responsibilities of each team member should be clear, ensuring that members have no questions about accountability and delegation. In case of an emergency, the person in charge may determine the safe area according to the emergency response manual and direct the person to evacuate to a safe place [15].

5 Conclusion

The circumstances of the incident described in this article are typical in China, and the incident reflects the common safety problems in small and medium-sized chemical enterprises. Managers should understand these lessons we learned from this incident. By using fish-bone diagram method, the causes of the explosion incident were obtained through qualitative analysis, including six aspects, such as people, design, management, environment, safety prevention and materials, and 26 factors influencing the incident were further analyzed. Then the analytic hierarchy process was used for quantitative analysis, and the results showed that lack of chemical industry experience, non-standard process design, and incomplete automatic control system were the main factors affecting the incident, so the enterprise should strictly control and prevent these factors. Finally, we summarized the causes of such incidents and put forward the measures from five aspects of process safety management, including process hazard analysis, pre-production safety inspection, operation procedures, training, and emergency plan and response plan. An incident brings not an only economic loss but also bad social influence to the factory. It is our hope that this paper will help prevent future accidents.

6 References

[1] Sichuan Provincial Emergency Management Office, Yibin Hengda Technology Co., LTD. "7.12" Major Explosion Fire Accident Investigation Report, Available at http://www.chinasafety.gov.cn/gk/jsts/201902/t20190214_224397.shtml, Accessed on February 14, 2019.

[2] Tongyuan Luo, Chao Wu and Lixiang Duan, Fishbone diagram and risk matrix analysis method and its application in safety assessment of natural gas spherical tank, Ind. Eng. J. 25(02), 2007, 138-141.

[3] Xing, W., QC Team Based Materials, China Society Press, Beijing, China, 2004.

[4] T Deng, X., Chen, Q. and Jiang, D., Application of fishbone chart analysis method in prevention for steam turbine overspeed incident, Guangdong Electr. Power, 20(02), 2013, 73-77.

[5] Changlong Zhu, Jiayong Zhu, Li Wang and M. Sam Mannan. Lessons learned from analyzing a VCE incident at a chemical plant, Journal of Loss Prevention in the Process Industries,50,2017, 397-402.

[6] Dennis P. Nolan, Chapter 5: Specialized Reviews—CHAZOP, EHAZOP, Bow-Tie Analysis, Layers of Protection Analysis, Safety Integrity Level, FishboneDiagram, and Cyber Security Vulnerability Analysis, Safety and Security Review for the Process Industries (Fourth Edition), Gulf Publishing Company,2015,Pages 17-27.

[7] Xinhua News Agency, Emergency management department: 7·12 incident enterprises have no law and discipline, and government supervision is lax, Available at http://www.gov.cn/xinwen/2018-07/17/content_5307196.htm?_zbs_baidu_bk, Accessed on July 17, 2018.

[8] Seong KonLee, GentoMogi and Jong WookKim. Decision support for prioritizing energy technologies against high oil prices: A fuzzy analytic hierarchy process approach, Journal of Loss Prevention in the Process Industries,22(6), 2009, 915-920.

[9] Saaty T.L., A scaling method for priorities in hierarchical structures, Journal of Mathematical Psychology, 15 (3),1977, 234-281.

[10] Heda Zhang and Xiaoping Zheng, Characteristics of hazardous chemical incidents in China: A statistical investigation, Journal of Loss Prevention in the Process Industries, 25(4), 2012, 686-693.

[11] Hao WU, Igor PEÑARRUBIA, Lin CUI and Jinsong ZHAO, Process safety management considerations for biofuel production, Higher Education Press, 4(3), 2017, 357-367.

[12] Yuan Gao, Yunxiao Fan, Jing Wang, Xi Li and Jingjing Pei, The mediating role of safety management practices in process safety culture in the Chinese oil industry, Journal of Loss Prevention in the Process Industries, 57, 2019, 223-230.

[13] Paul R.Amyotte and Cathleen S.Lupien, Chapter Three - Elements of Process Safety Management, Methods in Chemical Process Safety, 1, 2017,87-148.

[14] T. Kontogiannis, M.C. Leva and N. Balfe, Total Safety Management: Principles, processes and methods, Safety Science, 100(B), 2017, 128-142.

[15] Hao WU, Igor PEÑARRUBIA, Lin CUI and Jinsong ZHAO, Process safety management considerations for biofuel production, Front. Eng. Manag, 4(3), 2017, 357- 367.

Taming the Wild River Rapids – How Process Safety Can Apply Outside the Process Industries

Trish Kerin
IChemE Safety Centre
Level 7, 455 Bourke St, Melbourne, Victoria, Australia
tkerin@icheme.org

Prepared for Presentation at
American Institute of Chemical Engineers
2019 Spring Meeting and 15th Global Congress on Process Safety
New Orleans, LA
March 31 – April 3, 2019

—

Taming the Wild River Rapids – How Process Safety Can Apply Outside the Process Industries

Trish Kerin
IChemE Safety Centre
Level 7, 455 Bourke St, Melbourne, Victoria, Australia
tkerin@icheme.org

Keywords: lessons learned, process safety, design, maintenance

Abstract

On the 25th October 2016 the Thunder River Rapids ride at Dreamworld on Australia's Gold Coast suffered a pump failure, resulting in the tragic deaths of four riders. This incident was investigated by the police and later a coronial inquest to determine how a tragedy like this could occur. As a process safety professional, it was interesting to follow the coronial inquest findings and then think about how process safety management may have helped to prevent this incident. Assuming that a process safety event can be defined as a high consequence or potential consequence low likelihood loss of control event [1], this incident had all the hallmarks. Therefore, it is possible to apply the IChemE Safety Centre (ISC) framework [2] for process safety to identify where the system failed and understand how it could be averted in future. This paper will explore various elements of the incident and how process safety management techniques may have helped, including that the ride had been modified to be different from all other similar rides, that the staff did not know how to stop the ride in an emergency and that all means to stop the ride required manual detection and intervention, with no alarms or shut down systems installed.

1 The Incident

The Thunder River Rapids ride at Dreamworld in Queensland, Australia, was a water ride, simulating white water rafting. Up to six riders sit in a circular raft, which flows along a turbulent river, including travelling up a conveyor to ride down a waterfall. The water flow is driven by pumps and made turbulent by the installation of underwater obstacles. It was in full operation on 25th October 2016, when the pump managing the water flow tripped out. This resulted in the water level on the ride dropping. This drop in water level caused a raft to become stuck on the rails at the conveyor leading to the disembarkation platform. This raft remained stationary for some 57 seconds while the ride continued to run before it was impacted by the next raft in the sequence. This following raft contained 6 passengers, four adults and two children. When this raft struck the stationary one, it flipped over, killing the four adult passengers. The two children managed to escape the raft [3]. This incident is

—

currently under a Coronial Inquest. As such this paper is written from several media outlets that have been reporting the proceedings and based on the evidence reported, not the findings of the Coroner. The Coronial Report is due to be published in mid 2019.

2 History of the Ride

The Thunder River Rapids ride was completed in 1986 and was likely an in-house design and construction [4]. Sometime after construction, the ride was modified to remove some of the timber slats that form the base of the conveyor [4]. Two in every three slats were removed, though is unclear why this was done. This did create significant gaps in the base of the conveyor belt. The ride was also refurbished some months prior to the incident. [5].

Over the history of the ride there had been a number of incidents that had occurred with the rafts and the pumps. This in itself is not unusual, in that pumps can trip out from time to time, however there had been an incident in 2001 where the pump had tripped, and a raft became stuck [6]. This was then struck by a second raft, which flipped over. In this incident there were no passengers in the raft, and therefore no injuries. The investigation found that the ride operator had failed to adequately monitor the ride and he was dismissed. Details of this incident were not shared broadly and the operators and maintainers of the ride in 2017 had no knowledge of it. At the time an email was written by a park engineer discussing the incident and stated "I shudder when I think if there had been guests on the ride" [3]. This demonstrates that someone at some time recognized the significance of this incident, but it was then lost in history.

3 Technical and Organizational Causes of the Incident

There is one key immediate cause of this incident, however there are many different and wide-ranging causal factors. The immediate cause has been determined to be the tripping of the water pump, which in turn allowed the water level to drop, causing the first raft to become stuck. This provided the obstacle that the second raft, with six passengers, struck, causing it to then flip over. But let us examine the broader aspects of this incident, from the pump reliability to the safety systems and organizational capability. The ISC framework [2] provides a useful template to examine the causes. This framework in represented by Figure 1. Fundamentally leadership underpins all aspects of process safety. There are six functional elements that must be robust in an organization to manage major hazard safety. These are; knowledge & competence; engineering & design; systems & procedures; assurance; human factors and culture. In this incident there were clear failures in each of these elements.

—

Figure 1. ISC Framework for Process Safety

3.1 Knowledge & Competence

The training of the ride operators took only 90 minutes and was unstructured with no evidence of competence consistently recorded [7]. There was a procedure manual however it was thought to be confusing by the ride operators [8]. The operator running the ride on 25th October 2016 had been trained for 90 minutes that morning, as it was her first day running the Thunder River Rapids. The junior operator stated that the trainer "showed me the front page of the manual and just flipped through it. I had to go through it myself. She said just sign off on it when you are happy." [9] The trainer stated that she sometimes had to "invent" training techniques because of the complexity of the control panel [8]. There was no clear guidance on how and when to use the emergency stop for the ride. The new operator testified that she was told "don't worry about it, you don't need to use it" with regard to the emergency stop button [10]. This is disputed by the person who delivered the training. The new operator believed it was the role of the senior operator to shut the ride down. It should be noted that the emergency stop would shut down the moving equipment with in two seconds, where the normal shutdown took around eight seconds. The ride operator that day stated that she "would have waited for the senior operator to tell her to hit the e stop" [10]. The senior operator that day was "surprised" to learn there was an emergency stop that would have shut the ride down within two seconds [7].

3.2 Engineering & Design

Intamin Amusement Rides designed and built the first river rapids rides in the world, and following this tragedy released a statement stating that the ride was "neither built by Intamin nor supplied by Intamin". The ride at Dreamworld was a custom build, and this was said to be common for rides built in the 1980's in the Gold Coast[4].

3.2.1 The Timber Slats

—

A few years after the ride was opened, a modification was made to the timber slats in the conveyor system [4]. The slats appear to serve at least two safety purposes. The first was to provide a solid platform for a passenger if they fell from the raft, essentially isolating them for the moving machinery and providing a platform to stand on [11]. The second was so that the rafts did not get caught in gaps and become stuck or indeed flip over. The Thunder River Rapids ride had two of every three slats removed. This provided large gaps between the remaining slats. There does not appear to be a theme park anywhere else that has made this change. There also appears to be no evidence as to why the change was made, nor was there a considered Management of Change (MoC) review undertaken. It is not clear if the organization had a structured and systematic MoC process. The safety regulator did not approve the modification and believed it added in additional hazards [4, 12]. In this incident, the raft became lodged in the gap, and as the conveyor continued to move forward, the raft was driven up vertically. When the passengers then fell from the raft, there was no solid base and they landed in the mechanism of the conveyor. There had also been conveyor framework modifications made to the ride in early 2016 in an attempt to stop rafts flipping over [13]. It is unclear what this modification was or indeed why it was made.

Figure 2. Aerial view of Thunder River Rapids showing the large gaps in the timber slat conveyor after the incident [10].

Figure 3. Aerial view of Infinity Falls, Orlando, showing close spacing of the boards on the conveyor [14].

3.2.2 The Pump

The water pumps had failed three times in the days leading up to the incident, and then twice on 25th October 2016, before the final failure that lead to the tragedy [15]. The pump's purpose was to keep the water flowing through the ride to keep the rafts moving. If the pumps stopped the water not only stopped flowing, it also drained back, and the water level dropped such that rafts could become stuck on the underwater obstacles that produced the rapids [16]. The reason for the pumps tripping out was not determined, the protocol seemed to be to just reset them to get the ride operating again [3]. There also is some confusion as to whether an alarm was sounded when the pumps tripped out, with an old procedure having the comment about an alarm crossed out [17]. At the time of the incident

—

the pumps were approximately 10 years old, and the manufacturer stated that they had a life expectancy of about a decade. There did not appear to be any plans to replace the pumps, however there was a maintenance request into the pump supplier in the week prior, but this visit did not eventuate [18].

3.2.3 Low Water

Low water in the ride was a critical hazard, as the rafts could become stuck or indeed flip. There had been previous incidents where low water level hazard led to rafts flipping, notably in 2001. There was no low water level switch that would shut down the ride if the water level became too low [3]. The only indication the operators had was to notice the water stain level on the wall of the ride canal to determine if the water was too low [19]. When the water level dropped on 25th October, one empty raft became stuck on the rails near the base of the conveyor. As the second raft containing four adults and two children engaged with the conveyor it was then driven into the stuck raft. A safety review conducted after the event stated that the primary cause was a lack of a safety rated water level detection system. It was believed that such a device could have been fitted for as little as AUD 3,000. The ride had been refurbished some seven months prior to the incident, but a device was not fitted. The nearby log ride at the park had such a safety device fitted [18].

3.2.4 Emergency Stop

Around two years prior to the incident an external safety report was written suggesting the emergency stop system should be simplified, however this was rejected by the company as the emergency system in place was seen to be an "acceptable risk". The stopping procedure for the ride contained several individual steps. There was a single button which would stop all components of the ride, however this was located at the unloading area and not the control panel where the operators worked. [20]

There was a great amount of confusion regarding the emergency stop button and what it actually did. The senior operator stated that they were "surprised" to learn there was an emergency stop that would stop all components and the junior operator on her first day was unsure what the button did [7]. She stated she was told to not worry about the button because she did not need to use it and testified, she "would have waited for my senior operation to tell me to press it". She also did not know the button would stop the conveyor from moving. [10] The maintenance workers who performed the daily checks on the ride also stated they did not know what the emergency stop button did as they never tested it [21].

3.2.5 Maintenance

Seven months prior to the incident a directive to cut all repairs and maintenance spending was ordered by the owners. At the time of the incident the facility was approximately AUD125,000 over maintenance budget for the year to date. Capital expenditure could continue to be spent during this time [16]. One maintainer stated that he performed comprehensive daily, monthly and annual checks on the ride, which include consumable

—

replacements such as nuts and bolts [22]. When the regulator inspected the ride after the incident, he noted it was in "generally poor condition" with excessive corrosion and rust on the electrical cables [23].

The maintenance planner noted that if spending could be deferred it would be. He also noted that the ride was shut annually for review, but on the nine occasions he witnessed a shutdown, he never saw an engineer inspect the ride nor were any risk assessment carried out [24].

3.2.6 Seatbelts

The Thunder River Rapids rafts had seat belts fitted for the safety of the riders. It should be noted that in May 2017, some seven months after the Dreamworld incident, that an 11 year old girl died on a similar ride, Splash Canyon, at Drayton Manor in the United Kingdom. The Splash Canyon ride rafts had no seatbelts installed at all, and as the child stood to swap seats with a fellow rider she fell overboard and struck her head. She died days later in hospital. This was not the first time a rider had fallen from the Splash Canyon ride [25]. This incident highlights the importance of proper restraints on these types of rides.

The seatbelts provided on the Thunder River Rapids rafts consisted of Velcro straps. It was vital that these were engaged correctly to provide the necessary restraint. Unfortunately, the Velcro was worn, and this made correct engagement more difficult. It is thought that at least one of the four riders may have survived the incident if the rafts had a better quality seatbelt, such as a mechanical buckle type that are on aircraft. This is because as the raft was shaken by the mechanism the belt failed to hold and the rider dropped to the conveyor system below [23, 26].

3.3 Systems & Procedures

Dreamworld has a series of operating procedures for the Thunder River Rapids ride however they were thought to be confusing [8]. During the investigation into the 2001 incident where empty rafts flipped, it was determined that the operator failed to follow the startup and emergency procedures. The investigation also reviewed the operating procedures and determined that the chance of a similar incident occurring with passengers on the rafts "is nil" [6]. A senior engineer believed that the cause of the 2016 incident was the operator not following procedures and that if procedures were followed the "system is a safe system" [27, 28].

There was also a lack of adequate emergency response procedures. Operators had discussed wanting better procedures and undertaking emergency drills, however, even two years after the incident, these have not been conducted at all [29].

3.4 Assurance

—

Over the history of the ride there had been a number of third-party reviews, inspections or audits carried out. Three years prior the incident a safety audit was conducted, and it recommended replacing the complicated emergency shut down process with a single emergency stop button, but this recommendation was never actioned [30].

There was also a legal requirement for the rides to have a "certificate to operate" and the Thunder River Rapids ride certificate was due to expire in January 2016. The company sought at extension as they were unable to find a "competent person" as defined under the legislation to inspect the ride [16]. Eight day before the incident the ride was inspected by a third-party engineer. He declared the ride as "structurally sound and properly functioning" even though he did not complete the full safety audit and merely did a visual inspection. He also did not inspect the log book as was required by law in a push to have the ride certified before it expired, though he admitted that he could have refused to certify the ride because of a "lack of documentation" [31, 32].

3.5 Human Factors

The control panel for the ride was confusing [3] and the operators needed to perform 36 different tasks in a one minute period, consistently [10]. This made for a system that encouraged human error, as it was not possible for the operators to be effective under these conditions. In addition to this, the engineering team believed the operators needed to simply follow procedures for the ride to be safe, not taking into account how difficult the procedures were or the actual workload. A senior engineer stated "In my view if procedures are followed the system is a safe system, you may as well not open the park. Procedures are there to be followed, if we don't follow procedures in everyday life, things happen." This blind focus on procedures does not take into account how the human interfaces with the machine and how the system advantages or disadvantages human performance.

3.6 Culture

A maintenance worker stated that he believed that the "safety culture at Dreamworld is very sound" [22] however this is in contrast to other workers who testified that familiarity had bred contempt with respect to the Thunder River Rapids ride [16], where new attractions were given more focus that existing ones [17]. Junior operators believed there was a culture of fear and had engaged with their union over their working conditions and the demands put on them regarding lack of rest breaks [33].

The culture also appeared to not learn from incidents, with the 2001 incident of flipping rafts said to be a one off and having "nil" chance of occurring with riders on board [6]. even though the consequence appeared to be understood at the time with an engineer writing in an email "I shudder when I think if there had been guests on the ride." [3].

The park relied on its apparent good safety history rather than challenging how and when an incident could occur [11]. They did not have an effective incident tracking system to

—

help with the learning process. They introduced a database but there was "general reluctance to use what was a fairly complex database" by the operators [34].

Safety did not appear to be a value in the company culture. The former safety manager stated that "everyone thought it was someone else's responsibility". He believed that deficiencies in the corporate structure led to a lot of problems. In 2018 his corporate safety role was made redundant.

Around five months prior to the incident, a memo was issued to staff to 'remain vigilant when monitoring conveyor movements" for items that might interfere with the rafts. This suggests there was some understanding of hazards and consequences, but it does to appear to have driven any improvement activities or modifications to prevent an incident [24].

4 Ramifications

The full ramifications of this incident are not yet clear. The Queensland government has initiated some legislative improvements to manage theme parks, but there is likely to be more recommendations coming from the Coronial Inquest when it releases its findings. Aside from the tragic loss of four lives, from the Dreamworld perspective, their guest numbers have remained low since the incident occurred, and it is struggling to attract patrons. This has led to their parent company suffering financial losses for the past two years and 50% decreases in the share price over the two year period [35].

The incident also provided an interesting lesson in crisis management. The company was scheduled to hold their annual general meeting in the days following the incident. The meeting went ahead and in it the CEO was awarded a performance bonus of $840,000 including both shares and a cash component despite the four deaths days before [36]. From a public relations perspective this was not a successful event. The CEO also stated they had been in contact with the victim's families, but this was publicly disputed by the families [37]. This further damaged the company reputation at a difficult time.

This incident will not only have ramifications for the families of the deceased, but for the survivors and the staff who responded due to the nature of injuries sustained by the victims. Four former staff members are suing Dreamworld for psychological damage [38].

These ramifications show how a process safety type incident can have significant impact on a business and its ability to continue operations in a positive way.

5 Could the Principles of Process Safety Have Helped?

As had been shown, there were clear and systematic failures across the whole organization that led to this tragedy. Starting with knowledge and competence, which relates to

—

organizational capability, had the operators been trained better in the safe operation of the ride as well as in emergency response, their performance that day may have been improved. From an engineering and design perspective, the design and installation of the ride was inadequate, it underwent undocumented changes which introduced hazards, and it was not adequately maintained. The systems and procedures for the operation and maintenance the ride was confusing and did not appear to contain necessary information. The assurance processes conducted did not seem to be thorough, rather they focused on keeping the ride operating and recommendations were not actioned. The ride was not designed with human factors in mind, from the high workload to the confusing operating panel. The organization culture did not seem to place a value on safety, it only focused on cost reduction. It also did not appear to be a learning culture, where lessons from the past are understood and embedded. Lastly at the time of crisis, the overall leadership displayed by the parent company was publicly found to be lacking.

There are transferrable learnings for the process industries from this incident, with some clear questions to ask. How are you managing your competency program to ensure you have the right capability? Are your designs adequate, maintained and any changes managed? Are your procedures thorough, identifying hazards, consequences and controls and easy to comprehend? How are you monitoring the operation of your facility? Does the design and operation of your facility enhance human performance or make it more difficult? Does your culture support safe operations?

6 Conclusion

A focus on the elements of process safety can help identify what is important in a business. In this instance there had been recommendations for a low-cost instrumented safety system to be installed that would have prevented the incident from occurring. The importance of this was not understood by the organization, and the opportunity to upgrade the ride during its recent refurbishment was ignored. It is clear that the hazard and potential consequences and controls were not understood. As is often the case, when the hazard is not understood and therefore not managed, people suffer tragic consequences.

7 References

[1] Kerin, T. *Managing Process Safety. In the Core Body of Knowledge for Generalist OHS Professionals*. Safety Institute of Australia, Tullamarine, VIC. 2017

[2] ISC (IChemE Safety Centre). Safety Centre Framework, Available at https://www.icheme.org/knowledge/safety-centre/framework/, Accessed on January 2, 2019

—

[3] ABC News, Available at https://www.abc.net.au/news/2018-06-18/dreamworld-fatal-accident-ride-had-broken-down-inquest-told/9871898, Accessed on January 2, 2019

[4] Australian Financial Review, Available at https://www.afr.com/news/dreamworlds-rapids-ride-different-from-others-around-the-world-20161101-gsfd07, Accessed on January 2, 2019

[5] 9 News, Video 11, Available at https://www.9news.com.au/2018/12/06/03/33/dreamworld-inquest-enters-final-days, Accessed on January 2, 2019

[6] news.com.au, Available at https://www.news.com.au/national/queensland/courts-law/coroner-james-mcdougall-requests-proof-all-dreamworld-rides-were-certified/news-story/e77275b97b0bbee9c1c541bda6df0035, Accessed on January 2, 2019

[7] ABC News, Available at https://www.abc.net.au/news/2018-06-19/dreamworld-operator-on-first-day-when-four-people-died-on-ride/9885214, Accessed on January 2, 2019

[8] ABC News, Available at https://www.abc.net.au/news/2018-06-27/dreamworld-trainer-promoted-since-four-killed-on-ride/9914188, Accessed on January 2, 2019

[9] 9 News, Video 5, Available at https://www.9news.com.au/2018/12/06/03/33/dreamworld-inquest-enters-final-days, Accessed on January 2, 2019

[10] ABC News, Available at https://www.abc.net.au/news/2018-06-20/dreamworld-operator-impossible-load-responsibilities-manage-ride/9889776, Accessed on January 2, 2019

[11] 9 News, Video 1, Available at https://www.9news.com.au/2018/12/06/03/33/dreamworld-inquest-enters-final-days, Accessed on January 2, 2019

[12] 9News, Available at https://www.9news.com.au/2018/11/20/03/32/dreamworld-ride-passed-wphs-safety-audit, Accessed on January 2, 2019

[13] ABC News, Available at https://www.abc.net.au/news/2018-10-08/dreamworld-inquest-hears-total-failure-to-ensure-safety/10350034, Accessed on January 2, 2019

[14] Intamin Worldwide, Available at https://www.intaminworldwide.com/infinity-falls-orlando/ , Accessed on December 12, 2018

—

[15] 9 News, Video 8, Available at https://www.9news.com.au/2018/12/06/03/33/dreamworld-inquest-enters-final-days, Accessed on January 2, 2019

[16] ABC News, Available at https://www.abc.net.au/news/2018-06-25/dreamworld-inquest-told-of-ride-maintenance-cutbacks/9906214, Accessed on January 2, 2019

[17] Australian Financial Review, Available at https://www.afr.com/real-estate/commercial/hotels-and-leisure/dreamworld-prioritised-new-attractions-over-maintenance-of-old-rides-inquest-20180626-h11urb, Accessed on January 2, 2019

[18] 9 News, Video 10, Available at https://www.9news.com.au/2018/12/06/03/33/dreamworld-inquest-enters-final-days, Accessed on January 2, 2019

[19] ABC News, Available at https://www.abc.net.au/news/2018-06-21/dreamworld-ride-operator-highly-distressed-give-evidence-inquest/9893308, Accessed on January 2, 2019

[20] news.com.au, Available at https://www.news.com.au/national/queensland/courts-law/dreamworld-engineer-tells-families-i-apologise-for-not-being-to-make-eye-contact-with-you/news-story/40545dc747c95eb96b0293c888d77d96, Accessed on January 2, 2019

[21] Sunshine Coast Daily, Available at https://www.sunshinecoastdaily.com.au/news/dreamworld-thunder-river-rapids-button-wasnt-regul/3543997/, Accessed on January 2, 2019

[22] ABC News, Available at https://www.abc.net.au/news/2018-06-26/dreamworld-ride-operator-says-conveyor-was-never-stopped/9910192, Accessed on January 2, 2019

[23] news.com.au, Available at https://www.news.com.au/national/queensland/courts-law/safety-report-shows-simple-change-on-dreamworld-ride-could-have-saved-lives/news-story/9c1fa2c195efd12f86a3e8ad516ea1b8, Accessed on January 2, 2019

[24] 9 News, Video 20, Available at https://www.9news.com.au/2018/12/06/03/33/dreamworld-inquest-enters-final-days, Accessed on January 2, 2019

[25] news.com.au, Available at https://www.news.com.au/travel/travel-updates/incidents/girl-11-dies-after-falling-into-the-water-on-splash-canyon-ride-during-school-trip-to-uk-theme-park/news-story/f83330b0ce12ea11df2d089ece676519, Accessed on January 2, 2019

---

[26] 9News, Available at https://www.9news.com.au/2018/12/06/03/33/dreamworld-inquest-enters-final-days, Accessed on January 2, 2019

[27] 9News, Available at https://www.9news.com.au/2018/10/16/14/43/dreamworld-inquest-thunder-river-rapids-ride-engineers-give-evidence, Accessed on January 2, 2019

[28] 9News, Available at https://www.9news.com.au/2018/10/17/10/49/dreamworld-inquest-engineer-to-appear-again, Accessed on January 2, 2019

[29] Australian Financial Review, Available at https://www.afr.com/news/dreamworld-ride-trainer-promoted-after-tragedy-staff-wanted-emergency-training-20180628-h11yke, Accessed on January 2, 2019

[30] 9 News, Video 12, Available at https://www.9news.com.au/2018/12/06/03/33/dreamworld-inquest-enters-final-days, Accessed on January 2, 2019

[31] ABC news, Available at https://www.abc.net.au/news/2018-10-09/dreamworld-ride-declared-safe-days-before-tragedy/10352164, Accessed on January 2, 2019

[32] ABC News, Available at https://www.abc.net.au/news/2018-10-17/dreamworld-supervisor-sympathy-has-families-in-tears-inquest/10387186, Accessed on January 2, 2019

[33] 9 News, Video 9, Available at https://www.9news.com.au/2018/12/06/03/33/dreamworld-inquest-enters-final-days, Accessed on January 2, 2019

[34] 9 News, Video 16, Available at https://www.9news.com.au/2018/12/06/03/33/dreamworld-inquest-enters-final-days, Accessed on January 2, 2019

[35] Australian Financial Review, Available at https://www.afr.com/research-tools/AAD/share-prices?days=1095, Accessed on November 26, 2018

[36] The Guardian, Available at https://www.theguardian.com/australia-news/2016/oct/27/deborah-thomas-set-to-receive-840000-in-bonuses-despite-dreamworld-deaths, Accessed on December 19, 2018

[37] ABC News, Available at https://www.abc.net.au/news/2016-10-27/dreamworlds-shareholders-ask-parent-company-to-support-victims/7970104, Accessed on January 2, 2019

—

[38] news.com.au, Available at https://www.news.com.au/lifestyle/real-life/news-life/dreamworld-first-responders-to-sue-theme-park-for-psychological-damage/news-story/e8fd7399eec9d165c657dbdb3ac35968, Accessed on January 2, 2019

Fatal Toxic Chemical Release at DuPont

Tamara Qureshi
U.S. Chemical Safety and Hazard Investigation Board
1750 Pennsylvania Ave, NW Suite 910 Washington, DC 20006
tamara.qureshi@csb.gov

Dan Tillema, Steve Cutchen, Lauren Grim, and Chris Lyon
U.S. Chemical Safety and Hazard Investigation Board
1750 Pennsylvania Ave, NW Suite 910 Washington, DC 20006

Prepared for Presentation at
American Institute of Chemical Engineers
2019 Spring Meeting and 15th Global Congress on Process Safety
New Orleans, LA
March 31 – April 3, 2019

Fatal Toxic Chemical Release at DuPont

Tamara Qureshi
U.S. Chemical Safety and Hazard Investigation Board
1750 Pennsylvania Ave, NW Suite 910 Washington, DC 20006
tamara.qureshi@csb.gov

Dan Tillema, Steve Cutchen, Lauren Grim, and Chris Lyon
U.S. Chemical Safety and Hazard Investigation Board

Keywords: DuPont, toxic release, gas detection, alarm

Abstract

On November 15, 2014, four workers were killed after approximately 24,000 pounds of highly toxic methyl mercaptan released from an insecticide production unit (Lannate® Unit) at the E.I. du Pont de Nemours chemical manufacturing facility in La Porte, Texas. This paper will discuss the events leading to the methyl mercaptan release, the deficiencies in DuPont La Porte's toxic gas detection system, and offer lessons on how to prevent some of the deficiencies in the toxic gas detection system from occurring. An effective toxic gas detection system can protect workers by alerting them of a potentially dangerous situation and preventing them from unknowingly entering a potential toxic chemical release.

1 Introduction

On November 15, 2014, approximately 24,000 pounds of highly toxic methyl mercaptan released from an insecticide production[1] unit (Lannate® Unit) at the E.I. du Pont de Nemours chemical manufacturing facility in La Porte, Texas. Four workers at the DuPont La Porte facility entered the Lannate® Unit's manufacturing building (manufacturing building) and the highly toxic methyl mercaptan release. Three of these workers were responding to a coworker's distress call from inside the manufacturing building. The fourth worker entered after all the workers inside the manufacturing building stopped communicating with workers inside the control room. The Lannate® Unit's methyl mercaptan toxic gas detection system did not effectively warn workers of the large chemical release occurring within the manufacturing building. Additionally, this toxic gas detection system did not provide emergency responders with sufficient information to effectively evaluate the potential harm to the public.

---

[1] The Insecticide Business Unit (IBU) was divided into two primary insecticide product lines: Lannate®, which is DuPont's trade name for methomyl; and Agricultural Products Intermediates (API), which produced insecticides and nematicides known as Vydate®, the trade name for oxamyl [3, p. 11].

The DuPont La Porte methyl mercaptan toxic gas detection system had several deficiencies. First, the methyl mercaptan detector system did not meet its design intent. The alarm set point for methyl mercaptan concentration and the method of conveying alarms did not forewarn workers of the immediately dangerous to life and health (IDLH) conditions they would encounter by entering the manufacturing building. The methyl mercaptan gas detectors only alarmed on the Lannate® control room operator's computer screen. There was no audible or visible alarm provided in the field for outside personnel. Second, the placement of the methyl mercaptan toxic gas detectors did not fully portray the extent of the November 15, 2014 methyl mercaptan release. Third, the DuPont La Porte site normalized the hazards associated with methyl mercaptan. Finally, the alarms did not prompt protective action or investigations into the sources of alarms. During the hours prior to the November 15, 2014 incident, multiple highly toxic chemical gas detectors alarmed (sounded). However, the DuPont emergency response team was not notified nor was the area cleared of personnel. In addition, methyl mercaptan releases on November 13 and 14, 2014, were identified by methyl mercaptan detectors, but were never reported as releases nor investigated as serious process safety incidents. Personnel believed they knew the source of the releases and normalized their responses to these alarms. Consequently, there was a delayed awareness of the release by operations, and workers who entered the manufacturing building after the release began were not alerted of nor prepared for the lethal concentrations of methyl mercaptan being released, resulting in the death of two of these workers.

The DuPont La Porte incident is the CSB's third investigation of a fatal DuPont incident in five years.[2] On September 30, 2015, the CSB issued a presentation, safety video, and an Interim Recommendations report [1],[3] which detailed needed safety improvements at the facility and issued recommendations to the DuPont La Porte facility to be implemented prior to resuming operations of the Lannate® Unit.[4] Due to the amount of methyl mercaptan stored at the DuPont La Porte site, the Lannate® process was covered by Occupational Safety Health Administration's (OSHA) Process Safety Management (PSM) program and the Environmental Protection Agency's (EPA) Risk Management Plan (RMP) Rule [2].

An effective toxic gas detection system can make the difference between personnel being prepared for a potentially dangerous situation or entering a potential toxic chemical release. To be effective, a toxic gas detection system needs to consider the chemical's hazards, the gas detector's

---

[2] The CSB also investigated an incident at a DuPont plant in Belle, West Virginia (three incidents that occurred in a 33-hour period over January 22-23, 2010, resulting in one fatality), and another in Buffalo, New York (a single incident that occurred on November 9, 2010, resulting in one fatality and one serious injury) [16], [17].

[3] In the Interim Recommendations, the CSB identified significant process safety management deficiencies at the DuPont La Porte facility, including delayed maintenance of safety-critical equipment, lack of written procedures, poor hazard analysis practices, lack of implementation of important inherently safer design concepts, and a lack of hazard recognition.

[4] In response to the CSB findings and recommendations, DuPont committed to making changes at the DuPont La Porte facility. DuPont La Porte, however, was not able to address all of these findings and recommendations, because in the spring of 2016 DuPont announced that it would not reopen the Lannate® Unit. The company had determined that "significant changes in market conditions during the period of the shutdown [would] persist over the long term [14]" and the cost required for the restart "[was] not a long-term viable and cost-efficient option for the DuPont Crop Protection business [15]." In 2017, DuPont dismantled and removed the buildings and equipment from the DuPont La Porte facility that were associated with its insecticide and herbicide business units, which included the Lannate® Unit.

capabilities, the location of the gas detectors, and the alarms associated with the gas detectors. This paper will discuss some of the events leading to the methyl mercaptan release, the deficiencies in DuPont La Porte's toxic gas detection system and offer lessons on how to prevent some of the deficiencies in the toxic gas detection system from occurring.

2 Events Leading to the Incident

On Monday, November 10, 2014, an inadvertent chemical dilution caused operating difficulties that forced a shutdown of DuPont La Porte's Lannate® Unit [3, pp. 20-21]. On Wednesday, November 12, 2014, operators tried to restart the Lannate® process. During the shutdown, the methyl mercaptan piping to the reaction section of the process plugged due to the formation of a clathrate hydrate (hydrate),[5] halting restart. At that time, operators did not know the source of the plugging and began troubleshooting the process to try to clear the piping, unaware that the solid hydrate had formed.

Two days later, on Friday, November 14, 2014, the troubleshooting efforts to clear the plugging were still ongoing. That morning, a technical team composed of engineers and other employees experienced in the Lannate® process, met with operations personnel to discuss troubleshooting options to clear the plugging. This meeting identified the likely scenario that water had entered the methyl mercaptan system and a solid hydrate formed as a result. Based on their understanding of the hydrate,[6] the technical team asked operators to put hot water on the outside of the methyl mercaptan piping, under the insulation, to warm the piping and its contents, to break-up (dissociate[7]) the plugging. The team also realized that when heated, methyl mercaptan would expand, requiring a safe place to vent to avoid over-pressuring the piping. To address this concern, operations personnel opened valves between the methyl mercaptan piping and a waste gas vent header, not recognizing that this alignment created a pathway for liquid methyl mercaptan to eventually release from the piping inside the manufacturing building when drain valves were opened. In addition, when operations personnel opened other valves while troubleshooting, such as drain valves, methyl mercaptan released to the atmosphere causing a strong methyl mercaptan odor[8] that could be smelled by site personnel.

---

[5] At low temperatures (≤ 52 °F) water and methyl mercaptan form a solid, ice-like material called methyl mercaptan clathrate hydrate [10]. Temperatures in the Houston area for the 24 hours preceding the incident averaged approximately 40 °F and had been consistently below 55 °F since Tuesday, November 11, 2014. Although the potential for methyl mercaptan hydrate formation had been identified in DuPont's methyl mercaptan technical standard and in a process hazard analysis years earlier, DuPont never implemented safeguards, such as heat tracing, or developed a procedure to dissociate the hydrate safely [3, p. 22].

[6] DuPont's methyl mercaptan technical standard identifies the potential to form a hydrate at low temperatures. The technical standard states, "[methyl mercaptan] will form a hydrate with water, which is a solid below 40 deg F [degrees Fahrenheit] per information provided by a [methyl mercaptan] supplier."

[7] Methyl mercaptan clathrate hydrates dissociate back to the liquid phases of methyl mercaptan and water at approximately 52 °F [10].

[8] Methyl mercaptan has a low odor threshold. It gives off a smell like rotten cabbage. It can be smelled at only two parts per billion (ppb) [9, p. 1]. OSHA's permissible exposure limit (PEL), however, is 10 parts per million (ppm), which is 5,000 times the minimum concentration it can be smelled. The low odor threshold of mercaptans makes them useful as fuel gas odorants [9, p. 2].

During this troubleshooting process, methyl mercaptan released both outside and inside the manufacturing building, triggering 32 methyl mercaptan gas alarms on the control panel throughout the 17 hours preceding the incident. While methyl mercaptan is a toxic and flammable chemical, and at high enough concentrations lethal, operations personnel did not respond in accordance with the nature of the emergency circumstances, because they associated the alarms with ongoing troubleshooting efforts. They did not perceive the methyl mercaptan alarms as signifying a serious hazard because they normalized[9] both the methyl mercaptan odor within the Lannate® Unit as well as the detector alarms associated with them.

Early in the morning on Saturday, November 15, 2014, the hot water from the hoses warmed the hydrate plugging, causing it to dissociate, clearing the plugging. Due to the valve alignment, liquid methyl mercaptan flowed into the waste gas vent header, located inside the manufacturing building [3, pp. 25-26]. This vent header, however, was not intended for liquid methyl mercaptan.

At 2:51 am alarms began to sound on the control system indicating high pressure in equipment located inside the manufacturing building. Operations personnel did not realize that liquid methyl mercaptan was causing the high pressure. Instead, they attributed the high pressure to a common, long-standing problem with process condensate (that DuPont La Porte personnel believed to be mostly water) accumulating in the vent header piping [3, p. 28]. The Lannate® control room operator (Board Operator) contacted the night shift supervisor (Shift Supervisor) and another operator (Operator 1) separately, asking them to help troubleshoot the high-pressure situation. The Shift Supervisor and Operator 1 each went separately to the manufacturing building, likely with the intent to drain liquid from the vent header piping manually.[10] Sometime between 3:01 am and 3:13 am, two sets of drain valves on the vent header piping, located on the third floor of the manufacturing building were opened (Figure 1). But instead of being comprised mostly of water as workers believed, liquid methyl mercaptan that had filled the piping escaped out of the valves, vaporized, and killed the Shift Supervisor.

Between 3:24 am and 3:26 am, three methyl mercaptan detectors inside the manufacturing building sensed at least 25 parts per million (ppm) methyl mercaptan[11] and triggered alarms at the control panel. The Board Operator—still focused on reducing the high pressure—did not realize a major chemical release inside the manufacturing building was occurring.

---

[9] Normalization of deviance is a phenomenon in which individuals or work teams gradually accept a lower standard of performance until the lower standard becomes the norm. It is typically the result of conditions slowly changing and eroding over time [11, p. 4].

[10] In 2011, DuPont installed a NOx (nitrogen oxides) reducing scrubbed (NRS) incinerator system. The waste gas vent header piping to the NRS Incinerator was designed without sufficient consideration of liquid accumulation. It contained low points where liquid could accumulate, and there was no engineered equipment provided, such as a knock out drum, to remove liquid safely from the waste gas vent header piping system. The liquid accumulation in the piping would create pressure in process equipment. Pressure buildup within the equipment was resolved by manually draining liquid from the waste gas vent header. See DuPont La Porte, Texas Chemical Facility Toxic Chemical Release: Interim Recommendations [3, p. 18].

[11] The methyl mercaptan detectors alarm at 25 ppm–well above the National Institute of Occupational Safety and Health (NIOSH) recommended exposure limit (REL) of 0.5 ppm and the OSHA enforceable (PEL) ceiling limit of 10 ppm. OSHA recommends that employers consider using the alternative occupational exposure limits because the agency believes that exposures above some of these alternative occupational exposure limits may be hazardous to workers, even when the exposure levels are in compliance with the relevant PELs [5]. The Cal-OSHA permissible exposure limit (PEL) is an 8-hour time-weighted average (TWA) of 0.5 ppm and NIOSH recommended exposure limit (REL) is a 15- minute ceiling limit of 0.5 ppm [7], [6].

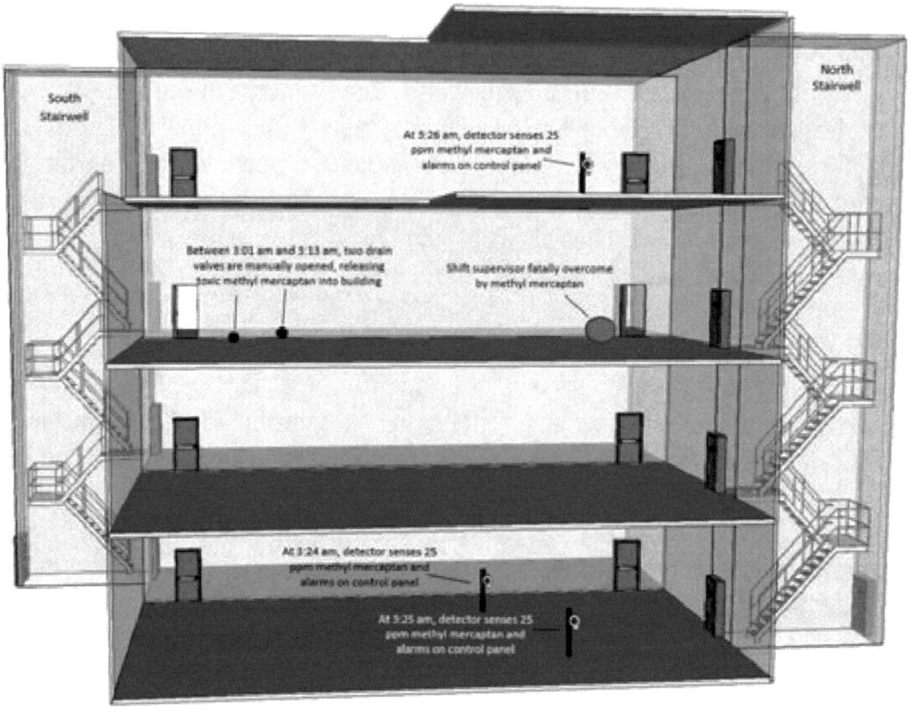

Figure 1. Depiction Showing Location Where Drain Valves Were Opened. These drain valves released toxic methyl mercaptan into the manufacturing building. Methyl mercaptan detectors on the first and fourth floors detected high concentrations of methyl mercaptan shortly after the release began. This depiction is a simplified graphic of the manufacturing building and does not show the location of Operator 1. Source: CSB

At about 3:30 am, Operator 1 made an urgent call for help over the radio. Various personnel interpreted her communication in different ways, generally as either "We need help!" or "I need help on the fourth floor!" The Board Operator tried to get more information through radio communication, but neither the Shift Supervisor nor Operator 1 ever responded [3, p. 29].
Two operators (Operator 2 and Operator 3) who were in the control room and heard the distress call ran to the manufacturing building to help. Another operator (Operator 4) saw Operator 2 and Operator 3 running to the manufacturing building, and he followed them. None of these operators knew there was a major release of toxic methyl mercaptan inside of the manufacturing building, and therefore they did not wear any respiratory protection when they ran into the manufacturing building. The manufacturing building lacked automatic visual or audible alarms to alert field workers or prevent them from entering a potentially toxic atmosphere.

Operator 2, Operator 3, and Operator 4 each entered the south stairway and took different routes inside the manufacturing building to try to find Operator 1. Operator 2 went to the third floor, where he was fatally overcome by methyl mercaptan [3, p. 29]. Operator 4 went to the second floor. He walked about ten feet and hit what he later described as a "wall" of methyl mercaptan, but he managed to retreat to the stairwell [3, p. 29]. Operator 3 went to the fourth floor but did not find anyone. He then announced on the Lannate® Unit's public-address system that he did not see anyone on the fourth floor. The Board Operator responded that the Shift Supervisor and Operator 1 may be on the third floor. Operator 3 then began to feel light headed. He made his way to the stairwell and lost consciousness while descending from the fourth floor [3, p. 29].

When Board Operator could not reach people inside the manufacturing building, an operator in the control room, Operator 6[12] (the brother of Operator 2), then grabbed three 5-minute escape respirators. Other operators in the control room warned him not to enter the manufacturing building because they did not know what was going on or where the non-responsive operators were. Operator 6, however, took the escape respirators and rushed to the manufacturing building [3, p. 29].

At about 3:40 am, Operator 6 encountered Operator 4 in the south stairway of the manufacturing building and put an escape respirator on him. The breathing air helped Operator 4 recover, and he exited the manufacturing building safely. A worker—likely Operator 6—then manually activated the manufacturing building fume release alarm, an alarm intended to alert area workers of a toxic chemical release in the building. Operator 6 then went to the third floor of the manufacturing building. At some point inside the manufacturing building after helping Operator 4, Operator 6 retrieved a self-contained breathing apparatus (SCBA). He tried to rescue his brother (Operator 2) by putting one of the escape respirators on his face. Operator 6 was then fatally overcome by methyl mercaptan before he could connect his own respirator mask to his SCBA tank.

3 Lessons from the DuPont La Porte Incident

3.1 Ensure Toxic Gas Detector Systems Operate as Intended

DuPont La Porte's methyl mercaptan technical standard stated, "[a]utomatic or continuous specific chemical detection systems *must* be installed where a [methyl mercaptan] release could have any off-site impact [...] [Emphasis in the original]." DuPont La Porte's methyl mercaptan toxic gas detection system consists of methyl mercaptan detectors, alarms, distributed control system (DCS), control panel, and a control room operator. The manufacturing building is equipped with gas detectors for methyl mercaptan in various locations. The third floor, where the release occurred, however, did not have any methyl mercaptan detectors. The detectors inside the manufacturing building were also set to alarm on the control panel when they detected a 25 ppm concentration of methyl mercaptan and were intended to protect workers from exposure to methyl mercaptan.[13]

DuPont used hydrogen sulfide (H_2S) detectors to identify methyl mercaptan. DuPont records indicate that the manufacturer has stated the detector should be capable of detecting methyl mercaptan within 1 ppm.[14] The alarms activated when the detectors sensed the hydrogen sulfide equivalent to 25 ppm methyl mercaptan.[15] The implementation of the methyl mercaptan toxic gas

[12] Operator 6 was a board operator for another process in the Lannate® Unit.

[13] The 25 ppm alarm set point is 50 times the NIOSH's REL of 0.5 ppm and more than double OSHA's ceiling limit of 10 ppm [7]. Therefore, alarming at this detection level could not protect workers from exposure to methyl mercaptan.

[14] DuPont La Porte's detectors for methyl mercaptan are calibrated for 0-50 ppm using a hydrogen sulfide (H_2S) calibration gas.

[15] The methyl mercaptan detectors were evaluated post-incident. Although the detectors inside the building have the ability to span 0-50 ppm, these detectors were set to span 0-25 ppm. Post-incident, the span was increased to 50 ppm. This configuration of the alarms limited the amount of information communicated to DuPont La Porte personnel.

detection system inside the manufacturing building, however, did not meet the design intent, federal safety regulations, or DuPont corporate standards.

The methyl mercaptan detectors outside and inside the manufacturing building were installed for two different purposes—the outside detectors were geared towards an offsite release and the inside detectors were meant to protect workers from a hazardous environment. DuPont La Porte documentation stated that the basis for the alarms outside the manufacturing building was to provide an early warning mechanism for a potential offsite methyl mercaptan release. The alarm setpoint was based on the ERPG (Emergency Response Planning Guidelines)-2[16] concentration[17] for methyl mercaptan which is 25 ppm. DuPont La Porte installed the methyl mercaptan detectors inside the manufacturing building after the methyl mercaptan detectors outside the building. Unlike the outside detectors, there was no documentation for the basis for setting the alarm setpoint to 25 ppm for the methyl mercaptan detectors inside the manufacturing building nor was there evidence that these setpoints were set at 25 ppm to prevent methyl mercaptan alarms from becoming a nuisance. It appears that the setpoints were copied from the outside methyl mercaptan detectors even though the inside detectors served a different purpose—worker protection. The methyl mercaptan detectors alarmed at 25 ppm—well above the NIOSH recommended exposure limit (REL) of 0.5 ppm and the OSHA enforceable PEL ceiling limit of 10 ppm.

Depending on the purpose of a toxic gas detection system alarm, there are different sets of values for facilities to consider. For instance, ERPG values are designed for the protection of the public during an emergency and not intended for the everyday routine protection of workers [4].[18] For worker protection, OSHA recommends that employers consider using the lower alternative occupational exposure limits because the agency believes that exposures above some of these alternative occupational exposure limits may be hazardous to workers, even when the exposure levels are in compliance with the relevant PELs [5]. For example, the Cal-OSHA permissible exposure limit (PEL) is an 8-hour time-weighted average (TWA) of 0.5 ppm and NIOSH's REL is a 15- minute ceiling limit of 0.5 ppm [6], [7].

Additionally, information from the methyl mercaptan detectors inside the manufacturing building only displayed on the DCS monitor and was not recorded by the site's DCS and data historian [3, p. 37]. CSB investigators, however, were able to obtain information from alarm logs about the detectors.

DuPont La Porte's management of change documentation for installation of methyl mercaptan detectors inside the manufacturing building stated the condition prompting the change was the

[16] ERPG-2 is the maximum airborne concentration below which nearly all individuals could be exposed for up to 1 hour without experiencing or developing irreversible or other serious health effects or symptoms which could impair an individual's ability to take protective action. At concentrations above ERPG-2, some members of the community may experience significant health effects which could impair them from taking protective action. Emergency responders often alert the community to take protective action in areas susceptible to the ERPG-2 concentration and above [19].

[17] The ERPG values for methyl mercaptan in parts per million (ppm) are: ERPG-1, 0.005; ERPG-2, 25; ERPG-3, 100 [18, p. 6].

[18] Emergency Response Planning Guidelines (ERPGs) are chemical concentration levels that are used during emergency response and when preparing for emergencies. They are designed to anticipate public health effects not occupational health effects from exposure to airborne chemicals [4].

possibility that personnel could enter a potentially hazardous location without knowing methyl mercaptan had been released:

> *Present Conditions and Description of Proposed Change:*
>
> *Personnel can enter a potentially hazardous location without knowing MeSH [methyl mercaptan] has been released to the environment.*

The methyl mercaptan gas detectors, however, only alarmed on the Lannate® control room operator's computer screen.

DuPont La Porte site did not have audible or visual alarms on or in the manufacturing building to alert field personnel of a methyl mercaptan release automatically. DuPont relied on verbal communication of alarms that automatically displayed on a continuously manned control board. Without being made aware of the major chemical release inside the building, Operator 2, Operator 3, and Operator 4 entered without respiratory protection. As a result, Operator 2 was killed from methyl mercaptan exposure, and Operator 3 and Operator 4 were exposed to dangerous concentrations of methyl mercaptan.

Facilities should make strategic decisions when choosing a toxic gas detector to ensure that a toxic gas detection system operates as intended. Considerations should include regulatory requirements, NIOSH recommendations, chemical interferences, and utility (worker protection, release modeling, etc.). The methyl mercaptan detectors at the DuPont La Porte facility had a span of 0-50 ppm. This range may be appropriate for early detection of a toxic leak when coupled with alarms and clear instructions to evacuate the premises at lower concentrations. If the goal is to understand other aspects of a release, like potential lethal concentration (150 ppm is methyl mercaptan's IDLH concentration) or modeling an offsite release, then a wider concentration range should be considered.

Additionally, facilities should also consider the use of automatic alarms designed to alert personnel of hazardous conditions—such as a release of a toxic chemical or explosive atmosphere—which can relay critical safety information immediately to personnel, without potential delays inherent in a system that solely relies on a control room operator to verbally communicate safety alarms to operators in the field.

3.2 Ensure Toxic Gas Detector Placement is Strategic

Although the DuPont La Porte site had 13[19] fixed methyl mercaptan detectors inside and outside the manufacturing building, there were no fixed methyl mercaptan detectors located on the third floor inside the manufacturing building where a majority of the November 15, 2014 methyl mercaptan release occurred. Additionally, no workers were provided with either portable or personal methyl mercaptan monitors. Part of the Lannate® process occurred inside the closed manufacturing building, a four-story building containing a large amount of piping and process equipment, dividing walls, multiple stairwells, and an elevator. The Lannate® Unit had methyl

[19] The site had a total of 15 methyl mercaptan detectors—there were two additional methyl mercaptan detectors installed in another process building adjacent to the manufacturing building.

mercaptan process piping both inside and outside the manufacturing building. A facility siting study and a process hazards analysis had identified the necessity of having methyl mercaptan detectors in the Lannate® manufacturing building. These studies identified the need for protecting workers from toxicity and explosivity hazards related to potential methyl mercaptan releases.

DuPont La Porte installed 10 methyl mercaptan detectors near process piping outside the manufacturing building. The design basis for the outside methyl mercaptan detectors was to provide an early warning of a methyl mercaptan release. DuPont La Porte placed these detectors around its methyl mercaptan storage tank and the area where methyl mercaptan could be unloaded. There were, however, no fixed methyl mercaptan detectors along the fence line that could relay whether there was an offsite release.

DuPont La Porte installed three methyl mercaptan detectors inside the manufacturing building. The stated design purpose for the methyl mercaptan detectors inside the building was worker protection. The inside detectors were located within the building on the first and fourth floors. These detector locations were determined based on potential release scenarios from a 2001 process hazard analysis (PHA).

The methyl mercaptan release occurred on the third floor where there were no methyl mercaptan detectors. Although the 2001 PHA did not identify methyl mercaptan detector locations on the third floor, DuPont La Porte routed the waste gas vent header piping along the third floor and this piping was directly connected to the top of the methyl mercaptan storage tank. During abnormal or upset conditions, the waste gas vent header could have been a potential source for a methyl mercaptan release. For example, ineffective hazard management while troubleshooting the plugged methyl mercaptan feed piping during the November 15th incident allowed liquid methyl mercaptan to flow into the waste gas vent header piping towards the NRS incinerator—a location where it was never intended to go.

Additionally, the routine draining of liquid from the waste gas vent header should have triggered a management of change review and considered whether additional safeguards like methyl mercaptan detectors were necessary. As discussed earlier, DuPont La Porte never fully resolved liquid accumulation in the waste gas vent header piping to the NRS incinerator with an engineered solution. Instead, operators were directed to drain the system by area management through written instructions in a daily instruction log book to operations staff [8, p. 18].[20] This routine draining presented potential dangers because DuPont La Porte personnel never sampled or analyzed the liquid workers routinely drained from this system [8, p. 18]. Daily instructions before the incident requested the operators to sample the uncharacterized liquid. DuPont La Porte management wanted samples of the liquid for a subsequent laboratory analysis to identify its chemical composition. Workers reported that the liquid had a strong, noxious odor of sulfur compounds [8, p. 18]. Even though DuPont La Porte was aware that the liquid likely contained more than just water and was unaware of the actual chemical composition [8, p. 18], instructions to workers did

[20] Beginning January 14, 2014, and continuing until the incident, the daily instructions directed operators to drain the methyl mercaptan waste gas vent header's "low point" on the 3rd floor once per shift. After the installation of a flexible hose from the waste gas vent header drain valve to the floor, the instructions changed to include directing the liquid through the hose under a running safety shower until the liquid stops flowing from the hose [8, p. 18].

not specify that additional precautions—like wearing additional PPE—were needed when draining the liquid from the waste gas vent header to the floor inside the manufacturing building.

Consequently, DuPont La Porte workers dealt with the common problem of liquid accumulation in the waste gas vent header on a routine basis by draining the liquid (line breaking) without engineered solutions or ensuring the use of safety procedures or personal protective equipment. On November 15, 2014, when the liquid drain valves were opened this time, flammable and highly toxic methyl mercaptan flowed onto the floor and filled the manufacturing building with toxic vapor and there no methyl mercaptan detectors on the third floor to alert personnel of the hazardous conditions that existed there.

As with toxic detector selection, the placement of toxic gas detectors must be strategic in order to capture the potential areas where a release could occur and the extent of the release. The methyl mercaptan detector placement both externally and internally at the DuPont La Porte facility did not operate as intended in part because the detectors did not fully cover the areas where the majority of the methyl mercaptan was released. For example, there was no real time data for emergency responders to use to evaluate the potential for offsite concentrations during the methyl mercaptan release. Although emergency responders conducted offsite monitoring hours after the initial release, the information obtained from air monitoring could not give an accurate portrayal of the methyl mercaptan release. Furthermore, the methyl mercaptan monitors inside the building did not provide accurate information about the release because they were not located near the open drain valves.

Facilities can adopt a variety of methods to select the location of toxic gas monitors with the understanding that they should be prepared for scenarios where the monitoring location may not be effective. One technique that a facility could use is to develop a series of release scenarios and based on those scenarios place monitors in the path of those releases (including by the potential sources of a leak). This technique, however, may not cover all potential release areas. Additionally, facilities should evaluate additional locations for toxic gas detectors if there is a change in process that creates new potential release scenarios. Another technique, if focusing on worker protection, is to give workers personal monitors when they enter toxic gas hazardous areas, especially if these areas are not normally occupied or congested with equipment. At DuPont La Porte, commercially available personal hydrogen sulfide monitors could have been issued to workers. Additionally, if using fixed toxic gas monitors for personnel protection, they should be installed near the breathing zone.

3.3 Ensure Personnel Understand Toxic Gas Hazards

DuPont La Porte personnel did not fully appreciate the effects of methyl mercaptan's physical characteristics. Although methyl mercaptan[21] is a highly toxic, flammable chemical, its low odor

[21] Personnel at DuPont La Porte would refer to methyl mercaptan as "MeSH." MeSH is shorthand based upon methyl mercaptan's chemical formula: CH_3SH. "Me" represents the methyl group (CH_3) and "SH" represents the mercaptan group (SH)—and when combined, these initialisms forms MeSH.

threshold[22] and ability to cause olfactory fatigue in personnel can generate a false sense of security when working with the chemical.

Although DuPont's methyl mercaptan technical standard states, "[a]ll DuPont sites *will* protect the site and neighboring businesses and communities from the escape of a highly toxic material [,]" methyl mercaptan detectors provided essentially no value to DuPont personnel [Emphasis in the original]. DuPont operations staff became accustomed to the odor of methyl mercaptan being present in their work environment. The low odor threshold of methyl mercaptan resulted in a false perception of safety. Consequently, the workers did not associate detecting the odor of methyl mercaptan as representing an immediate hazard.

> *"And I smell MeSH [methyl mercaptan] all the time ... but it never ... I smelled MeSH but I never felt I couldn't breathe. Like whether it's, there's -- whether it's a small leak or anything, it just never -- I didn't think MeSH was that dangerous. I thought it, you know, but it is."*

As seen by this comment from a DuPont La Porte worker, smelling methyl mercaptan at low concentrations without having any observable, adverse effects can potentially normalize the danger of a highly toxic chemical.

Furthermore, extended exposure to methyl mercaptan could potentially eliminate the perception of a methyl mercaptan release because olfactory fatigue will prevent workers from being able to smell the chemical. The odor threshold for methyl mercaptan is approximately two parts per billion (2 ppb). However, olfactory fatigue is known to occur with methyl mercaptan, resulting in the exposed person to no longer being able to detect the chemical by smell. As a result, using odor as an indicator of a release may not provide adequate warning of hazardous concentrations [9].

In the events leading to the major methyl mercaptan release, there were many smaller releases, filling the area with methyl mercaptan odor. In part because personnel believed they understood the source of releases and normalized the risk associated with these smaller releases, personnel did not apply additional safeguards like clearing the area or wearing the required personal protective equipment required by the site when dealing with a methyl mercaptan release.

Facilities should ensure that employees understand the hazards of chemicals that they are working with and may be potentially exposed to if there is a release. In cases where there are multiple hazardous chemicals, such as methyl isocyanate[23] (MIC) and methyl mercaptan (as was the case at DuPont La Porte), it is important that companies effectively train workers on the hazards of each chemical at their facility. Furthermore, companies should train workers that handle multiple highly toxic chemicals, such as methyl mercaptan and MIC, to treat these chemicals with equal

[22] Odor threshold, in general, is the lowest concentration of a gas or a material's vapor that can be detected by odor. Odor threshold values are not fixed physiological parameters or physical constants but are statistical points representing the best estimate value from a group of individual responses [12, p. 2].

[23] Methyl isocyanate (MIC) is a raw material used in the production of insecticides at La Porte including the Lannate® process. In December 3, 1984, a methyl isocyanate release at the Union Carbide insecticide plant in Bhopal, India resulted in an estimated 3,800 people that died within days, and tens of thousands that were injured. Eventually, the release killed tens of thousands of people. See CSB Safety Message on 30th anniversary of fatal chemical release that killed thousands in Bhopal, India [13].

importance. This training is important for workers to help ensure they have a working knowledge of those chemical hazards.

3.4 Ensure Response to a Toxic Gas Detector Alarm Is Appropriate

The toxic gas detection system at DuPont La Porte had a history of false readings. Because of these false readings, personnel supplanted their ability to observe and detect a leak over the toxic gas detection system. Personnel adapted their responses to a potential toxic gas release to include determining whether the toxic gas detectors were functioning properly by using their perception of smell before implementing the site's leak response procedures.

For example, the Lannate® Unit used chlorine in addition to methyl mercaptan. Workers could observe from a distance a chlorine leak by the green color of the gas or the immediate adverse effects (i.e., difficulty breathing) of the gas. Workers, similarly, would also use their sense of smell to detect methyl mercaptan. Methyl mercaptan gas, however, is colorless and the odor is not a reliable indicator of a release (for the reasons noted above). In both situations where workers used their physical observations to determine where there was a leak, workers were at risk for being overcome by a toxic chemical release because they were not using additional safeguards like personal protective equipment (PPE).

DuPont's highly toxic material standards require that toxic gas detection systems be treated the same as safety interlocks. DuPont defines a safety interlock as a system or function that detects an out-of-limits (abnormal) condition or improper sequence and brings it to a safe condition. In this case, an alarm should trigger a predetermined response from operators to bring the system to a safe condition. In practice, the methyl mercaptan detectors were not treated as a safety interlock.

Multiple DuPont employees informed the CSB that odor was the primary method used to locate a methyl mercaptan leak. DuPont personnel routinely responded to methyl mercaptan detector alarms without any additional respiratory protection. When they responded, if a significant leak was identified, these personnel consistently stated they would leave to get a SCBA and then return.

On the night of the incident, multiple methyl mercaptan detector alarms activated, but because the workers were aware that methyl mercaptan was being released from their activities associated with clearing the hydrate, these alarms were normalized and not treated as a safety interlock. The alarms were not treated as a near miss and there was no actual investigation into the source or how to prevent these releases. No additional personal protective equipment, such as air-purifying respirators or SCBAs was used. In addition, DuPont did not order a unit evacuation and did not notify the emergency response team until hours later, after Operator 1 made the distress call.

When a toxic detection system has a history of false readings it creates an opportunity for workers to normalize toxic gas detector readings and their response, making the toxic gas detection system ineffective. Facilities should take steps to address the potential for false readings and reduce opportunities for workers to normalize false readings and their alarms. One method is to treat toxic gas detector alarms as a near miss and investigate them even if there are no injuries or the readings are determined to be false. This method should trigger an effort to understand alarm management and safety management system issues and address them. Additionally, facilities should ensure that workers are using leak response procedures to prevent personnel from entering into potential hazardous situations without proper safeguards, like PPE.

4 Conclusions

The Lannate® Unit's toxic gas detection system did not effectively warn workers of the November 15, 2014 methyl mercaptan release occurring inside the manufacturing building. The manufacturing building, including the building's overall construction, detectors, and alarms, was not designed such that personnel could be readily notified of a major chemical release in the manufacturing building. Additionally, the methyl mercaptan toxic gas system did not provide real-time, accurate information to warn of potential offsite consequences.

There are several lessons that facilities can learn from the incident to help improve their toxic gas detection systems. Gas detectors must operate as intended to prevent workers from normalizing the hazard the detectors are protecting workers from. For toxic gas detection system to be effective, the detector selection and placement have to be strategic to identify and quantify releases. Finally, personnel have to understand and appreciate the hazards associated with the toxic gases they work with to help ensure they respond appropriately to potential toxic gas releases.

5 References

[1] U.S. Chemical Safety and Hazard Investigation Board (USCSB), "DuPont La Porte Facility Toxic Chemical Release," [Online]. Available: https://www.csb.gov/dupont-la-porte-facility-toxic-chemical-release-/. [Accessed 1 March 2019].

[2] U.S. Environmental Protection Agency (USEPA), "List of Regulated Substances under the Risk Management Plan (RMP) Program," [Online]. Available: https://www.epa.gov/rmp/list-regulated-substances-under-risk-management-plan-rmp-program. [Accessed 2017 August 28].

[3] U.S. Chemical Safety and Hazard Investigation Board (USCSB), "DuPont La Porte, Texas Chemical Facility Toxic Chemical Release: Interim Recommendations," 30 September 2015. [Online]. Available: https://www.csb.gov/assets/1/20/dupont_la_porte_interim_recommendations_2015-09-30_final1.pdf?15526. [Accessed 28 February 2019].

[4] National Oceanic and Atmospheric Administration (NOAA) Office of Response and Restoration, "Emergency Response Planning Guidelines (ERPGs)," [Online]. Available: https://response.restoration.noaa.gov/oil-and-chemical-spills/chemical-spills/resources/emergency-response-planning-guidelines-erpgs.html. [Accessed 10 March 2019].

[5] United States Department of Labor (USDOL) Occupational Safety and Health Administration (OSHA), "Permissible Exposure Limits - Annotated Tables," [Online]. Available: https://www.osha.gov/dsg/annotated-pels/. [Accessed 28 February 2019].

[6] United States Department of Labor (USDOL) Occupational Safety and Health Administration (OSHA), "Annotated Table Z-1," [Online]. Available: https://www.osha.gov/dsg/annotated-pels/tablez-1.html. [Accessed 28 February 2019].

[7] Centers for Diseases Control and Prevention National Institute of Occupational Safety and Health (NIOSH), "NIOSH Pocket Guide to Chemical Hazards - Methyl Mercaptan," 11 April 2016. [Online]. Available: https://www.cdc.gov/niosh/npg/npgd0425.html. [Accessed 28 February 2019].

[8] U.S. Chemical Safety and Hazard Investigation Board (USCSB), "DuPont La Porte, Texas Chemical Facility Toxic Chemical Release: Interim Recommendations," 30 September 2015. [Online]. Available: https://www.csb.gov/assets/1/20/dupont_la_porte_interim_recommendations_2015-09-30_final1.pdf?15526.

[9] Agency for Toxic Substances and Disease Registry (ATSDR), "Methyl Mercaptan," [Online]. Available: https://www.atsdr.cdc.gov/MHMI/mmg139.pdf. [Accessed 28 February 2019].

[10] A. H. Mohammadi and D. Richon, "Equilibrium Data of Sulfur Dioxide and Methyl Mercaptan Clathrate Hydrates," *Journal of Chemical & Engineering Data,* vol. 56, pp. 1666-1668, 2011.

[11] Center for Chemical Process Safety (CCPS), Recognizing Catastrophic Incident Warning Signs in the Process Industries, Hoboken: Wiley, 2012.

[12] A. H. Lehocky, S. S. Murnane and D. Patrick, Odor Thresholds for Chemicals with Established Occupational Health Standards, Falls Church: AIHA, 2013.

[13] USCSB, "On 30th Anniversary of Fatal Chemical Release that Killed Thousands in Bhopal, India, CSB Safety Message Warns it Could Happen Again," [Online]. Available: http://www.csb.gov/on-30th-anniversary-of-fatal-chemical-release-that-killed-thousands-in-bhopal-india-csb-safety-message-warns-it-could-happen-again-/?pg=4.

[14] K. Baker, "DuPont Closing Texas Plant Where 4 Died," The News Journal, 1 April 2016. [Online]. Available: https://www.usatoday.com/story/money/2016/04/01/dupont-closing-texas-plant-where-4-died/82512270/. [Accessed 29 June 2017].

[15] DuPont, "Form 8-K," 31 March 2016. [Online]. Available: https://www.sec.gov/Archives/edgar/data/30554/000003055416000115/dd33116form8-k.htm.

[16] U.S. Chemical Safety and Hazard Investigation Board (USCSB), "E.I. DuPont de Nemours & Co., Inc. Belle, West Virgina Investigative Report," September 2011. [Online]. Available: https://www.csb.gov/assets/1/20/csb%20final%20report.pdf?13966. [Accessed 1 March 2019].

[17] U.S. Chemical Safety and Hazard Investigation Board (USCSB), "E. I. DuPont De Nemours Co. Fatal Hotwork Explosion," [Online]. Available: https://www.csb.gov/e-i-dupont-de-nemours-co-fatal-hotwork-explosion/. [Accessed 1 March 2019].

[18] American Industrial Hygiene Association (AIHA), "Current ERPG Values," 2014. [Online]. Available: https://www.aiha.org/get-involved/AIHAGuidelineFoundation/EmergencyResponsePlanningGuidelines/Documents/2014%20ERPG%20Values.pdf.

[19] F. Cavender, S. Phillips and M. Holland, "Development of Emergency Response Planning Guidelines (ERPGs)," *Journal of Medical Toxicology,* vol. 4, no. 2, pp. 127-131, 2008.

553063 Hurricane Harvey 20 Months Later - Insights on the Arkema Crosby Flooding

On August 24, 2017, Hurricane Harvey, a Category 4 hurricane, made landfall in southeast Texas producing an unprecedented amount of rainfall. Hurricane Harvey killed 68 people and flooded over 300,000 structures, forcing roughly 40,000 people to evacuate their homes. Hurricane Harvey required a large emergency response effort supported by local, state, and Federal officials.

At Arkema's Crosby site, Harvey delivered an unprecedented amount of rainfall. The National Oceanic and Atmospheric Administration estimates show that the rainfall during Harvey had less than a one-in-one-thousand-year likelihood of occurring. Harris County has estimated that at Crosby, Harvey was a 5,000 year to 20,000 year event. The plant's hurricane preparedness procedure was followed and resulted in a safe shutdown and hurricane preparedness of the manufacturing units. However, finished goods inventory remaining in the warehouses was overwhelmed by high water and overcame multiple layers of protection used to safely store organic peroxide finished goods.

This presentation will explain the sites preparedness ahead of time, the events leading up to the flooding, the incident itself, and the emergency response.

The presentation will focus on the impacts, the main lessons learned and the key recommendations from the CSB report.

AUTHOR INDEX

AUTHOR INDEX